Strategies for Bioremediation of Organic and Inorganic Pollutants

Strategies for Bioremediation of Organic and Inorganic Pollutants

Editors

María S. Fuentes
Planta Piloto de Procesos Industriales Microbiológicos (PROIMI-CONICET)
Avenida Belgrano y Pasaje Caseros
Tucumán
Argentina

Verónica L. Colin
Planta Piloto de Procesos Industriales Microbiológicos (PROIMI-CONICET)
Avenida Belgrano y Pasaje Caseros
Tucumán
Argentina

Juliana M. Saez
Planta Piloto de Procesos Industriales Microbiológicos (PROIMI-CONICET)
Avenida Belgrano y Pasaje Caseros
Tucumán
Argentina

CRC Press
Taylor & Francis Group
Boca Raton London New York

CRC Press is an imprint of the
Taylor & Francis Group, an **informa** business
A SCIENCE PUBLISHERS BOOK

Cover illustrations reproduced by kind courtesy of Drs. Petra Susan Kidd, Verónica Leticia Colin and María Soledad Fuentes.

CRC Press
Taylor & Francis Group
6000 Broken Sound Parkway NW, Suite 300
Boca Raton, FL 33487-2742

Printed on acid-free paper
Version Date: 20171124

International Standard Book Number-13: 978-1-1386-2637-9 (Hardback)

Library of Congress Cataloging-in-Publication Data

Names: Fuentes, Marâia S., editor. | Colin, Verâonica L., editor. | Saez, Juliana M., editor.
Title: Strategies for bioremediation of organic and inorganic pollutants / editors, Marâia S. Fuentes (Planta Piloto de Procesos Industriales Microbiolâogicos (PROIMI-CONICET), Tucumâan, Argentina), Verâonica L. Colin (Planta Piloto de Procesos Industriales Microbiolâogicos (PROIMI-CONICET), Tucumâan, Argentina), Juliana M. Saez (Planta Piloto de Procesos Industriales Microbiolâogicos (PROIMI-CONICET), Tucumâan, Argentina).
Description: Boca Raton, FL : CRC Press, 2018. | "A science publishers book." | Includes bibliographical references and index.
Identifiers: LCCN 2017046821 | ISBN 9781138626379 (hardback : alk. paper)
Subjects: LCSH: Bioremediation. | Hazardous wastes--Biodegradation. | In situ bioremediation. | Sewage--Purification--Biological treatment.
Classification: LCC TD192.5 .S77 2018 | DDC 628.5--dc23
LC record available at https://lccn.loc.gov/2017046821

Visit the Taylor & Francis Web site at
http://www.taylorandfrancis.com

and the CRC Press Web site at
http://www.crcpress.com

Preface

The increased awareness concerning to environmental protection have prompted the development of eco-friendly technologies. This book provides a wide battery of versatile and innovative techniques for environmental clean-up of organic and inorganic pollutants generated from different anthropogenic activities. These technologies are based upon the use of diverse biological agents, including bacteria, yeast, fungi, algae, and plants or products derived from them for designing strategies for environmental remediation of pollutants such as agrochemicals, heavy metals and industrial wastes. Varied and efficient strategies used in bioremediation processes such as biosorption, immobilization of both biomass and metabolites, use of surfactants, biobeds technology, combined biological systems (microorganism-plant), and use of mixed microbial cultures are described.

The spectrum of strategies involved, and in large extent the successful combination of them, improves the removal of pollutants from diverse nature generated from anthropogenic activities with minimal environmental impact. This approach will be interesting for professionals involved with environmental technology and waste management.

Contents

1

Agro-industrial Wastewaters Bioremediation by Ligninolytic Macrofungi

Pablo M. Ahmed,[1,*] *María del M. Rosales Soro*,[2,a] *Lucía I.C. de Figueroa*[2,3,b] and *Hipólito F. Pajot*[2,c]

Introduction

Industrial activity has always resulted in some class of contamination, either solid waste, wastewater or gaseous pollution. The agro-industrial wastes consist of many and varied residues from agriculture and food industry and are worldwide produced at an estimated rate of thousand million tons per year. Although several agro-industrial residues can be disposed of safely in the environment due to its biodegradable nature, the vast quantities in which they are generated makes it necessary to look for disposition mechanisms involving the production of goods or services. Biotechnology offers many feasible alternatives to the disposal of agro-industrial wastes, allowing seeing the problems of waste disposal under a new light, as a source of valuable resources for the production of fuels, feeds, medical, pharmaceutical and industrial products.

Microbial processes are being examined as viable remediation technologies to fight environmental pollution, thus a variety of cleanup technologies have been put into practice and novel methods of bioremediation for the treatment of agro-industrial wastes are currently being worked out. An emerging field is the exploitation of waste's nutritive potential for the production of various high-value compounds.

This chapter presents a summary of the most studied macrofungi in the bioremediation of agro-industrial wastes, the processes involved as well as the main characteristics of important agro-industrial wastewaters and the simultaneous use of its nutritional potential

[1] Instituto de Tecnología Agroindustrial del Noroeste Argentino, dependiente de la Estación Experimental Agroindustrial Obispo Colombres y del Consejo Nacional de Investigaciones Científicas y Técnicas ITANOA-EEAOC-CONICET. Av. William Cross 3150 (T4101XAC), Las Talitas, Tucumán, Argentina.
[2] Planta Piloto de Procesos Industriales Microbiológicos PROIMI-CONICET. Av. Belgrano y Caseros (T4000INI), S.M. de Tucumán, Tucumán, Argentina.
[a] Email: milagrorosales@live.com.ar
[b] Email: proimiunt@gmail.com
[c] Email: hipolito_pajot@yahoo.com; hipolitopajot@hotmail.com
[3] Microbiología Superior, Facultad de Bioquímica, Química y Farmacia, Universidad Nacional de Tucumán, Ayacucho 471 (T4000INI), S.M. de Tucumán, Tucumán, Argentina.
[*] Corresponding author: pabloma@live.com.ar; pahmed@eeaoc.org.ar

for the production of several high-value compounds. The role of fungal enzymes in the degradation and transformation of phenols, lignin and related compounds in olive mill wastewater, pulp and paper mill effluents, and sugar crops stillages are discussed in detail. The most recent knowledge on these liquid wastes, in terms of their composition, obtaining and treatment methods, with special emphasis on the bioremediation, employing a specific fungi group—ligninolytic macrofungi—are also presented.

Mycotransformation, an Efficient Pollutant Removal Biotechnology

Bioremediation, also known as biotreatment, bio-reclamation or bio-restoration, is a process that utilizes the metabolic potential of living organisms such as bacteria, fungi, algae, and plants, to clean up contaminated environments and detoxify, degrade or remove environmental pollutants. Bioremediation is a promising, relatively efficient and cost-effective technology for the treatment of highly polluted industrial wastewaters with a positive environmental impact (Asamudo et al. 2005, Pant and Adholeya 2007a). According to Kulshreshtha (2012), current approaches about bioremediation are based on two main principles: metabolism or absorption of pollutants by living organisms. The organisms used in bioremediation may be autochthonous (indigenous), allochthonous (non-indigenous), or genetically modified organisms. Microorganisms can be used for bioremediation purposes in both *in situ* and *ex situ* conditions. With *in situ* techniques, the polluted site is treated in place; however, in *ex situ* techniques, samples from polluted sites are collected and transferred to the laboratory for its treatment (Rhodes 2014).

Fungi are a diverse group of organisms ubiquitous in the environment. They can exist and survive in almost every habitat and may play vital roles in all ecosystems. Fungi have the ability to regulate the flow of nutrients and energy through their mycelia networks and are known to degrade, or deteriorate, a wide variety of materials and compounds in processes referred as myco-degradation and myco-deterioration (Tiwari 2015). Fungal degradative activities against different kinds of wood, stored paper, textiles, plastics, leather, and electron insulating and various wrapping materials are well known. In fact, these are the unique microorganisms that can be employed in the complete mineralization of several wastes and wastewaters (Singh 2006). Filamentous fungi are also used to entrap and immobilize organic and inorganic particles of stillages, forming larger pellets or flocs, which enhance the subsequent separation and biodegradation (Alam and Razi 2003). In a similar way, fungi modify the structure of biosolids enhancing their bioseparation, dewaterability, and filterability (Mannan et al. 2005).

Environmental engineers all over the world have to solve problems of wastes and wastewaters. Fungal treatment of wastes in nature has been known for centuries. Due to their ubiquitous presence, fungi are capable of acclimatizing to some types of wastes, and growing in variable environmental conditions including the use of different carbon and nitrogen sources, different inoculum doses and time, as well as in static or agitated culture conditions. Some molds, yeasts, and filamentous fungi are highly tolerant to acidic or alkaline environment, while others are psychrophilic growing at temperatures near or below 0°C, whereas certain psychrotolerant fungi have the ability to survive at the very low temperature of –40°C. On the contrary, thermophilic fungi can grow above 40°C and may be cultivated in minimal media with growth yields and metabolic rates comparable to those of mesophilic fungi. Thus, fungi could thrive in the harsh conditions that characterize most industrial wastewaters and have been extensively used to remove heavy metals from liquid effluents and to mineralize phenols, petroleum hydrocarbons, polycyclic aromatic hydrocarbons, polychlorinated biphenyls, chlorinated pesticides, dyes, biopolymers, and other substances (Anastasi et al. 2013, Martorell 2014, Singh 2015, Haritash and Kaushik 2016).

Valorization of Agro-industrial Wastewaters

Industries of olive oil, starch, sugar cane, cotton bleaching, pulp and paper processing, and fruit packaging, among others, produce several billion liters of colored, toxic and harmful wastewaters over the world annually (Coulibaly et al. 2003). Generally, these industries generate wastes with similar features and in some cases at the same time of the year. For this reason, a common process to treat their residues is of great interest. There is a growing concern around the world regarding environmental pollution caused by this kind of agro-industrial wastewaters. In the last decade, the European Commission was established to influence the environmental regulations and technical aspects of the regulation of industrial wastewaters discharge, while, the United States Environmental Protection Agency (EPA) developed and implemented related politics and regulations more than a decade earlier (Singh 2006).

As a result of the implementation of protectionist environmental policies, the conversion of industrial wastewaters to valuable materials and energy is emerging as a promising trend that pursues the safe reuse of wastes for sustainable development as a manner of stopping the depletion of natural resources and the global climate change.

Agro-industrial wastewaters can provide us with environmental and economic benefits (nee' Nigam et al. 2009). Liquid effluents could be biotechnologically transformed into biofuels, microbial biomass or chemicals, or used as cheap and abundant sources of substances, such as polyphenols, exopolysaccharides, organic acids, catalysts, oligosaccharides, etc., for the cosmetic, nutraceutical, food conservation, packaging, pharmaceutical and medical industries (Zhang 2008, Sánchez 2009, ElMekawy et al. 2014).

Agriculture wastewaters such as sugar crops molasses, distilleries vinasses and the residues of milling processes are rather promising substrates for the production of added-value by-products because besides being cheap and abundant, they require minimal supplements to allow microbial growth. This kind of residues serves as a storehouse of carbon, energy and other macronutrients such as nitrogen, phosphorous, and sulfur, and growth factors as minerals and vitamins (Vandamme 2009, Santana-Méridas et al. 2012).

Molasses are produced by the sugar industry from cane or beet crops; the product is actually the mother liquor separated from the crystallized sucrose. Typical molasses have 50–55% total fermentable sugar by weight and it is used extensively in the production of bulk goods such as yeast, ethanol, poly-γ-glutamic acid, citric and succinic acids, industrial enzymes, and many others (Liu et al. 2008, Gopal and Kammen 2009, Zhang et al. 2012).

Vinasse is a liquid waste generated in the distillation of ethanol after fermentation of the molasses or juice from sugar crops, such as sugarcane and beet. Despite having variable concentration of potassium, phosphates, sulfates, calcium, iron, sodium, chlorides, carbon sources, nitrogen and other trace elements, several investigations have shown that vinasses can be used as culture media component for the production of microbial biomass proteins, lipids or enzymes, due to its nutritional characteristics (Aguiar et al. 2010, España-Gamboa et al. 2011).

Olive mill wastewater contains a variety of assimilable carbon sources, vitamins and minerals and therefore it can be regarded as fermentation medium for the production of high-added value products from industrial importance, e.g., enzymes, organic acids, exopolysaccharides, single cell oils, etc. (Yousuf et al. 2010, Bellou et al. 2014, Arous et al. 2016).

Pulping and bleaching plant effluents can be tranformed in valuable compounds (biological metabolites, or others like lignin and lignin-derivates), which have antioxidants properties or utility in pesticide and pharmaceutical industry (Arun and Eyini 2011, Lavoie et al. 2011).

Wastewaters Fungal Bioremediation

Fungi have the ability to mineralize, release and store several materials that are toxic to other microorganisms. In addition, fungi are a common source of secondary metabolites such as protein and valuable biochemicals products for medicinal, agricultural and industrial importance including antibiotics, immunosuppressants, anticancerous agents, enzymes and organic acids, among others (Tiwari 2015). Yeasts and filamentous fungi are also used extensively to reduce the chemical and biochemical oxygen demand of a wide variety of food and agriculture-processing wastewaters with the concomitant production of protein or fodder yeasts and fungi. Fungal treatment of wastewaters dated back to the '60s (Singh 2006). Fungal bioremediation not only converts the wastewater organics into added-value products such as amino acids, enzymes, dyes, organic acids, organic alcohols, and others (van Leeuwen et al. 2003), but it also produces highly dewaterable fungal biomass, which can be used as a source of animal feed and potentially in human diets (Guest and Smith 2002, Zheng et al. 2005).

In any ecosystem, fungi are among the major decomposers of plant polymers such as cellulose, hemicelluloses, and lignin. Fungal ligninolytic enzymes like manganese peroxidase, lignin peroxidase and laccase have been used to biodegrade recalcitrant compounds such as phenolic compounds, dyes, polycyclic aromatic hydrocarbons, explosives, and drugs among others, through nonspecific enzymatic oxidation reactions (D'Annibale et al. 2005, Chen and Ting 2015).

Ethanol Distilleries Vinasse

A variety of agro-industrial residues is nowadays being fermented worldwide to produce ethanol, used as fuel. The production of ethanol as a biofuel has increased in the latest years. The fuel ethanol market grew from less than a billion liters in 1975 to more than 39 billion liters in 2006. In 2008, global production of ethanol reached 79 billion liters and it is expected to reach more than 100 billion in the next years. Approximately 95% of this production was produced by fermentation-distillation (Krzywonos et al. 2008, Mussatto et al. 2010).

Nowadays, the American continent is the biggest worldwide producer of ethanol, with the United States and Brazil representing an important role in this sector. Although ethanol production in South America is concentrated in Brazil, other countries of the region as Colombia, Paraguay, and Argentina are actively developing the biofuel industry. In other Latin American countries, including Ecuador, Peru, Uruguay and Venezuela although promotion programs are in progress, the ethanol market is still very limited (Tomei and Upham 2009, Janssen and Rutz 2011). In the '70s, Brazil and Argentina were engaged in plans to promote the ethanol production, mainly from sugarcane molasses. In 1975, Brazil's government created the National Ethanol Program (PROÁLCOOL) during the oil crisis, as an alternative to petroleum products (Christofoletti et al. 2013). Similarly, the use of ethanol as a fuel was aimed by Argentina in the Alconafta plan, when the country was reached by the oil crisis. However, due to the expansion of the sugar-ethanol industry, refineries began to play an important role in environmental pollution.

The liquid waste of the distillation processes usually referred as vinasse sometimes is also called stillage or molasses spent wash. Vinasse is produced as the main by-product of ethanol generation from different feedstocks including sugarcane in South America (Hannon and Trenkle 1990), beet, wine, and fruits in Europe, corn and tequila in North America (Gianchini and Ferraz 2009, España-Gamboa et al. 2011). A standard ethanol distillery generates on average between 9–15 l vinasse per each liter of ethanol, depending on the distillery equipment and the efficiency of the overall process (España-Gamboa et al. 2015).

Aquatic ecosystems are the main receptors of vinasse. The main concern fought by the sugar-alcohol industries is the huge volume of vinasse generated, which represents a potential pollutant to water bodies when it is inadequately managed (Ferreira 2009, Ferreira et al. 2011). Unrestricted discharge of vinasse into nature causes severe damage to the environment, due to its high contents of suspended solids, ions, color and odor, low pH, high corrosivity and excessive organic load (Freire and Cortez 2000, Silva et al. 2007).

Vinasse Physicochemical Properties

Vinasse is a dark-brown effluent, with a typical odor and a strongly acidic nature (pH 3.5–5). It has high ash content and concentration of mineral salts (Algur and Gokalp 1991). Wilkie et al. (2000) reported that ethanol production from various crops, such as sugar crops (beet, sugarcane, molasses, etc.), starch crops (corn, wheat, rice, cassava, etc.), and/or cellulosic material (sugarcane bagasse, harvesting and wood residues), reaches considerable volumes of vinasse with high contents of perilous potential. According to España-Gamboa et al. (2011) the distinctive characteristics of vinasse depend on the feedstock (biomass) used for the production of ethanol. Table 1 summarizes some of the main parameters that define the chemical composition and yield of vinasses resulting from the fermentation and distillation of different raw materials.

Major organic compounds present in sugarcane vinasse are: glycerol, ethanol, lactic acid, acetic acid, oxalate, malate, carbohydrates and a high content of phenolic compounds, mainly tannic and humic acids (Pant and Adholeya 2007a, Parnaudeau et al. 2008, Mohana et al. 2009). The presence of phenols in sugarcane vinasse is due partially to degradation of lignin from lignocellulosic fibers remaining in the cane juice after sugarcane pressing (Syaichurrozi and Sumardiono 2013).

The intense brown color and the recalcitrant nature of the vinasse are usually attributed to the presence of compounds with different molecular weights, structures, and properties. These are formed during the sugar crops industrialization as a result of caramelization reactions, pH changes, thermal effects, and the reactions between amino compounds and carbohydrates (Coca et al. 2004, Arimi et al. 2014). The main colorants determined in vinasse are melanoidin (resulting from the Maillard amino-carbonyl reaction), caramels, alkaline degradation products (derived from the superheating of sugars at high temperatures) and melanins (Wilkie et al. 2000, Pant and Adholeya 2007b, España-Gamboa et al. 2011). These compounds show antioxidants properties and may reduce or even inhibit microbiological activity both in soil and in aquatic environments (Naik et al. 2008, Mohana et al. 2009), affecting vital biogeochemical cycles (e.g., carbon and nitrogen cycles). Likewise, melanoidins are inhibitory to fermentation and are toxic to many microorganisms present in wastewaters treatment processes (Chandra et al. 2008).

Table 1. Physicochemical composition and yield of distilleries vinasses from several raw materials (modified of España-Gamboa et al. 2011).

Parameter	Sugar source					
	Cane juice	**Cane molasses**	**Grapes (wine)**	**Agave (tequila)**	**Sweet sorghum**	**Beet molasses**
BOD [g l^{-1}]	16.7	39.5	14.54–16.3	20.6	46	27.5–44.9
COD [g l^{-1}]	30.4	84.9–95	26–50.2	55.2–66.3	79.9	55.5–91.1
N_T [mg l^{-1}]	102–628	153–1,230	104.9–650	na	800	1800–4750
P_T [mg l^{-1}]	71–130	1–190	65–118.4	41	1990	160–163
K [mg l^{-1}]	1733–1952	4893–11000	118–800	240–345	na	10000–10030
S_T [mg l^{-1}]	1356	1500–3480	120	780–880	na	3500–3720
pH	4.04–4.6	4.46–4.8	3–4.2	3.4	4.5	4.3–5.35
Cu [mg l^{-1}]	4	0.27–1.71	0.2–3.26	0.36–4	37	2.1–5*
Cd [mg l^{-1}]	na	0.04–1.36	0.05–0.08	0.01–0.2	na	< 1*
Pb [mg l^{-1})	na	0.02–0.48	0.55–1.34	0.065–0.5	na	< 5*
Fe [mg l^{-1}]	16	12.8–157.5	0.001–0.077	35.2–45	317	203–226*
Phenols [mg l^{-1}]	na	34	29–474	44–81	na	450*
V_Y [l $l_{ethanol}^{-1}$]	13	12–20	11.6	10–12	14.3–16	9–15
References	Wilkie et al. 2000, Salomon and Lora 2009	Pathak et al. 1999, Hutnan et al. 2003, Kannan and Upreti 2008, Chidankumar et al. 2009	Wilkie et al. 2000, Bustamante et al. 2005, Vlyssides et al. 2005	Orendain and Flores 2004, Méndez-Acosta et al. 2010, Buitrón and Carvajal 2010, López-López et al. 2010	Cail and Barford 1985, Wilkie et al. 2000, Gnansounou et al. 2005	Madejón et al. 2001, Decloux et al. 2002, Jiménez et al. 2006, Tejada and Gonzalez 2006

BOD: biochemical oxygen demand; COD: chemical oxygen demand; N_T: total nitrogen; P_T: total phosphorus; S_T: total sulfate; V_Y: vinasse yield; na: data not available.

*Unit is mg kg^{-1}

Distillery Wastewaters Treatment by Ligninolytic Macrofungi

Fungal transformation of vinasse usually employs submerged processes in shaken flasks, fermentors and bioreactors. Ligninolytic macrofungi that were investigated for their ability to remediate distillery spent wash are summarized in Table 2.

España-Gamboa et al. (2015) evaluated total phenols and color removal from sugarcane vinasse, using an air-pulsed bioreactor inoculated with *Trametes versicolor*. The batch operation of the bioreactor removed 71% of total phenols, 18% of color and 40% of chemical oxygen demand (COD) and the maximum laccase (LAC) activity achieved was 428 U l^{-1}. The air-pulsed bioreactor was subsequently operated in continuous for a period of 25 d, removing 80% of total phenols, 17% of color and 60% of COD with a LAC activity ranging from 956 to 1630 U l^{-1}. Ferreira et al. (2010) performed sugarcane vinasse discoloration assays with the macrofungus *Pleurotus sajor-caju* CCB 020, demonstrating that color removal and degradation of this complex effluent occurs concomitantly with the increase of the LAC activity, reaching 400 to 450 U l^{-1} between the 9th and 10th day of growth, and manganese peroxidase (MnP) activity at the 12th day of cultivation

Table 2. Ligninolytic macrofungi used in treatments of distilleries effluents (based on Mohana et al. 2009).

Ligninolytic macrofungus	Discoloration [%]	Treatment conditions	References
Coriolus hirsutus	80	A large amount of glucose was required to decolorize the effluent. The addition of peptone reduced the discoloring ability of the fungus	Miyata et al. 2000
Coriolus hirsutus IF04917	45	Vinasse was subjected to sequential batch discoloration operation by the immobilized fungal cells	Fujita et al. 2000
Coriolus sp. no. 20	80	First macrofungal strain applied in the discoloration of sugar cane vinasse. Discoloring activity was found when glucose or sorbose [0.5% (w/v)] was added to the medium as a carbon source	Watanabe et al. 1982
Coriolus versicolor no. 20	34.5	Trial was performed on a 10% vinasse dilution (v/v) with addition of 2% glucose (w/v) as a carbon source	Chopra et al. 2004
Coriolus versicolor Ps4a	80	Two types of enzymes (dependent and independent of sugar) were found responsible for the decolorizing activity	Ohmomo et al. 1985
Flavodon flavus	80	It was possible to biodegrade melanoidins and to remove 68% benzo[a]pyrene present in sugar cane vinasse	Raghukumar and Rivonkar 2001, Raghukumar et al. 2004
Funalia trogii	57	Vinasse sugarcane on a 30% concentration (v/v) was decolorized, using cotton fiber as an additional carbon source	Kahraman and Yeşilada 2003
Panerochaete chrysosporium	53.5	It required a readily available carbon source to discolorise. Nitrogen source had no effect. The maximum discoloration rate was observed at 6.25% wastewater dilution (v/v)	Kumar et al. 1998
Panerochaete chrysosporium ATCC 24775	55	This macrofungal strain was immobilized in different supports. The best removal percentages with polyurethane foam were obtained when it worked on a rotating biological contactor (RBC)	Raghukumar et al. 2004, Guimarães et al. 2005
Panerochaete chrysosporium JAG-40	80	Culture medium was supplemented with glucose and peptone	Dahiya et al. 2001
Panerochaete chrysosporium NCIM 1073	0	Vinasse discoloration was evaluated under submerged culture conditions, with and without agitation	Thakkar et al. 2006
Panerochaete chrysosporium NCIM 1106	82		
Panerochaete chrysosporium NCIM 1197	76		
Pleurotus florida Eger EM1303	86.3	A distillery effluent pretreated by a hydroponic culture was discolored	Pant and Adholeya 2009
Pleurotus ostreatus 3022	50	It was worked with a 50% dilution (v/v) of sugar cane vinasse	Rodríguez Pérez et al. 2006
Pleurotus ostreatus 3024	-		

Table 2 contd. ...

...*Table 2 contd.*

Ligninolytic macrofungus	Discoloration [%]	Treatment conditions	References
Pleurotus pulmonarius	43	As with *F. trogii*, the greatest discoloration of 30% (v/v) sugarcane vinasse was achieved by using cotton fiber as an additional carbon source	Kahraman and Yeşilada 2003
Pleurotus sajor-caju	> 90	Media with molasses spent wash were enriched with glucose, macro and micronutrients	Ferreira et al. 2011
Pycnoporus coccineus	60	The immobilized mycelium removed 50% more color than in the free state	Chairattanamanokorn et al. 2005
Trametes versicolor	80	Chemical oxygen demand (COD) and amines (N-NH4) decreased in presence of sucrose and KH_2SO_4 as nutrients	Benito et al. 1997
Trametes sp. I-62 (CECT 20197)	73	There was no color associated with the mycelium, nor with the polysaccharides secreted by the macrofungus. The discoloration was attributed to an enzymatic biodegradation mechanism	González et al. 2000

(60 to 100 U l^{-1}). One year later, the same authors used the system *P. sajor-caju*/sugarcane vinasse in a biodegradation study and achieved reductions of 82.8% in COD, 75.3% in biochemical oxygen demand (BOD), 99.2% of color and 99.7% in turbidity. A significant vinasse toxicity reduction was further observed when bioassays with *Pseudokirchneriella subcapitata, Daphnia magna, Daphnia similis* and *Hydra attenuate* were performed (Ferreira et al. 2011). Kahraman and Yeşilada (2003) treated vinasse from sugar fermentation with the macrofungi *Coriolus versicolor, Funalia trogii, Phanerochaete chrysosporium* and *Pleurotus pulmonarius* in the presence of cotton stalk. Decolorizing and COD removal capability of the four fungi were determined in the wastewater concentrations range of 10%, 20%, and 30%. *Coriolus versicolor* and *F. trogii* decolorized effectively the effluent. It was also demonstrated that cotton stalk enhanced discoloration, by acting as both, support and source of nutrients. Verma et al. (2011) reported a three-step sequential treatment of four vinasses from sugarcane molasses, by using a combination of ultrasound-induced acoustic cavitation, whole-fungal culture-treatment using the marine fungus NIOCC #2a, and the biosorption of the residual color with heat-inactivated wet biomass of the fungus. The sonication step reduced the odor and turbidity of the effluents, increasing their biodegradability in the second stage. Laccase synthesis in the presence of the four vinasses was directly correlated with their discoloration. At the end of this sequential treatment, a removal of 60–80% in color, 50–70% in COD and 60–70% in total phenols were achieved. These results were then confirmed through bioassays against *Artemia* sp. larvae, in which the toxicity of vinasses was reduced by two to five-fold. Strong (2010) conducted lab-scale treatment assays on vinasses from an alcoholic beverage, derived from the distillation of fermented marula fruit and employing various ligninolytic macrofungi: *Trametes pubescens* MB 89, *Pycnoporus cinnabarinus, P. chrysosporium* and *Ceriporiopsis subvermispora*. The marula vinasse showed a high phenols content (866 mg l^{-1}), high suspended solids (10.5 g l^{-1}) and 27 g l^{-1} of COD at pH 3.8. The full-strength effluent was treated in shake-flask inoculated with pure cultures of all studied macrofungi. *Trametes pubescens* MB 89 performed best in terms of phenolic compounds, COD and color removal. In a second trial, the authors evaluated the treatment of six concentrations of the wastewater (10%, 20%, 30%, 40%, 60%, and 80%), using an axenic culture of *T. pubescens* MB 89 at pH 4.5. The COD and phenols

removal efficiencies did not vary with effluent concentrations, getting between 71–77% and 87–92%, respectively. However, the highest LAC activity (1063 U l^{-1}) was quantified in the 80% wastewater concentration.

Enzymes Involved on Fungal Bioremediation of Vinasse

Biorefinery integrates biological processes for the production of fuels, energy and chemical by-products derived from biomass in a process analogous to an oil refinery (Cherubini 2010). Anaerobic digestion is suitable for a sugar biorefinery, not only because it provides a proper management of the wastes and wastewaters produced (vinasse), but also because it generates biogas as an internal fuel capable of satisfying any energy requirement into the own industry (Rajeshwari et al. 2000, Rocha et al. 2012). However, vinasse is difficult to biodegrade and decolorize by anaerobic treatment, thus there is an increasing demand from the public and private sectors for research and develop aerobic discoloration alternatives of this effluent (Singh 2006). In recent years, different aerobic microorganisms, such as bacteria (Krzywonos and Seruga 2012), algae (Marques et al. 2013) and fungi (Nair and Taherzadeh 2016) have been studied for their use in vinasse degradation processes.

The use of fungi is advantageous as they produce a wide array of biochemicals and enzymes, which tend to be more effective in degrading complex polymers than the ones produced by other microorganisms (Singh and Chen 2008, Bugg et al. 2011). Ligninolytic macrofungi are among the most significant degraders in nature and their use for bioremediation was firstly proposed by Bumpus and Aust (1987), after the discovery of the extracellular oxidative enzymes of the white-rot fungus *P. chrysosporium*. In the '90s, many researchers used this kind of fungi for the discoloration of distilleries spent wash. Despite the enzymatic system related with bioremediation of vinasse is yet to be completely understood, it seems largely connected with the expression of fungal ligninolytic enzymes. Lignin peroxidase (LiP, E.C. 1.11.1.14), MnP (E.C. 1.11.1.13) and phenol oxidases such as LAC (1.10.3.2), among other extracellular oxidases, are capable of degrading several compounds, such as melanoidins and polyaromatic compounds that cannot be degraded by other microorganisms (Benito et al. 1997).

Discoloration of sugar refinery vinasse by *P. chrysosporium* coincided in the time with the detection of MnP in the culture medium. Miyata et al. (1998) reported a complete enzymatic study on the discoloration of melanoidins by *Coriolus hirsutus*. Pellets of this ligninolytic macrofungus decolorized melanoidins by the participation of MnP, manganese-independent peroxidase (MIP) and the extracellular H_2O_2 synthesized by glucose oxidase along with the partial participation of LAC. Raghukumar et al. (2004) proposed that in the discoloration of molasses spent wash by free mycelia of the macrofungus *Flavodon flavus*, glucose oxidase activity leads to H_2O_2 formation, which acts as a bleaching agent on the vinasse bioremediation. For all the above exhibited, the contribution of MnP, MIP, LiP, LAC, the intracellular enzymatic H_2O_2-producing system, etc., for the degradation, discoloration, and detoxification of vinasse are required.

Other enzymes are produced simultaneously to the above mentioned, such as other peroxidases, enzymes producing H_2O_2 required by peroxidases (e.g., glyoxal oxidase and superoxide dismutase), enzymes linked to lignocellulose degradation pathways (glucose oxidase and aril alcohol oxidase) and others such as polysaccharide monooxygenases (PMOs), whose role is not yet fully understood (Kubicek 2013, Isaksen et al. 2014, McCotter et al. 2016).

In addition, various reports declared that intracellular enzymes as sorbose oxidase and glucose oxidase could also decolorize melanoidins. Aoshima et al. (1985) reported the induction of this type of enzymes by *Coriolus versicolor* Ps4a in the presence of melanoidins and concluded that two-thirds of the enzymes produced were sugar dependent enzymes, which seemed to be the same as sorbose-oxidase from *Coriolus* sp. no. 20 or a comparable type of sugar-oxidase (Watanabe et al. 1982).

Immobilization Strategies

Immobilization of biocatalysts, especially microbial adhesion to surfaces, has been studied intensely for stillages and polluted water treatment (Bhamidimarri 1990). Fungi can grow in submerged culture in different forms including dispersed mycelia (filamentous growth) and the formation of filamentous pellets. According to Truong et al. (2004), fungal pellets also have a potential application as immobilized cell systems because they do not require cross-linking or entrapment. Immobilized fungi demonstrated high mechanical withstand and stability against media conditions, thus several immobilization techniques have been employed in bioremediation of effluents with fungi as transformer agents.

Phanaerochaete chrysosporium immobilized in calcium-alginate beads decolorized molasses spent wash between 0 and 2 d, resulting in a faster discoloration than the one achieved with free cells, nevertheless, the color reduction decreased from 85% with free cells to 59% in the immobilized system after 10 d (Fahy et al. 1997). Fitzgibbon et al. (1995) decreased the COD (reaching a total reduction of 77%) of pretreated vinasse during 10 d with *Geotrichum candidum* at 30°C, employing *C. versicolor* immobilized on nylon cubes in a packed-bed bioreactor. *Flavodon flavus* cells immobilized on polyurethane foam cubes was successfully used to bioremediate fresh 10% sugarcane vinasse, for three consecutive cycles, reducing about 98% of the toxicity of this wastewater according to assays using an estuarine fish, *Oreochromis massambicus*, as bioindicator (Raghukumar et al. 2004). This was the first study reporting the biodegradation of vinasse from molasses and the simultaneous detoxification and removal in the benzo[a]pyrene content of the effluent (68% after 5 d).

Added-Value Products Obtained from Vinasses Fungal Bioremediation

The use of several residues in the production of enzymes, polymers, pigments, plant hormones, etc., has received notable attention in recent years. Macrofungi produce a double benefit during the biotransformation of distillery wastewaters. In an earlier report, Yürekli et al. (1999) informed about the production of indole acetic, gibberellic, and abscisic acids and cytokinin by *F. trogii* and *T. versicolor* from the 25% molasses spent wash treatment.

Conversion of the organic loadings of ethanol vinasses to microbial biomass has special relevance to countries where the supplementation of food and feeds is an urgent necessity and it has been reported by several researchers. Pant et al. (2006) cultivated *Pleurotus florida*, *P. pulmonarius*, and *P. sajor-caju* on wheat straw and sugarcane bagasse amended with biomethanated sugarcane vinasse. The macrofungus *P. florida* achieved the highest biomass yield at 50% vinasse concentration in the presence of wheat straw, which was found to be a better support and substrate than bagasse.

Enzymes are among the most relevant molecules obtained from macrofungi. Aguiar et al. (2010) attempted to produce LAC and MnP by growing three lignocellulolytic fungi, *P. sajor-caju*, *Pleurotus ostreatus* and *Pleurotus ostreatotoroseus*, in chemically treated sugarcane bagasse plus sugarcane vinasse. After 30 d of cultivation in such medium, mycelial growth was higher than the growth measured in mineral medium. *Pleurotus ostreatus* produced more than 300 U l^{-1} of LAC and 27.69 U l^{-1} of MnP at the 21st day of cultivation, whereas

P. sajor-caju showed maximum LAC (10.53 U l^{-1}) and MnP (17.93 U l^{-1}) activities at the 6th and 12th day of cultivation, respectively.

The above-discussed data reveal the viability of using vinasse amended media for the production of enzymes for its use in industrial processes as biofuel production, among others. During the last years, many processes related to wastewater bioremediation has been considerably improved, however the study of different fungal consortia, their ligninolytic enzymes (LiP, MnP, and LAC) and their immobilization is still poorly developed (Iqbal and Asgher 2013, Asgher et al. 2014). Accordingly, different immobilization strategies are being used for the development of thermostable, recoverable and reusable immobilized ligninolytic macrofungi cultures, their oxidative and hydrolytic enzymes, high yields of edible fungal biomass, or secondary fungal metabolites, for its commercial applications.

Olive Mill Wastewater

About 98% of the global production of olive oil is located in Mediterranean countries (McNamara et al. 2008). The biggest olive oil producer country is Spain, followed by Italy, Greece, Turkey, Tunisia, Portugal, Morocco and Algeria (Dourou et al. 2016). Outside the Mediterranean region, olives are cultivated in the Middle East, USA, Argentina, and Australia.

Olive oil is produced from olive trees; each olive tree yielding between 15 and 40 kg of olives per yr. Large quantities of olive mill wastewater (OMW) are produced during the manufacture of olives by the traditional mill and press processes, with 1–2 t of OMW being generated during the processing of 1 t of olives (García García et al. 2000, Paraskeva and Diamadopoulos 2006). Olive mill wastewater is a stable emulsion composed of vegetation water of olives, washing and process water, soft tissues from the olive pulp and traces of olive. In numerous cases, direct disposal of olive mill effluents into lakes, rivers and water streams resulted in disastrous environmental consequences because of its pollutant capacity (Tsonis et al. 1989, Di Giovacchino et al. 2002).

Composition and Characteristics

The characteristics of OMW are variable, depending on factors such as type and maturity of olives, region of origin, method of extraction, climatic conditions, cultivation/processing procedures, type of process (batch or continuous) involved in obtaining oil, etc. (Fountoulakis et al. 2002, Paraskeva and Diamadopoulos 2006, Dourou et al. 2016). Commonly OMW is characterized by containing high COD, BOD and high amounts of total phenols, carbohydrates, polysaccharides, fatty acids, polyalcohols, pectins and tannins (Lesage-Meessen et al. 2001, Fountoulakis et al. 2002, Paraskeva and Diamadopoulos 2006, Dourou et al. 2016). This effluent is harmful to sewage treatment plants due to the large amounts of organic and suspended matter, especially because of its oil content (Rytwo et al. 2013). In addition to its high-polluting power, OMW usually exhibits a high level of phytotoxic and antibacterial activity, due to the presence of various phenolic compounds, which make it difficult to treat by biological technologies and soil microbial communities (Karpouzas et al. 2010, Ouzounidou et al. 2012).

Physicochemical processes have been employed for OMW treatment; however, their implementation is often associated with large-scale feasibility and cost-efficiency issues which may be further accompanied by other technical or environmental troubles related to emission of air pollutants, membrane fouling, toxicity induced by radical species and the formation of large amounts of toxic sludge, among others. It is known that biological treatment of OMW can lead to the synthesis of various value-added products, e.g., enzymes and natural antioxidants, microbial biomass, polymers, energy, soil conditioners and fertilizers (Aviani et al. 2010, Padovani et al. 2013, Zervakis et al. 2013, Romero-García et al. 2014).

OMW Treatment by Ligninolytic Macrofungi and their Enzymes

Ligninolytic macrofungi constitute an alternative tool for achieving OMW detoxification since they demonstrated the ability to degrade recalcitrant aromatic compounds like the ones present in this wastewater through the production of nonspecific oxidative enzymes, e.g., LAC, LiP and MnP (Cerrone et al. 2011, Ntougias et al. 2013).

Kissi et al. (2001) described a strain of *Phanaerochaete chrysosporium* isolated from Moroccan OMW with the ability to degrade OMW under diverse conditions. This macrofungus removed more than 50% of the color and phenolic compounds from OMW within 6 d, while *Pleurotus ostreatus* reached similar results in 12 d. According to Blánquez et al. (2002), *Pleurotus flavido-alba* is also able to decolorize OMW in static or semi-static cultures, reducing the aromatic compounds and toxicity simultaneously, through the synthesis of ligninolytic enzymes as MnP and LAC. Similarly, *Pleurotus sajor-cajú*, *Lentinus (Lentinula) tigrinus*, *Coriolopsis polyzona* and *Pycnoporus coccineus* were able to decolorize and to remove COD from crude OMW in only 20 d, without any additional carbon source (Jaouani et al. 2003).

Aggelis et al. (2002) selected *P. ostreatus* out of eight ligninolytic macrofungi strains employed to treat OMW. *Pleurotus ostreatus* reached the highest discoloration and total phenolics removal efficiency (49% and 52%, respectively) after one-month cultivation. The main ligninolytic activities measured were LAC, followed by MIP, and a substantial MnP activity was also observed, while neither LiP or veratryl alcohol oxidase activities could be detected. Similarly, Martirani et al. (1996) informed that the phenol-detoxifying activity of *P. ostreatus* was associated with increasing phenol oxidase production. Likewise, Vinciguerra et al. (1995) demonstrated a very significant correlation between color, total organic carbon, and total phenols reduction, using the ligninolytic macrofungus *Lentinus edodes*. Through nuclear magnetic resonance analysis, the authors confirmed the bioconversion of phenolic and aliphatic compounds of OMW into their lower metabolic products by enzymatic activities (phenol oxidase and MnP at the seventh day of treatment), concluding that the ligninolytic system of this macrofungus participates in the OMW degradation.

In a recent study, Ntougias et al. (2015) evaluated a total of 49 macrofungal extracts in terms of their ability to degrade, decolorize and detoxify OMW. Almost all of them achieved high total phenolics (> 60%) and color removal (≤ 70%), while COD and phytotoxicity decreased to a lesser extent. *Agrocybe cylindracea*, *Inonotus andersonii*, *P. ostreatus* and *Trametes versicolor* culture extracts remained unaffected in the presence of commercial catalase, indicating no interaction of the latter with fungal enzymes and no competition for H_2O_2. However, the addition of H_2O_2 resulted in a drastic improvement on OMW's discoloration, without noticeable effects on the total phenolic content, suggesting that the enzyme-mediated oxidation transforms colored components, but not necessarily phenolics. When fungal extracts were heat-treated, the phenols content of OMW did not decrease, confirming, therefore, their enzymatic rather than physicochemical oxidation. When LAC enzyme was added to OMW-based media, it was reversibly inhibited by the effluent's high phenolic load, while peroxidases activities remained stable and active during the entire process.

Immobilization Strategies

The treatment of OMW by immobilized fungal biomass has been widely studied in the recent years (Sayadi et al. 1996, D'Annibale et al. 1999, 2000, Ahmadi et al. 2006). *Phanaerochaete chrysosporium* could decolorize OMW, removing more than 70% COD when it was immobilized on polyurethane foam. The process could be repeated up to three times without sporulation or cell lysis (Sayadi et al. 1996). Ten years later, Ahmadi et al. (2006) reported the use of *P. chrysosporium* immobilized on loofa sponge for the bioremediation of OMW. Working with 20%, 30%, and 50% OMW and the addition of basal saline medium, they reported a reduction of 50% of total phenolic compounds and on COD content. The color of 50% OMW solution decreased 40–50% after 7 d of incubation. D'Annibale et al. (1999) reported that LAC of *L. edodes,* firstly immobilized on chitosan and then cross-linked with glutaraldehyde, eliminated total phenols, and ortho-phenols from OMW, after 24 hr treatment. A LAC from *L. edodes* was further immobilized on oxirane, showing the ability to remove about 70% of total phenols, 88% of ortho-phenols and 22% of color from OMW after 2 hr-incubation, in a treatment operated in a batch recirculation mode (D'Annibale et al. 2000).

Duarte et al. (2014) performed a biotreatment for the removal of phenolic compounds in three different samples: OMW, OMW plus phenolic compounds (OMWS) and water supplemented with phenolic compounds (WS). The process was based on the use of alive or heat-killed *P. sajor-caju,* immobilized on silica-alginate (biocomposites) that were added to batch reactors containing OMW, OMWS or WS. After 28 d of treatment at 25°C, biocomposites active of *P. sajor-caju* showed ability to remove between 64.6 and 88.4% of phenolic compounds from OMW and OMWS and between 91.8% and 97.5% from WS. Furthermore, a removal of 30–38.1% of fatty acids, 68.7% of the sterol and 35% of COD was achieved in the case of OMW.

Assessment of OMW for the Obtention of Value-Added Products

Since OMW contains various carbon sources, organic compounds, and minerals, it can be considered as an exploitable resource, so it should be valorized as a substrate in several biotechnological processes (Lanciotti et al. 2005).

Several kinds of fungi can grow on OMW, producing high-added value products, such as enzymes, organic acids, exopolysaccharides, etc. (Sarris et al. 2011, Bellou et al. 2014, Mateo and Maicas 2015, Arous et al. 2016). Fenice et al. (2003) utilized an OMW based medium (2-fold diluted OMW supplemented with 0.5% sucrose and 0.1% yeast extract) for the growth of *Panus trigrinus* CBS 577.79 with the consequent production of LAC and MnP. The highest activity levels achieved were 4600 ± 98 U l^{-1} and 370 ± 15 U l^{-1}, respectively, in a stirred-tank reactor and 4300 ± 23 U l^{-1} and 410 ± 22 U l^{-1}, respectively, in an airlift reactor.

In a study performed by Zerva et al. (2016), the OMW's oxidative capacity of two ligninolytic macrofungi species, *Pleurotus citrinopileatus* LGAM 28684 and *Irpex lacteus* LGAM 238, were tested. The degradation of OMW (25% v/v) was investigated for several culture conditions, namely pH, agitation speed, nitrogen-based supplements and their concentration. The selected parameters were pH 6, agitation rate 150 rpm, 30 g l^{-1} corn steep liquor as a nitrogen source for *P. citrinopileatus* and 20 g l^{-1} diammonium tartrate for *I. lacteus*. Employing OMW as substrate, the production of biotechnologically valuable enzymes, such as LAC (1048.9 ± 2.9 U l^{-1} for *P. citrinopileatus* and 57.4 ± 2.2 U l^{-1} for *I. lacteus*), MnP (303.7 ± 15.2 U l^{-1} in the *P. citrinopileatus* culture medium and 100.2 ± 5.0 U l^{-1} for *I. lacteus*) and MIP (735.0 ± 4.27 U l^{-1} for *P. citrinopileatus* and 674.9 ± 33.0 U l^{-1} for *I. lacteus*), as well as glucans (14.05% and 10.93% of total glucan content for *P. citrinopileatus* and *I. lacteus*, respectively), were demonstrated with the simultaneous effluent degradation (more than 90% color and phenols reduction within a 24 d cultivation period). These results indicate the potential valorization of OMW, which can be used for the industrial production of enzymes and/or β-glucans deriving from the fungal biomass.

Koutrotsios et al. (2016) evaluated the suitability of OMW (12.5%, 25%, and 50% v/v) as a substrate for the production of *Hericium erinaceus* biomass and fungal derivates, such as LAC and MnP. During OMW bio-treatment, *H. erinaceus* produced abundant mycelial biomass (154.80 ± 8.45 mg per 100 ml in OMW 50%) and the obtained enzymes profile revealed a high LAC activity (134 U l^{-1} at 28 d in OMW 50%). Although to a lesser extent, MnP was also excreted during the first week of treatment and reached a peak during the second week for two out of three treatments (more than 20 U l^{-1} and ≥ 10 U l^{-1} in OMW 12.5% and 50%, respectively). Manganese-independent peroxidase was also generated toward the end of growth, but with low activities (~ 15 U l^{-1} in OMW 12.5%). *Hericium erinaceus* biomass proved to produce polysaccharides (especially β-D-glucans) with immunomodulatory, antihyperglycemic, antihyperlipidemic, anticancer, neuroprotective, and other health-promoting properties (Friedman 2015). In their work, Koutrotsios et al. (2016) also obtained successful results in the production of β-glucan and total glucan content in this ligninolytic macrofungus cultivated on olive pruning residues, another derivate from the olive processing industry.

Olive mill wastewater was also employed in the production of high-quality compost (Tomati et al. 1996), and it was demonstrated that a pretreatment of OMW with *P. ostreatus* enhances the biogas production by methanogenic bacteria (Fountoulakis et al. 2002).

Pulp and Paper Mill Wastewaters

Pulp and paper (P&P) industry is considered one of the major users of wood resources, producing plenty of biomass-based energy and materials (Svensson and Berntsson 2014). Cellulose pulp is the main raw material in the production of different types of paper and paperboard. The wood pulp production increased to about 50% worldwide, from 120 million t in 1979 to nearly 180 million t in 2013. The global P&P market has changed intensely in the last decade. Globalization has led to increased competitiveness in this international market, and new players have emerged both as producers and consumers of wood pulp (Diesen 2007). The main P&P producers are in North America and Europe with a wood pulp production of 38% and 28%, respectively, even though in 2013, Asia and South America produced 17% and 13% of the wood pulp, respectively. Brazil, an emerging country in the P&P market, went from being a pulp consumer to a world leader in wood pulp production, and since 2008, it has been the fourth largest pulp producer in the world. In both traditional and emerging P&P producers, such as United States (Schneider 2011), China (Zhu et al. 2012) and India (Afroz and Singh 2014), P&P factories are considered like the main source of environmental pollutants.

Wood pulp quality depends on the pulping process. According to Thompson et al. (2001), the wood pulp can be prepared by three main processes: mechanical, semi-chemical and chemical pulping. Each process consumes large amounts of water and produces effluents containing high amounts of toxic organic and inorganic compounds. In the mechanical process, wood chips are mechanically grounded by abrasive action and the fibers are stripped off and suspended in water.

Chemical pulping employs sufficient quantities of chemicals to break down the wood in the presence of pressure and heat. In this process, a significant part of the wood components (mainly lignin) is chemically dissolved to obtain a solid compound with high cellulose fiber content. There are two principal methods of chemical pulping: sulfite pulping and sulfate (kraft) pulping. The first one uses aqueous sulfur dioxide and a base of calcium, sodium, magnesium or ammonium. On the other hand, the kraft process employs a mixture of sodium hydroxide and sodium sulfide, known as white liquor, at a high pressure and temperature. The spent liquor is then either recycled or disposed of by burning for heat recovery.

Finally, a combination of both processes (a semi-chemical pulping) could be carried out. This combines the chemical and mechanical methods, where wood is earliest partially softened by chemicals and the excess of the pulping proceeds with mechanical force (Cardoso et al. 2009, Zheng et al. 2013).

Pulp and paper plants are recognized as heavily polluting industries worldwide, discharging a variety of liquid, gaseous and solid wastes to the environment. Black liquor is the main wastewater of P&P industry. It is considered a pollutant because it contains a toxic mixture of pulping residues (like hemicelluloses and chemicals from the kraft process), and about 50% of lignin (Zaied and Bellakhal 2009). Its high organic load and solid content affect the aquatic ecosystems in several ways, causing oxygen depletion in large areas and numerous changes in fish reproduction and physiology. The release of this effluent into nature without any previous treatment is responsible for several damages for the environment and constitutes a threat to human health (Lara et al. 2003).

Properties of Pulp and Paper Wastewaters and Treatment Methods

One of the major issues regarding P&P effluents discharge in water bodies is its color, which may vary from light tan to deep brown or black. Other major concern is the large amount of chemicals released to the environment, thus 1 t of pulp processing originates about 100 kg of color responsible compounds and from 2–4 kg of organochlorines in the bleach plant wastewaters (Nagarathnamma et al. 1999).

Paper and pulp effluents can be different, depending on the type of process, type of wood, technology applied, management practices, internal recirculation of the effluents recovery, and the amount of water used. The P&P wastewaters commonly have high COD and BOD (Table 3, Kamali and Khodaparast 2015), low biodegradability (relation BOD_5/COD), and more than 200–300 different organic and inorganic compounds (Karrasch et al. 2006). Moreover, these effluents normally include wood debris and soluble wood materials, such as resin acids (which naturally occur in the resin of tree wood), unsaturated fatty acids (such as oleic, linoleic, and linolenic acids), diterpene alcohols, chlorinated resin acids, colored compounds, etc. (Pokhrel and Viraraghavan 2004, Buyukkamaci and Koken 2010).

Bleaching of wood pulp is another high pollutant process that uses chlorine-based agents, thus due to the presence of chlorinated organic compounds, P&P liquors show a significant toxic effect on human beings (Deshmukh et al. 2009). These organochlorine compounds are commonly known as adsorbable organic halides (AOX), and about 500 different AOX have been identified in P&P mill effluents (Savant et al. 2006). In many countries, the environmental control authorities set strict restrictions on the discharges of AOX into the aquatic environments. Whereas a discharge of up to 1.5 kg of AOX per t of cellulose pulp is permitted in several countries of Europe, the United State of America government has imposed a total ban on AOX discharge (Taseli and Gokcay 1999, Requejo et al. 2012).

Table 3. Nature and characteristics of effluents from P&P production processes (Kamali and Khodaparast 2015).

Unit operations	pH	COD (mg l^{-1})	BOD (mg l^{-1})	BOD_5/COD	TSS (mg l^{-1})	References
Wood yard and chipping[a]	7	1275	556	-	7150	Avsar and Demirer 2008
Thermo-mechanical pulping	4.0–4.2	3343–4250	-	-	330–510	Qu et al. 2012
Chemical thermo-mechanical pulping[b]	7.43	7521	3000	-	350	Liu et al. 2011
Kraft cooking section	13.5	1669.7	460	0.27	40	Wang et al. 2007
Pulping process operations[c]	5.5	9065	2440	-	1309	Avsar and Demirer 2008
Bleaching[d]	8.2	3680	352	-	950	Kansal et al. 2008
Paper machine	6.5	1116	641	-	645	Avsar and Demirer 2008
Integrated pulp and paper mill	6.5	3791	1197	-	1241	Avsar and Demirer 2008
Recycled paper mill	6.2–7.8	3380–4930	1650–2565	0.488–0.52	1900–3138	Zwain et al. 2013

[a]Pulpwood storage, debarking, and chipping; [b]Alkaline peroxide mechanical pulping; [c]Pulping, pulp screening, pulp washing and thickening, bleaching, and kraft repulping; [d]A combination of chlorination and alkaline extraction stages.

According to data previously mentioned, black liquor can cause serious environmental pollution problems, thus several methods for its treatment were developed. However, pollutants continue to be found in this wastewater (Orrego et al. 2010). Physical and chemical technologies employed for the reduction of color and organic materials in black liquor include sedimentation, floatation, coagulation and precipitation (Eskelinen et al. 2010, Razali et al. 2011). Other technologies were also assayed with interesting results, including membranes technologies (Li and Zhang 2011, Chanworrawoot and Hunsom 2012), adsorption with various adsorbents (Pokhrel and Viraraghavan 2004, Ciputra et al. 2010), oxidation processes (Lucas et al. 2012) as well as new catalytic materials (Herney-Ramirez et al. 2011) and rapid land infiltration (Camberato et al. 2006). Unfortunately some limitations and the expensive nature of these technologies leaded to investigate about other treatment alternatives, including fungal degradation (Asgher et al. 2014, Kamali and Khodaparast 2015, Rajwar et al. 2016).

Bioremediation of P&P Mill Wastewaters by Ligninolytic Macrofungi

Livernoche et al. (1983) isolated and screened 15 strains of lignocellulolytic macrofungi and informed that *Trametes versicolor*, *Phanerochaete chrysosporium*, *Pleurotus ostreatus*, and *Polyporus versicolor* were capable of decolorizing P&P mill effluent. Raghukumar et al. (1996) also selected an unidentified macrofungus (fungus no. 312) for its ability to decolorize 74% of bleach plant effluent at pH 8.2 and 98% at pH 4.5. A secreted LAC enzyme was responsible for the wastewater effective discoloration. As reported by Prasad and Gupta (1997), *T. versicolor* and *P. chrysosporium* proved to have a substantial ability to reduce the color and COD content from a P&P mill effluent. Saxena and Gupta (1998) demonstrated that the ligninolytic macrofungus *P. chrysosporium* was able to remove COD, color and lignin content in P&P mill effluents, especially when it was combined with other white-rot fungi, such as *Pycnoporus sanguineus*, *Pleurotus ostreatus* or *Heterobasidion annosum*, and with surfactants. Choudhury et al. (1998) also demonstrated that lignin, BOD, COD, and color content were removed achieved to the extent of 77%, 76.8%, 60%, and 80%, respectively, when kraft mill effluent was treated by *P. ostreatus*. Lara et al. (2003) evaluated the polymerization and depolymerization of lignin upon remediation of black liquor by *Trametes elegans*, demonstrating that ligninolytic activity could be detected even in the absence of nutrients.

Freitas et al. (2009) reported that the ligninolytic macrofungi *Pleurotus sajor-caju*, *T. versicolor*, and *P. chrysosporium*, among others, have the potential to reduce color and COD in a wastewater derived from the secondary treatment of a bleached kraft pulp mill processing of *Eucalyptus globulus*. *Pleurotus sajor-caju* was one of the most effective in the biodegradation of organic compounds present in this P&P effluent, achieving reductions of relative absorbance (25% and 72% at 250 and 465 nm, respectively), and in COD level (64% and 72–77% after 6 and 13 d of incubation respectively). The synthesis of LAC and LiP were correlated with the degradation of organic compounds in this bleached kraft effluent. Jaganathan et al. (2009) also employed *P. chrysosporium* for the treatment of P&P mill effluent, and concluded that the color, COD, and BOD were removed in 86.4%, 78.8%, and 70.5%, respectively, after 3 d in shake-flask batch experiments. The authors observed the fragmentation of the fungal mycelial mass by the shaking action, which increased the contact area, and probably this was responsible for a better performance in the batch experiments. Prasongsuk et al. (2009) isolated three thermotolerant ligninolytic fungi with the capability of decolorizing P&P mill effluents: *Daedaleopsis* sp., *Schizophyllum commune* PT and *S. commune* SL. Among them, the macrofungus *Daedaleopsis* sp. exhibited the highest decolorization efficiencies, achieving 52% decolorization of wastewater from the pulping process and ~ 81% in wastewater from combined pulping and paper recycling processes.

Immobilization Strategies

Earlier immobilization systems used in the treatment of P&P effluents, including RBCs and trickling filters, utilized passive immobilization strategies involving adhesion of free cells to a solid support. For instance, an RBC inoculated with *P. chrysosporium* was used by Eaton et al. (1982) for treating an alkaline extraction stage effluent. The reactor was operated continuously during 35 d without a noticeable loss in discoloration activity. However, the discoloration process with immobilized *P. chyrisosporium* whole cells was patented later (Chang et al. 1987). Subsequently, the alkaline extraction stage effluent was treated in a continuous bioreactor also inoculated with *P. chyrisosporium* but immobilized on polyurethane foam, reaching 70% decoloration, 64% reduction in total phenols and 7% reduction in organic chlorine, after 5.8 hr (Cammarota and Sant'Anna 1992). Immobilization enhanced decoloration and phenol removal, perhaps by inducing ligninolytic enzymes synthesis.

Anaerobically treated P&P effluents have been also inoculated with immobilized fungi. Robledo-Narvaez et al. (2006) employed a mixed culture of *T. versicolor* and *Lentinus edodes* immobilized on oak sawdust and activated carbon to remediate this effluent, and reported a significant increase in the overall removal of organic matter, color, lignin, and COD. However, the use of mixed cultures proved to be unnecessary since pure cultures of *T. versicolor* immobilized on oak sawdust produced even higher discoloration and lignin degradation (69%), but with a lesser reduction of COD (32%) (Ortega-Clemente et al. 2007).

These researchers concluded that in the anaerobic phase, more than 90% of the organic load was removed from the effluent, while in the macrofungal treatment step, about 60% remaining organic matter was removed.

Biofilms have been also used to enhance fungal density in the treatment of industrial wastewaters. Wu et al. (2005) demonstrated the capacity of *P. chrysosporium*, *P. ostreatus*, *L. edodes*, *T. versicolor* and the unidentified fungi S22 to degrade lignin from black liquor when growing on a porous plastic support media. Under such conditions, more than 71% lignin degradation and 48% COD removal were achieved. Several factors as pH and the concentration of carbon, nitrogen, and trace elements in the wastewater, proved to have significant effects in the process.

Production of Value-Added Goods from P&P Mill Wastewaters

Fungal wastewater treatment with concomitant by-products recovery has various advantages, as indicated in the previous sections. One additional benefit of the concomitant fungal bioremediation is that a single species could be used for deriving multiple valuable biological compounds, such as oxidative enzymes and others potentially marketable like lignin or lignin-derivates.

White rot fungi synthesize extracellular ligninolytic enzymes during the degradation of bleach plant effluents with high concentrations of toxic phenolic compounds (Minussi et al. 2007). Laccase is the major secreted enzyme during the discoloration and detoxification of effluents from P&P industry (D'Souza et al. 2006, Font et al. 2006), either alone or in combination with MnP (Driessel and Christov 2001, Ortega-Clemente et al. 2009).

In the last decade, studies have been conducted in search of low molecular weight compounds generated by lignin bioconversion from black liquor, such as phenols (Sánchez 2009, Arun and Eyini 2011). Vanillin, gallic acid, and catechols are the major monomers produced from the depolymerization of black-liquors lignins (Ribbons 1987). However, actual processes yield a mixture of monomers, dimers, and trimers (Beauchet et al. 2012). Such products have a wide array of application in pesticide and pharmaceutical synthesis (Lavoie et al. 2011) or may also have antioxidants properties, which make them high added-value products (Egüés et al. 2014). In this sense, Negrão et al. (2015) cultivated *Bjerkandera adusta* (UAMH 8258) in 10% organosolv black-liquor with nutrients, in order to obtain low molecular weight lignin. After fungal treatment (15 d), the researchers concluded that organosolv black liquor has economic importance to industries, since it is an excellent substrate for *B. adusta* cultivation, with the concomitant biological modification of the remaining lignin, which could, in turn, be used in the generation of flavouring compounds like vanillic acid, among others (Zamzuri et al. 2013).

Concluding Remarks

Most agro-industrial wastewaters are toxic and have intense color. One of the main goals of their treatment is to discharge colorless and non-harmful effluents to receiving water bodies. Diverse physicochemical technologies have been employed to clean-up these effluents, however, most of them are inappropriate to industrial scales, due to their high cost and to the generation of secondary pollution.

Considerable success has been reached using fungi in the biological treatment of agro-industrial residues, such as vinasses, olive mill, and P&P mill wastewaters. Fungal bioremediation have proved to be useful in the decrease of color, BOD, COD, and phenolic compounds content, among others.

Whereas the mechanisms involved in the transformation of agro-industrial wastewaters are not yet fully understood, ligninolytic enzymes as laccase and manganese peroxidase clearly play a crucial role in fungal bioremediation. The use of several bioreactor configurations was successfully used in the remediation of agro-industrial wastewaters. Immobilized systems greatly enhance the treatment performance, extending the lifetime and the mechanical and chemical stability of the biocatalysts.

In addition, macrofungal treatment of agro-industrial wastes and wastewaters couples the production of high added-value products in great demand, with the remediation of wastewaters. In this sense, these liquid wastes could be considered valuable industrial by-products that can inexpensively substitute synthetic culture media in the production processes. The potential supply of edible microbial biomass is especially important for countries that have a scarcity of protein resources. The production of enzymes could also be of economic importance since they may be utilized in several biotechnology sectors and industrial applications, with important health, environmental and economic benefits.

References Cited

Afroz, Z. and A. Singh. 2014. Impact of pulp and paper mill effluent on water quality of river Aami and its effect on aquatic life (fish). Glob. J. Pharmaco. 8(2): 140–149.

Aggelis, G., C. Ehaliotis, F. Nerud, I. Stoychev, G. Lyberatos and G. Zervakis. 2002. Evaluation of white-rot fungi for detoxification and decolorization of effluents from the green olive debittering process. Appl. Microbiol. Biotechnol. 59: 353–360.

Aguiar, M.M., L.F.R. Ferreira and R.T.R. Monteiro. 2010. Use of vinasse and sugarcane bagasse for the production of enzymes by lignocellulolytic fungi. Braz. Arch. Biol. Technol. 53(5): 1245–1254.

Ahmadi, M., F. Vahabzadeh, B. Bonakdarpour, M. Mehranian and E. Mofarrah. 2006. Phenolic removal in olive oil mill wastewater using loofah-immobilized *Phanerochaete chrysosporium*. World J. Microbiol. Biotechnol. 22(2): 119–127.

Alam, M.Z. and A. Fakhru'l-Razi. 2003. Enhanced settleability and dewaterability of fungal treated domestic wastewater sludge by liquid state bioconversion process. Water Res. 37: 1118–1124.

Algur, O.F. and H.Y. Gokalp. 1991. Batch culture of *Rhizopus arrhizus* and *Actinomucor elegans* on vinasse medium for biomass production and BOD reduction. Doga Tr. J. Biol. 5: 198–205.

Anastasi, A., V. Tigini and G.C. Varese. 2013. The bioremediation potential of different ecophysiological groups of fungi. pp. 29–49. *In*: Goltapeh, E.M., Y.R. Danesh and A. Varma [eds.]. Fungi as Bioremediators. Springer, Berlin, Heidelberg.

Aoshima, I., Y. Tozawa, S. Ohmomo and K. Ueda. 1985. Production of decolorizing activity for molasses pigment by *Coriolus versicolor* Ps4a. Agric. Biol. Chem. 49(7): 2041–2045.

Arimi, M.M., Y. Zhang, G. Götz, K. Kiriamiti and S.U. Geißen. 2014. Antimicrobial colorants in molasses distillery wastewater and their removal technologies. Int. Biodeterior. Biodegradation 87: 34–43.

Arous, F., S. Azabou, A. Jaouani, H. Zouari-Mechichi, M. Nasri and T. Mechichi. 2016. Biosynthesis of single-cell biomass from olive mill wastewater by newly isolated yeasts. Environ. Sci. Pollut. Res. 23(7): 6783–6792.

Arun, A. and M. Eyini. 2011. Comparative studies on lignin and polycyclic aromatic hydrocarbons degradation by basidiomycetes fungi. Bioresour. Technol. 102(17): 8063–8070.

Asamudo, N.U., A.S. Daba and O.U. Ezeronye. 2005. Bioremediation of textile effluent using *Phanerochaete chrysosporium*. Afr. J. Biotechnol. 3(13): 1548–1553.

Asgher, M., M. Shahid, S. Kamal and H.M.N. Iqbal. 2014. Recent trends and valorization of immobilization strategies and ligninolytic enzymes by industrial biotechnology. J. Mol. Cat. B Enzym. 101: 56–66.

Aviani, I., Y. Laor, S. Medina, A. Krassnovsky and M. Raviv. 2010. Co-composting of solid and liquid olive mill wastes: management aspects and the horticultural value of the resulting composts. Bioresour. Technol. 101(17): 6699–6706.

Avsar, E. and G.N. Demirer. 2008. Cleaner production opportunity assessment study in SEKA Balikesir pulp and paper mill. J. Clean. Prod. 16: 422–431.

Beauchet, R., F. Monteil-Rivera and J.M. Lavoie. 2012. Conversion of lignin to aromatic-based chemicals (L-chems) and biofuels (L-fuels). Bioresour. Technol. 121: 328–334.

Bellou, S., A. Makri, D. Sarris, K. Michos, P. Rentoumi, A. Celik et al. 2014. The olive mill wastewater as substrate for single cell oil production by Zygomycetes. J. of Biotechnol. 170: 50–59.

Benito, G.G., M.P. Miranda and D.R. de los Santos. 1997. Decolorization of wastewater from an alcoholic fermentation process with *Trametes versicolor*. Bioresour. Technol. 61(1): 33–37.

Bhamidimarri, S.R. 1990. Adsorption and Attachment of Microorganisms to Solid Supports. Wastewater Treatment by Immobilized Cells. CRC Press, Inc., Boca Ratón, FL, USA.
Blánquez, P., G. Caminal, M. Sarra, M.T. Vicent and X. Gabarrell. 2002. Olive oil mill waste waters decoloration and detoxification in a bioreactor by the white rot fungus *Phanerochaete flavido-alba*. Biotechnol. Prog. 18(3): 660–662.
Bugg, T.D., M. Ahmad, E.M. Hardiman and R. Rahmanpour. 2011. Pathways for degradation of lignin in bacteria and fungi. Nat. Prod. Rep. 28(12): 1883–1896.
Buitrón, G. and C. Carvajal. 2010. Biohydrogen production from Tequila vinasses in an anaerobic sequencing batch reactor: Effect of initial substrate concentration, temperature and hydraulic retention time. Bioresour. Technol. 101(23): 9071–9077.
Bumpus, J.A. and S.D. Aust. 1987. Biodegradation of environmental pollutants by the white rot fungus *Phanerochaete chrysosporium*: Involvement of the lignin degrading system. BioEssays 6(4): 166–170.
Bustamante, M.A., C. Paredes, R. Moral, J. Moreno-Caselles, A. Pérez-Espinosa and M.D. Pérez-Murcia. 2005. Uses of winery and distillery effluents in agriculture: Characterisation of nutrient and hazardous components. Water Sci. Technol. 51(1): 145–151.
Buyukkamaci, N. and E. Koken. 2010. Economic evaluation of alternative wastewater treatment plant options for pulp and paper industry. Sci. Total Environ. 408: 6070–6078.
Cail, R.G. and J.P. Barford. 1985. An evaluation of the performance of an upflow floc (tower) digester treating sugar beet and sweet sorghum stillages. Biomass 7(4): 279–286.
Camberato, J.J., B. Gagnon, D.A. Angers, M.H. Chantigny and W.L. Pan. 2006. Pulp and paper mill by-products as soil amendments and plant nutrient sources. Can. J. Soil. Sci. 86: 641–653.
Cammarota, M.C. and G.L. Sant'Anna Jr. 1992. Decolorization of kraft bleach plant E1 stage effluent in a fungal bioreactor. Environ. Technol. 13(1): 65–71.
Cardoso, M., É.D. de Oliveira and M.L. Passos. 2009. Chemical composition and physical properties of black liquors and their effects on liquor recovery operation in Brazilian pulp mills. Fuel 88(4): 756–763.
Cerrone, F., P. Barghini, C. Pesciaroli and M. Fenice. 2011. Efficient removal of pollutants from olive washing wastewater in bubble-column bioreactor by *Trametes versicolor*. Chemosphere 84(2): 254–259.
Chairattanamanokorn, P., T. Imai, R. Kondo, M. Sekine, T. Higuchi and M. Ukita. 2005. Decolorization of alcohol distillery wastewater by thermotolerant white rot fungi. Appl. Biochem. Microbiol. 41(6): 583–588.
Chandra, R., R.N. Bharagava and V. Rai. 2008. Melanoidins as major colourant in sugarcane molasses based distillery effluent and its degradation. Bioresour. Technol. 99(11): 4648–4660.
Chang, H.M., T.W. Joyce and T.W. Kirk. 1987. Process of treating effluent from a pulp or papermaking operation. U.S. Patent # 4 655, 926.
Chanworrawoot, K. and M. Hunsom. 2012. Treatment of wastewater from pulp and paper mill industry by electrochemical methods in membrane reactor. J. Environ. Manage. 113: 399–406.
Chen, S.H. and A.S.Y. Ting. 2015. Biodecolorization and biodegradation potential of recalcitrant triphenylmethane dyes by *Coriolopsis* sp. isolated from compost. J. Environ. Manage. 150: 274–280.
Cherubini, F. 2010. The biorefinery concept: Using biomass instead of oil for producing energy and chemicals. Energy Conver. Manage. 51(7): 1412–1421.
Chidankumar, C.S., S. Chandraju and R. Nagendraswamy. 2009. Impact of distillery spent wash irrigation on the yields of top vegetables (Creepers). World Appl. Sci. J. 6(9): 1270–1273.
Chopra, P., D. Singh, V. Verma and A.K. Puniya. 2004. Bioremediation of melanoidin containing digested spent wash from cane-molasses distillery with white rot fungus *Coriolus versicolor*. Indian J. Microbiol. 44: 197–200.
Choudhury, S., M. Rohella Manthan and N. Sahoo. 1998. Decolorization of kraft mill effluent by white rot fungi. Indian J. Microbiol. 38: 221–224.
Christofoletti, C.A., J.P. Escher, J.E. Correia, J.F.U. Marinho and C.S. Fontanetti. 2013. Sugarcane vinasse: Environmental implications of its use. Waste Manag. 33(12): 2752–2761.
Ciputra, S., A. Antony, R. Phillips, D. Richardson and G. Leslie. 2010. Comparison of treatment options for removal of recalcitrant dissolved organic matter from paper mill effluent. Chemosphere 81: 86–91.
Coca, M., M.T. García, G. González, M. Peña and J.A. García. 2004. Study of coloured components formed in sugar beet processing. Food Chem. 86(3): 421–433.
Coulibaly, L., G. Germain and A. Spiros. 2003. Utilization of fungi for biotreatment of raw wastewaters. Afr. J. Biotechnol. 2: 620–630.
D'Annibale, A., M. Ricci, V. Leonardi, D. Quaratino, E. Mincione and M. Petruccioli. 2005. Degradation of aromatic hydrocarbons by white-rot fungi in a historically contaminated soil. Biotechnol. Bioeng. 90(6): 723–731.
D'Annibale, A., S.R. Stazi, V. Vinciguerra, E. Di Mattia and G.G. Sermanni. 1999. Characterization of immobilized laccase from *Lentinula edodes* and its use in olive-mill wastewater treatment. Process Biochem. 34(6): 697–706.
D'Annibale, A., S.R. Staz, V. Vinciguerra and G.G. Sermanni. 2000. Oxirane-immobilized *Lentinula edodes* laccase: stability and phenolics removal efficiency in olive mill wastewater. J. Biotechnol. 77(2): 265–273.
D'Souza, D.T., R. Tiwari, A.K. Sah and C. Raghukumar. 2006. Enhanced production of laccase by a marine fungus during treatment of colored effluents and synthetic dyes. Enzyme Microb. Technol. 38: 504–511.

Dahiya, J., D. Singh and P. Nigam. 2001. Decolourisation of synthetic and spent wash melanoidins using the white-rot fungus *Phanerochaete chrysosporium* JAG-40. Bioresour. Technol. 78(1): 95–98.
Decloux, M., A. Bories, R. Lewandowski, C. Fargues, A. Mersad, M.L. Lameloise et al. 2002. Interest of electrodialysis to reduce potassium level in vinasses. Preliminary experiments. Desalination 146(1-3): 393–398.
Deshmukh, N.S., K.L. Lapsiya, D.V. Savant, S.A. Chiplonkar, T.Y. Yeole, P.K. Dhakephalkar et al. 2009. Upflow anaerobic filter for the degradation of adsorbable organic halides (AOX) from bleach composite wastewater of pulp and paper industry. Chemosphere 75: 1179–1185.
Di Giovacchino, L., S. Sestili and D. Di Vincenzo. 2002. Influence of olive processing on virgin olive oil quality. Eur. J. Lipid Sci. Technol. 104(9-10): 587–601.
Diesen, M. 2007. Economics of the Pulp and Paper Industry, Book 1. 2nd Ed. Finnish Paper Engineers' Association/ Paperi ja Puu Oy, Helsinki, Finland.
Dourou, M., A. Kancelista, P. Juszczyk, D. Sarris, S. Bellou, I.E. Triantaphyllidou et al. 2016. Bioconversion of olive mill wastewater into high-added value products. J. Clean. Prod. 139: 957–969.
Driessel, B.V. and L. Christov. 2001. Decolorization of bleach plant effluent by mucoralean and white-rot fungi in a rotating biological contactor reactor. J. Biosci. Bioeng. 92: 271–276.
Duarte, K.R., C. Justino, T. Panteleitchouk, A. Zrineh, A.C. Freitas, A.C. Duarte et al. 2014. Removal of phenolic compounds in olive mill wastewater by silica-alginate-fungi biocomposites. Int. J. Environ. Sci. Technol. 11(3): 589–596.
Eaton, T.K., H.M. Chang, T.W. Joyce, T.W. Jeffries and T.K. Kirk. 1982. Method obtains fungal reduction of the color of extraction-stage kraft bleach effluents. TAPPI J. 65: 89–92.
Egüés, I., C. Sanchez, I. Mondragon and J. Labidi. 2014. Antioxidant activity of phenolic compounds obtained by autohydrolysis of corn residues. Ind. Crops Prod. 36(1): 164–171.
ElMekawy, A., L. Diels, L. Bertin, H. DeWever and D. Pant. 2014. Potential biovalorization techniques for olive mill biorefinery wastewater. Biofuel Bioprod. Bior. 8: 283–293.
Eskelinen, K., H. Särkkä, T.A. Kurniawan and M.E.T. Sillanpää. 2010. Removal of recalcitrant contaminants from bleaching effluents in pulp and paper mills using ultrasonic irradiation and Fenton-like oxidation, electrochemical treatment, and/or chemical precipitation: a comparative study. Desalination 255: 179–187.
España-Gamboa, E., J. Mijangos-Cortés, L. Barahona-Perez, J. Dominguez-Maldonado, G. Hernández-Zarate and L. Alzate-Gaviria. 2011. Vinasses: characterization and treatments. Waste Manag. Res. 29(12): 1235–1250.
España-Gamboa, E., T. Vicent, X. Font, J. Mijangos-Cortés, B. Canto-Canché and L. Alzate. 2015. Phenol and color removal in hydrous ethanol vinasse in an air-pulsed bioreactor using *Trametes versicolor*. J. Biochem. Technol. 6(3): 982–986.
Fahy, V., F.J. Fitzgibbon, G. McMullan, D. Singh and R. Marchant. 1997. Decolorization of molasses spent wash by *Phanerochaete chrysosporium*. Biotechnol. Lett. 19(1): 97–99.
Fenice, M., G.G. Sermanni, F. Federici and A. D'Annibale. 2003. Submerged and solid-state production of laccase and Mn-peroxidase by *Panus tigrinus* on olive mill wastewater-based media. J. Biotechnol. 100(1): 77–85.
Ferreira, L.F.R. 2009. Fungi biodegradation of vinasse from industrial sugarcane processing. Ph.D. Thesis, University of Sao Paulo–ESALQ, Brazil.
Ferreira, L.F.R., M. Aguiar, G. Pompeu, T.G. Messias and R.R. Monteiro. 2010. Selection of vinasse degrading microorganisms. World J. Microbiol. Biotechnol. 26(9): 1613–1621.
Ferreira, L.F.R., M.M. Aguiar, T.G. Messias, G.B. Pompeu, A.M.Q. Lopez, D.P. Silva et al. 2011. Evaluation of sugar-cane vinasse treated with *Pleurotus sajor-caju* utilizing aquatic organisms as toxicological indicators. Ecotoxicol. Environ. Saf. 74(1): 132–137.
Fitzgibbon, F.J., P. Nigam, D. Singh and R. Marchant. 1995. Biological treatment of distillery waste for pollution-remediation. J. Basic Microbiol. 35(5): 293–301.
Font, X., G. Caminal, X. Gabarrell and T. Vicent. 2006. Treatment of toxic industrial wastewater in fluidized and fixed-bed batch reactors with *Trametes versicolor*: influence of immobilization. Environ. Technol. 27: 845–854.
Fountoulakis, M.S., S.N. Dokianakis, M.E. Kornaros, G.G. Aggelis and G. Lyberatos. 2002. Removal of phenolics in olive mill wastewaters using the white-rot fungus *Pleurotus ostreatus*. Water Res. 36(19): 4735–4744.
Freire, W.J. and L.A. Cortez. 2000. Vinhaça de cana-de-açúcar. Dissertation, University of Piracicaba–ESALQ, Brazil.
Freitas, A.C., F. Ferreira, A.M. Costa, R. Pereira, S.C. Antunes, F. Gonçalves et al. 2009. Biological treatment of the effluent from a bleached kraft pulp mill using basidiomycete and zygomycete fungi. Sci. Total Environ. 407(10): 3282–3289.
Friedman, M. 2015. Chemistry, nutrition and health-promoting properties of *Hericium erinaceus* (Lion's mane) mushroom fruiting bodies and mycelia and their bioactive compounds. J. Agric. Food Chem. 63: 7108–7123.
Fujita, M., A. Era, M. Ike, S. Soda, N. Miyata and T. Hirao. 2000. Decolorization of heat-treatment liquor of waste sludge by a bioreactor using polyurethane foam-immobilized white rot fungus equipped with an ultramembrane filtration unit. J. Biosci. Bioeng. 90(4): 387–394.

García García, I., P.R. Jiménez Pena, J.L. Bonilla Venceslada, A. Martín Martín, M.A. Martín Santos and E. Ramos Gómez. 2000. Removal of phenol compounds from olive mill wastewater using *Phanerochaete chrysosporium, Aspergillus niger, Aspergillus terreus* and *Geotrichum candidum*. Proc. Biochem. 35(8): 751–758.

Gianchini, C.F. and M.V. Ferraz. 2009. Benefícios da utilização de vinhaça em terras de plantio de cana-de-açúcar. Revisão de Literatura. Rev. Cient. Electron. Agron. 3: 1–15.

Gnansounou, E., A. Dauriat and C.E. Wyman. 2005. Refining sweet sorghum to ethanol and sugar: Economic trade-offs in the context of North China. Bioresour. Technol. 96(9): 985–1002.

González, T., M.C. Terrón, S. Yagüe, E. Zapico, G.C. Galletti and A.E. González. 2000. Pyrolysis/gas chromatography/mass spectrometry monitoring of fungal-biotreated distillery wastewater using *Trametes* sp. I-62 (CECT 20197). Rapid Commun. Mass Spectrom. 14(15): 1417–1424.

Gopal, A.R. and D.M. Kammen. 2009. Molasses for ethanol: the economic and environmental impacts of a new pathway for the lifecycle greenhouse gas analysis of sugarcane ethanol. Environ. Res. Lett. 4(4): 5.

Guest, R.K. and D.W. Smith. 2002. A potential new role for fungi in a wastewater MBR biological nitrogen reduction system. J. Environ. Eng. Sci. 1(6): 433–437.

Guimarães, C., P. Porto, R. Oliveira and M. Mota. 2005. Continuous decolourization of a sugar refinery wastewater in a modified rotating biological contactor with *Phanerochaete chrysosporium* immobilized on polyurethane foam disks. Process Biochem. 40(2): 535–540.

Hannon, K. and A. Trenkle. 1990. Evaluation of condensed molasses fermentation solubles as a nonprotein nitrogen source for ruminants. J. Anim. Sci. 68(9): 2634–2641.

Haritash, A.K. and C.P. Kaushik. 2016. Degradation of low molecular weight polycyclic aromatic hydrocarbons by microorganisms isolated from contaminated soil. Int. J. Environ. Sci. 6(5): 646–656.

Herney-Ramirez, J., A.M.T. Silva, M.A. Vicente, C.A. Costa and L.M. Madeira. 2011. Degradation of acid orange 7 using a saponite-based catalyst in wet hydrogen peroxide oxidation: kinetic study with the Fermi's equation. Appl. Catal. B Environ. 101: 197–205.

Hutnan, M., M. Hornak, I. Bodik and V. Hlavacka. 2003. Anaerobic treatment of wheat stillage. Chem. Biochem. Eng. Q 17(3): 233–242.

Iqbal, H.M.N. and M. Asgher. 2013. Characterization and decolorization applicability of xerogel matrix immobilized manganese peroxidase produced from *Trametes versicolor* IBL-04. Prot. Pept. Lett. 20(5): 591–600.

Isaksen, T., B. Westereng, F.L. Aachmann, J.W. Agger, D. Kracher, R. Kittl et al. 2014. A C4-oxidizing lytic polysaccharide monooxygenase cleaving both cellulose and cello-oligosaccharides. J. Biol. Chem. 289(5): 2632–2642.

Jaganathan, B., S.M. Hossain, K.M.S. Begum and N. Anantharaman. 2009. Aerobic pollution abatement of pulp mill effluent with the white rot fungus *Phanerochaete chrysosporium* in three-phase fluidized bed bioreactor. Chem. Eng. Res. Bull. 13(1): 13–16.

Janssen, R. and D.D. Rutz. 2011. Sustainability of biofuels in Latin America: risks and opportunities. Energy Pol. 39(10): 5717–5725.

Jaouani, A., S. Sayadi, M. Vanthournhout and M.J. Penninckx. 2003. Potent fungi for decolourisation of olive oil mill wastewaters. Enzym. Microb. Technol. 33(6): 802–809.

Jiménez, A.M., R. Borja, A. Martín and F. Raposo. 2006. Kinetic analysis of the anaerobic digestion of untreated vinasses and vinasses previously treated with *Penicillium decumbens*. J. Environ. Manage. 80(4): 303–310.

Kahraman, S. and O. Yeşilada. 2003. Decolorization and bioremediation of molasses wastewater by white-rot fungi in a semi-solid-state condition. Folia Microbiol. 48(4): 525–528.

Kamali, M. and Z. Khodaparast. 2015. Review on recent developments on pulp and paper mill wastewater treatment. Ecotoxicol. Environ. Saf. 114: 326–342.

Kannan, A. and R.K. Upreti. 2008. Influence of distillery effluent on germination and growth of mung bean (*Vigna radiata*) seeds. J. Hazard Mat. 153(1–2): 609–615.

Kansal, S.K., M. Singh and D. Sud. 2008. Effluent quality at kraft/soda agro-based paper mills and its treatment using a heterogeneous photocatalytic system. Desalination 228: 183–190.

Karpouzas, D.G., S. Ntougias, E. Iskidou, C. Rousidou, K.K. Papadopoulou, G.I. Zervakis et al. 2010. Olive mill wastewater affects the structure of soil bacterial communities. Appl. Soil Ecol. 45(2): 101–111.

Karrasch, B., O. Parra, H. Cid, M. Mehrens, P. Pacheco, R. Urrutia et al. 2006. Effects of pulp and paper mill effluents on the microplankton and microbial self-purification capabilities of the Biobío River. Chile Sci. Total Environ. 359: 194–208.

Kissi, M., M. Mountadar, O. Assobhei, E. Gargiulo, G. Palmieri, P. Giardina et al. 2001. Roles of two white-rot basidiomycete fungi in decolorisation and detoxification of olive mill waste water. Appl. Microbiol. Biotechnol. 57(1): 221–226.

Koutrotsios, G., E. Larou, K.C. Mountzouris and G.I. Zervakis. 2016. Detoxification of olive mill wastewater and bioconversion of olive crop residues into high-value-added biomass by the choice edible mushroom *Hericium erinaceus*. Appl. Biochem. Biotechnol. 180(2): 195–209.

Krzywonos, M., E. Cibis, T. Miśkiewicz and C.A. Kent. 2008. Effect of temperature on the efficiency of the thermo- and mesophilic aerobic batch biodegradation of high-strength distillery wastewater (potato stillage). Bioresour. Technol. 99(16): 7816–7824.

Krzywonos, M. and P. Seruga. 2012. Decolorization of sugar beet molasses vinasse, a high-strength distillery wastewater, by lactic acid bacteria. Pol. J. Environ. Stud. 21(4): 943–948.

Kubicek, C.P. 2013. Fungi and Lignocellulosic Biomass, 1st Ed. Wiley-Blackwell, Oxford, UK.

Kulshreshtha, S. 2012. Current trends in bioremediation and biodegradation. J. Bioremed. Biodeg. 3: e114. doi: 10.4172/2155–6199.1000e114.

Kumar, V., L. Wati, P. Nigam, I.M. Banat, B.S. Yadav, D. Singh et al. 1998. Decolorization and biodegradation of anaerobically digested sugarcane molasses spent wash effluent from biomethanation plants by white-rot fungi. Process Biochem. 33(1): 83–88.

Lanciotti, R., A. Gianotti, D. Baldi, R. Angrisani, G. Suzzi, D. Mastrocola et al. 2005. Use of Yarrowia lipolytica strains for the treatment of olive mill wastewater. Bioresour. Technol. 96(3): 317–322.

Lara, M.A., A.J. Rodríguez-Malaver, O.J. Rojas, O. Holmquist, A.M. González, J. Bullón et al. 2003. Black liquor lignin biodegradation by Trametes elegans. Int. Biodeterior. Biodegradation 52(3): 167–173.

Lavoie, J.M., W. Baré and M. Bilodeau. 2011. Depolymerization of steam-treated lignin for the production of green chemicals. Bioresour. Technol. 102(7): 4917–4920.

Lesage-Meessen, L., D. Navarro, S. Maunier, J.C. Sigoillot, J. Lorquin, M. Delattre et al. 2001. Simple phenolic content in olive oil residues as a function of extraction systems. Food Chem. 75(4): 501–507.

Li, S. and X. Zhang. 2011. The study of PAFSSB on RO pre-treatment in pulp and paper wastewater. Proc. Environ. Sci. 8: 4–10.

Liu, T., H. Hu, Z. He and Y. Ni. 2011. Treatment of poplar alkaline peroxide mechanical pulping (APMP) effluent with *Aspergillus niger*. Bioresour. Technol. 102: 7361–7365.

Liu, Y.P., P. Zheng, Z.H. Sun, Y. Ni, J.J. Dong and L.L. Zhu. 2008. Economical succinic acid production from cane molasses by *Actinobacillus succinogenes*. Bioresour. Technol. 99(6): 1736–1742.

Livernoche, D., L. Jurasek, M. Desrochers, J. Dorica and I.A. Veliky. 1983. Removal of color from kraft mill wastewaters with cultures of white-rot fungi and with immobilized mycelium of *Coriolus versicolor*. Biotechnol. Bioeng. 25(8): 2055–2065.

López-López, A., G. Davila-Vazquez, E. León-Becerril, E. Villegas-García and J. Gallardo-Valdez. 2010. Tequila vinasses: generation and full scale treatment processes. Rev. Environ. Sci. Biotechnol. 9(2): 109–116.

Lucas, M.S., J.A. Peres, C. Amor, L. Prieto-rodríguez and M.I. Maldonado. 2012. Tertiary treatment of pulp mill wastewater by solar photo-Fenton. J. Hazard. Mater. 225–226: 173–181.

Madejón, E., R. López, J.M. Murillo and F. Cabrera. 2001. Agricultural use of three (sugar-beet) vinasse composts: effect on crops and chemical properties of a Cambisol soil in the Guadalquivir river valley (SW Spain). Agric. Ecosyst. Environ. 84(1): 55–65.

Mannan, S., A.F. Razia and M.Z. Alam. 2005. Use of fungi to improve bioconversion of activated sludge. Water Res. 39: 2935–2943.

Marques, S.S.I., I.A. Nascimento, P.F. de Almeida and F.A. Chinalia. 2013. Growth of *Chlorella vulgaris* on sugarcane vinasse: the effect of anaerobic digestion pretreatment. Appl. Biochem. Biotechnol. 171(8): 1933–1943.

Martirani, L., P. Giardina, L. Marzullo and G. Sannia. 1996. Reduction of phenol content and toxicity in olive oil mill waste waters with the ligninolytic fungus *Pleurotus ostreatus*. Water Res. 30(8): 1914–1918.

Martorell, M.M. 2014. Participación de mecanismos enzimáticos en biodecoloración con levaduras aisladas de ambientes naturales. Aspectos básicos y aplicación biotecnológica. Doctoral Thesis, Universidad Nacional de Tucumán, Argentina.

Mateo, J.J. and S. Maicas. 2015. Valorization of winery and oil mill wastes by microbial technologies. Food Res. Int. 73: 13–25.

McCotter, S.W., L.C. Horianopoulos and J.W. Kronstad. 2016. Regulation of the fungal secretome. Curr. Genet. 62(3): 533–545.

McNamara, C.J., C.C. Anastasiou, V. O'Flaherty and R. Mitchell. 2008. Bioremediation of olive mill wastewater. Int. Biodeter. Biodegradation 61(2): 127–134.

Méndez-Acosta, H.O., R. Snell-Castro, V. Alcaraz-González, V. González-Álvarez and C. Pelayo-Ortiz. 2010. Anaerobic treatment of Tequila vinasses in a CSTR-type digester. Biodegradation 21(3): 357–363.

Minussi, R.C., G.M. Pastore and N. Durán. 2007. Laccase induction in fungi and laccase/N-OH mediator systems applied in paper mill effluent. Bioresour. Technol. 98(1): 158–164.

Miyata, N., T. Mori, K. Iwahori and M. Fujita. 2000. Microbial decolorization of melanoidin-containing wastewaters: Combined use of activated sludge and the fungus *Coriolus hirsutus*. J. Biosci. Bioeng. 89(2): 145–150.

Miyata, N., K. Iwahori and M. Fujita. 1998. Manganese-independent and -dependent decolorization of melanoidin by extracellular hydrogen peroxide and peroxidases from *Coriolus hirsutus* pellets. J. Ferment. Bioeng. 85(5): 550–553.

Mohana, S., B.K. Acharya and D. Madamwar. 2009. Distillery spent wash: Treatment technologies and potential applications. J. Hazard. Mater. 163(1): 12–25.
Mussatto, S.I., G. Dragone, P.M. Guimarães, J.P.A. Silva, L.M. Carneiro, I.C. Roberto et al. 2010. Technological trends, global market, and challenges of bio-ethanol production. Biotechnol. Adv. 28(6): 817–830.
Nagarathnamma, R., P. Bajpai and P.K. Bajpai. 1999. Studies on decolourization and detoxification of chlorinated lignin compounds in kraft bleaching effluents by *Ceriporiopsis subvermispora*. Process Biochem. 34: 939–948.
Naik, N.M., K.S. Jagadeesh and A.R. Alagawadi. 2008. Microbial decolorization of spent wash: A review. Indian J. Microbiol. 48(1): 41–48.
Nair, R.B. and M.J. Taherzadeh. 2016. Valorization of sugar-to-ethanol process waste vinasse: A novel biorefinery approach using edible ascomycetes filamentous fungi. Bioresour. Technol. 221: 469–476.
nee' Nigam, P.S., N. Gupta and A. Anthwal. 2009. Pre-treatment of agro-industrial residues. pp. 13–33. *In*: Singh nee' Nigam, P. and A. Pandey [eds.]. Biotechnology for Agro-industrial Residues Utilization. Springer, Netherlands.
Negrão, D.R., M. Sain, A.L.L.L. Leão, J. Sameni, R. Jeng, J.P.F. de Jesus et al. 2015. Fragmentation of lignin from organosolv black liquor by white rot fungi. Bioresources 10(1): 1553–1573.
Ntougias, S., F. Gaitis, P. Katsaris, S. Skoulika, N. Iliopoulos and G.I. Zervakis. 2013. The effects of olives harvest period and production year on olive mill wastewater properties—Evaluation of *Pleurotus* strains as bioindicators of the effluent's toxicity. Chemosphere 92(4): 399–405.
Ntougias, S., P. Baldrian, C. Ehaliotis, F. Nerud, V. Merhautová and G.I. Zervakis. 2015. Olive mill wastewater biodegradation potential of white-rot fungi-Mode of action of fungal culture extracts and effects of ligninolytic enzymes. Bioresour. Technol. 189: 121–130.
Ohmomo, S., N. Itoh, Y. Watanabe, Y. Kaneko, Y. Tozawa and K. Ueda. 1985. Continuous decolorization of molasses waste water with mycelia of *Coriolus versicolor* Ps4a. Agric. Biol. Chem. 49(9): 2551–2555.
Orendain, E.H. and V.R. Flores. 2004. Process for the treatment of stillage generated by distillation in the tequila industry. U.S. Patent #10/580,684.
Orrego, R., J. Guchardi, R. Krause and D. Holdway. 2010. Estrogenic and anti-estrogenic effects of wood extractives present in pulp and paper mill effluents on rainbow trout. Aquat. Toxicol. 99: 160–167.
Ortega-Clemente, A., M.T. Ponce-Noyola, M.C. Montes-Horcasitas, M.T. Vicent, J. Barrera-Cortés and H.M. Poggi-Varaldo. 2007. Semi-continuous treatment of recalcitrant anaerobic effluent from pulp and paper industry using hybrid pellets of *Trametes versicolor*. Water Sci. Technol. 55(6): 125–133.
Ortega-Clemente, A., S. Caffarel-Méndez, M.T. Ponce-Noyola, J. Barrera-Córtes and H.M. Poggi-Varaldo. 2009. Fungal post-treatment of pulp mill effluents for the removal of recalcitrant pollutants. Bioresour. Technol. 100(6): 1885–1894.
Ouzounidou, G., S. Ntougias, M. Asfi, F. Gaitis and G.I. Zervakis. 2012. Raw and fungal-treated olive-mill wastewater effects on selected parameters of lettuce (*Lactuca sativa* L.) growth—The role of proline. J. Environ. Sci. Health B 47(7): 728–735.
Padovani, G., C. Pintucci and P. Carlozzi. 2013. Dephenolization of stored olive-mill wastewater, using four different adsorbing matrices to attain a low-cost feedstock for hydrogen photo-production. Bioresour. Technol. 138: 172–179.
Pant, D., U.G. Reddy and A. Adholeya. 2006. Cultivation of oyster mushrooms on wheat straw and bagasse substrate amended with distillery effluent. World J. Microbiol. Biotechnol. 22(3): 267–275.
Pant, D. and A. Adholeya. 2007a. Biological approaches for treatment of distillery wastewater: A review. Bioresour. Technol. 98(12): 2321–2334.
Pant, D. and A. Adholeya. 2007b. Identification, ligninolytic enzyme activity and decolorization potential of two fungi isolated from a dstillery effluent contaminated site. Water Air Soil Pollut. 183(1-4): 165–176.
Pant, D. and A. Adholeya. 2009. Nitrogen removal from biomethanated spent wash using hydroponic treatment followed by fungal decolorization. J. Environ. Eng. Sci. 26(3): 559–565.
Paraskeva, P. and E. Diamadopoulos. 2006. Technologies for olive mill wastewater (OMW) treatment: a review. J. Chem. Technol. Biotechnol. 81(9): 1475–1485.
Parnaudeau, V., N. Condom, R. Oliver, P. Cazevieille and S. Recous. 2008. Vinasse organic matter quality and mineralization potential, as influenced by raw material, fermentation and concentration processes. Bioresour. Technol. 99(6): 1553–1562.
Pathak, H., H.C. Joshi, A. Chaudhary, R. Chaudhary, N. Kalra and M.K. Dwiwedi. 1999. Soil amendment with distillery effluent for wheat and rice cultivation. Water Air Soil Poll. 113(1-4): 133–140.
Pokhrel, D. and T. Viraraghavan. 2004. Treatment of pulp and paper mill wastewater—a review. Sci. Total Environ. 333: 37–58.
Prasad, G.K. and R.K. Gupta. 1997. Decolourization of pulp and paper mill effluent by two white-rot fungi. Indian J. Environ. Health 39(2): 89–96.
Prasongsuk, S., P. Lotrakul, T. Imai and H. Punnapayak. 2009. Decolourization of pulp mill wastewater using thermotolerant white rot fungi. Science Asia 35: 37–41.

Qu, X., W.J. Gao, M.N. Han, A. Chen and B.Q. Liao. 2012. Integrated thermophilic submerged aerobic membrane bioreactor and electrochemical oxidation for pulp and paper effluent treatment—towards system closure. Bioresour. Technol. 116: 1–8.
Raghukumar, C., D. Chandramohan, F.C. Michel and C.A. Redd. 1996. Degradation of lignin and decolorization of paper mill bleach plant effluent (BPE) by marine fungi. Biotechnol. Lett. 18(1): 105–106.
Raghukumar, C. and G. Rivonkar. 2001. Decolorization of molasses spent wash by the white-rot fungus *Flavodon flavus*, isolated from a marine habitat. Appl. Microbiol. Biotechnol. 55(4): 510–514.
Raghukumar, C., C. Mohandass, S. Kamat and M. Shailaja. 2004. Simultaneous detoxification and decolorization of molasses spent wash by the immobilized white-rot fungus *Flavodon flavus* isolated from a marine habitat. Enzyme Microb. Technol. 35(2-3): 197–202.
Rajeshwari, K.V., M. Balakrishnan, A. Kansal, K. Lata and V.V.N. Kishore. 2000. State-of-the-art of anaerobic digestion technology for industrial wastewater treatment. Renew. Sustain. Energy Rev. 4(2): 135–156.
Rajwar, D., S. Joshi and J.P.N. Rai. 2016. Ligninolytic enzymes production and decolorization potential of native fungi isolated from pulp and paper mill sludge. Nat. Environ. Poll. Technol. 15(4): 1241–1248.
Razali, M.A.A., Z. Ahmad, M.S.B. Ahmad and A. Ariffin. 2011. Treatment of pulp and paper mill wastewater with various molecular weight of polyDADMAC induced flocculation. Chem. Eng. J. 166: 529–535.
Requejo, A., A. Rodríguez, J.L. Colodette, J.L. Gomide and L. Jiménez. 2012. TCF bleaching sequence in kraft pulping of olive tree pruning residues. Bioresour. Technol. 117: 117–123.
Rhodes, C.J. 2014. Mycoremediation (bioremediation with fungi)-growing mushrooms to clean the earth. Chem. Spec. Bioavailab. 26(3): 196–198.
Ribbons, D.W. 1987. Chemicals from lignin. Phil. Trans. R Soc. Lond. A 321(1561): 485–494.
Robledo-Narvaez, P.N., M.C. Montes-Horcasitas, M.T. Ponce-Noyola and H.M. Poggi-Varaldo. 2006. Production of biocatalysts of *Trametes versicolor* for the treatment of recalcitrant effluents of the pulp and paper industry. Battelle. ISBN: 1-57477-145-0.
Rocha, G.J.M., A.R. Gonçalves, B.R. Oliveira, E.G. Olivares and C.E.V. Rossell. 2012. Steam explosion pretreatment reproduction and alkaline delignification reactions performed on a pilot scale with sugarcane bagasse for bioethanol production. Ind. Crops Prod. 35(1): 274–279.
Rodríguez Pérez, S., R.C. Bermúdez Savón, M.S. Díaz and A. Kourouma. 2006. Selección de cepas de *Pleurotus ostreatus* para la decoloración de efluentes industriales. Revista Mexi Micol. 23: 9–15.
Romero-García, J.M., L. Niño, C. Martínez-Patiño, C. Álvarez, E. Castro and M.J. Negro. 2014. Biorefinery based on olive biomass. State of the art and future trends. Bioresour. Technol. 159: 421–432.
Rytwo, G., R. Lavi, Y. Rytwo, H. Monchase, S. Dultz and T.N. König. 2013. Clarification of olive mill and winery wastewater by means of clay-polymer nanocomposites. Sci. Total Environ. 442: 134–142.
Sánchez, C. 2009. Lignocellulosic residues: Biodegradation and bioconversion by fungi. Biotechnol. Adv. 27(2): 185–194.
Salomon, K.R. and E.E.S. Lora. 2009. Estimate of the electric energy generating potential for different sources of biogas in Brazil. Biomass Bioenergy 33(9): 1101–1107.
Santana-Méridas, O., A. González-Coloma and R. Sánchez-Vioque. 2012. Agricultural residues as a source of bioactive natural products. Phytochem. Rev. 11(4): 447–466.
Sarris, D., M. Galiotou-Panayotou, A.A. Koutinas, M. Komaitis and S. Papanikolaou. 2011. Citric acid, biomass and cellular lipid production by *Yarrowia lipolytica* strains cultivated on olive mill wastewater-based media. J. Chem. Technol. Biotechnol. 86(11): 1439–1448.
Savant, D.V., R. Abdul-Rahman and D.R. Ranade. 2006. Anaerobic degradation of adsorbable organic halides (AOX) from pulp and paper industry wastewater. Bioresour. Technol. 97(9): 1092–1104.
Saxena, N. and R.K. Gupta. 1998. Decolourization and delignification of pulp and paper mill effluent by white rot fungi. Indian J. Exp. Biol. 36: 1049–1051.
Sayadi, S. and R. Ellouz. 1996. Decolorization of olive mill waste-water by free and immobilized *Phanerochaete chrysosporium* cultures. Appl. Biochem. Biotechnol. 56: 265–276.
Schneider, T.E. 2011. Is environmental performance a determinant of bond pricing? Evidence from the U.S. pulp and paper and chemical industries. Contemp. Account Res. 28: 1537–1561.
Silva, M.A.S., N.P. Griebeler and L.C. Borges. 2007. Uso de vinhaça e impactos nas propriedades do solo e lençol freático. Rev. Bras. Eng. Agríc. 11(1): 108–114.
Singh, D. and S. Chen. 2008. The white-rot fungus *Phanerochaete chrysosporium*: conditions for the production of lignin-degrading enzymes. Appl. Microbiol. Biotechnol. 81(3): 399–417.
Singh, H. 2006. Mycoremediation: Fungal Bioremediation. John Wiley & Sons Inc., New Jersey, USA.
Singh, S.N. 2015. Microbial Degradation of Synthetic Dyes in Wastewaters. Springer International Publishing, Switzerland.
Strong, P.J. 2010. Fungal remediation of Amarula distillery wastewater. World J. Microbiol. Biotechnol. 26(1): 133.
Svensson, E. and T. Berntsson. 2014. The effect of long lead times for planning of energy efficiency and biorefinery technologies at a pulp mill. Renew. Energy 61: 12–16.

Syaichurrozi, I. and S. Sumardiono. 2013. Predicting kinetic model of biogas production and biodegradability organic materials: Biogas production from vinasse at variation of COD/N ratio. Bioresour. Technol. 149: 390–397.

Taseli, B.K. and C.F. Gokcay. 1999. Biological treatment of paper pulping effluents by using a fungal reactor. Water Sci. Technol. 40(11-12): 93–99.

Tejada, M. and J.L. Gonzalez. 2006. Effects of two beet vinasse forms on soil physical properties and soil loss. CATENA 68(1): 41–50.

Thakkar, A.P., V.S. Dhamankar and B.P. Kapadnis. 2006. Biocatalytic decolourisation of molasses by *Phanerochaete chrysosporium*. Bioresour. Technol. 97(12): 1377–1381.

Thompson, G., J. Swain, M. Kay and C.F. Forster. 2001. The treatment of pulp and paper mill effluent: a review. Bioresour. Technol. 77(3): 275–286.

Tiwari, K. 2015. The future products: Endophytic fungal metabolites. J. Biodivers Biopros Dev. 2: 145. doi: 10.4172/2376–0214.1000145.

Tomati, U., E. Galli, F. Fiorelli and L. Pasetti. 1996. Fertilizers from composting of olive-mill wastewaters. Int. Biodeterior. Biodegradation 38(3-4): 155–162.

Tomei, J. and P. Upham. 2009. Argentinean soy-based biodiesel: An introduction to production and impacts. Energy Pol. 37(10): 3890–3898.

Truong, T.Q., N. Miyata and K. Iwahori. 2004. Growth of *Aspergillus oryzae* during treatment of cassava starch processing wastewater with high content of suspended solids. J. Biosci. Bioeng. 97(5): 329–335.

Tsonis, S.P., V.P. Tsola and S.G. Grigoropoulos. 1989. Systematic characterization and chemical treatment of olive oil mill wastewater. Toxicol. Environ. Chem. 20-21: 437–457.

van Leeuwen, J.H., Z. Hu, T. Yi, A.L. Pometto III and B. Jin. 2003. Kinetic model for selective cultivation of microfungi in a microscreen process for food processing wastewater treatment and biomass production. Eng. Life Sci. 23(2-3): 289–300.

Vandamme, E.J. 2009. Agro-industrial residue utilization for industrial biotechnology products. pp. 3–11. *In*: Singh nee' Nigam, P. and A. Pandey [eds.]. Biotechnology for Agro-Industrial Residues Utilisation. Springer, Netherlands.

Verma, A.K., C. Raghukumar and C.G. Naik. 2011. A novel hybrid technology for remediation of molasses-based raw effluents. Bioresour. Technol. 102(3): 2411–2418.

Vinciguerra, V., A. D'Annibale, G. Delle Monache and G.C. Sermanni. 1995. Correlated effects during the bioconversion of waste olive waters by *Lentinus edodes*. Bioresour. Technol. 51(2-3): 221–226.

Vlyssides, A.G., E.M. Barampouti and S. Mai. 2005. Wastewater characteristics from Greek wineries and distilleries. Water Sci. Technol. 51(1): 53–60.

Wang, B., L. Gu and H. Ma. 2007. Electrochemical oxidation of pulp and paper making wastewater assisted by transition metal modified kaolin. J. Hazard Mater. 143: 198–205.

Watanabe, Y., R. Sugi, Y. Tanaka and S. Hayashida. 1982. Enzymatic decolorization of melanoidin by *Coriolus* sp. No. 20. Agric. Biol. Chem. 46(6): 1623–1630.

Wilkie, A.C., K.J. Riedesel and J.M. Owens. 2000. Stillage characterization and anaerobic treatment of ethanol stillage from conventional and cellulosic feedstocks. Biomass Bioenergy 19(2): 63–102.

Wu, J., Y.Z. Xiao and H.Q. Yu. 2005. Degradation of lignin in pulp mill wastewaters by white-rot fungi on biofilm. Bioresour. Technol. 96(12): 1357–1363.

Yousuf, A., F. Sannino, V. Addorisio and D. Pirozzi. 2010. Microbial conversion of olive oil mill wastewaters into lipids suitable for biodiesel production. J. Agric. Food Chem. 58(15): 8630–8635.

Yürekli, F., O. Yesilada, M. Yürekli and S.F. Topcuoglu. 1999. Plant growth hormone production from olive oil mill and alcohol factory wastewaters by white rot fungi. World J. Microbiol. Biotechnol. 15(4): 503–505.

Zaied, M. and N. Bellakhal. 2009. Electrocoagulation treatment of black liquor from paper industry. J. Hazard Mat. 163(2): 995–1000.

Zamzuri, N.A., S. Abd-Aziz, R.A. Rahim, L.Y. Phang, N.B. Alitheen and T. Maeda. 2013. A rapid colorimetric screening method for vanillic acid and vanillin-producing bacterial strains. J. Appl. Microbiol. 116(4): 903–910.

Zerva, A., G.I. Zervakis, P. Christakopoulos and E. Topakas. 2016. Degradation of olive mill wastewater by the induced extracellular ligninolytic enzymes of two wood-rot fungi. J. Environ. Manage (in press).

Zervakis, G.I., G. Koutrotsios and P. Katsaris. 2013. Composted versus raw olive mill waste as substrates for the production of medicinal mushrooms: an assessment of selected cultivation and quality parameters. BioMed. Research Int., vol. 2013, pp. 1–13.

Zhang, D., X. Feng, Z. Zhou, Y. Zhang and H. Xu. 2012. Economical production of poly (γ-glutamic acid) using untreated cane molasses and monosodium glutamate waste liquor by *Bacillus subtilis* NX-2. Bioresour. Technol. 114: 583–588.

Zhang, Y.H.P. 2008. Reviving the carbohydrate economy via multi-product lignocellulose biorefineries. J. Ind. Microbiol. Biotechnol. 35: 367–375.

Zheng, S., M. Yang and Z. Yang. 2005. Biomass production of yeast isolate from salad oil manufacturing wastewater. Bioresour. Technol. 96(10): 1183–1187.

Zheng, Y., L.I. Chai, Z.H. Yang, C.J. Tang, Y.H. Chen and Y. Shi. 2013. Enhanced remediation of black liquor by activated sludge bioaugmented with a novel exogenous microorganism culture. Appl. Microbiol. Biotechnol. 97(14): 6525–6535.

Zhu, X.L., J. Wang, Y.L. Jiang, Y.J. Cheng, F. Chen and S.B. Ding. 2012. Feasibility study on satisfing standard of water pollutants for pulp and paper industry. Appl. Mech. Mater. 178-181: 637–640.

Zwain, H.M., S.R. Hassan, N.Q. Zaman, H.A. Aziz and I. Dahlan. 2013. The start-up performance of modified anaerobic baffled reactor (MABR) for the treatment of recycled paper mill wastewater. J. Environ. Chem. Eng. 1: 61–64.

2

Bioremediation of Lignocellulosic Waste Coupled to Production of Bioethanol

María del M. Rosales Soro,[1,*] *Pablo M. Ahmed,*[2]
Lucía I. Castellanos de Figueroa[1,3] and *Hipólito F. Pajot*[1]

Introduction

There remains a continuous interest in the use of renewable energy sources (Schubert 2006). Ethanol is a renewable fuel oil which over the next 20 years could become one of the main types of biofuel in the transport sector (Hahn-Hägerdal et al. 2006). Ethanol has a high octane number, has oxygen in its chemical structure and has a higher heat of vaporization than gasoline, which makes alcohol a perfect fuel for hybrid automobiles of the next-generation. Ultimately, this practice leads to better fuel oxidation and thus reduces the exhaust gas (MacLean and Lave 2003).

The first-generation ethanol employed nowadays is produced from comestible stocks such as corn and sugarcane, raising several concerns about world hunger and making the search for new raw materials a priority concern (Tenenbaum 2008). Second generation ethanol is produced from lignocellulosic biomass generated as an agricultural and industrial waste. According to the data of the United States Department of Energy (http://www.energy.gov/), ethanol from lignocellulose could decrease the release of greenhouse gasses by 85%, as compared to the use of gasoline. Such waste does not have alternative uses. However, producing ethanol from lignocellulosic biomass is quite a complicated process involving multiple processes and profitable technologies which have not been developed yet (Kumar et al. 2008). In the saccharification process, monomeric sugars are released

[1] Planta Piloto de Procesos Industriales Microbiológicos PROIMI-CONICET. Av. Belgrano y Caseros, San Miguel de Tucumán, T4001MVB, Tucumán, Argentina.
Email: hipolitopajot@hotmail.com
[2] Instituto de Tecnología Agroindustrial del Noroeste Argentino, dependiente de la Estación Experimental Agroindustrial Obispo Colombres y del Consejo Nacional de Investigaciones Científicas y Técnicas ITANOA-EEAOC-CONICET. Av. William Cross 3150, Las Talitas, T4101XAC, Tucumán, Argentina.
Email: pabloma@live.com.ar
[3] Microbiología Superior, Facultad de Bioquímica, Química y Farmacia, Universidad Nacional de Tucumán, Ayacucho 450, San Miguel de Tucumán, T4001MVB, Tucumán, Argentina.
Email: proimiunt@gmail.com
* Corresponding author: milagrorosales@live.com.ar

from celluloses and hemicelluloses leading mainly to the accumulation of glucose and xylose. The fermentation of such monomeric sugars in pilot factories as IoGen, Abengoa, is performed by *Saccharomyces cerevisiae*. The technology for the production of ethyl alcohol from glucose or sucrose with baker's yeast has been around for centuries. However, *S. cerevisiae* ferments only hexoses leaving xylose unconsumed. Therefore, the search of yeast strains capable of fermenting all sugars released from lignocellulose has received special attention recently (Li et al. 2015).

Metabolic engineering permits improving fermentation characteristics of yeasts (Jeffries and Jin 2004). Significant progress has been made in the heterologous gene expression of bacterial and fungal xylose reductases (XR) and xylose isomerases (XI) in *S. cerevisiae* cells to increase ethanol yield from xylose (Moyses et al. 2016). One of the most promising approaches is to obtain a modified and robust industrial strain of *S. cerevisiae* by genetic engineering and subsequently taking adaptive evolution with selection pressure for optimal xylose utilization (dos Santos et al. 2016).

This chapter discusses the advantages of using lignocellulosic biomass to produce ethanol explaining the need for biomass pretreatment to enhance the hydrolysis of cellulose by the rigid association of cellulose with lignin. Particular emphasis will be placed on the cellulose hydrolysis to produce fermentescible sugars, the convertion of these sugars into ethanol, and the improvement of engineered *S. cerevisiae* strains for xylose fermentation.

World Energy Sources

Energy availability is considered a prominent factor in global economic activity (Kalogirou 2013). In the 21st century, energy consumption is rising dramatically due to the increase of the world population, the improvements in the quality of life, and the industrialization in the developing nations. However, the three main energy sources available today, nuclear power, fossil fuels, and renewable energies, are far from being considered perfect. The ideal world energy source must be economic, socially accepted, environmentally friendly and safe (Demirbas 2005). In this context, nuclear power plants are worldwide perceived as a potential risk for human health and the environment, especially after the Chernobyl, the Three Mile Island and the Fukushima Daiichi accidents (Bromet 2016). The utilization of fossil fuels not only leads to the depletion of reserves but to the increase of greenhouse gas (GHG) emissions, which in turn is correlated with the appearance of severe environmental problems as global climate change (Panwar et al. 2011, Stigka et al. 2014). One of the strategies to reduce dependence on fossil resources is to reduce energy consumption by applying energy saving programs focused on energy demand reduction and energy efficiency in the industrial and domestic spheres. Another strategy consists of using renewable energy sources (RES), not only for large-scale energy production but also for stand-alone systems (Baños et al. 2011). These types of energies are also called alternative energy sources and are those resources which can be used to produce energy again and again with decreased environmental pollution (Panwar et al. 2011).

Unfortunately, renewable energy is still expensive as compared to fossil-fuel-generated energy. Thus, a worldwide research in the field of renewable energy sources (RES) has been carried out during the last two decades concluding that it could supply 50% of the world energy demand by 2040 (Amponsah et al. 2014). Main renewable energy sources are given in Fig. 1.

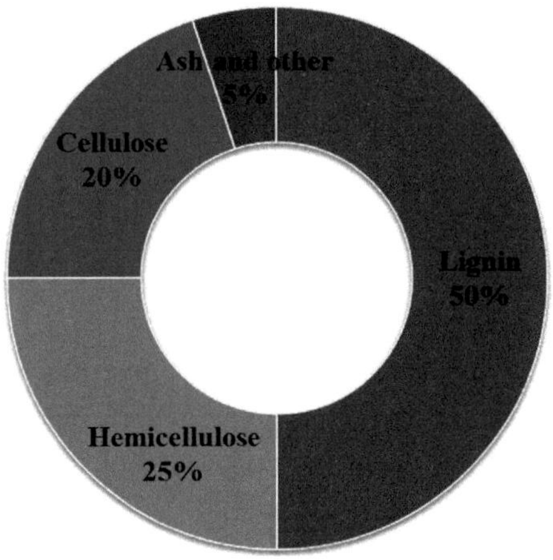

Figure 1. Summary of renewable energy sources.

Description of Renewable Energy Sources

Geothermal energy is extracted from the earth's interior, where it is stored in rocks or trapped as steam or liquid water (Ellabban et al. 2014). Geothermal energy resources offer many benefits, including, but not limited to, highly efficient, leveled energy costs, low GHG emissions, and a small environmental footprint when compared with fossil fuels (Thompson and Macdonald 2014). However, the main advantage of geothermal energy is that this source can provide 24-hour power, without the problems associated with other renewable resources such as the wind and solar energies (Baños et al. 2011).

Geothermal energy is used for the generation of electricity or for house and greenhouses heating (Demirbas 2005). Heat could be used directly when the energy flows naturally on earth's surface or indirectly through heat pumps, where water is injected into the rock bed in the horizontal or vertical coil of pipes to absorb heat. Heat pumps can also be employed to transfer heat out of the buildings and into the cooler soil surrounding them (Bose 2014).

The main disadvantage of geothermal energy is that it is currently very limited from a technological point of view. In addition, when using this type of energy there is a risk of releasing sulfur or other dangerous gases into the atmosphere. On the other hand, geothermal energy is more expensive than fossil fuels and hydropower (Schilling and Esmundo 2009).

Hydroelectric power alludes to the energy that can be captured from flowing water and is generated through a turbine, due to the mechanical conversion of energy into electricity in a highly efficient manner (Manohar and Adeyanju 2009). This energy is one of the oldest forms of renewable energy and has been employed in the form of water wheels since 200–400 b.c. in the near east (Reynolds 2002). Since rivers and streams are primarily filled with snow and rain, this energy source will remain forever while the rain exists (Kömürcü and Akpinar 2010). However, classical hydroelectric plants should be designed according to the river system, limiting their usage as they are site-specific (Ellabban et al. 2014). The newest forms of hydroelectric power are those that take advantage of the energy of waves and tides. Such technologies bring the promise of generating huge amounts of energy near in an eco-friendly and socially accepted way. The disadvantage of hydroelectric power plants is that local acceptance of energy infrastructure is often a factor that could restrict tidal power plants to unpopulated locations because they are detrimental to the local environment by replacing river ecosystems with lakes and disrupts fish migrations altering their natural habitats (Bertsch et al. 2016).

Solar energy is radiant energy that is produced by the sun. Solar energy collectors are special kind of heat ex-changers that transform solar radiation energy to internal energy of the transport medium. It is available in both direct and indirect forms. There is a vast scope to utilize the available solar energy for thermal applications such as cooking, water heating, crop drying, etc. Solar cooking is the most direct and convenient application of solar energy (Wentzel and Pouris 2007).

Although solar power has a very large potential, its commercial capacity is very limited and this is due to the high cost of this type of renewable energy, which is its main disadvantage (Wilkinson et al. 2016).

Renewable energy available in the wind has been used for centuries to generate electricity or mechanical power through the use of wind turbines and mills. This source of energy is nonpolluting and freely available in many areas. However, the power available increases with the wind speed, thus high wind areas are much more attractive for energy generation. This is the most striking characteristic of the wind resource, its variability. The wind is very variable, both temporally and geographically. Furthermore, this variability persists in a wide range of scales, both in time and space, this being one of the main disadvantages of this type of energy (Burton et al. 2001). Another disadvantage of wind power is that wind farms are often unsightly so wind sites are not close to urban areas, requiring an investment in transmission lines (Schilling and Esmundo 2009).

Windmills have been used for many years, principally for pumping water, grinding grain and in sailing ships. Many of the wind farms are now being replaced with modern wind turbines. These wind turbines incorporate sophisticated control systems in complex aerodynamic, mechanical and electrical machines (Tavner et al. 2007).

Biomass Energy

The term biomass energy refers to any source of heat produced from non-fossil, biological materials and as such has been employed for cooking and heating since the beginning of humanity (Field et al. 2008). Biomass is produced by green plants converting sunlight into plant material through photosynthesis and includes woody plants, herbaceous plants/grasses, aquatic plants, and manures, as well as all organic wastes, non-recyclable paper, and switch grass. Biomass is nowadays among the most precious and versatile resources on earth since it can be used in similar ways to fossil fuels. Moreover, it also exhibits a reasonable cost level in comparison to other renewable energies (Demirbas 2005, Moyses et al. 2016). In the developing world, biomass is still the major source of energy. Thus, the first biomass-driven power plant was established in 1,897 in India, a nation where biomass power represents 12% of the installed renewable energy capacity. In developed countries as Germany, the contribution of biomass power is also significant, rising to 31% by 2013. However, new biomass driven power plants in the proximities of populated centers are still highly resisted in both countries, on the basis of landscape modification, pollution and olfactory impacts (Patel and Rao 2016, Bertsch et al. 2016).

Biomass can produce energy by different processes, namely thermochemical and biochemical/biological. The thermochemical processes include Combustion, Gasification, and pyrolysis. Biomass combustion is a series of chemical reactions by which biomass energy is converted into heat, where carbon is oxidized to carbon dioxide, and hydrogen is oxidized to water.

Gasification, on the other hand, is the process of converting biomass into a combustible gas mixture of hydrogen, carbon monoxide, and methane that can be burnt to produce steam and heat, or used to obtain electricity in gas turbines cycles. If biomass integrated gasification combined gas-steam cycles are utilized, higher conversion efficiencies may be reached (Srivastava 2013).

Pyrolysis is defined as the thermal destruction of biomass, in the absence of oxygen, to produce liquid fraction (bio-oil), a solid fraction (charcoal) and a gaseous fraction.

Biochemical processes, on the other hand, include Anaerobic Digestion and Fermentation. The first is the conversion of biomass into biogas, which is a mixture of gases that is mainly composed of CH_4 40–70%, CO_2 30–60%, and other gases 1–5%, by means of bacterial microorganisms in the absence of oxygen. This technology is widely used for treating high moisture content biomass such as municipal solid waste (Caputo et al. 2005, Panwar et al. 2011). The second is the conversion of biomass into ethanol, also called alcoholic fermentation, which involves the breaking down of sugars (mainly glucose) producing energy, ethanol, and carbon by yeasts or bacteria.

The suitability of plant biomass (lignocellulose) for subsequent processing as energy crops is determined by the relative amounts of cellulose, hemicellulose, and lignin that compose it. Cellulose is a glucose polymer, consisting of linear chains of (1, 4)-D-glucopyranose units, in which the units are linked 1–4 in the b-configuration. Hemicellulose, in turn, is a copolymer with an average molecular weight of < 30,000 Da composed mainly by the five-carbon monosaccharides xylose and arabinose (McKendry 2002). Lignin is regarded as a group of amorphous compounds, with a high-molecular-weight, chemically related and non-water soluble. This amorphous heteropolymer consists of three different phenylpropane units that are held together by different types of linkages: coniferyl, *p*-coumaryl, and sinapyl alcohol. The proportions of each unit depend on the source of the polymer. The key objective of lignin in the plant is to provide structural support, resistance against microbial attack and oxidative stress and impermeability (Hendriks and Zeeman 2009).

Cellulose constitutes about 40–50% of the biomass by weight representing the largest fraction of the biomass; the hemicellulose portion represents 20–35% of the material by weight and lignin between 15–20%. Depending on the growing conditions, many types of perennial grasses (sugarcane, wheat, maize) have widely different yields. For example, sugarcane can be grown only in moist climatic and warm conditions when wheat can be grown with a wide range of rainfall in both hot and temperate climates (McKendry 2002).

Ethanol is produced mainly from corn grains in The United States, while in Brazil most of the production is based on sugarcane. However, with the increasing demand for ethanol, the production of fuel from raw materials that are also used for animal feed and human needs could interfere with food supply and prices (Demirbas 2006). In addition, ethanol production is higher from non-food lignocellulosic crops such as maize leaf litter, barley straw, bagasse compared to food crops (Gupta and Verma 2015). In this context, the problems of food security could be overcome by the exploitation of abundant lignocellulose feedstocks such as agricultural, forestry or agro-food residues and by-products (Schubert 2006, Hahn-Hägerdal et al. 2007). All these materials are widespread and are usually cheap as they do not have further industrial applications. Wastes from food industries have some advantages, such as being available in large quantities at the industrial sites (Fernandes et al. 2016). Lignocellulosic biomass becomes so lucrative and important for ethanol production because of its abundance and low cost. Obtaining ethanol as a transport fuel from lignocellulosic biomass represents a great opportunity to improve price stability, thereby reducing the trade deficit, improving energy security, and reducing GHG emissions into the atmosphere (Singh et al. 2014).

Bioethanol

The potential growth of bioethanol started with the oil crises in 1973 when bioethanol emerged as an alternative fuel produced from corn in the United States and from sugarcane in Brazil (Elghali et al. 2016).

In 1975, Brazil launched the National Alcohol Program, based on the production of ethanol from sugarcane (Lemos and Mesquita 2016). The significant investments in bioethanol production for over 40 years soon placed Brazil as one of the world's largest producers and consumers of bioethanol. Thus, in 2014, 88.2% of Brazilian domestic vehicles used flex-fuel motors (Anfavea 2015).

In the US, the sharp increase in bioethanol production started in 2002, with an increase in the number of bioethanol plants, from 61 to 209, in just one decade. Brazil and the United States are the world leading producers of bioethanol, accounting for about 62% of world production, followed by China which produces bioethanol from sugarcane, cassava, and yam.

In Europe, the feedstock used for bioethanol is predominately wheat, sugar beet and waste from the wine industry. Contrary to the general situation in the world, bioethanol production in Europe is by far less important than the production of another biofuel such as biodiesel. However, the potential of cost reduction is more significant for generation of bioethanol. Therefore, at present, the development of lignocellulosic bioethanol is highly important (Gnansounou 2010). According to the Renewable Energy Directive, the use of biofuels in the European Union will lead to a 35% overall reduction of green house gases. In addition to Brazil, biofuel production in Latin America is becoming an attractive sector mainly in Colombia and Argentina, whereas in other Latin American countries, the biofuel market is still very confined. On the other hand, the fuel ethanol industry of Central America is at an early stage (Janssen and Rutz 2011, Lemos and Mesquita 2016).

Fermentation and Production Process

People have been converting grain and fruit sugars into ethanol for thousands of years (Schubert 2006). Regardless of the raw material, the process is usually mediated by *S. cerevisiae*, the baker's yeast, which has been selected for ethanol production from glucose or sucrose for the last 5000 years (Lee 1997). Ethanol from lignocellulose is also produced by the same process, but the complex matrix of biomass has to be disintegrated first and the cellulose hydrolyzed to form glucose (Singh et al. 2014).

Lignocellulose is the main component of the plant cell wall and has evolved to resist degradation. Wood, for example, typically consists of 40% cellulose, 25% hemicelluloses, 20% lignins and 5% ash and other, with glucose being the principal hexose while D-xylose and L-arabinose are the main pentose sugars (Fig. 2, Moyses et al. 2016).

Ethanol production from lignocellulose is a four stage process that includes pretreatment, hydrolysis, fermentation and distillation. The first step in the production of ethanol from lignocellulose is the breakage of the polymeric matrix into its respective polymers and is usually known as the pre-treatment. Then, the liberated polymers, especially the cellulose, need to be hydrolyzed to yield fermentable sugars (Harmsen et al. 2010). Once glucose has been produced, it should be fermented to yield yeast biomass,

Figure 2. Composition of lignocellulosic biomass from wood.

ethanol, and other metabolites. Finally, ethanol must be separated from the blend in the distillation step.

Despite its immense potential and the abundance of lignocellulosic biomass, the use of this material for bioethanol production faces three main hindrances:

1. The pre-treatment step, which involves the use of extreme conditions, and the use of hydrolytic enzymes in massive amounts to release fermentable sugars such as glucose;
2. Some compounds derived from the pre-treatment steps (e.g., furaldehydes, formate, acetate, phenolic derivatives) could inhibit fermentation;
3. Pentoses are not readily fermented by *S. cerevisiae* (Moyses et al. 2016).

Some of these obstacles are discussed in more detail below.

Pretreatment

The pretreatment of lignocellulose is a complex step that must modify the fibrous structure of the biomass, de-crystallize the celluloses and depolymerize hemicelluloses, breaking the reticulated matrix of hemicelluloses and lignin and preserving the integrity of cellulose. Pretreatment must also forbid the generation of fermentation inhibitors, allow the recovery of value-added products such as lignin, consume a low amount of energy, and be a cost-effective process (Chaturvedi and Verma 2013, Anwar et al. 2014).

A number of pretreatment technologies based on physical, chemical, physicochemical and biological methods have been developed. Concentrated strong acids such as H_2SO_4 and HCl were, for example, employed for treating lignocellulosic materials because of their hydrolytic potential. However, the process has been overlooked by the industry since concentrated strong acids are toxic, hazardous and corrosive, and they must be recovered after hydrolysis in order to make the process cost-effective (Sivers and Zacchi 1995, Sun and Cheng 2002). Nowadays, diluted acid pretreatments at low or high temperatures are being used in pilot plants around the world since they are relatively inexpensive and produce similar cellulose yields than concentrated acid treatments (Kim and Holtzapple 2005, Mood et al. 2013, Cannella and Jørgensen 2014, Kroppam et al. 2014, Khoo 2015). Unfortunately, concentrated and diluted acid treatments are more expensive than physical treatments since require a neutralization step prior to the enzymatic treatment of the biomass and are prone to produce fermentation inhibitors (Sun and Cheng 2002).

Fermentation inhibitors originated during the pretreatment can be classified into (1) aliphatic acids like acetic, mainly caused by the deacetylation of hemicelluloses, and formic and levulinic acids from the dissociation of furfural and 5-hydroxy-2-methyllfurfural (HMF), respectively; (2) furan derivatives such as 5-hydroxy-2-methyllfurfural (HMF) and 2-furaldehide (furfural) from the thermal decomposition of pentoses and hexoses; and (3) phenolic compounds released during the treatment of lignin (Ravindran and Jaiswal 2016).

Furaldehydes are especially known for their biological effects and their roles as inhibitors in fermentation processes (Almeida et al. 2009, Matsushika et al. 2009a). However, aliphatic weak-acids, furfural, HMF, and phenolic compounds could inhibit enzymatic activity or impair microbial metabolism, acting synergistically to reduce ethanol yields.

Acid pretreatments at higher temperatures generate lower amounts of fermentation inhibitors; moreover, the furfural-type inhibitors produced could also degrade into other aliphatic acids with less negative effects.

Different approaches have been studied in order to produce lower concentrations of these inhibitory compounds during pre-treatment, for example optimizing the temperature and retention time, or avoiding the use of some chemicals (Mupondwa et al. 2017). However, the use of these strategies implicates an additional cost in the process.

Alkaline treatments have been extensively studied since they promise to generate fewer fermentation inhibitors and to extract lignin in relatively mild conditions. Deacetylation is one of the most studied treatments for alkaline conditions and it is able to remove up to 90% of acetate from hemicellulose. However, only 10–20% of lignin is solubilized in the process, thus deacetylation is used as a first step prior to the mild acid pretreatment of biomass. Besides reducing the formation of acetic acid, deacetylation greatly reduces the formation of furfural and 5-hydroxymethlyfurfural; and increases xylose yields in the acidic step and glucose yields in the saccharification step. More recently the soda-anthraquinone treatment received considerable attention since it allows the extraction of considerable amounts of lignin in a single step, maximize cellulose and hemicellulose retention, and greatly reduces the production of inhibitors of the fermentation step, making further acid treatments unnecessary (Karp et al. 2014).

Hydrolysis

All sugars that form a part of celluloses and hemicelluloses should be released before being converted into the desired product(s) (Harner et al. 2015). The enzymatic hydrolysis of cellulose is called saccharification, and is nowadays the preferred strategy to produce free glucose from lignocellulosic biomass since it requires low energy and causes less pollution than other methods. This hydrolysis step is mainly catalyzed by cellulases, hemicellulases, and b-glucosidases; however, a number of accessory enzymes are also required for a functional process. Thus, commercial production of bioethanol from lignocellulosic biomass requires large quantities of efficient cellulolytic enzymes (Jourdier et al. 2013, Singhania et al. 2014).

There are two strategies that have been applied to perform the enzymatic hydrolysis and fermentation of lignocellulosic biomass: (1) separate hydrolysis and fermentation (SHF) in which hydrolysis and fermentation are run in separate steps at optimal conditions of temperature and pH and (2) simultaneous saccharification and fermentation (SSF) where hydrolysis and fermentation are run in the same step, usually by using a consortium of microorganisms with cellulolytic and an ethanol-fermenting capacity. The separation of hydrolysis and fermentation steps, even when conceptually easier, could lead to the inhibition of cellulases because of glucose and cellobiose accumulation in the hydrolysis step (Tomás-Pejó et al. 2008). In the simultaneous process, however, the glucose released by the action of cellulases is converted rapidly into ethanol by the fermenting microorganism minimizing the end-product inhibition on hydrolytic activity. Moreover, the presence of ethanol in the medium helps to avoid undesired microbial contaminations, thus resulting in higher rates of saccharification and ethanol yields in the process. However, the growth of cellulolytic microorganisms should be restricted to avoid the excessive consumption of cellulose as carbon and energy source (Singhania et al. 2014).

Cellulase Enzymes

Cellulases enzymes belong to two groups of enzymes known as endoglucanases (EG) and exocellulases or cellobiohydrolases (CBH), respectively. The enzymatic hydrolysis of cellulose is initiated when EG randomly attacks the internal glycosidic bonds of cellulose chain, to create nonreducing free ends that are susceptible to attack by an exo-type cellulase or CBH to cleave glucose dimers (cellobiose) outside (Neha and Rekha 2016).

For many years, the cost of cellulolytic enzymes has been one of the main constraints to its commercialization. Even though the cost of enzymes has been reduced over the last decade, it is still considerable and contributes to the overall cost of the bioconversion process. This could be solved either by improving properties such as activity, pH stability, and thermal stability or by optimizing the proportions of the different enzymes in commercial preparations (Koppram et al. 2014). As described above, low enzymatic hydrolysis yields and rates have a great impact on the overall efficiency of the cellulose-to-ethanol conversion and have been identified as one of the bottlenecks of bioethanol production processes during the past decade (Fernandes et al. 2016).

As the cellulose content increases (high solids levels), the efficiency in the enzymatic hydrolysis decreases; thus, up to one-third of the total sugars are obtained as oligomers or polymers that cannot be further fermented by *S. cerevisiae* (Xiros et al. 2011). Part of this inhibition is caused by cellobiose accumulation. The addition of b-glucosidases, which hydrolyzes cellobiose into glucose, could help to avoid this end-product inhibition (Hahn-Hägerdal et al. 2007). A new generation of cellulase preparations with higher saccharification yields and better performances in the breakdown of complex lignocellulosic substrates has been developed and is commercially available to date (Rosgaard et al. 2006, Cannella and Jørgensen 2014). These preparations present higher β-glucosidase activities plus a new class of enzymes termed lytic polysaccharide monooxygenases, or LPMO's (previously known as GH61), which act synergistically with endoglucanase and cellobiohydrolase. In the presence of oxygen, LPMO activity generates oxidative cuts on the lignocellulose surface, generating new entry sites for exocellulases, thus increasing the total amount of released products (Leggio et al. 2012, Cannella and Jørgensen 2014, Koppraam et al. 2014).

Fermentation

In contrast to the production of ethanol from sucrose and starch, the production from lignocellulose is the fermentation of mixed sugars in the presence of inhibitory compounds formed during the pretreatment and/or hydrolysis of the raw material, as described above (Dogaris et al. 2013). A high ethanol yield process is essential for the successful production of biofuels on a commercial scale; thus, it is necessary that both hexoses and pentoses generated during the hydrolysis step could be converted to ethanol efficiently. Pentoses, especially xylose, result from the breakdown of hemicellulose, which comprises 20–35% of lignocellulosic biomass and are not readily converted into ethanol by most yeasts.

S. cerevisiae presents numerous advantages over other microorganisms used for the industrial production of ethanol (Moyses et al. 2016). *S. cerevisiae* shows high ethanol productivity, high tolerance to ethanol and tolerance to the inhibitory compounds present in the hydrolysates of lignocellulosic biomass. Additionally, unlike its prokaryotic counterparts, *S. cerevisiae* withstands low pH and is insensitive to bacteriophage infection, which is particularly relevant in large industrial processes (Hahn-Hägerdal et al. 2006, Olsson et al. 2006). Unfortunately, *S. cerevisiae* is unable to produce ethanol from xylose, despite the presence in its genome of xylose and arabinose metabolizing enzymes (Wimalasena et al. 2014).

Numerous yeasts able to ferment xylose and produce ethanol have been identified so far. Among these, *Scheffersomyces (Pichia) stipitis*, *Pachysolen tannophilus* and *Scheffersomyces (Candida) shehatae*, are able to convert glucose or xylose individually into ethanol with reasonable productivity and yields (Ho et al. 1998, Neha and Rekha 2016). However, such yeasts are susceptible to the fermentation inhibitors generated during the pretreatment of lignocellulose and are unable to ferment xylose and glucose blends efficiently, since they tend to ferment glucose earlier, producing ethanol and inhibiting further ethanol production from xylose. In addition, xylose fermentation leads to the formation of by-products such as xylitol, ribitol, arabitol and acetic acid, which leads to a low yield of ethanol (Harner et al. 2014). Consequently, such yeasts have not been considered for large-scale processes and most of the research efforts have been devoted to the development of *S. cerevisiae* strains capable of efficiently fermenting xylose. This remains one of the challenges in the fermentation of lignocellulosic biomass (Olsson et al. 2006).

Metabolic Engineering of Xylose Fermentation

The improvement of yeasts capable of fermenting a five-carbon sugar as xylose is the essential step to achieve a viable production of second-generation ethanol (dos Santos et al. 2016).

Initially, several groups attempted to develop metabolic engineering strategies in order to efficiently convert xylose into ethanol by *S. cerevisiae* (Hahn-Hägerdal et al. 2007, Tomás-Pejó et al. 2008). Several key points emerged from these first experiences, including the need for reducing redox imbalances and xylitol formation, development of pathways for the metabolism of xylose in *S. cerevisiae*, and the necessity of modulating the expression of genes of the non-oxidative pentose phosphate pathway, which will be discussed later (Demeke et al. 2013). Xylose catabolism occurs through different fungal and bacterial pathways (Fig. 3, Moyses et al. 2016).

The xylose is assimilated by fungi through the oxidative-reductive pathway also called XRXDH, which involve two main enzymes: the NAD(P)H-dependent xylose reductase (XR, EC 1.1.1.21) encoded by XYL1, which converts D-xylose into Xilitol, and the NADP+-dependent xylitol dehydrogenase (XDH, EC 1.1.1.9) encoded by XYL2, which in turn transforms Xilitol into D-Xylulose (Lee 2014). Finally, xylulokinase (XK, EC 2.7.1.17) phosphorylates D-xylulose into 5P-xylulose, which is further metabolized through the pentose phosphate pathway (PPP) and glycolysis (Bruinenberg et al. 1984, Rizzi et al. 1988, Matsushika et al. 2009b, Moyses et al. 2016). These reactions depend on the potential of phosphorylation of the cell (ATP), therefore, of its energy charge (Matsuchika 2009a). Bacteria, on the other hand, metabolizes xylose trough the xylose isomerase pathway (XI). Thus, D-xylose is directly isomerized to D-xylulose by the xylose isomerase enzyme (XI, EC 5.3.1.5) which does not require cofactors and therefore, does not have the redox bottleneck associated with the XR-XDH pathway. In addition, since it has a lower accumulation of

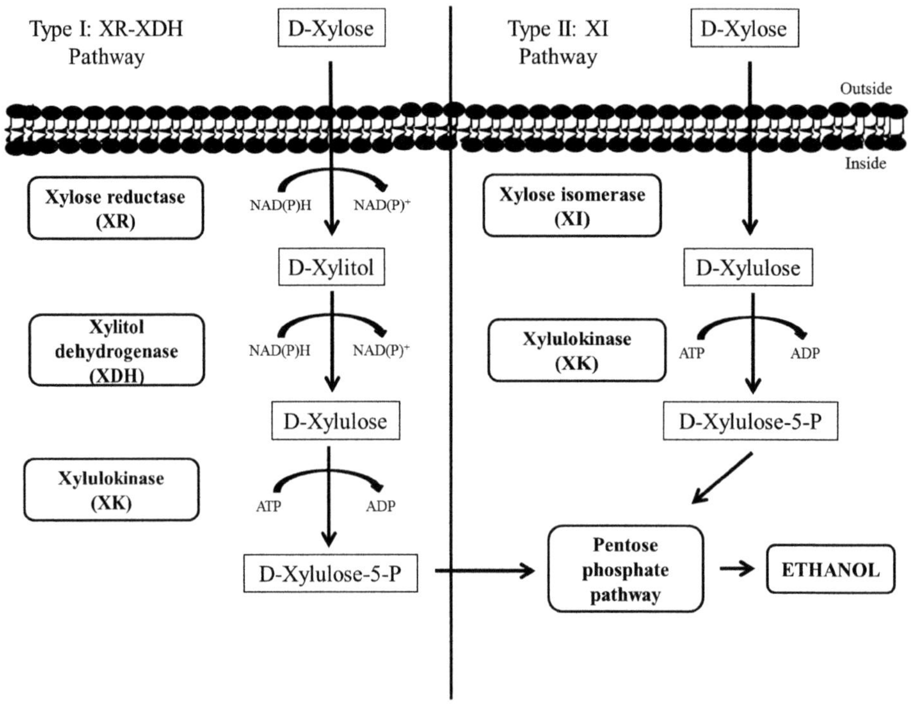

Figure 3. D-xylose metabolic pathway in fungi and bacteria (Modified according to Matsushika et al. 2009b).

by-products, the total amount of ethanol that can be produced is higher (Kuyper et al. 2004). The synthesis of these xylose-metabolizing enzymes requires the presence of xylose for induction and is also totally or at least partially repressed by the presence of glucose.

Engineering Saccharomyces cerevisiae with the XR/XDH Pathway

Despite of being demonstrated that the genome of several *Saccharomyces* spp. strains, if not all of them, contain encoding functional XR, XDH, and XK genes, the enzymes are produced at very low levels and the yeasts are not able to grow on xylose.

The *S. stipitis* XR/XDH pathway is one of the most frequently used to engineer baker's yeast for xylose fermentation (Kötter et al. 1990). However, the XR/XDH pathway has an important limitation; while XR preferentially uses NADPH as a cofactor, XDH exclusively uses NAD^+ (Hahn-Hägerdal et al. 2007); therefore, this pathway leads to a NADH/NADPH cofactor imbalance in the cell, resulting in the significant production of the by-product xylitol under anaerobic conditions. The production of xylitol, in turn, leads to a reduction in carbon assimilation, diminishing ethanol yields in recombinant strains (Latimer et al. 2014). Moreover, in several strains, xylitol is also produced from the activity of an unspecific aldose reductase (GRE3) which reduces D-xylose to xylitol using NADPH as a cofactor, causing a further decrease in ethanol yields (Träff et al. 2001, Nogué and Karhuma 2015).

However, *S. stipitis* is one of the few yeast types that have specific XR cofactors of NADPH and NADH, thus the yeast possesses the ability to excrete less xylitol during xylose fermentation. *P. tannophilus* also possesses this dual specific XR but produces more xylitol than *S. stipitis* (Cadete et al. 2016).

A number of strategies have been devised to lower xylitol formation. Enzymes with the preference of cofactor XR and XDH (mutant enzymes) have been developed to solve this asymmetry, however, these mutations led to an alteration in the use of cofactors and therefore in enzymatic kinetics (Jeppsson et al. 2006, Latimer et al. 2014). It has also been shown that increasing the level of XDH relative to XR decreases the xylitol yield and therefore, increases the ethanol yield (Karhumaa et al. 2007). Other metabolic engineering strategies have been approached such as expression of an $NADP^+$ dependent D-glyceraldehyde-3-phosphate dehydrogenase to facilitate NADPH regeneration because the resulting strain was able to ferment D-xylose to ethanol at a higher rate and yield than the corresponding strain without the mutation since the level of the secondary product xylitol was lowered (Verho et al. 2003). Another strategy was the deletion of the endogenous nonspecific NADPH-dependent aldose reductase GRE3 to decrease the production of xylitol (Träff et al. 2001, Karhumaa et al. 2007). Unfortunately, the aldose reductase, encoded by GRE3, plays a role in stress protection and its deletion is not desirable in industrial yeast strains (Aguilera and Prieto 2001).

Engineering Saccharomyces with the XI Pathway

The successful expression of a functional XI in *S. cerevisiae* would prevent the occurrence of the NADH/NADPH cofactor imbalance, reducing xylitol formation while increasing ethanol yields. However, expression of functional bacterial XI in *S. cerevisiae* has proven to be a difficult task. Heterologous expression of XI genes from *Bacillus subtilis* (Amore et al. 1989) and *Streptomyces rubiginosus* (Gárdonyi and Hahn-Hägerdal 2003) generated inactive and insoluble proteins. Unsuccessful heterologous expression in *S. cerevisiae* was attributed to several reasons, including protein misfolding, post-translational modifications, disulfide bridge formation, sub-optimal internal pH and absence of specific metal ions (Matsushika et al. 2009a). However, functional recombinants have indeed been created. The first one carried an XI enzyme from *Thermus thermophilus* and proved to ferment xylose poorly, possibly due to the low specific activity of the enzyme (Walfridsson et al. 1996). The first recombinant *S. cerevisiae* strain demonstrating high activity of prokaryotic XI was based on the expression of an enzyme from *Clostridium phytofermentans* (Brat et al. 2009). This enzyme is weakly inhibited by xylitol but the rate of D-xylose consumption and ethanol production is still low (Demeke et al. 2013).

Some anaerobic cellulolytic fungi do not have the usual eukaryotic pathway of xylose metabolism, lacking xylose reductases and xylitol dehydrogenases. Thus, the expression of recombinant fungal XI in *S. cerevisiae* is comparatively easier. Recombinant strains expressing the eukaryotic xylA from the anaerobic fungus *Piromyces* sp. (Kuyper et al. 2003) and *Orpinomyces* (Madhavan et al. 2009) showed high XI activities, but the growth of *S. cerevisiae* on xylose based media was slow in aerobic conditions and null under anaerobiosis. To overcome these difficulties, recombinant strains were adapted to xylose under anaerobic conditions, resulting in the isolation of *Piromyces* sp. strain (RWB202-AFX) with greater specific growth rate and ethanol yields (Kuyper et al. 2003). However, XI is strongly inhibited by xylitol, which is still generated by the action of GRE3, making the recombinant strain unsuitable for industrial scale production (Toivari et al. 2004). Genetic modifications, including the deletion of GRE3, the overexpression of genes from the non-oxidative phase of PPP, and the overexpression of xylose transporters proved to enhance both the rate of D-xylose consumption and ethanol production XI expressing *S. cerevisiae* (Karhumaa et al. 2007, Kuyper et al. 2005, Demeke et al. 2013, Moyses et al. 2016).

To address these difficulties, the genes encoding XK, ribulose 5 phosphate epimerase, ribulose 5-phosphate isomerase, transketolase, and transaldolase were overexpressed in strain RWB202-AFX and GRE3 was removed. The strain obtained (RWB218) showed higher specific growth rate and ethanol yield (Hahn-Hägerdal et al. 2007).

In the same way, the recombinant fungus *Orpinomyces* recently mentioned was further modified by overexpressing XKS1 (which encodes XK) and a heterologous sugar transporter (PsSUT1) following evolutionary engineering. Resulting strain showed an increment on the growth rate, ethanol production, and low xylitol production (Madhavan et al. 2009).

Despite significant efforts that have been made in modifying heterologous enzymes to increase the consumption rates and ethanol yields from xylose, a major obstacle for efficient and fast fermentation of lignocellulosic hydrolysates is that yeasts are unable to simultaneously consume glucose and pentoses (Farwick et al. 2014), mainly because most xylose transporters are competitively inhibited by glucose (Parachin et al. 2011, Young et al. 2012, Kim et al. 2013).

Xylose Uptake

In *S. cerevisiae,* xylose has been found to be transported by facilitated diffusion mainly through non-specific hexose transporters (Kötter and Ciriacy 1993). Unfortunately, the affinity of those transporters for xylose is approximately 200 fold lower than for glucose, thus xylose transport is competitively inhibited by glucose. There are 18 hexose transporters in *S. cerevisiae's* genome (Lin and Li 2011). The transporters can be divided into three classes, according to their affinities for glucose. There are high-affinity transporters Hxt6, Hxt7, and Gal2; intermediate affinity transporters Hxt2, Hxt4, and Hxt5; and low-affinity transporters Hxt1 and Hxt3 (Moyses et al. 2016). Studies have indicated that the high- and intermediate-affinity hexose transporters are responsible for most of xylose uptake. Furthermore, it has been shown that a low (but non-zero) glucose concentration is needed in the medium for efficient xylose uptake (Oloffson et al. 2008).

Naturally, if the internal metabolic pathway of xylose is improved, its transport will become an increasing limitation. Accordingly, the new approaches have to achieve optimization of transport and metabolism together (Young et al. 2010).

One possible solution to overcome this bottleneck is the overexpression of *S. cerevisiae's* high-affinity transporters. Thus, HXT5 and HXT7 genes are only expressed when *S. cerevisiae* grows on xylose as the sole carbon source; nevertheless, despite attempts to enhance the growth and xylose fermentation by overexpression of Hxt7 and Gal2 transporters, no significant effects have been observed so far (Hamacher et al. 2002, Matsuchika et al. 2009b).

Another solution leads to the metabolic engineering of xylose-assimilating *S. cerevisiae* strains in order to reduce the affinity of native transporters for glucose or to raise the affinity for xylose. Thus, *S. cerevisiae* native carriers have been subjected to adaptive evolution and mutagenesis studies to obtain fast uptake of xylose, resulting in high expression of hexose transporters in the strain (Wahlbom et al. 2003, Moyses et al. 2016). Using a strategy for transporter evolution, a chimeric Hxt36 transporter constructed by fusing Hxt3 Hxt6 proteins was also subjected to directed evolution, using a severe selection pressure including just 0.5%–1% xylose in media with up to 10% glucose. The resultant strain (N367A) was unable to grow on glucose, but showed a slow growth on xylose amended media, since the transporter affinity for glucose was lower than for xylose (Nijland et al. 2014).

The adaptive evolution of other genes also allows a rapid improvement in fermentative yield. Dos Santos et al. (2016) modified a robust industrial strain of *S. cerevisiae* by the addition of essential genes encoding the enzymes for pentose metabolism: ribose-5-phosphate ketol-isomerase, transaldolase, ribulose-5-phosphate 3-epimerase, and transketolase. All these genes were from *Orpinomyces* sp. and were expressed under the control of constitutive promoters of genes from the glycolytic pathway of *S. cerevisiae*. Subsequently, taken through cycles of adaptive evolution, strains able to efficiently ferment xylose (as the only source of carbon) into ethanol could be selected. A third possible solution is the expression of heterologous xylose transporters. Nevertheless, up until now, there are few functional heterologous xylose-transporters expressed in *S. cerevisiae*, probably due to structural or functional barriers for the expression of heterologous membrane permeases in yeasts (Leandro et al. 2006, Gonçalves et al. 2014). Thus, the genes encoding for transporters from Arabidopsis thaliana (At5g59250, Hector et al. 2008), *Trichoderma reesei* (Xlt1, Saloheimo et al. 2007), *Debaryomyces hansenii* (Xylh, Ferreira et al. 2013), and *Meyerozyma guilliermondii* (MgT05196, Wang et al. 2015) have been expressed in HXT-null *S. cerevisiae*. However, less than 50 percent of these sugar transporters were functional in *S. cerevisiae*, probably due to misfolding or improper localization, as already suggested above (Moyses et al. 2016).

Yeasts like *S. stipitis*, *S. shehatae*, and *P. tannophilus*, have efficient xylose transport systems, including proton symporters and facilitated diffusion transporters, while it is true that most of them present greater affinity for glucose (Harner et al. 2015). Genes coding for sugar transporters from *S. stipitis*: SUT1, SUT2, XUT1, belonging to the subfamily of glucose transporters with high-affinity glucose and xylose uptake, but with higher Vmax for xylose uptake compared with hexoses (Weierstall et al. 1999, Katahira et al. 2008) have already been expressed in *S. cerevisiae* as well as genes from *Candida intermedia* (GXS1 and GXF1, Young et al. 2010). The expression of GXF1, a glucose/xylose H^+ symporter able to transport xylose against a concentration gradient, produced faster xylose uptake and greater ethanol production in *S. cerevisiae* recombinant strains (Leandro et al. 2006, 2008, Runquist et al. 2009). Both transporters (GXS1 and GXF1) are efficient xylose transporters despite having a high affinity for glucose (Young et al. 2010).

Xylose transport is an obstacle in hydrolysates fermentation, and certainly, is aggravated with improvements of intracellular xylose metabolism (Farwick et al. 2014). Nevertheless, the expressions of these heterologous xylose transporters provide a great promise for future development of xylose-fermenting *S. cerevisiae* strains possessing high rates of xylose utilization (Matsushika et al. 2009a).

Concluding Remarks

Environmental and energy crises in the world today led to the search for alternative uses of renewable natural resources, using clean technologies. In this sense, the microbial production of ethanol from lignocellulosic biomass is a potential way to partially replace the oil raw materials.

Lignocellulosic biomass is the most promising raw material considering its high availability and low cost. Lignocellulosic biomass can be converted into bioethanol by hydrolysis and subsequent fermentation by yeast *S. cerevisiae*, but this type of yeast has a limited range of substrate assimilation. The lignocellulosic hydrolysate contains not only glucose but also pentose sugars (mainly xylose), which requires a microorganism capable of efficiently fermenting both glucose and xylose for a successful industrial production of bioethanol.

Together with the genetic engineering, it is important to use other approaches, such as adaptive evolution and expression of heterologous sugar carriers discussed above to obtain engineered *S. cerevisiae* strains that efficiently ferment xylose and, consequently, improve the ethanol production.

There is no doubt that the development of studies on process strategies for the development of yeast capable of efficiently fermenting pentoses will continue in order to establish an economically viable industrial process.

Acknowledgments

This work was supported by Agencia Nacional de Promoción Científica y Tecnológica and Fondo para la Investigación Científica y Tecnológica FONCYT (PICT 1154/2013; PICT 3639/2015); Consejo Nacional de Investigaciones Científicas y Técnicas (PIO CONICET YPF 133201301000022CO); Secretaría de Ciencia, Arte e Innovación Tecnológica de la Universidad Nacional de Tucumán SCAIT UNT (PIUNT D509).

References Cited

Aguilera, J. and J. Prieto. 2001. The *Saccharomyces cerevisiae* aldose reductase is implied in the metabolism of methylglyoxal in response to stress conditions. Curr. Genet. 39: 273–283.

Almeida, J.R., M. Bertilsson, M.F. Gorwa-Grauslund, S. Gorsich and G. Lidén. 2009. Metabolic effects of furaldehyde and impacts on biotechnological processes. Appl. Microbiol. Biotechnol. 82: 625–638.

Amore, R., M. Wilhelm and C.P. Hollenberg. 1989. The fermentation of xylose—an analysis of the expression of Bacillus and Actinoplanes xylose isomerase genes in yeast. Appl. Microbiol. Biotechnol. 30: 351–357.

Amponsah, N.Y., M. Troldborg, B. Kington, I. Aalders and R.L. Hough. 2014. Greenhouse gas emissions from renewable energy sources: A review of lifecycle considerations. Renew. Sust. Energ. Rev. 39: 461–475.

Anfavea (National association of Brazilian car builders). 2015. Annual statistics. Brazil. Available from: <http://www.anfavea.com.br/tabelas.html>

Anwar, Z., M. Gulfraz and M. Irshad. 2014. Agro-industrial lignocellulosic biomass a key to unlock the future bio-energy: a brief review. J. Radiat. Res. Appl. Sci. 7: 163–173.

Baños, R., F. Manzano-Agugliaro, F.G. Montoya, C. Gil, A. Alcayde and J. Gómez. 2011. Optimization methods applied to renewable and sustainable energy: A review. Renew. Sust. Energ. Rev. 15: 1753–1766.

Bertsch, V., M. Hall, C. Weinhardt and W. Fichtner. 2016. Public acceptance and preferences related to renewable energy and grid expansion policy: Empirical insights for Germany. Energy. 114: 465–477.

Bose, B.K. 2014. Energy, global warming and impact of power electronics in the present century. pp. 1–26. *In*: Abu-Rub, H., M. Malinowski and K. Al-Haddad [eds.]. Power Electronics for Renewable Energy Systems, Transportation and Industrial Applications. Wiley-IEEE Press. New York, USA.

Brat, D., E. Boles and B. Wiedemann. 2009. Functional expression of a bacterial xylose isomerase in *Saccharomyces cerevisiae*. Appl. Environ. Microbiol. 75: 2304–2311.

Bromet, E.J. 2016. Emotional consequences of three Mile Island and Chernobyl: Lessons learned for Fukushima. pp. 67–82. *In*: Shigemura, J. and R.K. Chhem [eds.]. Mental Health and Social Issues Following a Nuclear Accident. Springer Japan. Tokio, Japan.

Bruinenberg, P.M., P.H.M. de Bot, J.P. van Dijken and W.A. Scheffers. 1984. NADH-linked aldose reductase: the key to anaerobic fermentation of xylose by yeasts. Appl. Microbiol. Biotechnol. 19: 256–260.

Burton, T., D. Sharpe, N. Jenkins and E. Bossanyi. 2001. Wind Energy Handbook. John Wiley and Sons.

Cadete, R.M., A.M. de las Heras, A.G. Sandström, C. Ferreira, F. Gírio, M.F. Gorwa-Grauslund et al. 2016. Exploring xylose metabolism in *Spathaspora* species: XYL1.2 from *Spathaspora passalidarum* as the key for efficient anaerobic xylose fermentation in metabolic engineered *Saccharomyces cerevisiae*. Biotechnol. Biofuels 9: 167.

Cannella, D. and H. Jørgensen. 2014. Do new cellulolytic enzyme preparations affect the industrial strategies for high solids lignocellulosic ethanol production? Biotechnol. Bioeng. 111: 59–68.

Caputo, A.C., M. Palumb, P.M. Pelagagge and F. Scacchia. 2005. Economics of biomass energy utilization in combustion and gasification plants: effects of logistic variables. Biomass Bioenerg. 28: 35–51.

Chaturvedi, V. and P. Verma. 2013. An overview of key pretreatment processes employed for bioconversion of lignocellulosic biomass into biofuels and value added products. Biotechnology 3(5): 415–431.

Demeke, M.M., H. Dietz, Y. Li, M.R. Foulquié-Moreno, S. Mutturi, S. Deprez et al. 2013. Development of a D-xylose is fermenting and inhibitor tolerant industrial *Saccharomyces cerevisiae* strain with high performance in lignocellulose hydrolysates using metabolic and evolutionary engineering. Biotechnol. Biofuels 6: 89.

Demirbaş, A. 2005. Potential applications of renewable energy sources, biomass combustion problems in boiler power systems and combustion related environmental issues. Prog. Energy Combust. Sci. 31: 171–192.

Demirbaş, A. 2006. Global renewable energy resources. Energy Sources, Part A 28: 779–792.

Dogaris, I., D. Mamma and D. Kekos. 2013. Biotechnological production of ethanol from renewable resources by *Neurospora crassa*: an alternative to conventional yeast fermentations. Appl. Microbial. Biotechnol. 97: 1457–1473.

dos Santos, L.V., M.F. Carazzolle, S.T. Nagamatsu, N.M.V. Sampaio, L.D. Almeida, R.A.S. Elghali, L. et al. 2016. Developing a sustainability framework for the assessment of bioenergy systems. Energy Policy 35: 6075–83.

Ellabban, O., H. Abu-Rub and F. Blaabjerg. 2014. Renewable energy resources: Current status, future prospects and their enabling technology. Renew. Sust. Energ. Rev. 39: 748–764.

Farwick, A., S. Bruder, V. Schadeweg, M. Oreb and E. Boles. 2014. Engineering of yeast hexose transporters to transport D-xylose without inhibition by D-glucose. Proc. Natl. Acad. Sci. U. S. A. 111: 5159–5164.

Fernandes, M.C., I. Torrado, F. Carvalheiro, V. Dores, V. Guerra, P.M. Lourenço et al. 2016. Bioethanol production from extracted olive pomace: dilute acid hydrolysis. Bioethanol. 2.

Ferreira, D., A. Nobre, M.L. Silva, F. Faria-Oliveira, J. Tulha, C. Ferreira et al. 2013. XYLH encodes a xylose/H+ symporter from the highly related yeast species *D. fabryi* and *D. hansenii*. FEMS Yeast Res. 13: 585–596.

Field, C.B., J.E. Campbell and D.B Lobell. 2008. Biomass energy: the scale of the potential resource. Trends Ecol. Evol. 23: 65–72.

Gárdonyi, M. and B. Hahn-Hägerdal. 2003. The *Streptomyces* rubiginosus xylose isomerase is misfolded when expressed in *Saccharomyces cerevisiae*. Enzyme Microb. Tech. 32: 252–259.

Gnansounou, E. 2010. Production and use of lignocellulosic bioethanol in Europe: Current situation and perspectives. Bioresour. Technol. 101: 4842–4850.

Gonçalves, D.L., A. Matsushika, B. Belisa, T. Goshima, E.P. Bon and B.U. Stambuk. 2014. Xylose and xylose/glucose co-fermentation by recombinant *Saccharomyces cerevisiae* strains expressing individual hexose transporters. Enzyme Microb. Technol. 63: 13–20.

Gupta, A. and J.P. Verma. 2015. Sustainable bio-ethanol production from agro-residues: a review. Renew. Sust. Energ. Rev. 41: 550–567.

Hahn-Hägerdal, B., M. Galbe, M.F. Gorwa-Grauslund, G. Lidén and G. Zacchi. 2006. Bio-ethanol—the fuel of tomorrow from the residues of today. Trends Biotechnol. 24(12): 549–556.

Hahn-Hägerdal, B., K. Karhumaa, C. Fonseca, I. Spencer-Martins and M.F. Gorwa-Grauslund. 2007. Towards industrial pentose-fermenting yeast strains. Appl. Microbial. Biotechnol. 74: 937–953.

Hamacher, T., J. Becker, M. Gárdonyi, B. Hahn-Hägerdal and E. Boles. 2002. Characterization of the xylose transporting properties of yeast hexose transporters and their influence on xylose utilization. Microbiology 148: 2783–2788.

Harmsen, P.F.H., W. Huijgen, L. Bermudez and R. Bakker. 2010. Literature review of physical and chemical pretreatment processes for lignocellulosic biomass (No. 1184). Wageningen UR Food and Biobased Research.

Harner, N.K., P.K. Bajwa, M.B. Habash, J.T. Trevors, G.D. Austin and H. Lee. 2014. Mutants of the pentose fermenting yeast *P. tannophilus* tolerant to hardwood spent sulfite liquor and acetic acid. Anton. Leeuw. 105: 29–43.

Harner, N.K., X. Wen, P.K. Bajwa, G.D. Austin, C.Y. Ho, M.B. Habash et al. 2015. Genetic improvement of native xylose-fermenting yeasts for ethanol production. J. Ind. Microbiol. Biotechnol. 42: 1–20.

Hector, R.E., N. Qureshi, S.R. Hughes and M.A. Cotta. 2008. Expression of a heterologous xylose transporter in a *Saccharomyces cerevisiae* strain engineered to utilize xylose improves aerobic xylose consumption. Appl. Microbiol. Biotechnol. 80: 675–684.

Hendriks, A.T.W.M. and G. Zeeman. 2009. Pretreatments to enhance the digestibility of lignocellulosic biomass. Bioresour. Technol. 100: 10–18.

Ho, N.W., Z. Chen and A.P. Brainard. 1998. Genetically engineered *Saccharomyces yeast* capable of effective cofermentation of glucose and xylose. Appl. Environ. Microbiol. 64: 1852–1859.

Jeffries, T.W. and Y.S. Jin. 2004. Metabolic engineering for improved fermentation of pentoses by yeasts. Appl. Microbiol. Biotechnol. 63: 495–509.

Jeppsson, M., O. Bengtsson, K. Franke, H. Lee, B. Hahn-Hägerdal and M.F. Gorwa-Grauslund. 2006. The expression of a *Pichia stipitis* xylose reductase mutant with higher K(M) for NADPH increases ethanol production from xylose in recombinant *Saccharomyces cerevisiae*. Biotechnol. Bioeng. 93: 665–673.

Jourdier, E., C. Cohen, L. Poughon, C. Larroche, F. Monot and F.B. Chaabane. 2013. Cellulase activity mapping of *Trichoderma reesei* cultivated in sugar mixtures under fed-batch conditions. Biotechnol. Biofuels 6: 79.

Kalogirou, S.A. 2013. Solar energy engineering: processes and systems. Academic Press.

Karhumaa, K., R. Fromanger, B. Hahn-Hägerdal and M.F. Gorwa-Grauslund. 2007. High activity of xylose reductase and xylitol dehydrogenase improves xylose fermentation by recombinant *Saccharomyces cerevisiae*. Appl. Microbiol. Biotechnol. 73: 1039–1046.
Karp, E.M., B.S. Donohoe, M.H. O'Brien, P.N. Ciesielski, A. Mittal, M.J. Biddy et al. 2014. Alkaline pretreatment of corn stover: bench-scale fractionation and stream characterization. ACS Sustainable Chem. Eng. 2: 1481–1491.
Katahira, S., M. Ito, H. Takema, Y. Fujita, T. Tanino, T. Tanaka et al. 2008. Improvement of ethanol productivity during xylose and glucose co-fermentation by xylose-assimilating *S. cerevisiae* via expression of glucose transporter Sut1. Enzyme Microb. Technol. 43: 115–119.
Khoo, H.H. 2015. Review of bio-conversion pathways of lignocellulose-to-ethanol: Sustainability assessment based on land footprint projections. Renew. Sust. Energ. Rev. 46: 100–119.
Kim, S. and M.T. Holtzapple. 2005. Lime pretreatment and enzymatic hydrolysis of corn stover. Bioresour. Technol. 96: 1994–2006.
Kim, S.R., Y.C. Park, Y.S. Jin and J.H. Seo. 2013. Strain engineering of *Saccharomyces cerevisiae* for enhanced xylose metabolism. Biotechnol. Adv. 31: 851–861.
Kömürcü, M.İ. and A. Akpinar. 2010. Hydropower energy versus other energy sources in Turkey. Energ. Sources Part B 5: 185–198.
Koppram, R., E. Tomás-Pejó, C. Xiros and L. Olsson. 2014. Lignocellulosic ethanol production at high-gravity: challenges and perspectives. Trends Biotechnol. 32: 46–53.
Kötter, P., R. Amore, C.P. Hollenberg and M. Ciriacy. 1990. Isolation and characterization of the *Pichia stipitis* xylitol dehydrogenase gene, XYL2, and construction of a xylose-utilizing *Saccharomyces cerevisiae* transformant. Curr. Genet. 18: 493–500.
Kötter, P. and M. Ciriacy. 1993. Xylose fermentation by *Saccharomyces cerevisiae*. Appl. Microbiol. Biotech. 38: 776–783.
Kumar, R., S. Singh and O.V. Singh. 2008. Bioconversion of lignocellulosic biomass: biochemical and molecular perspectives. J. Ind. Microbiol. Biotechnol. 35: 377–391.
Kuyper, M., H.R. Harhangi, A.K. Stave, A.A. Winkler, M.S. Jetten, W.T. de Laat et al. 2003. High-level functional expression of a fungal xylose isomerase: the key to efficient ethanolic fermentation of xylose by *Saccharomyces cerevisiae*? FEMS Yeast Res. 4: 69–78.
Kuyper, M., A. Aaron, J.P. van Dijken and J.T. Pronk. 2004. Minimal metabolic engineering of *Saccharomyces cerevisiae* for efficient anaerobic xylose fermentation: a proof of principle. FEMS Yeast Res. 4: 655–664.
Kuyper, M., M.J. Toirkens, J.A. Diderich, A.A. Winkler, J.P. van Dijken and J.T. Pronk. 2005. Evolutionary engineering of mixed-sugar utilization by a xylose-fermenting *Saccharomyces cerevisiae* strain. FEMS Yeast Res. 5: 925–934.
Latimer, L.N., M.E. Lee, D. Medina-Cleghorn, R.A. Kohnz, D.K. Nomura and J.E. Dueber. 2014. Employing a combinatorial expression approach to characterize xylose utilization in *Saccharomyces cerevisiae*. Metab. Eng. 25: 20–29.
Leandro, M.J., P. Gonçalves and I. Spencer-Martins. 2006. Two glucose/xylose transporter genes from the yeast *C. intermedia*: first molecular characterization of a yeast xylose-H^+ symporter. Biochem. J. 395: 543–549.
Leandro, M.J., I. Spencer-Martins and P. Gonçalves. 2008. The expression in *Saccharomyces cerevisiae* of a glucose/xylose symporter from *C. intermedia* is affected by the presence of a glucose/xylose facilitator. Microbiology 154: 1646–1655.
Lee, J. 1997. Biological conversion of lignocellulosic biomass to ethanol. J. Biotechnol. 56: 1–24.
Lee, S.M., T. Jellison and H.S. Alper. 2014. Systematic and evolutionary engineering of a xylose isomerase-based pathway in *Saccharomyces cerevisiae* for efficient conversion yields. Biotech. Biofuels 7: 122.
Leggio, L.L., D. Welner and L. De Maria. 2012. A structural overview of GH61 proteins–fungal cellulose degrading polysaccharide monooxygenases. Comput. Struct. Biotechnol. J. 2: 1–8.
Lemos, P. and F.C. Mesquita. 2016. Future of global bioethanol: An appraisal of results, risk and uncertainties. pp. 221–237. *In*: Luiz, M.S.F.S. and B.C.L. Augusto [eds.]. Global Bioethanol: Evolution, Risks, and Uncertainties. Elseiver Inc. Ámsterdam, Países Bajos.
Li, W., M. Li, L. Zheng, Y. Liu, Y. Zhang, Z. Yu et al. 2015. Simultaneous utilization of glucose and xylose for lipid accumulation in black soldier fly. Biotechnol. Biofuels 8: 117.
Lin, Z. and W.H. Li. 2011. Expansion of hexose transporter genes was associated with the evolution of aerobic fermentation in yeasts. Mol. Biol. Evol. 28: 131–42.
MacLean, H.L. and L.B. Lave. 2003. Evaluating automobile fuel/propulsion system technologies. Prog. Energy Combust. Sci. 29: 1–69.
Madhavan, A., S. Tamalampudi, A. Srivastava, H. Fukuda, V.S. Bisaria and A. Kondo. 2009. Alcoholic fermentation of xylose and mixed sugars using recombinant *Saccharomyces cerevisiae* engineered for xylose utilization. Appl. Microbiol. Biotechnol. 82: 1037–1047.
Manohar, K. and A.A. Adeyanju. 2009. Hydropower energy resources in Nigeria. J. Eng. Appl. Sci. 4: 68–73.

Matsushika, A., H. Inoue, S. Watanabe, T. Kodaki, K. Makino and S. Sawayama. 2009a. Efficient bioethanol production by a recombinant flocculent *Saccharomyces cerevisiae* strain with a genome-integrated NADP+-dependent xylitol dehydrogenase gene. Appl. Environ. Microbiol. 75: 3818–3822.

Matsushika, A., H. Inoue, T. Kodaki and S. Sawayama. 2009b. Ethanol production from xylose in engineered *Saccharomyces cerevisiae* strains: current state and perspectives. Appl. Environ. Microbiol. 84: 37–53.

McKendry, P. 2002. Energy production from biomass (part 1): overview of biomass. Bioresour. Technol. 83: 37–46.

Mood, S.H., A.H. Golfeshan, M. Tabatabaei, G.S. Jouzani, G.H. Najafi, M. Gholami et al. 2013. Lignocellulosic biomass to bioethanol, a comprehensive review with a focus on pretreatment. Renew. Sust. Energ. Rev. 27: 77–93.

Moyses, D.N., V.C.B. Reis, J.R.M.D. Almeida, L.M.P.D. Moraes and F.A.G. Torres. 2016. Xylose fermentation by *Saccharomyces cerevisiae*: challenges and prospects. Int. J. Mol. Sci. 17: 207.

Mupondwa, E., X. Li, L. Tabil, S. Sokhansanj and P. Adapa. 2017. Status of Canada's lignocellulosic ethanol: Part I: Pretreatment technologies. Renew. Sust. Energ. Rev. 72: 178–190.

Neha, N. and R. Rekha. 2016. Production of bioethanol from lignocellulosic feedstock, as raw material through 2 step enzymatic process: An alternative energy fuel. Int. J. Eng. Res. Dev. 12: 24–32.

Nijland, J.G., H.Y. Shin, R.M. de Jong, P.P. De Waal, P. Klaassen and A.J. Driessen. 2014. Engineering of an endogenous hexose transporter into a specific D-xylose transporter facilitates glucose-xylose co-consumption in *Saccharomyces cerevisiae*. Biotechnol. Biofuels 7: 168.

Nogué, V.S. and K. Karhumaa. 2015. Xylose fermentation as a challenge for commercialization of lignocellulosic fuels and chemicals. Biotechnol. Lett. 37: 761–772.

Olofsson, K., M. Bertilsson and G. Lidén. 2008. A short review on SSF—an interesting process option for ethanol production from lignocellulosic feedstocks. Biotechnol. Biofuels 1: 7.

Olsson, L., H.R. Soerensen, B.P. Dam, H. Christensen, K.M. Krogh and A.S. Meyer. 2006. Separate and simultaneous enzymatic hydrolysis and fermentation of wheat hemicellulose with recombinant xylose utilizing *Saccharomyces cerevisiae*. Appl. Biochem. Biotechnol. 129: 117–129.

Panwar, N.L., S.C. Kaushik and S. Kothari. 2011. Role of renewable energy sources in environmental protection: a review. Renew. Sust. Energ. Rev. 15: 1513–1524.

Parachin, N.S., B. Bergdahl, E.W. van Niel and M.F. Gorwa-Grauslund. 2011. Kinetic modelling reveals current limitations in the production of ethanol from xylose by recombinant *Saccharomyces cerevisiae*. Metab. Eng. 13: 508–517.

Patel, S. and K.V.S. Rao. 2016. Social acceptance of a biomass plant in India. Power and Energy Systems: Towards Sustainable Energy (PESTSE). Biennial International Conference on Power and Energy Systems: Towards Sustainable Energy. Bengalore, India. pp. 1–6.

Ravindran, R. and A.K. Jaiswal. 2016. A comprehensive review on pre-treatment strategy for lignocellulosic food industry waste: challenges and opportunities. Bioresour. Technol. 199: 92–102.

Reynolds, T.S. 2002. Stronger than a hundred men: a history of the vertical water wheel, JHU Press.

Rizzi, M., P. Erlemann, N.A. Bui-Thanh and H. Dellweg. 1988. Xylose fermentation by yeasts. 4. Purification and kinetic studies of xylose reductase from *Pichia stipitis*. Appl. Microbiol. Biotechnol. 29: 148–154.

Rosgaard, L., S. Pedersen, J.R. Cherry, P. Harris and A.S. Meyer. 2006. Efficiency of new fungal cellulase systems in boosting enzymatic degradation of barley straw lignocellulose. Biotechnol. Prog. 22: 493–498.

Runquist, D., C. Fonseca, P. Radstrom, I. Spencer-Martins and B. Hahn-Hägerdal. 2009. Expression of the Gxf1 transporter from *C. intermedia* improves fermentation performance in recombinant xylose-utilizing *Saccharomyces cerevisiae*. Appl. Microbiol. Biotechnol. 82: 123–130.

Saloheimo, A., J. Rauta, V. Stasyk, A.A. Sibirny, M. Penttilä and L. Ruohonen. 2007. Xylose transport studies with xylose-utilizing *Saccharomyces cerevisiae* strains expressing heterologous and homologous permeases. App. Microbiol. Biotechnol. 74: 1041–1052.

Schilling, M.A. and M. Esmundo. 2009. Technology S-curves in renewable energy alternatives: Analysis and implications for industry and government. Energy Policy 37: 1767–1781.

Schubert, C. 2006. Can biofuels finally take center stage? Nat. Biotechnol. 24: 777–784.

Singh, R., A. Shukla, S. Tiwari and M. Srivastava. 2014. A review on delignification of lignocellulosic biomass for enhancement of ethanol production potential. Renew. Sust. Energ. Rev. 32: 713–728.

Singhania, R.R., J.K. Saini, R. Saini, M. Adsul, A. Mathur, R. Gupta et al. 2014. Bioethanol production from wheat straw via enzymatic route employing *Penicillium janthinellum* cellulases. Bioresour. Technol. 169: 490–495.

Sivers, V.M. and G. Zacchi. 1995. A techno-economical comparison of three processes for the production of ethanol from pine. Bioresour. Technol. 51: 43–52.

Srivastava, T. 2013. Renewable energy (gasification). Adv. Electron. Electr. Eng. 3: 1243–1250.

Stigka, E.K., J.A. Paravantis and G.K. Mihalakakou. 2014. Social acceptance of renewable energy sources: A review of contingent valuation applications. Renew. Sust. Energ. Rev. 32: 100–106.

Sun, Y. and J. Cheng. 2002. Hydrolysis of lignocellulosic materials for ethanol production: a review. Bioresour. Technol. 83: 1–11.

Tavner, P.J., J. Xiang and F. Spinato. 2007. Reliability analysis for wind turbines. Wind Energy 10: 1–18.

Tenenbaum, D.J. 2008. Food vs. fuel: diversion of crops could cause more hunger. Environ. Health Perspect. 116: 254–257.
Thompson, A. and A. Macdonald. 2014. The State of the Geothermal Energy Industry in Canada. GSA Annual Meeting in Vancouver, British Columbia.
Toivari, M.H., L. Salusjarvi, L. Ruohonen and M. Penttila. 2004. Endogenous xylose pathway in *Saccharomyces cerevisiae*. Appl. Environ. Microbiol. 70: 3681–3686.
Tomás-Pejó, E., J.M. Oliva, M. Ballesteros and L. Olsson. 2008. Comparison of SHF and SSF processes from steam-exploded wheat straw for ethanol production by xylose-fermenting and robust glucose-fermenting *Saccharomyces cerevisiae* strains. Biotechnol. Bioeng. 100: 1122–1131.
Träff, K.L., R.R. Otero Cordero, W.H. van Zyl and B. Hahn-Hägerdal. 2001. Deletion of the GRE3 aldose reductase gene and its influence on xylose metabolism in recombinant strains of *Saccharomyces cerevisiae* expressing the xylA and XKS1 genes. Appl. Environ. Microbiol. 67: 5668–5674.
Verho, R., J. Londesborough, M. Penttilä and P. Richard. 2003. Engineering redox cofactor regeneration for improved pentose fermentation in *Saccharomyces cerevisiae*. Appl. Environ. Microbiol. 69: 5892–5897.
Wahlbom, C.F., R.R.C. Otero, W.H. van Zyl, B. Hahn-Hägerdal and L.J. Jönsson. 2003. Molecular analysis of a *Saccharomyces cerevisiae* mutant with improved ability to utilize xylose shows enhanced expression of proteins involved in transport, initial xylose metabolism, and the pentose phosphate pathway. Appl. Environ. Microbiol. 69: 740–746.
Walfridsson, M., X. Bao, M. Anderlund, G. Lilius, L. Bülow and B. Hahn-Hägerdal. 1996. Ethanolic fermentation of xylose with *Saccharomyces cerevisiae* harboring the *T. thermophiles* xylA gene, which expresses an active xylose (glucose) isomerase. Appl. Environ. Microbiol. 62: 4648–4651.
Wang, C., X. Bao, Y. Li, C. Jiao, J. Hou, J.Q. Zhang et al. 2015. Cloning and characterization of heterologous transporters in *Saccharomyces cerevisiae* and identification of important amino acids for xylose utilization. Metabol. Eng. 30: 79–88.
Weierstall, T., C.P. Hollenberg and E. Boles. 1999. Cloning and characterization of three genes (SUT1-3) encoding glucose transporters of the yeast *Pichia stipitis*. Mol. Microbiol. 31: 871–883.
Wentzel, M. and A. Pouris. 2007. The development impact of solar cookers: a review of solar cooking impact research in South Africa. Energy Policy 35: 1909–1919.
Wilkinson, S., P. Stoller, P. Ralph, B. Hamdorf, L.N. Catana and G.S. Kuzava. 2016. Exploring the feasibility of algae building technology in NSW. In SBE16 International High Performance Built Environments Conference. Australia.
Wimalasena, T.T., D. Greetham, M.E. Marvin, G. Liti, Y. Chandelia, A. Hart et al. 2014. Phenotypic characterisation of *Saccharomyces* spp. yeast for tolerance to stresses encountered during fermentation of lignocellulosic residues to produce bioethanol. Microb. Cell Fact. 13: 47.
Xiros, C., P. Katapodis and P. Christakopoulos. 2011. Factors affecting cellulose and hemicellulose hydrolysis of alkali treated brewers spent grain by *Fusarium oxysporum* enzyme extract. Bioresour. Technol. 102: 1688–1696.
Young, E., S.M. Lee and H. Alper. 2010. Optimizing pentose utilization in yeast: the need for novel tools and approaches. Biotechnol. Biofuels 3: 24.
Young, E.M., A.D. Comer, H. Huang and H.S. Alper. 2012. A molecular transporter engineering approach to improving xylose catabolism in *Saccharomyces cerevisiae*. Metab. Eng. 14: 401–411.

3

Recovery of Sugarcane Vinasse by Microbial Pathways

An Integral Approach

Macarena M. Rulli,[1,a] *Luciana M. Del Gobbo,*[1,b] *María J. Amoroso*[1,2] and *Verónica L. Colin*[2,*]

Introduction

Vinasse is the main effluent generated from alcohol distilleries, with an average volume of 12 l of vinasse per l of ethyl alcohol produced (Christofoletti et al. 2013). Vinasse composition varies according to the raw material (corn, wheat, rice, potatoes, sugar beets, sugarcane and sweet sorghum) and equipment utilized in the process of alcohol production (Romanholo Ferreira et al. 2011). However, the effluent is a brown liquid which contains basically water (93%) as well as organic solids and minerals (7%). Vinasse is usually characterized by a low pH (4.5 to 5.5), high salt content (30.5 to 45.2 d Sm^{-1}), and a chemical oxygen demand (COD) of around 80,000 to 100,000 mg l^{-1} while the biochemical oxygen demand (BOD) often range from 30 to 70% of the COD (Pant and Adholeya 2007). Vinasse also contains heavy metals in variable concentrations and numerous potential toxins that decrease effluent quality (Colin et al. 2016).

The sugar industry in Argentina is concentrated in Northwest of the country, particularly in Tucumán, Salta and Jujuy provinces. Sugarcane production has significant economic and social importance in the Tucumán province since it is one of its main agro-industrial activities. When comparing the average crop performance of Tucumán with the world average, productivity (measured in tons of cane per ha) is well above the world average (Perez et al. 2007). Thereby, in our province, vinasse is mainly produced from sugarcane juice or from sugarcane molasses, with both feedstocks rich in potassium, phosphate, sulfate, calcium, iron, sodium, and organic compounds such as glycerol, lactic

[1] Facultad de Bioquímica, Química y Farmacia, Universidad Nacional de Tucumán, Ayacucho 491, 4000, Tucumán, Argentina.
[a] Email: macarenarulli@gmail.com
[b] Email: lucianadelgobbbo@gmail.com
[2] Planta Piloto de Procesos Industriales Microbiológicos (PROIMI-CONICET), Av. Belgrano y Pje. Caseros, San Miguel de Tucumán, T4001MVB, Tucumán, Argentina.
Email: amoroso@proimi.org.ar
* Corresponding author: veronicacollin@yahoo.com.ar

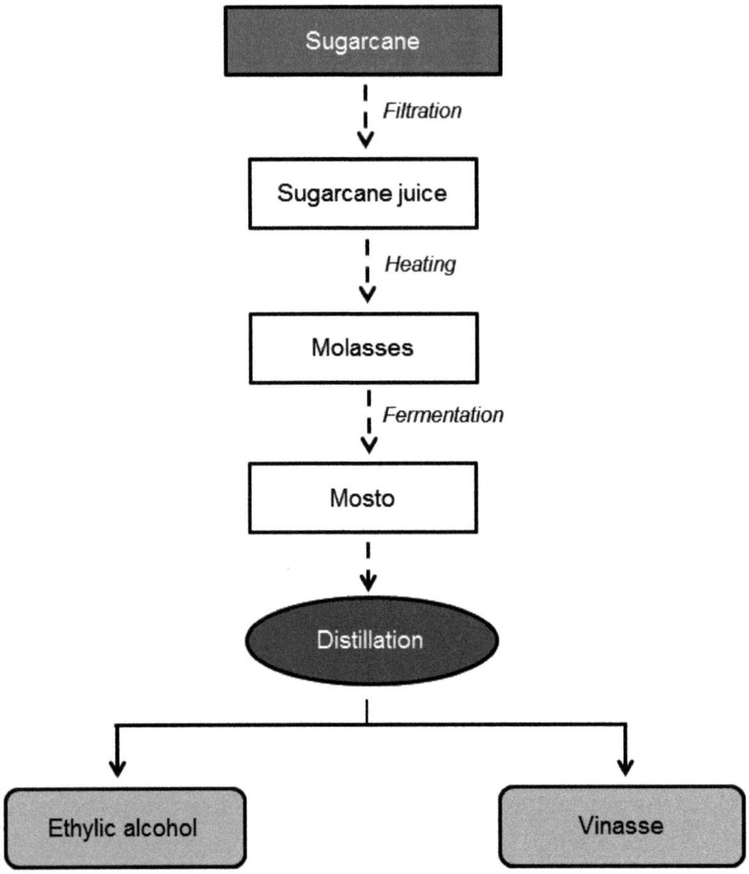

Figure 1. Flow chart of the ethanol production process and underproduction of sugarcane vinasse (Adapted from Christofoletti et al. 2013).

acid and acetic acid (Rajagopal et al. 2014). In Fig. 1, a flow chart summarizing the ethanol production process and underproduction of sugarcane vinasse is shown.

The vinasse is often released into the main canal that runs through to the sugar factory and meanders into the secondary channels around the cane fields. Aquatic ecosystems are the main receptors for toxic substances from industrial and agricultural activities. These ecosystems are able to assimilate and even neutralize the toxic substances as long as they do not exceed the depuration capacity of the water. Because of the chemical composition of vinasse, it has been observed that its continuous disposal in the waterway could result in the depletion of the oxygen supply and the accumulation of very high levels of heavy metals and other toxins, causing the canal to become not habitable for aquatic life. Finally, its detrimental impact could extend to human life (Clementson et al. 2016). The low pH, electric conductivity, and chemical elements present in sugarcane vinasse could also cause changes in the chemical and physical-chemical properties of soil (Escher et al. 2016). Soil is often considered as the direct receptor of pollutants, providing an important tool for genetic tests and environmental monitoring. In fact, different assays demonstrated that excessive amounts of vinasse applied in soils can have toxic, cytotoxic and genotoxic effects on plants (Srivastava and Jain 2010, Jain and Srivastava 2012, Juárez Cortes 2016).

One of the principles of green chemistry enunciates that it is preferable to avoid the generation of a residue than to try to clean it once formed (Anastas and Warner 1998). However, the Tucumán province produces just over 60% of Argentina's sugarcane alcohol in 11 sugar mills coupled with autonomous distilleries. The generation of large volumes of vinasse is, therefore, unavoidable making it necessary to look for alternatives to improve the management of this waste. The Experimental Agroindustrial Obispo Colombres Station is a regional center of Tucumán province that has worked for more than 25 years on this topic, proposing to consider vinasse as a sub-product with a potential utility (Mornadini and Quaia 2013). Among potential utilities, the reconversion of vinasse in products of biotechnological interest is considered as a highly promising alternative. In our research group, the generation of surface-active compounds (SACs) from actinobacteria was intensively assayed (Colin et al. 2013a,b, 2016). Although the SACs of microbial origin have similar properties than their synthetic counterparts, they present certain advantages (greater biodegradability, reduced toxicity, stability in extreme conditions of pH, temperature, and salinity, etc.); thereby, they tend to be more acceptable from a social point of view (Colin et al. 2013b). In the field of the bioremediation, the microbial SACs are considered an attractive group of compounds with potential utility for the recovery of petroleum and its derivatives (Pacwa-Płociniczak et al. 2011, Silva et al. 2014, Motevasel 2014, Montero-Rodriguez et al. 2015). However, the direct use of microbial SACs as soil washing agents, without the presence of the producer microorganisms in the cleaning process, is considered a methodology with greater pragmatism for the recovery of diverse pollutants.

Despite unquestionable advantages of SACs of natural occurrence, their large scale application is restricted by the high cost of production based upon use of synthetic substrates. As an example, it is convenient to cite to the surfactin, a glycopeptide produced by *Bacillus subtilis* which has an exceptional surface activity. Compared to some synthetic surfactants such as sodium dodecyl sulfate (SDS) and trimethyl ammonium bromide (BTA), surfactin has a higher capacity to reduce surface and interfacial tension and presents lower values of critical micellar concentration (CMC) (Cavalcante Barros et al. 2007). Although the surfactin is marketed in the pure state and in small amounts by some companies, its cost is too high (over 50 USD per milligram of product) as to be routinely used in environmental remediation technologies. This reflects the inability to compete with synthetic surfactants that are routinely used; thereby, the biosynthesis of microbial SACs from local wastes has emerged as an attractive practice.

Taking into account this background, the present chapter reviews the current advances related to the recovery of vinasse by microbial pathways. Their reconversion in microbial SACs is particularly considered as well as the application of these biomolecules in soil washing technologies.

Integral Approach for Vinasse Recovery by Microbial Pathways

The vinasse recovery by microbial pathways is feasible due to the high concentration of biodegradable organic carbon. Figure 2 shows an approach based on considering the vinasse as a byproduct of the sugar-alcohol industry. The approach aims at the integral use of vinasse as it is intended to take advantage of both the microbial biomass and the supernatant generated from the raw effluent. This approach is divided into three main blocks which are detailed as follows:

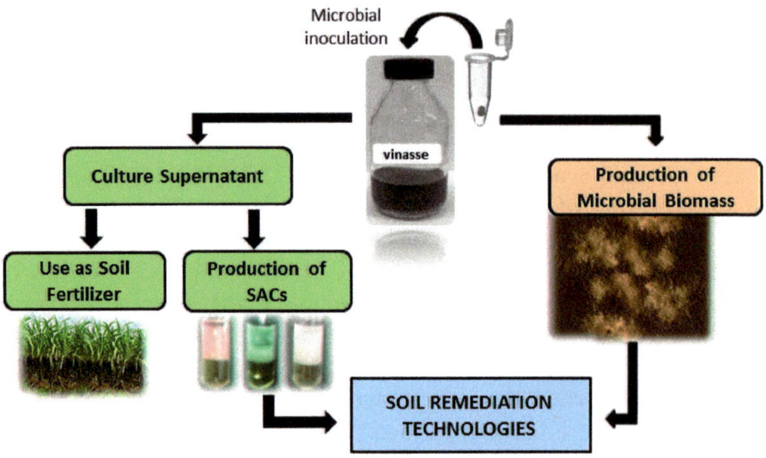

Figure 2. Flow chart of the experimental procedure purposed for the integral use of vinasse as a byproduct of the sugar-alcohol industry.

Vinasse as Feedstock for the Production of Biomass

It is known that the vinasse can be used as a source of nutrients for the production of microbial biomass. For example, Sartori et al. (2015) reported on the production of fungal mycelia of *Pleurotus sajor-caju*, *P. ostreatus*, *P. albidus* and *P. flabellatus* after cultivation in vinasse for 15 d. The mycelia produced were then lyophilized and used as a complementary diet for *Danio rerio* fish. Studies provided by Barrocal et al. (2010) also reported on the biomass production of cyanobacterium *Spirulina maxima* from vinasse, which can be used as an animal feed supplement.

In our particular case, biomass production aims its application as a bioremediation tool for contaminated environments. Chojnacka (2009) refers to two processes involving the use of microbial biomass to clean environment: one of them, known as bioaccumulation, involves the use of living biomass being it an active process mediated by transport systems that require cellular energy. The second process, known as biosorption, uses non-living biomass and refers to a passive and universal process, in which diverse materials of biological origin can be used as biosorbents. Both low-cost and high-cost biosorbents are available. The first group includes by-products or waste products from industrial processes, including seaweeds, yeast from wine or brewery industry, agricultural wastes, etc. (Volesky 2007). High-cost biosorbents include biomass, which should possess certain characteristics such as very high biosorption capacity (200 mg g^{-1}) and a high affinity to sorbate. The biomass of microalgae *Spirulina* sp. constitutes a typical example of a high-cost biosorbent (Chojnacka 2009).

Our research group has provided various studies based upon use of living actinobacteria biomass to remove both heavy metals and pesticides from soils and liquid systems (Colin et al. 2012, Álvarez et al. 2017). The actinobacteria are suited for their inoculation in contaminated systems as a consequence of their relatively rapid growth rates and ability to colonize substrates due to the mycelial formation (Ravel et al. 1998). However, the studies up to this date are mostly based on the use of biomass produced in commercial synthetic media (Polti et al. 2014). Therefore, the first step to ensuring the sustainability of the biotechnological process would consider the biomass generation from cheap feedstocks. With respect to this, Aparicio et al. (2017) referred to the ability of four actinobacteria to generate biomass from sugarcane vinasse. Interestingly, the biomass fractions recovered until three consecutive cycles of vinasse re-use were useful to remove lindane and hexavalent chromium from the soil. Based on these findings, the authors inferred that a volume from 10,000 to 12,000 l of vinasse could be consumed in order to produce the biomass necessary for the treatment of one hectare of contaminated soil. Also, it was inferred that this volume would be the equivalent of the amount of vinasse generated each 1,000 l of ethylic alcohol. Likewise, a removal more than 70% of the BOD was achieved at the end of the reused vinasse cycle, suggesting a proportional reduction in the effluent toxicity. Being a promising alternative to improve the management of waste from the sugar-alcohol industry, this approach could be one of the bases for the future production of biomass on a large scale.

Vinasse as Fertilizer

Organic residues generated as byproducts of the sugar and alcohol industry can be used as soil improvers and substitutes for inorganic phosphorus and potassium fertilizers (de Mello Prado et al. 2013). The alternatives of agricultural use of vinasse nowadays are the ones of greater application, not only in our province but also in the whole world. Even in the Tucumán province, this practice has been recognized as environmentally sustainable and incorporated into environmental legislation (Res. SEMA N° 040, 2011). It is important to point out that the agricultural use of vinasse is a low-cost investment alternative since it only requires the land available to carry out it. However, the use of the residues in agriculture requires some specific recommendations prior to its application, in order to prevent potential environmental damage (Soler da Silva et al. 2013).

The use of vinasse in crude form without a prior conditioning treatment can be detrimental to soil quality due to acid pH to the large amounts of organic matter, and to the high concentrations of suspended and volatile solids. Occasionally, other toxic compounds like heavy metals can be detected in the crude effluent at very variable concentrations. Besides, the irrigation of vinasse can cause emissions of greenhouse gases, such as carbon dioxide, nitrous oxide and methane (Moran-Salazar et al. 2016). To overcome these limitations, different modalities for the vinasse application to the soil were developed. Biocomposite is, for example, a modality that consists of the aerobic decomposition of organic matter by microbial action. Different organic residues from the sugar-alcohol industry (cachaça, bagasse, etc.) are used in mixtures where up to 30% of the produced vinasse can be incorporated (Mornadini and Quaia 2013). Additional microorganisms (inocula) prepared to accelerate the decomposition process may also be used.

A second modality consists in to apply the vinasse in a concentrated form either by evaporation or natural drying, and that aims to reduce the initial volume of the effluent. Concentrated vinasse is a syrup that can be transported at lower costs than the original liquid effluent. At a later stage, it is possible to dehydrate it completely, until a fine powder that retains its characteristics of organic fertilizer is obtained (Mornadini and Quaia 2013).

A third alternative, known as fertigation, makes direct use of raw vinasse (at a dilution 1:10 to 1:30 v/v) as soil improvers. This is a system of easy implementation and low cost, which has been applied for more than 15 yr in the Tucumán province (Mornadini and Quaia 2013). The use of diluted vinasse aims mainly to reduce the organic load and the concentration of other toxins, avoiding a detrimental effect on the environment. However, this alternative has the disadvantage of spending water resources, increasing the final volume of the handled effluent even more. An alternative to mitigate the vinasse toxicity and improve its quality as fertilizer, without spending water resources, consists in subjecting it to the microbial action in order to neutralize the pH and to balance its C/N ratio (Mornadini and Quaia 2013). With respect to this, we reported a first advance on the recovery of a sugarcane vinasse sample using an actinobacterium from our collection of cultures, *Streptomyces* sp. MC1 (Colin et al. 2016). A significant mitigation of the effluent toxicity subjected to microbial action during 4 d was effectively demonstrated with respect to the crude vinasse, using *Lactuca sativa* seedlings as bioremediation indicator.

In our laboratory, two strains of filamentous fungi were isolated from soil exposed to high sugarcane vinasse doses for long periods of time (Fig. 3). The strains denoted as V_1 (pink spores) and V_2 (green spores) (Fig. 3A), were isolated in a selective solid medium constituted by vinasse diluted in distilled water, without the addition of exogenous nutrients (unpublished data). Microscopical observations of these colonies revealed the presence of sparsely branched and septate hyphae as well as the presence of aspergillary heads (Fig. 3B).

Assays in liquid media revealed the ability of V_1 and V_2 strains to grow at high vinasse concentrations, removing a large percentage of the BOD in a short time period (unpublished data). This could be due to the that prolonged exposure of soil to high doses of vinasse allows the development of microorganisms capable of degrading this

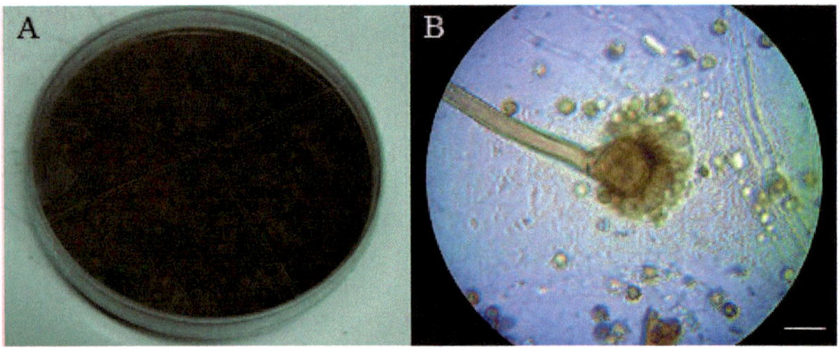

Figure 3. Colonies of filamentous fungi isolated in a vinasse-based solid medium (A). Aspergillary heads detected in microscopic preparations of both colonies with the bar representing 10 µm (B).

effluent (Carrillo Pérez et al. 2004). Based on these findings, we are currently designing new microbiological processes using these new isolates to provide a biotechnological tool committed to improving the quality of vinasse.

Vinasse as Feedstock for the Production of Microbial Surface-Active Compounds

The generation of products from economically viable feedstocks is another of the principles of green chemistry that must be considered to optimize any biotechnological process (Anastas and Warner 1998). This is of great importance, more so if it is considered that the feedstock can represent from 10–30% of total production costs (Makkar and Cameotra 2002). The wastewater of great regional availability like vinasse can be used as a low-cost substrate for producing metabolites by microbial fermentation. In fact, Aguiar et al. (2010) reported on the production of fungal lignocellulolytic enzymes using vinasse and sugarcane bagasse as cheap feedstock. Utilization of vinasse for the production of biopolymers by bacteria *Haloferax mediterranei* and *Haloarcula marismortui* was also successfully reported (Bhattacharyya et al. 2012, Pramanik et al. 2012). The anaerobic digestion of vinasse is also a process of high biotechnological value since it allows the generation of the methane biogas from a waste (Janke et al. 2014, Moraes et al. 2014).

Most of the SACs used for industrial applications are synthetically manufactured using petrochemicals as precursors (Dreja et al. 2012). This implies major limitations because they are derived from a non-renewable resource. Additionally, synthetic SACs are often associated with higher toxicity, and low biodegradability compared to their biologically-derived counterparts (Banat et al. 2010, Fracchia et al. 2012). During the last years, SACs of biological origin have been driven by changing government legislation that points towards the use of renewable and less toxic compounds (Marchant and Banat 2012a,b). Plant and animal sources used for supplying biological SACs are non-sustainable since they can be greatly affected by meteorological conditions. Besides, SACs derived from these sources are often required in large quantities in order to achieve optimal functionality. In this context, there is an increasing interest in the use of alternative and more reliable sources for the production of biological SACs. Microorganisms offer a considerable alternative to replace SACs derived from plants and animals. Additionally, microbial SACs are a group of structurally diverse molecules, which offer a good performance under extreme conditions of temperature, pH, pressure, salinity, even at low concentrations (Banat et al. 2010, Khopade et al. 2012, Colin et al. 2013a). Despite their potential advantages, the commercial availability of microbial SACs is currently limited, mainly due to economic obstacles. The production of new biomolecules, from local wastewaters like vinasse could, therefore, be the key to overcoming this challenge. Guerra de Oliveira and García Cruz (2013), for example, studied the production parameters and physicochemical properties of a biosurfactant synthesized by *Bacillus pumilus* when different concentrations of vinasse and waste frying oil were used as carbon sources. These authors detected the maximum production of crude biosurfactant at a concentration of 5% for both substrates. On the other hand, de Lima and Rodríguez de Souza (2014) focused on the use of vinasse to produce biosurfactant by *Bacillus subtilis* PC, in order to reduce SACs production costs and minimize the environmental impact generated by this effluent. The authors evaluated the effect of both vinasse concentration and inoculum concentration on the surface tension (ST). They noted a maximum reduction of the ST for a 55% (v/v) concentration of vinasse, and 20.4% (v/v) inoculum concentration. However, it is concluded that the inoculum concentration had a greater effect than vinasse concentration on the ST response variable. Also, Colin et al. (2016) evaluated the ability of *Streptomyces* sp. MC1 to produce bioemulsifiers in culture media containing vinasse in distilled water at variable concentrations (M_1 media). As an alternative, this strain was cultivated in a defined medium whose composition was described by Colin et al. (2013b), but using different vinasse concentrations as the carbon source (M_2 media), instead of glucose. Under such assay conditions, bioemulsifier production could not be detected in the M_1 culture media. However, when vinasse was used as a carbon source (M_2 culture media) at concentrations ranging 0.1 to 10% (v/v), a significant emulsifying activity was detected with a maximum emulsification index (63.0 ± 0.5%) for 1.0% v/v. At vinasse concentrations higher than 10%, no emulsifying activity could be detected probably due to the high content of total solids. These findings represented the first advances provided by our research group related to the reconversion of a regional effluent in microbial SACs.

Application of Microbial Surface-Active Compounds in Soil Washing Agents Technologies

Association between hydrocarbon biodegradation and SACs production is of great interest. This association encourages the applications of SACs-producing microorganisms for the Microbial Enhanced Oil Recovery (MEOR) (Nayak et al. 2009) or of hydrocarbon-contaminated environments (Reddy et al. 2010). However, production and purification of microbial SACs for their direct use in soil washing technologies results in an approach with greater pragmatism. Among the technologies of environmental restoration, the soil washing with the help of a detergent action agent is positioned as one of the most efficient techniques, and with a relatively moderate cost compared to other techniques. SACs are able to form complexes with pollutants attached to soil matrix, promoting their desorption towards the aqueous phase. Once in the aqueous phase, the hydrophobic pollutants are stabilized inside of the SACs micelles, which favor their solubility and subsequent removal during the washing process (Fig. 4). At the end of the process, the contaminated water is withdrawn to be processed in specific treatment plants (EPA 2002).

The washing technologies favor the significant reduction of the contaminated soil volume since fine particles are extracted from it during the process. Its efficiency is however limited by both pollutants characteristics and the soil itself. In general lines, soils with small particles such as clays (< 2.0 μm) and silts (2.0 to 50.0 μm) have greater absorbent power than sands (50.0 μm to 2.0 mm) and gravels (> 2.0 mm). Often, it is assumed that the soil with fine particles offers a larger surface tending to bind more contaminants than soil with large particles (Xu et al. 2014). In this way, the efficiency and cost of the process will be affected by the soil granulometry.

Two modalities for the washing soil are basically available. The *in situ* modality, that implies the direct injection of an extraction solution (e.g., surfactants) into the soil in order to dissolve and mobilize contaminants. The extraction solution can be injected through wells, trenches, infiltration galleries, etc. This modality is often used to complete or enhance the performance of traditional remediation techniques. As an alternative, the *ex situ* modality is also available, which involves the soil extraction from the contaminated area. The soil

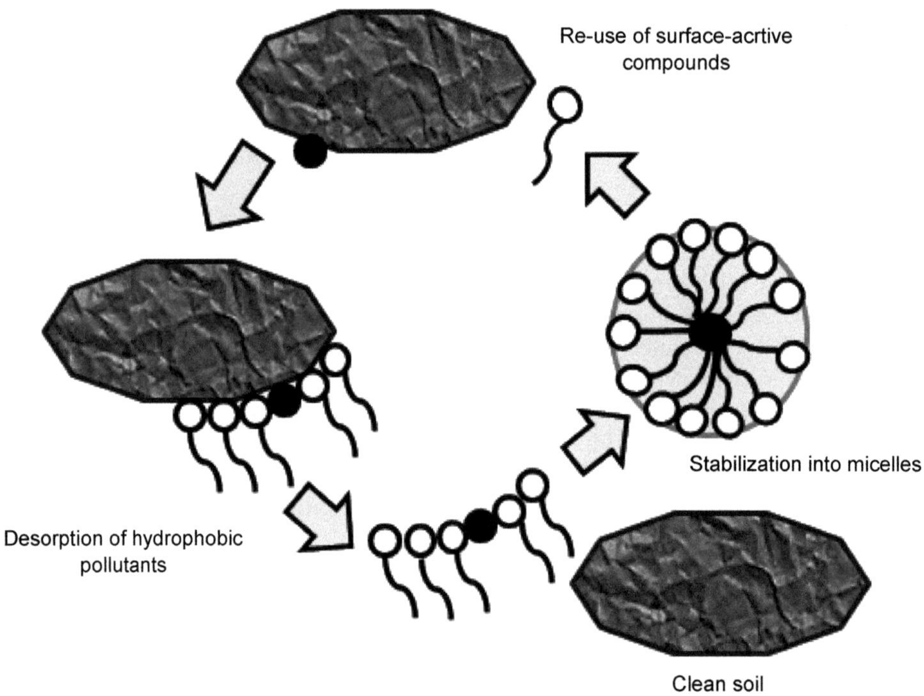

Figure 4. Flow chart summarizing the mechanism of the SACs to desorb hydrophobic pollutants from soil (Adapted from Pacwa-Płociniczak et al. 2011).

to clean is then placed in an automatic mixing machine, with a large proportion of the contaminants being removed by simple physical separation during the washing process. However, the efficiency of this process can be also improved by adding surfactants agents to the washing solution (Shah et al. 2016).

In the south of our country, a sanitation project based on the *ex situ* washing of an affected area, which had suffered intermittent hydrocarbon inputs during 20 yr has been carried out since 2006 (Pando et al. 2010). The contamination was attributed to the spillage of liquid effluents and the sprinkling of hydrocarbons from a gas and hydrocarbon plant located in the Río Negro province (Fernandez Oro Station). Pollution was not discovered until 2003, with average original values of concentrations around 50 mg kg^{-1} of soil for Total Petroleum Hydrocarbons (TPH), and peaks that can reach 200 mg kg^{-1}. One of the most important challenges of the *ex situ* process was the movement of soils towards the washing plant, which represented about 17% of the total budget. At the end of the treatment, the HTP in the decontaminated material was around 1,400 mg kg^{-1}, values well below the limit established by the Provincial Authority. This implied a high efficiency of the cleaning process, with an HTP removal of the 89%.

Despite the effectiveness of the soil washing technologies, surfactant selection guidelines are needed in order to aid in the implementation of agents that enhanced the washing process. Many synthetic surfactants traditionally used in this process are non-biodegradable so their persistence in the environments could affect the soil properties. In addition, synthetic surfactant-induced foaming is often one of the operating considerations that can also impact the surfactant selection. Total or partial replacement of synthetic molecules by those recovered from microbial sources could help to solve many of these problems, with similar performance in the contaminant's removal. Respect to this, Urum et al. (2006) carried out a comparative study to investigate the efficiency of different surfactants (synthetic and natural) to remove crude oil from contaminated soil by using the washing methodology. The authors reported a similar removal percentage of crude oil by the synthetic surfactant sodium dodecyl sulfate and rhamnolipid biosurfactants. Likewise, Kang et al. (2010) analyzed the performance of microbial surfactant sophorolipid, Tween 80/60/20 and Span 20/80/85 as washing agents to remove 2-methylnaphthalene from the artificially contaminated soil. A higher removal efficiency associated with the use of microbial surfactant, compared to any other tested synthetic surfactants except Tween 80, can be detected. These results encourage us to deepen the studies related to the use of microbial SACs for the development of cleaner and safer remediation technologies. Currently, most of our efforts are aimed at designing sustainable processes for the manufacture of microbial SACs from sugarcane vinasse.

Concluding Remarks

Numerous studies encourage us to use sugarcane vinasse as a byproduct of the sugar-alcohol industry. An integral approach to address the problem of contamination with this effluent is considered throughout this chapter, emphasizing on the search of microorganisms capable of degrading and/or reconverting the raw vinasse in products of biotechnological interest. The generation of SACs and microbial biomass itself would have an economic return as both products are applicable in environmental remediation technologies. Likewise, the reduction of vinasse toxicity subjected to the microbial action could be an adequate alternative to improve effluent quality, promoting its application as a soil amendment. This approach could constitute an adequate alternative to improve the management of a sugar-alcohol industry waste. Currently, complementary studies are in progress in order to select the more appropriate taxa for recovery of the sugarcane vinasse by microbial pathways.

Acknowledgments

This work was supported by Agencia Nacional de Promoción Científica y Tecnológica (ANPCyT) (PICT 2012 N° 2920) and (PICT 2013 N° 0141), Programa CAPES-CONICET-MINCYT 2014 (PCB II), and CONICET.

References Cited

Aguiar, M.M., L.F. Romanholo Ferreira and R.T. Rosim Monteiro. 2010. Use of vinasse and sugarcane bagasse for the production of enzymes by lignocellulolytic fungi. Braz. Arch. Biol. Technol. 53(5): 1245–1254.

Álvarez, A., J.M. Saez, J.S. Dávila Costas, V.L. Colin, M.S. Fuentes, S.A. Cuozzo et al. 2017. Actinobacteria: Current research and perspectives for bioremediation of pesticides and heavy metals. Chemosphere 166: 41–62.

Anastas, P.T. and J.C. Warner. 1998. Green Chemistry: Theory and Practice, Oxford University Press: New York, 1998, p. 30.

Aparicio, J.D., C.S. Benimeli, C.A. Almeida, M.A. Poli and V.L. Colin. 2017. Integral use of sugarcane vinasse for biomass production of actinobacteria: Potential application in soil remediation. Chemosphere 181: 478–484.

Banat, I.M., A. Franzetti, I. Gandolfi and R. Marchant. 2010. Microbial biosurfactants production, applications and future potential. Appl. Microbiol. Biotechnol. 87: 427–444.

Barrocal, V.M., M.T. García-Cubero, G. González-Benito and M. Coca. 2010. Production of biomass by *Spirulina maxima* using sugar beet vinasse in growth media. N. Biotechnol. 27(6): 851–6.

Bhattacharyya, A., A. Pramanik, S.K. Maji, S. Haldar, U.K. Mukhopadhyay and J. Mukherjee. 2012. Utilization of vinasse for production of poly-3-(hydroxybutyrate-co-hydroxyvalerate) by *Haloferax mediterranei*. AMB Express 2: 34.

Cavalcante Barros, F.F., C. Pereira de Quadros, M.R. Maróstica Júnior and G.M. Pastore. 2007. Surfactina: propiedades químicas, tecnológicas e funcionais para aplicações em alimentos. Quim. Nova 30(2): 409–414.

Chojnacka, K. 2009. Biosorption and Bioaccumulation in Practice. Nova Science Publishers, New York.

Christofoletti, C.A., J.P. Escher, J.E. Correia, J.F.U. Marinho and C.S. Fontanetti. 2013. Sugarcane vinasse: Environmental implications of its use. Waste Manage. 33(12): 2752–2761.

Clementson, C., B.N. Abrahim and O. Homenauth. 2016. An investigation of the spatial variability of elements due to vinasse disposal in waterways at the Albion Bioethanol Plant, Berbice Guyana. G.J.E.M.P.S. 5(3): 074–087.

Colin, V.L., L.B. Villegas and C.M. Abate. 2012. Indigenous microorganisms as potential bioremediators for environments contaminated with heavy metals. Int. Biodeter. Biodegr. 69: 28–37.

Colin, V.L., M.F. Castro, M.J. Amoroso and L.B. Villegas. 2013a. Production of bioemulsifiers by *Amycolatopsis tucumanensis* DSM 45259 and their potential application in remediation technologies for soils contaminated with hexavalent chromium. J. Hazard. Mater. 26: 577–583.

Colin, V.L., C.E. Pereira, L.B. Villegas, M.J. Amoroso and C.M. Abate. 2013b. Production and partial characterization of bioemulsifiers from a chromium-resistant actinobacteria. Chemosphere 90: 1372–1378.

Colin, V.L., A.A. Juárez Cortes, J.D. Aparicio and M.J. Amoroso. 2016. Potential application of a bioemulsifier-producing actinobacterium for treatment of vinasse. Chemosphere 144: 842–847.

de Lima, A.M. and R. Rodríguez de Souza. 2014. Use of sugar cane vinasse as substrate for biosurfactant production using *Bacillus subtilis* PC. Chem. Eng. Trans. 37: 673–678.

de Mello Prado, R., G. Caione and C.N. Silva Campos. 2013. Filter cake and vinasse as fertilizers contributing to conservation agriculture. Appl. Environ. Soil Sci. http://dx.doi.org/10.1155/2013/581984.

Dreja, M., I. Vockenroth and N. Plath. 2012. Biosurfactants—exotic specialties or ready for application? Tenside Surf. Det. 49: 10–17.

EPA. 2002. Guidelines for Ensuring and Maximizing the Quality, Objectivity, Utility, and Integrity of Information Disseminated by the Environmental Protection Agency.

Escher, J.P., C.A. Christofoletti, Y.A. Rodríguez and C.S. Fontanetti. 2016. Sugarcane vinasse, a residue of ethanol industry: toxic, cytotoxic and genotoxic potential using the *Allium cepa* test. J. Environ. Protect. 7: 602–612.

Fracchia, L., M. Cavallo, M.G. Martinotti and I.M. Banat. 2012. Biosurfactants and bioemulsifiers biomedical and related applications—present status and future potentials. pp. 325–370. *In*: Ghista, D.N. [ed.]. Biomedical Science, Engineering and Technology. INTECH open Science.

Guerra de Oliveira, J. and C.H. García-Cruz. 2013. Properties of a biosurfactant produced by *Bacillus pumilus* using vinasse and waste frying oil as alternative carbon sources. Braz. Arch. Biol. Technol. 56(1): 155–160.

Jain, R. and S. Srivastava. 2012. Nutrient composition of spent wash and its impact on sugarcane growth and biochemical attributes. Physiol. Mol. Biol. Plants 18(1): 95–99.

Janke, L., A.F. Leite, H. Wedwitschka, T. Schmidt, M. Nikolausz and W. Stinner. 2014. Biomethane production integrated to the Brazilian sugarcane industry: The case study of São Paulo state. In Proceedings of the 22nd European Biomass Conference and Exhibition, Hamburg, Germany, 23–26.

Juárez Cortes, A.A. 2016. Aplicación de procesos microbiológicos para el tratamiento de vinaza y la producción de bioemulsificantes por *Streptomyces* sp. MC1. Tesis de Grado, Facultad de Bioquimica, Quimica y Farmacia, Universidad Nacional de Tucuman, Argentina.

Kang, S.W., Y.B. Kim, J.D. Shin and E.K. Kim. 2010. Enhanced biodegradation of hydrocarbons in soil by microbial biosurfactant, sophorolipid. Appl. Biochem. Biotechnol. 160: 780–790.

Khopade, A., R. Biao, X. Liu, K. Mahadik, L. Zhang and C. Kokare. 2012. Production and stability studies of the biosurfactant isolated from marine *Nocardiopsis* sp. B4. Desalination 285: 198–204.

Makkar, R.S. and S.S. Cameotra. 2002. An update on the use of unconventional substrates for biosurfactants production and their new applications. Appl. Microbiol. Biotechnol. 58: 428–434.

Marchant, R. and I.M. Banat. 2012a. Microbial biosurfactants: challenges and opportunities for future exploitation. Trends Biotechnol. 30: 558–565.

Marchant, R. and I.M. Banat. 2012b. Biosurfactants: a sustainable replacement for chemical surfactants? Biotechnol. Lett. 34: 1597–1605.

Montero-Rodríguez, D., R.F.S. Andrade, D.L. Ramos Ribeiro, D. Rubio-Ribeaux, R.A. Lima, H.W.C. Araújo et al. 2015. Bioremediation of petroleum derivative using biosurfactant produced by *Serratia marcescens* UCP/ WFCC 1549 in low-cost medium. Int. J. Curr. Microbiol. App. Sci. 4(7): 550–562.

Moraes, B.S., T.L. Junqueira, L.G. Pavanello, O. Cavalett, P.E. Mantelatto, A. Bonomi et al. 2014. Anaerobic digestion of vinasse from sugarcane biorefineries in Brazil from energy, environmental, and economic perspectives: Profit or expense? Appl. Energy 113: 825–835.

Moran-Salazar, R.G., A.L. Sanchez-Lizarraga, J. Rodriguez-Campos, G. Davila-Vazquez, E.N. Marino-Marmolejo, L. Dendooven et al. 2016. Utilization of vinasses as soil amendment: consequences and perspectives. Springer Plus 5: 1007. DOI: 10.1186/s40064-016-2410-3.

Mornadini, M. and E. Quaia. 2013. Alternativas para el aprovechamiento de la vinaza como subproducto de la actividad sucroalcoholera. Avance Agroindustrial 34(2): EEAO-DOSSIER.

Motevasel, M. 2014. A study of surface biosurfactants applications on oil degradation. A.J.O.C.T. 2: 9.

Nayak, A.S., M.H. Vijaykumar and T.B. Karegoudar. 2009. Characterization of biosurfactant produced by *Pseudoxanthomonas* sp. PNK-04 and its application in bioremediation. Int. Biodeter. Biodegr. 63: 73–79.

Pacwa-Płociniczak, M., G.A. Płaza, Z. Piotrowska-Seget and S.S. Cameotra. 2011. Environmental applications of biosurfactants: recent advances. Int. J. Mol. Sci. 12: 633–654.

Pando, M.M., D. Rosa, S. Casabal and S. Wharton. 2010. Planta de lavado de suelos: saneamiento de pasivo ambiental en la Estación Fernández Oro, Río Negro. Petrotecnia 16–21.

Pant, D. and A. Adholeya. 2007. Biological approaches for treatment of distillery wastewater: a review. Bioresour. Technol. 98: 2321–2334.

Pérez, D., C. Fandos, J. Scandaliaris, L. Mazzone, F. Soria and P. Scandaliaris. 2007. Estado actual y evolución de la productividad del cultivo de caña de azúcar en Tucumán y el noroeste argentino en el período 1990–2007. Publicación Especial 34–EEAOC.

Polti, M.A., J.D. Aparicio, C.S. Benimeli and M.J. Amoroso. 2014. Simultaneous bioremediation of Cr(VI) and lindane in soil by actinobacteria. Int. Biodeter. Biodegr. 88: 48–55.

Pramanik, A., A. Mitra, M. Arumugam, A. Bhattacharyya, S. Sadhukhan, A. Ray et al. 2012. Utilization of vinasse for the production of polyhydroxybutyrate by *Haloarcula marismortui*. Folia Microbiol. 57(1): 71–79.

Rajagopal, V., S.M. Paramjit, K.P. Suresh, S. Yogeswar, R.D.V.K. Nageshwar and N. Avinash. 2014. Significance of vinasse waste management in agriculture and environmental quality-Review. Afr. J. Agric. Res. 9(38): 2862–2873.

Ravel, J., M.J. Amoroso, R.R. Colwell and R.T. Hill. 1998. Mercury resistant actinomycetes from Chesapeake Bay. FEMS Microbiol. Lett. 162: 177–184.

Reddy, M.S., B. Naresh, T. Leela, M. Prashanthi, N.C. Madhusudhan, G. Dhanasri et al. 2010. Biodegradation of phenanthrene with biosurfactant production by a new strain of *Brevibacillus* sp. Bioresour. Technol. 101: 7980–7983.

Romanholo Ferreira, L.F., M. Aguiar, T. Messias and R.T. Monteiro. 2011. Evaluation of sugar-cane vinasse treated with *Pleurotus sajor-caju* utilizing aquatic organisms as toxicological indicators. Ecotoxicol. Environ. Saf. 74(1): 132–137.

Sartori, S.B., L.F.R. Ferreira, T.G. Messias, G. Souza, G.B. Pompeua and R.T.R. Monteiro. 2015. *Pleurotus* biomass production on vinasse and its potential use for aquaculture feed. Mycology 6(1): 28–34.

Shah, S., S. Shahzad, A. Munir, M.N. Nadagouda, G.S. Khan, D.F. Shams et al. 2016. Micelles as soil and water decontamination agents. Chem. Rev. 116(10): 6042–6074.
Silva, R. de C.F.S., D.G. Almeida, R.D. Rufino, J.M. Luna, V.A. Santos and L.A. Sarubbo. 2014. Applications of biosurfactants in the petroleum industry and the remediation of oil spills. Int. J. Mol. Sci. 15: 12523–12542.
Soler da Silva, M.A., H.J. Kliemann, A. Borges De-Campos, B.E. Madari, J.D. Borge and J.M. Gonçalves. 2013. Effects of vinasse irrigation on effluent ionic concentration in Brazilian Oxisols. Afr. J. Agric. Res. 8(45): 5664–5677.
Srivastava, S. and R. Jain. 2010. Effect of distillery spent wash on cytomorphological behaviour of sugarcane settlings. J. Environ. Biol. 31(5): 809–812.
Urum, K., S. Grigson, T. Pekdemir and S.A. McMenamy. 2006. Comparison of the efficiency of different surfactants for removal of crude oil from contaminated soils. Chemosphere 62: 1403–1410.
Volesky, B. 2007. Biosorption and me. Water Res. 41(18): 4017–4029.
Xu, J., D.B. Kleja, H. Biester, A. Lagerkvist and J. Kumpiene. 2014. Influence of particle size distribution, organic carbon, pH and chorides on washing of mercury contaminated soil. Chemosphere 109: 99–105.

4

Use of Immobilized Biomass as Low-Cost Technology for Bioremediation of PAHs Contaminated Sites

Mauricio J. Alessandrello,[1,a] *María S. Juárez Tomás,*[1,b] *Paula Isaac,*[2] *Diana L. Vullo*[3] and *Marcela A. Ferrero*[1,*]

Introduction

Organic compounds have been produced for industrial applications for decades and they are some of the widest spread and most persistent pollutants found in air, water, and sediments (Ritter et al. 2002, Roots et al. 2010). Those so-called persistent organic pollutants (POPs) include polycyclic aromatic hydrocarbons (PAHs) comprising several of the most prioritized compounds on the US EPAs lists because of its detrimental biological effects, including toxicity, mutagenicity, and carcinogenicity (Peng et al. 2008) (Table 1). Its remediation is a priority.

PAHs are a group of organic compounds with two or more fused benzene rings, originating from natural as well as anthropogenic sources. They are widely distributed environmental contaminants and are not easily degraded under natural conditions. PAH persistence increases as the molecular weight increases. Although they are the main air pollutants, soil acts as final depository of these compounds (Usman et al. 2016).

PAHs released to the environment are removed *via* volatilization, photooxidation, chemical oxidation, adsorption to soil particles and leaching, the main removal processes being probably the transformation and degradation by microorganisms (Baboshin and Golovleva 2012). Because many of these compounds are extremely ubiquitous, numerous

[1] Planta Piloto de Procesos Industriales Microbiológicos (PROIMI-CONICET), Av. Belgrano y Pje. Caseros, San Miguel de Tucumán, T4001MVB, Tucumán, Argentina.

[a] Email: mauricio.alessandrello@gmail.com

[b] Email: sjuareztomas@proimi.org.ar

[2] Centro de Investigación y Transferencia-Villa María (CIT-UNVM) CONICET, Arturo Jauretche 1555, Villa María, 5900 Córdoba, Argentina.
Email: pauisaac86@yahoo.com.ar

[3] Universidad Nacional de General Sarmiento, Instituto de Ciencias, Área Química. J.M. Gutiérrez 1150, Los Polvorines, Buenos Aires, B1613GSX, Argentina.
Email: dvullo@ungs.edu.ar

* Corresponding author: mferrero@proimi.org.ar

Table 1. US EPA's 16 priority pollutant PAHs and selected physical-chemical properties (ATSDR 2005).

Polycyclic aromatic hydrocarbons	Structure (no. of rings)	Molecular weight (g mol^{-1})	Solubility (mg l^{-1})	Vapor pressure (mm Hg)
Naphthalene	2	128.17	31	8.89×10^2
Acenaphthene	3	154.21	3.8	3.75×10^3
Acenaphthylene	3	152.20	16.1	2.90×10^2
Anthracene	3	178.23	0.045	2.55×10^5
Phenanthrene	3	178.23	1.1	6.80×10^4
Fluorene	3	166.22	1.9	3.24×10^3
Fluoranthene	4	202.26	0.26	8.13×10^6
Benzo(a)anthracene	4	228.29	0.011	1.54×10^7
Chrysene	4	228.29	0.0015	7.80×10^9
Pyrene	4	202.26	0.132	4.25×10^6
Benzo(a)pyrene	5	252.32	0.0038	4.89×10^9
Benzo(b)fluoranthene	5	252.32	0.0015	8.06×10^8
Benzo(k)fluoranthene	5	252.32	0.0008	9.59×10^{11}
Dibenzo(a,h)anthracene	6	278.35	0.0005	2.10×10^{11}
Benzo(g,h,i)perylene	6	276.34	0.00026	1.00×10^{10}
Indeno[1,2,3-cd]pyrene	6	276.34	0.062	1.40×10^{10}

bacterial species have evolved through metabolic versatility, adapting their catabolic activities in order to metabolize those (Vandera et al. 2015). Therefore, indigenous bacterial communities are capable of metabolizing persistent organic pollutants but due to their low availability and activity, together with a lack of access to the contaminants and low available nutrients in the environment, these processes occur at low rates in terms of reducing their exposure in the aqueous environment (Petrie et al. 2003). However, a wide diversity of bacteria has been isolated from soils and sediments with the ability to use PAHs as their sole carbon and energy source (Cerniglia 1984, Bamforth and Singleton 2005, Haritash and Kaushik 2009).

The production of emulsifying molecules by PAHs degrading microorganisms is a relevant strategy to increase hydrocarbon bioavailability (Rosenberg and Ron 1999, Ron and Rosenberg 2002). The use of synthetic emulsifiers improved PAH biodegradation (Grimberg 1996, Willumsen et al. 2001); however, bioemulsifiers were confirmed to be more selective, biodegradable and friendly to the environment (Poremba et al. 1991).

Even though numerous bacteria capable of degrading a wide range of PAHs have been isolated from contaminated sites, not many genera have been reported be able to degrade both low molecular weight (LMW) and high molecular weight (HMW) polycyclic hydrocarbons (Song et al. 2011). Several mixed cultures of strains belonging to different bacterial genera and showing different degradation capabilities were proposed to favor the degradation of a PAH mixture (Mrozik 2003, Chavez et al. 2004, Janbandhu and Fulekar 2011, Zhong et al. 2011, Mikesková et al. 2012).

Bacterial communities have been utilized for many years to neutralize, degrade, and mineralize xenobiotic compounds in wastewater activated sludge (Byrns 2001, Bertin et al. 2007). Also, the importance of promoting biofilm formation to improve the detoxifying activity and the longevity of microbial diversity and abundance has been actively considered (Boon et al. 2003, Accinelli et al. 2012).

The objective of this chapter is to provide an overview of the strategies for bioremediation of PAHs by immobilized biomass and biotransformation that have been applied based on the benefits from the biofilm mode of growth. In the following sections, general concepts about some microbial immobilization processes will be explained. In later sections, different case studies concerning microbial immobilization systems used for hydrocarbon decontamination in waters and soils will be reviewed.

Microbial Immobilization Processes for Bioremediation of Hydrocarbons Contaminated Environments

In bioremediation processes employing microorganisms, one of the critical challenges is to achieve appropriate microbial cell persistence in natural habitats or complex sites of application (García-Delgado et al. 2015a,b). Usually, microbial biomass employed as inoculum is produced in bioreactors under optimal growth conditions, an important step for the bioaugmentation processes being the transference of such cultures to the interest sites (Nopcharoenkul et al. 2011). The complexity of different environment can cause abiotic and biotic stresses and a decrease of microbial viability in a short time, mainly if microorganisms are applied as free cell suspensions (Nopcharoenkul et al. 2011, Bayat et al. 2015).

Several studies have evidenced that different microbial immobilization strategies can provide a protector niche for microorganisms applied in bioremediation processes of various environmental pollutants (Rivelli et al. 2013, Han et al. 2014, García-Delgado et al. 2015a,b). Microbial immobilization favors a decrease of toxic compound concentrations in the cellular microenvironment, protects against depredation and competition, and improves the access to nutrients. The use of immobilized cultures could function as a mini-bioreactor in the environment, favoring a higher microbial retention and generating a higher resistance to diverse stress conditions (Nuñal et al. 2014, Bayat et al. 2015). In bioaugmentation technologies, the main advantages of microbial immobilization processes are the high microbial stability (both in viability and decontaminant activity), the easy separation, recovery, and re-use of systems, and the reduction of operating costs (Nopcharoenkul et al. 2013, KokKee et al. 2015, Shen et al. 2015).

For different biotechnological applications, microorganisms can be immobilized by binding to a surface (adhesion, biofilm formation) and by covering with an appropriate material (entrapment/encapsulation) (Muffler et al. 2014, Bayat et al. 2015, KokKee et al. 2015, Shen et al. 2015). Since different systems containing immobilized microbial cells present a higher useful life and stability, they have been widely proposed or employed in the treatment of wastewater and in the bioremediation of environments contaminated with hydrocarbons (Sheppard et al. 2014, Dellagnezze et al. 2016, Huang et al. 2016, Wang et al. 2016, Zhang et al. 2016).

Microbial Immobilization by Adhesion and Biofilm Formation

The first step in the immobilization by attachment phenomena is the contact between microbial cells and a surface. Binding of microorganisms to surfaces is a complex process that can be affected by several factors, such as support, environmental and microbial characteristics (Liao et al. 2015).

Among the support characteristics, the most relevant features in the attachment process can be size, surface area, porosity, roughness, chemical composition, surface charge and hydrophobicity (Vanysacker et al. 2014). Ionic strength, pH, temperature, flux rate, nutrients and presence of different chemicals are environmental factors that potentially influence microbial cell-surface interaction (Vanysacker et al. 2014, Liao et al. 2015). On the other hand, the initial cell-support interaction and the unspecific or specific adhesion to surfaces can be influenced by a wide variety of microbial characteristics. Some of them are the following: physiological state and genetic predisposition, cell shape and size, cell surface hydrophobicity and electrostatic charge, cellular surface structures—flagella, fimbriae and pili—presence of outer membrane proteins, lipopolysaccharides, S-layer and extracellular polymeric substances (EPS), biosurfactant production, and cellular communication systems—*quorum sensing*—among other factors (Vanysacker et al. 2014, Liao et al. 2015). Several microbial properties mentioned are strain-dependent.

In particular, cell surface hydrophobicity and biosurfactant production are microbial properties that can positively or negatively influence the attachment process to different surfaces, mainly in the initial adhesion steps (Liao et al. 2015). Moreover, in the specific case of hydrocarbon-degrading microorganisms, the cell hydrophobicity and biosurfactant production can favor, through different mechanisms, the interaction between no polar molecules and microbial cells. Thus, these properties are relevant because they can affect both the microbial adhesion and hydrocarbon removal ability (Nikolopoulou et al. 2013, Ron and Rosenberg 2014).

Microbial adhesion to a surface is a crucial but no predictive event in the biofilm formation. Biofilms are defined as structured microbial communities, which are adhered to a surface/interface or to other microbial cells, enclosed in a matrix of extracellular polymeric substances (Costerton et al. 1995, Donlan and Costerton 2002). Biofilms are heterogeneous biological systems in their structure and composition, the main characteristics being their higher persistence in the environment respecting planktonic microbial cells, and their higher tolerance to environmental toxic compounds and to different stress conditions (Abdallah et al. 2014). On the other hand, in marine environments, dynamic biofilms, dominated by hydrocarbonoclastic bacteria and diatoms, are fundamental for hydrocarbon removal in coastal mudflats (Coulon et al. 2012). Thus, biofilms represent systems of high hydrocarbon degradation activity, and the biofilm-based technologies constitute attractive options for the bioremediation of contaminated environments.

Numerous supports, organic or inorganic, natural or synthetic materials, can be employed for the development of adhesion/biofilm-based immobilization systems to be used in bioremediation processes (Bayat et al. 2015). The supports must fulfill the following requirements: to favor microbial adhesion, to be innocuous for organisms and environment, to present an appropriate mechanical resistance, and to be economical and easily available (KokKee et al. 2015). Some natural organic materials (e.g., corncob powder, waste mussel shells, sugarcane bagasse, etc.) are economically relevant, because they constitute low-cost by-products, are biodegradable and contain high concentrations of organic nutrients that can favor microbial growth (Rivelli et al. 2013, Sheppard et al. 2014, Liu et al. 2015).

Microbial Immobilization by Entrapment/Encapsulation

Immobilization by entrapment/encapsulation of microorganisms is another alternative and efficient strategy for environmental decontamination of hydrocarbons (Bayat et al. 2015). The main advantages of entrapment/encapsulation immobilization systems are the following: protection of substances or cells to immobilize against stress factors, stabilization of interest material, gradual release of immobilized material, and incorporation of a liquid material in a solid system (when applying some entrapment/encapsulation techniques), among others (Umer et al. 2011, Bayat et al. 2015).

According to Bayat et al. (2015), the entrapment is an immobilization technique that involves the capture of substances or cells within a support matrix or inside a hollow fiber. Instead, when the immobilization is achieved by enveloping interest components within various forms of semi-permeable membranes, the method is denominated encapsulation. The processes mentioned allow obtaining particles, whose characteristics depend on the specific technique applied, on the components to immobilize, on the cover/protector materials, and on the different immobilization, conditions used, among other factors (Umer et al. 2011, Martín et al. 2015).

Different cover materials have been proposed for their employment in entrapment/encapsulation systems with potential application in environmental bioremediation. For example, natural organic polymers (polysaccharides and proteins), semi-synthetic organic polymers (e.g., cellulosic derivatives), synthetic organic polymers (e.g., acrylic derivatives, polyesters, polyethers, etc.), inorganic polymers (e.g., silicon), and other inorganic materials (e.g., silicates) have been evaluated (Kuyukina et al. 2006, Maqbool et al. 2012, Han et al. 2014, Bayat et al. 2015). Cover materials must be cohesive, chemically compatible and not reactive with substances or cells to encapsulate, and they must provide appropriate coating properties, such as strength, flexibility, suitable permeability (semi-permeability) and stability (Umer et al. 2011).

Several methods for microbial immobilization by entrapment/encapsulation are available, such as extrusion-gelation (by thermal or ionic cross-linking), spray drying, fluid bed coating and freeze or vacuum drying, among other (Gasperini et al. 2014, Martín et al. 2015). In particular, the extrusion-ionic gelation is a frequently used technique, consisting of two main steps. In the first step, the mixture of the material to be encapsulated and the precursor material of the gel is extruded, resulting in the formation of droplets falling into a suitable hardening solution. In the second step, ionic gelation between charged macromolecules (e.g., polyanions such as alginate, a natural polysaccharide widely used) with oppositely charged ions (e.g., Ca^{2+} of the hardening solution) is achieved (Shen et al. 2015, Alessandrello et al. 2017a). This process leads to the formation of water-insoluble but permeable gel particles, which have a matrix structure (i.e., the interesting material dispersed in the protective material). The extrusion-ionic gelation method is a simple and economical technique, which does not require unfavorable conditions (such as toxic solvents or harmful temperatures) for microbial cells and allows an adequate particle size control (Martín et al. 2015, Kumar et al. 2016). However, some disadvantages are the relative difficulty for scaling, the susceptibility of polymers to structural damage and defects and the limited choice of cover material (Martín et al. 2015).

Bioremediation of Hydrocarbons Contaminated Soil by Immobilized Microorganisms

There are several reasons for using immobilized microorganisms for bioremediation of contaminated soils: immobilized microorganisms are often more resistant to stressful environmental conditions as can be a polluted site. They are also less prone to predation and protection against competition with autochthonous microorganisms. In addition to this, encapsulated microbial cells are surrounded by a microenvironment in which the entrance of toxic compounds is gradual, thus allowing the survival of the immobilized cells in a highly contaminated site (Kuyukina et al. 2013, Dellagnese et al. 2016). For its use in contaminated soil, the immobilization carrier should be economic, readily available, have a high specific surface area and slow but high biodegradability since it will not be recovered from the soil after the bioremediation process.

Zhang et al. (2016) studied the use of biochar as the immobilization carrier for a *Corynebacterium* strain, which was isolated from oil contaminated soil. Biochars are often selected as carriers due to their high affinity with organic pollutants rather than with soil organic matter. The immobilization method, mediated by bacterial cell adhesion, was simple and quick although a quantification of the immobilized biomass was not done. Immobilization efficiency was checked only by scanning electron microscopic analysis which showed a biofilm over the carrier surface. The degradation of a hydrocarbons' mixture by the immobilized cells was assayed in the M9 mineral medium. Assays with immobilized cells showed a 7% higher hydrocarbons' removal compared to the free cells counterparts. Overall removal by immobilized cells was 78.9%. Although the difference in the hydrocarbons' removal values between free and immobilized cells may not seem too big, it could be expected that the free cells would not reach those removal values in soil, since they would be unprotected against predation and the harsh environmental conditions. Using wastewaters as a culture medium for biomass production could improve this technology from an ecological and economical point of view.

Kuyukina et al. (2013) used in poly(vinyl alcohol) (PVA) immobilized *Rhodococcus* cells to amend oil polluted soils. This carrier is a cryogel that is relatively inexpensive compared to other synthetic polymers as polyacrylamide. The immobilization procedure requires low temperatures (–18°C) and this could be a problem for the up-scaling of this methodology due to the energy cost. However, this may be compensated by the higher mechanical strength and resistance to abiotic and biotic damage that have PVA immobilized cells. In that study, the bioremediation strategy consisted in applying, simultaneously, two *Rhodococcus* strains, that were immobilized in separate, to an oil contaminated soil. For comparison, a cell suspension of the two strains was added to another contaminated soil. The experiment lasted for 14 mon. Results showed that immobilized cells could remove 90% of the alkanes present while free cells removed 77%. Respirometry assays also reported higher metabolic activity in soils amended with immobilized cells than in soils incubated with free cells. Another very important aspect to consider when proposing the use of immobilized microorganisms for bioremediation is its storing capacity. In the case of an accidental oil spill, it is desirable that the immobilized microorganisms are ready to use and, for that, a stock should be maintained. The authors in that study determined the viability of PVA immobilized *Rhodococcus* strains after 14 mon storage at –20°C and found that these cryogel-entrapped bacteria could be preserved for that time with a viability loss of approximately 40% for both strains.

Wang et al. (2016) used the entrapment method with calcium alginate to immobilize bacterial cells of *Bacillus* adhered to pineapple peel biochar. This was done to enhance pollutant sorption and preconcentrate it near the immobilized cells improving biodegradation. When introduced in an artificial wetland co-contaminated with pyrene and Cr(VI), together with plants of *Kyllinga brevifolia Rottb*, a pyrene removal of approximately 63% was observed. When the immobilized cells were incubated alone, a pyrene removal of 55% was obtained. Although this immobilization method is not as simple as the method used by Zhang et al. (2016) (see above), the alginate entrapment provides more resistance to the bacterial cells. While Zhang et al. (2016) tested the biodegradation of oil in a sterile medium, Wang et al. (2016) studied the removal of pyrene in soil. It is probably this extra resistance that could be necessary for the soil environment.

A similar approach was used by Huang et al. (2016). They co-immobilized cinder with a mixed co-culture of *Pseudomonas taiwanensis* and *Acinetobacter baumannii* using modified PVA. The function of the cinder was the same as the biochar in the previous examples: the sorption of the contaminant to enhance biodegradation. When the bacterial-cinder beads were incubated in an oil-contaminated soil, a reduction of the pyrene and indeno(1,2,3-cd)pyrene concentration of, respectively, 70.73 and 80.92% was observed. The free cells removed approximately 50% of both polycyclic aromatic hydrocarbons (PAH). The higher removal of indeno(1,2,3-cd)pyrene was attributed to its higher adsorption on the cinder due to its higher hydrophobicity compared to that of pyrene. Again, since the biodegradation of these PAH was carried out in soil, an additional protection through the PVA coating was needed.

Bioremediation of Hydrocarbons Contaminated Water by Immobilized Microorganisms

When using immobilized microorganisms to remediate contaminated water, the same principles for bioremediation of contaminated soil apply. The immobilized microorganisms are protected against predation and the harmful environmental conditions. Additionally, the immobilized microorganisms remain concentrated in the polluted area and are not washed away as would be planktonic microorganisms. The carriers used, in this case, should have good buoyancy since the hydrocarbons accumulate on the surface of the water. Besides, since the carrier could be eventually separated from the water, there is no need for it to be biodegradable. For the bioremediation process to be more economic, low-cost culture media and carrier materials should be used. In this sense, Sheppard et al. (2014) used waste mussel shells as a carrier. They embedded dry shell grits in a solution containing nutrients (biostimulation solution) and in a suspension containing bacterial cells (bioaugmentation suspension). The immobilized nutrients and bacterial cells were used for remediation of 5,000 l hydrocarbon contaminated sea water. After 27 wk, a 53.3% oil concentration reduction was observed. According to the authors, the simplicity of the application and the recycling of marine materials are among the advantages of the bioremediation method.

Wang et al. (2015) used expanded graphite, expanded perlite and bamboo-charcoal as carriers for the immobilization of three oil-degrading bacterial strains separately. The immobilization method was biofilm formation and took seven days of incubation at 25°C with agitation. The culture medium for immobilization was sea salt solution (3.38% w/v) containing 800 mg l^{-1} glucose. The immobilized cells were examined using scanning electron microscopy. In presence of 0.2% v/v diesel oil, the immobilized cells showed a greater oil degradation percentage than the free cells. Oil removal percentages of the three strains were nearly 100% when immobilized on expanded graphite and in expanded perlite. In bamboo charcoal, only one strain achieved similar removal values while the other two reached approximately an 83% removal. According to the authors, the best carrier was expanded graphite for three reasons: (1) it has a greater surface area available for supporting microbial growth, (2) it has greater adsorbing capacity and (3) its lightweightedness and hydrophobicity provide it with excellent floating capacity and makes it easier to recover. The authors proposed an adsorption-synergic biodegradation of the diesel oil that would occur in two stages. First, the oil would be adsorbed on the carrier material thus preventing its spreading. Second, the immobilized microorganisms degrade the adsorbed oil recovering the carrier adsorption capacity. The authors found that the oil adsorption in the immobilized cells on expanded graphite was between 6–16%, indicating that, although the carrier was an excellent sorbent, oil did not accumulate on it due to the biodegradation activity of the immobilized cells.

Liu et al. (2016) also tested the adsorption-synergic biodegradation model. They used PVA to co-immobilize bacterial cells and activated carbon. The sorption capacity of the activated carbon improved hydrocarbon removal from sea water by 47%.

Dellagnezze et al. (2016) immobilized a bacterial consortium on chitosan and used it in a mesocosm study (3,000 l) for bioremediation of oil contaminated sea water. The study lasted 30 d. Alkane degradation of 90% could be observed at 10th d, even without adding the immobilized microorganisms. The authors suggest that the oil added to the water stimulated the growth of hydrocarbon degrading microorganisms explaining the high degradation rates obtained in the control experiment. When analyzing the degradation of PAH, however, higher removal values were obtained in the experiment with immobilized microorganisms. Since the chitosan immobilized cells did not float, the authors adhered traps containing the immobilized microorganisms to the walls of the mesocosm beneath the surface. Thus, they propose the use of this technology in enclosed systems.

Hydrocarbons Removal by Immobilized Bacterial Cells Using Low-Cost Culture Media for Biomass Production: A Case Study

All the works reviewed so far have used economic carrier materials for microorganism immobilization. However, none has addressed the use of economic culture media for hydrocarbon-degrading microbial growth. Therefore, our recent work focused on the use of waste or industrial by-products as culture media for PAH-degrading bacteria (Alessandrello et al. 2017a). In that work, corn steep liquor, sugar cane molasses, whey permeate and crude glycerol were used as carbon or nitrogen source in different media formulations. Corn steep liquor is produced in the food industry during the corn wet milling process. Sugar cane molasses is a by-product of sugarcane refining into sugar. Whey permeate is a by-product of the cheese manufacturer. Crude glycerol is co-produced with biodiesel. All these components have in common is that they are produced in a higher amount than they are demanded and have a limited market. Therefore, many times, these products have to be discarded increasing the cost of wastewater treatments. Culture media formulated with these products could provide a new use for them so that they no longer represent an economic loss and environmental threat.

In our study, the media used were (nitrogen source-carbon source): corn steep liquor-molasses, whey permeate-molasses, corn steep liquor-crude glycerol and whey permeate-crude glycerol. Different concentrations of each component were tested to obtain the optimal growth medium. The two strains used (*Pseudomonas monteilii* P26 and *Gordonia* sp. H19) showed high growth in all the media assayed (1×10^8 CFU ml^{-1}) after 48 hr culture.

Since corn steep liquor and molasses were readily available in our region, they were selected as culture medium (1% v/v corn steep liquor and 2% m/v molasses) for immobilization of a co-culture of the two strains. The carriers tested were a polyurethane foam, sand, and calcium alginate. The immobilization method in sand and polyurethane foam was by biofilm formation while in calcium alginate was by cell entrapment. For biofilm formation, the carriers were submerged in the culture medium and it was inoculated with both strains. The culture remained static and the air was supplied through an aeration pump. Every 48 hr (time in which both strains achieved maximum growth), the spent medium was replaced by fresh one. This procedure allowed the enrichment of immobilized biomass by removing the planktonic bacteria in every medium replacement and promoting the growth of the attached ones. The process lasted for 14 d. For cell entrapment, a co-culture of both strains was incubated for 48 hr and then the cells were harvested and concentrated. The suspension was mixed with a solution of calcium alginate and dripped over a $CaCl_2$ solution. The beads formed were allowed to rest in that solution at 4°C for 15 min, then washed and immediately used. The immobilized co-cultures in the different carriers were incubated for 12 d in presence of 0.2 mM naphthalene, phenanthrene, and pyrene in a liquid medium containing amino acids and yeast extract which provided essential nutrients.

Results showed a high PAH removal mediated by sorption on the carriers. Naphthalene and phenanthrene were removed almost 100% by sterile sand and calcium alginate beads. The immobilized cells on polyurethane foam removed, respectively, 60%, 28.9% and 30% more naphthalene, phenanthrene and pyrene than the sterile carrier. The pyrene removal by calcium alginate and sand immobilized cells was nearly 90% while the sterile carriers removed in average 55% in 12 d (Fig. 1).

These results highlight the importance of the hydrocarbons' sorption on the carriers in the removal process and show the feasibility of using low-cost culture media for hydrocarbon-degrading bacteria production. Both calcium alginate and sand immobilized cells presented similar PAH removal values. When comparing the two immobilization methods used, i.e., cell entrapment and biofilm formation, the time needed for cell immobilization and the costs should be considered. Although the cell entrapment in calcium alginate is costlier than the method of biofilm formation, it is less time-consuming. The final application of the immobilized cells will determine which method is more appropriate.

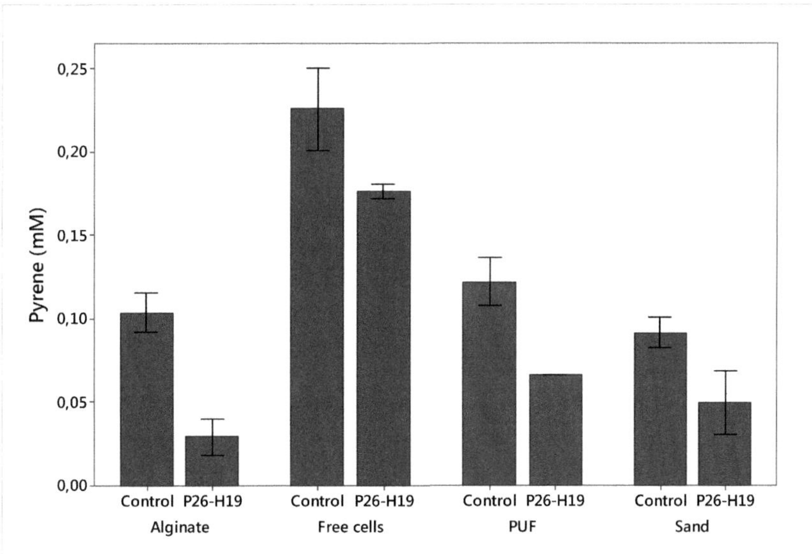

Figure 1. Remaining pyrene after 12 d by the different systems. The carriers used were alginate, polyurethane foam (PUF), sand and no carrier (free cells). P26-H19: co-culture of *Pseudomonas monteilii* P26 and *Gordonia* sp. H19. Control: assay done without bacterial cells.

Enhancing Biofilm Formation for Bacterial Immobilization: Case Studies

The immobilization of microorganisms by biofilm formation has the following main disadvantage compared with cell entrapment and encapsulation immobilization methods: it is not possible to control the number of bacterial cells that will be immobilized and often a higher concentration of immobilized cells is desired. Therefore, several studies aimed to find methods of enhancing biofilm formation to increase the number of immobilized cells and thus improving hydrocarbon biodegradation.

Isaac et al. (2017) tested the effect of PAH addition during the biofilm formation on glass of a mixed culture of PAH-degrading bacteria composed of *Pseudomonas* and actinobacteria strains. The defined mixed culture called C15 was formulated with *P. monteilii* P26, *Pseudomonas* sp. N3, *Gordonia* sp. H19, and *Rhodococcus* sp. F27. For biofilm formation in glass tubes, the consortium was incubated in JPP broth (% m/v: NaCl, 2; yeast extract, 0.1; meat peptone, 0.2; pH = 7) during three d at 30°C without agitation in the presence or absence of a mixture of pyrene and phenanthrene (0.2 mM). After incubation, the supernatant was removed and the biofilm formed on the walls of the tubes were stained with crystal violet for its quantification. They found that the biofilm formation was enhanced 78% when the mixture of PAH was added to the culture medium. When the formed biofilm was incubated in JPP broth containing 0.2 mM of pyrene and phenanthrene, the one that had been formed in the presence of PAH showed higher hydrocarbon removal values than the one that had been formed without PAH. In 3 d, the enhanced biofilm removed 100% phenanthrene and 69% pyrene while the non-stimulated biofilm removed a 65% and 28% of phenanthrene and pyrene, respectively, after 10 d of incubation (Fig. 2).

Additionally, the authors investigated the bioemulsifier production of both stimulated and non-stimulated biofilms during the PAH removal assay. It is well-known that emulsifier substances enhance hydrocarbon bioavailability and hence its biodegradation. Results showed that biofilms formed in presence of PAH produced 20% more emulsifier activity than biofilms formed in absence of PAH. When analyzing the biofilm formation kinetics in both conditions by confocal laser scanning microscopy, it was found that the presence of PAH accelerated biofilm maturation by 48 hr in comparison to the biofilm formation in the absence of PAH. No biofilm structural changes due to PAH addition were observed, indicating that the same was indeed resistant to PAH toxicity.

In conclusion, the addition of PAH in the biofilm formation stage enhanced its formation and hence bioemulsifier production and PAH removal. Probably, the biofilm formation by bacteria consortium was a stress response to the PAH toxicity. This method of biofilm formation enhancement, however, would not be suitable for the production of an immobilized cell system for its use in bioremediation because it implies the use of very toxic compounds during the production stage. Therefore, natural stressors are an alternative for increasing biofilm cell density.

Alessandrello et al. (2017b) purified annonaceous acetogenins (ACG) from *Annona cherimola* seeds and tested their biofilm formation stimulation activity. ACG are natural stressors that have been proposed for the treatment of tumors, as anti-parasitic compounds and pesticides due to their cytotoxic activity. The chemical structure of an ACG (itrabin) is shown in Fig. 3 as an example.

After purification, the ACG itrabin, jetein, squamocin and laherradurin, as well as a crude extract of *A. cherimola* were tested for biofilm formation stimulation of *Pseudomonas monteilii* P26, a PAH-degrading bacterium. The assays were conducted in microtiter plates where cultures of *P. monteilii* P26 in corn steep liquor were supplemented with different concentrations of ACG or the *A. cherimola* crude extract were incubated for 24 hr. After incubation, the supernatants were discarded and the biofilms were stained with crystal violet for its quantification. This simple assay allowed determining those ACG with the highest activity which was itrabin, jetein and the *A. cherimola* extract.

The selected ACG were then used for immobilization of *P. monteilii* P26 on polyurethane foam. The desired amount of carrier was submerged in corn steep liquor (10% v/v, pH = 7) and was inoculated with *P. monteilii* P26 in presence of the selected ACG or the

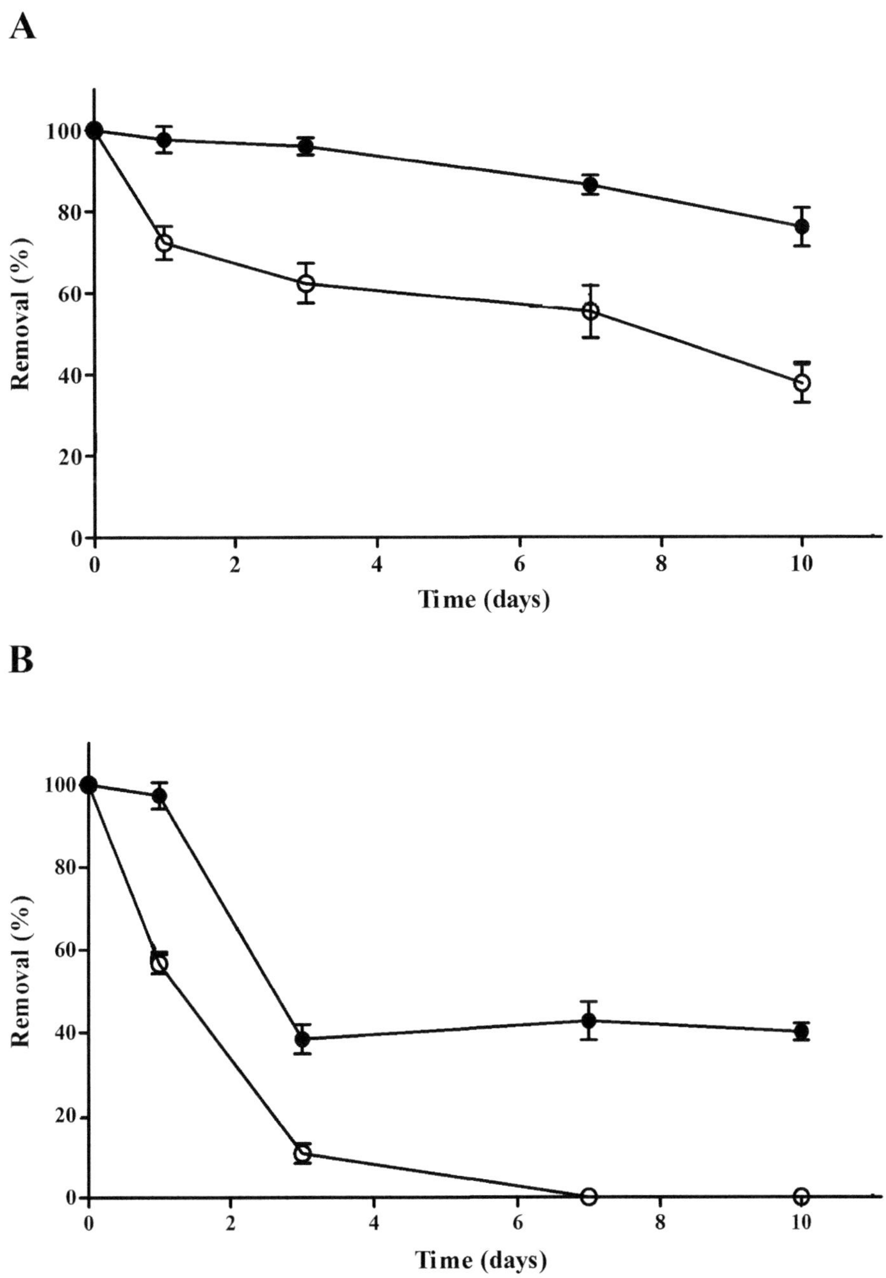

Figure 2. Kinetics of PAH removal by biofilm-immobilized mixed culture C15. Biofilms formed in the absence (**A**) or presence (**B**) of hydrocarbons. Residual phenanthrene (○) and pyrene (●) values were quantified considering hydrocarbon abiotic loss in all cases. Adapted from Isaac et al. (2017).

crude extract. Air was provided through a 0.2 μm nylon filter with an air pump. The system remained without agitation and mixing was provided by the air flow. After every 48 hr, the spent medium was replaced with fresh corn steep liquor. The process lasted 5 d after which the biofilm on the polyurethane foam was quantified. Polyurethane foam pieces with immobilized cells were washed to remove planktonic cells, cut into smaller parts and

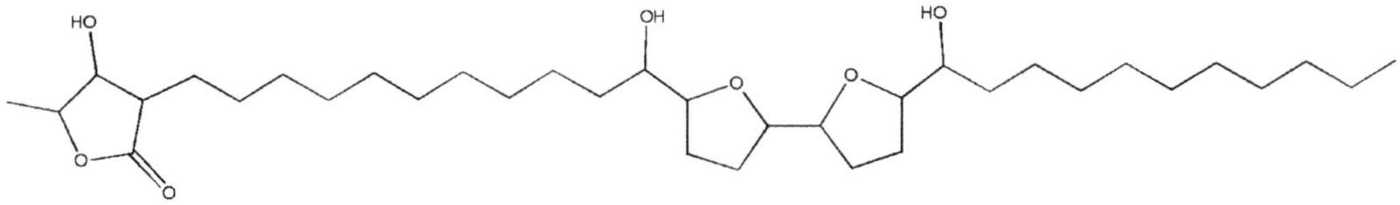

Figure 3. Chemical structure of itrabin.

vortexed in saline solution to detach the immobilized cells. The cell suspension was then plated and the CFU per g carrier determined.

Results showed that the addition of itrabin and the *A. cherimola* extract produced the highest biofilm density. The CFU per g carrier increased 100-fold in presence of these stressors compared to the biofilm formed without stimulation. However, when the stimulated biofilms were incubated in presence of PAH for seven days, the same removal values as with the non-stimulated biofilms were obtained. Almost 100% PAH removal was achieved independently of the stressor added during biofilm formation. PAH sorption on the carrier was the major factor responsible for the removal observed, accounting for 60% removal. It seemed that originally non-stimulated biofilms reached similar biofilm densities as stimulated ones during the PAH removal assay due to the stress caused by PAH. This would explain why there was no difference in the removal values between stimulated and non-stimulated biofilms. The study demonstrated the feasibility of using natural stressors (even without purification) for enhancing biofilm formation. In further studies, the PAH concentration in the removal assay will be increased and the PAH removal kinetics will be determined to corroborate if the biofilm enhancement really improves PAH biodegradation.

Concluding Remarks

A wide variety of low-cost carriers for microorganism immobilization and its use in bioremediation of hydrocarbons contaminated soils and waters have been proposed. The studies analyzed here showed bioremediation processes with different degrees of success. Although there are differences in the immobilization methods used, most studies agree in that hydrocarbon sorption on the carrier is the first step in the biodegradation process and improving contaminant sorption will improve its subsequent degradation. Strategies for enhancing biofilm formation should also improve hydrocarbon removal when using biofilm immobilized microorganisms.

For a bioremediation technology to be economically viable, not only the carrier has to be low-cost but also the culture media used to produce the microorganisms. Therefore, the use of low-value by-products or waste as alternative culture media should be further investigated. Currently, there is a lack of economic analysis studies concerning bioremediation technologies. Further, field and pilot scale studies are needed in which the economic, as well as the environmental impact, are assessed.

Finally, it is concluded that the selection of immobilization strategies (immobilization technique, support or cover material) would mainly depend on the characteristics of microbial cells to immobilize, and on the requirements of the product according to its specific application in bioremediation.

Acknowledgments

This work was supported by grants from the National Research Council of Argentina (CONICET). MJA is recipient of a fellowship from CONICET and the YPF Foundation, and PI is recipient of a fellowship from CONICET. MSJT, DLV and MAF are staff members of CONICET.

References Cited

Abdallah, M., C. Benoliel, D. Drider, P. Dhulster and N.E. Chihib. 2014. Biofilm formation and persistence on abiotic surfaces in the context of food and medical environments. Arch. Microbiol. 196: 453–472.
Accinelli, C., M.L. Sacca, M. Mencarelli and A. Vicari. 2012. Application of bioplastic moving bed biofilm carriers for the removal of synthetic pollutants from wastewater. Bioresour. Technol. 120: 180–186.
Alessandrello, M.J., M.S. Juárez Tomás, P. Isaac, D.L. Vullo and M.A. Ferrero. 2017a. PAH removal by immobilized bacterial cells-support systems using low-cost culture media for biomass production. Int. Biodeter. Biodegr. 120: 6–14.
Alessandrello, M.J., E.A. Parellada, M.S. Juárez Tomás, A. Neske, D.L. Vullo and M.A. Ferrero. 2017b. Polycyclic aromatic hydrocarbons removal by immobilized bacterial cells using annonaceous acetogenins for biofilm formation stimulation on polyurethane foam. J. Environ. Chem. Eng. 5: 189–195.
ATSDR. 2005. Toxicology profile for polyaromatic hydrocarbons. ATSDR's Toxicological Profiles on CD-ROM, CRC Press, Boca Raton, FL.
Baboshin, M.A. and L.A. Golovleva. 2012. Aerobic bacterial degradation of polycyclic aromatic hydrocarbons (PAHs) and its kinetic aspects. Microbiology 81: 639–650.
Bamforth, S.M. and I. Singleton. 2005. Bioremediation of polycyclic aromatic hydrocarbons: current knowledge and future directions. J. Chem. Technol. Biotechnol. 80: 723–736.
Bayat, Z., M. Hassanshahian and S. Cappello. 2015. Immobilization of microbes for bioremediation of crude oil polluted environments: a mini review. Open Microbiol. J. 9: 48–54.
Bertin, L., S. Capodicasa, F. Occulti, S. Girotti, L. Marchetti and F. Fava. 2007. Microbial processes associated to the decontamination and detoxification of a polluted activated sludge during its anaerobic stabilization. Water Res. 41: 2407–2416.
Boon, N., E.M. Top, W. Verstraete and S.D. Siciliano. 2003. Bioaugmentation as a tool to protect the structure and function of an activated-sludge microbial community against a 3-chloroaniline shock load. Appl. Environ. Microbiol. 69: 1511–1520.
Byrns, G. 2001. The fate of xenobiotic organic compounds in wastewater treatment plants. Water Res. 35: 2523–2533.
Cerniglia, C.E. 1984. Microbial metabolism of polycyclic aromatic hydrocarbons. Adv. Appl. Microbiol. 30: 31–71.
Chávez, F.P., H. Lunsdorf and C.A. Jerez. 2004. Growth of polychlorinated-biphenyl-degrading bacteria in the presence of biphenyl and chlorobiphenyls generates oxidative stress and massive accumulation of inorganic polyphosphate. Appl. Environ. Microbiol. 70: 3064–3072.
Costerton, J.W., Z. Lewandowski, D.E. Caldwell, D.R. Korber and H.M. Lappin-Scott. 1995. Microbial biofilms. Annu. Rev. Microbial. 49: 711–745.
Coulon, F., P.M. Chronopoulou, A. Fahy, S. Païssé, M. Goñi-Urriza, L. Peperzak et al. 2012. Central role of dynamic tidal biofilms dominated by aerobic hydrocarbonoclastic bacteria and diatoms in the biodegradation of hydrocarbons in coastal mudflats. Appl. Environ. Microbiol. 78: 3638–3648.
Dellagnezze, B.M., S.P. Vasconcellos, A.L. Angelim, V.M.M. Melo, S. Santisi, S. Cappello et al. 2016. Bioaugmentation strategy employing a microbial consortium immobilized in chitosan beads for oil degradation in mesocosm scale. Mar. Pollut. Bull. 107: 107–117.
Donlan, R. and J.W. Costerton. 2002. Biofilms: survival mechanisms of clinically relevant microorganisms. Clin. Microbiol. Rev. 15: 167–193.
García-Delgado, C., A. D'Annibale, L. Pesciaroli, F. Yunta, S. Crognale, M. Petruccioli et al. 2015a. Implications of polluted soil biostimulation and bioaugmentation with spent mushroom substrate (*Agaricus bisporus*) on the microbial community and polycyclic aromatic hydrocarbons biodegradation. Sci. Total Environ. 508: 20–28.
García-Delgado, C., I. Alfaro-Barta and E. Eymar. 2015b. Combination of biochar amendment and mycoremediation for polycyclic aromatic hydrocarbons immobilization and biodegradation in creosote-contaminated soil. J. Hazard Mater. 285: 259–266.
Gasperini, L., J.F. Mano and R.L. Reis. 2014. Natural polymers for the microencapsulation of cells. J. R. Soc. Interface. 11(100): 20140817.
Grimberg, S.J. 1996. Quantifying the biodegradation of phenanthrene by *Pseudomonas stutzeri* P16 in the presence of a nonionic surfactant. Appl. Environ. Microbiol. 62: 2387–2392.

Han, Y., W. Zhang, W. Lu, Z. Zhou, Z. Zhuang and M. Li. 2014. Co-immobilization of *Pseudomonas stutzeri* YHA-13 and *Alcaligenes* sp. ZGED-12 with polyvinyl alcohol-alginate for removal of nitrogen and phosphorus from synthetic wastewater. Environ. Technol. 35: 2813–2820.

Haritash, A.K. and C.P. Kaushik. 2009. Biodegradation aspects of polycyclic aromatic hydrocarbons (PAH): a review. J. Hazard. Mater. 169: 1–15.

Huang, R., W. Tian, Q. Liu, H. Yu, X. Jin, Y. Zhao et al. 2016. Enhanced biodegradation of pyrene and indeno(1,2,3-cd)pyrene using bacteria immobilized in cinder beads in estuarine wetlands. Mar. Pollut. Bull. 102: 128–33.

Isaac, P., M.J. Alessandrello, A.J. Macedo, M.C. Estévez and M.A. Ferrero. 2017. Pre-exposition to polycyclic aromatic hydrocarbons (PAHs) enhance biofilm formation and hydrocarbon removal by native multi-species consortium. J. Environ. Chem. Eng. (in press).

Janbandhu, A. and M.H. Fulekar. 2011. Biodegradation of phenanthrene using adapted microbial consortium isolated from petrochemical contaminated environment. J. Hazard. Mater. 187: 333–340.

KokKee, W., H. Hazaimeh, S.A. Mutalib, P.S. Abdullah and S. Surif. 2015. Self-immobilised bacterial consortium culture as ready-to-use seed for crude oil bioremediation under various saline conditions and sea water. Int. J. Environ. Sci. Technol. 12: 2253–2262.

Kumar, G., A. Mudhoo, P. Sivagurunathan, D. Nagarajan, A. Ghimire, C.H. Lay et al. 2016. Recent insights into the cell immobilization technology applied for dark fermentative hydrogen production. Bioresour. Technol. 219: 725–737.

Kuyukina, M.S., I.B. Ivshina, A.Y. Gavrin, E.A. Podorozhko, V.I. Lozinsky, C.E. Jeffree et al. 2006. Immobilization of hydrocarbon-oxidizing bacteria in poly(vinyl alcohol) cryogels hydrophobized using a biosurfactant. J. Microbiol. Methods 65: 596–603.

Kuyukina, M.S., I.B. Ivshina, T.N. Kamenskikh, M.V. Bulicheva and G.I. Stukova. 2013. Survival of cryogel-immobilized *Rhodococcus* strains in crude oil-contaminated soil and their impact on biodegradation efficiency. Int. Biodeter. Biodegr. 84: 118–125.

Liao, C., X. Liang, M.L. Soupir and L.R. Jarboe. 2015. Cellular, particle and environmental parameters influencing attachment in surface waters: a review. J. Appl. Microbiol. 119: 315–330.

Liu, J., S. Chen, J. Ding, Y. Xiao, H. Han and G. Zhong. 2015. Sugarcane bagasse as support for immobilization of *Bacillus pumilus* HZ-2 and its use in bioremediation of mesotrione-contaminated soils. Appl. Microbiol. Biotechnol. 99: 10839–10851.

Liu, P.G., D. Yang, J. Tang, H. Hsu, C. Chen and I. Lin. 2016. Development of a cell immobilization technique with polyvinylalcohol for diesel remediation in sea water. Int. Biodeter. Biodegr. 113: 397–407.

Maqbool, F., Z. Wang, Y. Xu, J. Zhao, D. Gao, Y.G. Zhao et al. 2012. Rhizodegradation of petroleum hydrocarbons by *Sesbania cannabina* in bioaugmented soil with free and immobilized consortium. J. Hazard. Mater. 237–238: 262–269.

Martín, M.J., F. Lara-Villoslada, M.A. Ruiz and M.E. Morales. 2015. Microencapsulation of bacteria: A review of different technologies and their impact on the probiotic effects. Innov. Food Sci. Emerg. Technol. 27: 15–25.

Mikesková, H., C. Novotný and K. Svobodová. 2012. Interspecific interactions in mixed microbial cultures in a biodegradation perspective. Appl. Microbiol. Biotechnol. 95: 861–870.

Mrozik, A. 2003. Bacterial degradation and bioremediation of polycyclic aromatic hydrocarbons. Polish J. Environ. Studies 12: 15–25.

Muffler, K., M. Lakatos, C. Schlegel, D. Strieth, S. Kuhne and R. Ulber. 2014. Application of biofilm bioreactors in white biotechnology. Adv. Biochem. Eng. Biotechnol. 146: 123–161.

Nikolopoulou, M., N. Pasadakis and N. Kalogerakis. 2013. Evaluation of autochthonous bioaugmentation and biostimulation during microcosm-simulated oil spills. Mar. Pollut. Bull. 72: 165–173.

Nopcharoenkul, W., P. Pinphanichakarn and O. Pinyakong. 2011. The development of a liquid formulation of *Pseudoxanthomonas* sp. RN402 and its application in the treatment of pyrene-contaminated soil. J. Appl. Microbiol. 111: 36–47.

Nopcharoenkul, W., P. Netsakulnee and O. Pinyakong. 2013. Diesel oil removal by immobilized *Pseudoxanthomonas* sp. RN402. Biodegradation 24: 387–397.

Nuñal, S.N., S.M. Santander-De Leon, E. Bacolod, J. Koyama, S. Uno, M. Hidaka et al. 2014. Bioremediation of heavily oil-polluted sea water by a bacterial consortium immobilized in cocopeat and rice hull powder. Biocontrol Sci. 19: 11–22.

Peng, R.H., A.S. Xiong, Y. Xue, X.Y. Fu, F. Gao, W. Zhao et al. 2008. Microbial biodegradation of polyaromatic hydrocarbons. FEMS Microbiol. Rev. 32: 927–955.

Petrie, L., N.N. North, S.L. Dollhopf, D.L. Balkwill and J.E. Kostka. 2003. Enumeration and characterization of iron(III)-reducing microbial communities from acidic subsurface sediments contaminated with uranium(VI). Appl. Environ. Microbiol. 69: 7467–7479.

Poremba, K., W. Gunkel, S. Lang and F. Wagner. 1991. Toxicity testing of synthetic and biogenic surfactants on marine microorganisms. Environ. Toxicol. Water Qual. 6: 157–163.

Ritter, L., K. Solomon, P. Sibley, K. Hall, P. Keen, G. Mattu et al. 2002. Sources, pathways, and relative risks of contaminants in surface water and groundwater: a perspective prepared for the Walkerton inquiry. J. Toxic Environ. Health A 65: 1–142.
Rivelli, V., A. Franzetti, I. Gandolfi, S. Cordoni and G. Bestetti. 2013. Persistence and degrading activity of free and immobilized allochthonous bacteria during bioremediation of hydrocarbon-contaminated soils. Biodegradation 24: 1–11.
Ron, E.Z. and E. Rosenberg. 2002. Biosurfactants and oil bioremediation. Curr. Opin. Biotechnol. 13: 249–252.
Ron, E.Z. and E. Rosenberg. 2014. Enhanced bioremediation of oil spills in the sea. Curr. Opin. Biotechnol. 27: 191–194.
Roots, O., A. Roose, A. Kull, I. Holoubek, P. Cupr and J. Klanova. 2010. Distribution pattern of PCBs, HCB and PCB using passive air and soil sampling in Estonia. Environ. Sci. Pollut. Res. Int. 17: 740–749.
Rosenberg, E. and E.Z. Ron. 1999. High- and low-molecular-mass microbial surfactants. Appl. Microbiol. Biotechnol. 52: 154–162.
Shen, T., Y. Pi, M. Bao, N. Xu, Y. Li and J. Lu. 2015. Biodegradation of different petroleum hydrocarbons by free and immobilized microbial consortia. Environ. Sci. Process Impacts 12: 202–233.
Sheppard, P.J., K.L. Simons, E.M. Adetutu, K.K. Kadali, A.L. Juhasz, M. Manefield et al. 2014. The application of a carrier-based bioremediation strategy for marine oil spills. Mar. Pollut. Bull. 84: 339–46.
Song, H., S. Payne, C. Tan and L. You. 2011. Programming microbial population dynamics by engineered cell–cell communication. Biotechnol. J. 6: 837–849.
Umer, H., H. Nigam, A.S. Tamboli and M.S.M. Nainar. 2011. Microencapsulation: Process, techniques and applications. I.J.R.P.B.S. 2: 474–481.
Usman, M., K. Hanna and S. Haderlein. 2016. Science of the total environment fenton oxidation to remediate PAHs in contaminated soils : A critical review of major limitations and counter-strategies. Sci. Total Environ. 569–570: 179–190.
Vandera, E., M. Samiotaki, M. Parapouli, G. Panayotou and A.I. Koukkou. 2015. Comparative proteomic analysis of *Arthrobacter phenanthrenivorans* Sphe3 on phenanthrene, phthalate and glucose. J. Proteomics 113: 73–89.
Vanysacker, L., B. Boerjan, P. Declerck and I.F. Vankelecom. 2014. Biofouling ecology as a means to better understand membrane biofouling. Appl. Microbiol. Biotechnol. 98: 8047–8072.
Wang, C., L. Gu, X. Liu, S. Ge, T. Chen and X. Hu. 2016. Removal of pyrene in simulated wetland by joint application of *Kyllinga brevifolia Rottb* and immobilized microbes. Int. Biodeter. Biodegr. 114: 228–233.
Wang, X., X. Wang, M. Liu, Y. Bu, J. Zhang, J. Chen et al. 2015. Adsorption–synergic biodegradation of diesel oil in synthetic sea water by acclimated strains immobilized on multifunctional materials. Mar. Pollut. Bull. 92: 195–200.
Willumsen, P.A., U. Karlson, E. Stakebrandt and R.M. Kroppenstedt. 2001. *Mycobacterium frederiksbergense* sp. nov., a novel polycyclic aromatic hydrocarbon-degrading *Mycobacterium* species. Int. J. Syst. Evol. Microbiol. 51: 1715–1722.
Zhang, H., J. Tang, L. Wang, J. Liu, R.G. Gurav and K. Sun. 2016. A novel bioremediation strategy for petroleum hydrocarbon pollutants using salt tolerant *Corynebacterium variabile* HRJ4 and biochar. J. Environ. Sci. 47: 7–13.
Zhong, Y., T. Luan, L. Lin, H. Liu and N.F.Y. Tam. 2011. Production of metabolites in the biodegradation of phenanthrene fluoranthene and pyrene by the mixed culture of *Mycobacterium* sp. and *Sphingomonas* sp. Bioresour. Technol. 102: 2965–2972.

5

Biosorption of Dyes by Brown Algae

Josefina Plaza Cazón and *Edgardo Donati**

Introduction

A wide variety of synthetic and natural organic dyes are used by different industries (textile dyeing industries, for inks and tinting, in paper and pulp, adhesives, ceramics, construction, food, glass, polymers, etc.) (Yagub et al. 2014). Dyes can be classified mainly according to their solubility or the nature of their chromophore (Iscen et al. 2007, Yagub et al. 2014). The azo compound class (containing the azo –N=N– group) comprise more than 60% of all dyes.

Although the dyes contribute only with a minor fraction of organic load, their removal (and/or their decolorization) is required not only for esthetic reasons but in many cases also due to the toxicity, mutagenicity, and carcinogenicity of them and/or of their degradation products (Lim et al. 2011). There are numerous techniques used to treat textile industrial wastewater and can be classified into physical, chemical and biological approaches (El-Zawahry et al. 2016). Unfortunately, most of the existing methods, such as sorption using activated carbon, coagulation, membrane filtration, and irradiation, require high costs and/or are non-environmentally friendly (Yagub et al. 2014). Furthermore, some of them produce high amounts of chemical sludge that require secondary treatments (Jain et al. 2014).

Reactive dyes are chemically stable and most cannot be easily decomposed by natural agents including microorganisms. In spite of that, there are many studies showing the ability of different microorganisms to efficiently degrade dyes. For example, Lim et al. (2011) showed the decolorization of azo dye Reactive Black 5 (RB5) using *Paenibacillus* biofilm achieving a high rate of decolorization (91%) and removal of chemical oxygen demand (79%), although some metabolites such as sulphanilic acid were also detected in the final effluent. In the same way, Adnan et al. (2015) used a strain of the fungus *Trichoderma atroviride* to treat RB5 reaching a high biodegradation (91.1%) and without the generation of secondary aromatic amines. Pajot et al. (2011) showed the degradation but also the accumulation of several textile dyes by the yeast *Trichosporon akiyoshidainum.*

Centro de Investigación y Desarrollo en Fermentaciones Industriales, CINDEFI (CCT La Plata-CONICET, UNLP), Facultad de Ciencias Exactas, 50 y 115, (1900), La Plata. Argentina.
Email: joplaca@hotmail.com
* Corresponding author: donati@quimica.unlp.edu.ar

Although these results support the chance to use biodegradation processes as a technology to treat dyes, some problems can be identified: cells are negatively affected by dyes, the treatment requires a continuous supply of nutrients, and cells need to be immobilized to be reused in several cycles; in addition, dyes are not recovered to be used again and, in many cases, some secondary and toxic products are produced.

An alternative methodology to avoid the last two disadvantages would be the removal of dyes keeping intact their chemical structure. Biosorption emerged as a promising technology to achieve such purpose and it can be defined as the sorption of solutes onto metabolically inactive and inexpensive biological materials. Biosorption is a metabolism-independent process easy to operate and with several inherent advantages, including low cost, operation over a wide range of conditions and the possible reuse of biosorbents (Vijayaraghavan and Yun 2007). An economic biosorbent is defined as one which is abundant in nature, or is a by-product or waste from industry and requires little processing (Hamzeh et al. 2012). Different kinds of low cost biomaterial such as algae, fungus, biomass fly ash, and agricultural wastes have been used to remove dyes from the solution (Pengthamkeerati et al. 2008). Algae biomass is an abundant resource that is widely available in both fresh and salt waters. Many reports have shown that algal biomass can be an efficient biosorbent. The algal biosorption capacity is attributed to their high surface area as well as high binding affinity. Algal cell wall has different functional groups, e.g., hydroxyl, carboxylate, amino and phosphate, which can play an important role in the pollutant removal from wastewater. Brown algae have proven to be highly effective (even more than other type of algae) to remove different kind of pollutants from aqueous solutions (Davis et al. 2003, Romera et al. 2007); although not many studies about dye biosorption have been reported.

In this chapter, RB5 dye sorption from artificial solutions using biomasses of two brown algae, *Macrocystis pyrifera* and *Undaria pinnatifida*, is reported.

Case Study

Macrocystis pyrifera and *Undaria pinnatifida* belong to Phaeophyceae (brown algae) group. *M. pyrifera* is a native species broadly distributed in Argentina from Chubut to Tierra del Fuego. Great amounts of these species are deposited on the beach of Bahía de Camarones causing unpleasant odors and a negative impact on local tourism. *Undaria pinnatifida* is an invasive species and it was firstly detected at Puerto Madryn in 1992; today it occupies more than 1,000 km of Patagonic coast. Recent studies have shown negative impacts on native macroalgal assemblages and reef fish (Plaza Cazón et al. 2014). This case study describes the biosorption of RB5 dye using *M. pyrifera* and *U. pinnatifida* biomasses.

Materials and Methods

Dye description

Reactive Black 5 dye belongs to the group of azo. Figure 1 shows the chemical structure of RB5 dye. Vilmafix®Black B-V was used throughout this study. Powdered dyestuff was dissolved in distilled water in order to prepare 1000 mg l^{-1} solution that was later filtered through 0.22 μm membranes.

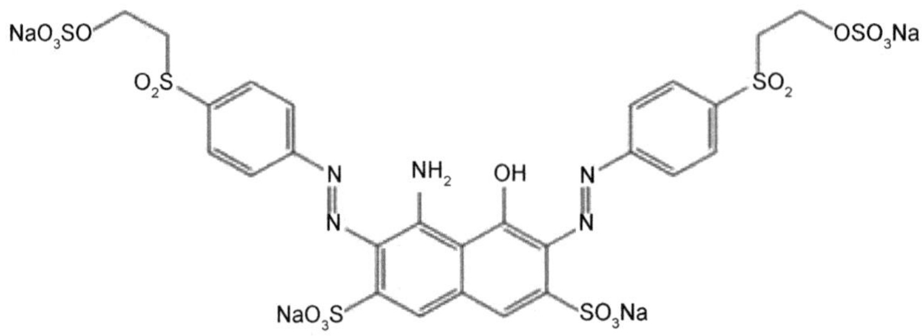

Figure 1. Reactive Black 5: chemical structure.

Biological material: pretreatment

Algae biomass was ground and sieved to select the 10–16 mesh fraction (1.18–2 mm particle size). Biomass was washed several times with distilled water and dried into an oven at 50°C for 48 hr. Part of the biomass was treated for 24 hr with 0.2 M $CaCl_2$ solution at pH 5.0 and another part was treated with 0.1 M HCl for 24 hr. Then, both biomasses were repeatedly washed with distilled water and dried in an oven at 50°C for 48 hr.

Choosing the pre treatment

The efficiency of those two different treatments on the RB5 dye removal was tested in batch experiments. For that, individual solutions of RB5 dye at a concentration of 50 mg l^{-1} (prepared by dilution of the 1000 mg l^{-1} stock solution), were placed in flasks Erlenmeyers containing 0.1 g of untreated or treated (using $CaCl_2$ or HCl) biomass. The pH of the solutions was adjusted at 1.0. Flasks were maintained under agitation at 180 rpm and 30°C. Samples were collected after 24 hr.

Choosing the optimal pH

Assays were carried out to choose the optimal pH for RB5 removal by algal biomass. Selected from the previous assays, only $CaCl_2$-treated biomass was used for these experiments. 100 ml of 50 mg l^{-1} RB5 solution were prepared at different pH values (from 1.0 to 4.0) and placed in flasks containing 0.1 g algal treated biomass. The flasks were kept in agitation at 180 rpm and 30°C.

Determining the optimum dosage of biomass for RB5 biosorption

Different concentrations of biomass (in g, 0.1; 0.2; 0.4; 0.6; 0.8; 1.0; 2.0; 4.0) were tested in order to select the optimal biomass for biosorption process. Biomass was put in contact with 50 mg l^{-1} RB5 solution, pH 1.0 (selected from the results of the experiments to choose the optimum pH value). The flasks were kept in agitation at 180 rpm in a controlled-temperature room for 24 hr.

Kinetic studies and equilibrium studies

Sorption kinetics is needed to determine the equilibrium time, the main sorption mechanism, and the rate-controlling steps (resistance to film diffusion, intraparticle diffusion, or chemical reaction rates) for batch adsorption experiments. Experiments were carried out at initial pH 1.0 adding 0.6 g of biomass previously treated with $CaCl_2$

(biosorbent) to 100 ml of 50 mg l^{-1} RB5 solution. Flasks were kept at 20°C in a shaker at 180 rpm. Samples were collected at different time intervals to analyze the concentration of RB5 remaining in solution.

Reactive Black 5 sorption isotherms were obtained by mixing 6.0 g of biosorbent with 100 ml RB5 solution (initial concentration varying from 10 to 400 mg l^{-1}) at 20°C in a shaker (180 rpm). The initial pH (1.0) and the contact time (6 hr) were selected according to the previous experiments. At the end of the experiments, RB5 and also Ca^{2+} concentrations were analyzed in the solution.

Dye desorption studies

After 6 hr of contact time with 400 mg l^{-1} of RB5, biomass was recovered and washed several times with distilled water. Later the biomass was re-suspended in 100 ml of aqueous solutions with different pH values (from 1 to 14). Samples were taken at different times to measure RB5 concentration.

Analysis of RB5 degradation products using HPLC

Filtered samples from the control (distilled water at pH 1 with biosorbent) and 400 mg l^{-1} RB5 systems were taken after the biosorption process. 20 μl of each sample were injected in HPLC with Autosampler (717 Water) using a Symmetric C 18 column at a flow rate of 1 ml min^{-1} (Water 1525 Binary HPLC Pump). An isocratic method by a mixture of 90% of 15 mM acid phosphoric and 10% pure methanol as a mobile phase was utilized. The detection was done at wavelength 266 nm (Waters 2996 PAD). The data were analyzed with a Software Empoware.

Analytical methods

Initial and final dye concentrations in solution were measured by UV/visible spectrophotometer (TGO PG Instruments) at λ_{max} = 594 nm. Ca^{2+} concentration was determined by absorption atomic spectrophotometry (Shimadzu AA-CC66-50).

Equations and models

The amount of RB5 adsorbed was calculated by the mass balance equation:

$$q = \frac{V(c_i - c_f)}{m} \tag{1}$$

where q is the solute uptake (mg g^{-1}); C_i and C_f are the initial and final solute concentrations in solution (mg l^{-1}), respectively; V is the solution volume (l) and m is the mass of biosorbent (g).

In order to investigate the mechanism of biosorption and the potential rate controlling step, the following kinetic models have been used.

The integrated pseudo first order equation (PFORE). This equation can be expressed as:

$$\log\left(\frac{q_{eq} - q(t)}{q_{eq}}\right) = \frac{-k_1}{2.303}t \tag{2}$$

where q_{eq} (mg g^{-1}) and $q_{(t)}$ are the sorption capacities at equilibrium (experimental value) and at time t, respectively, and k_1 (min^{-1}) is the pseudo-first order rate constant.

The integrated pseudo-second order rate equation (PSORE) can be expressed by the following equation:

$$\frac{t}{q(t)} = \frac{1}{k_2 q_{eq}^2} + \frac{1}{q_{eq}} t \tag{3}$$

where q_{eq} (mg g^{-1}) and $q_{(t)}$ are the sorption capacities at equilibrium (calculated value from experimental data) and at time t, respectively, and k_2 (g mg^{-1} min^{-1}) is the pseudo-second order rate constant.

The prediction of the rate-limiting step is an important factor to be considered in the adsorption process. The solid-liquid sorption process can be divided in three steps: (i) mass transfer of the adsorbate from the bulk liquid to the particle surface (boundary layer diffusion), (ii) adsorption at an exterior site, (iii) intraparticle diffusion. The external mass transfer coefficient, β_L (m min^{-1}) of RB5 in the liquid film boundary can be evaluated using the equation proposed by Gupta et al. (2001):

$$\ln\left(\frac{C_t}{C_o} - \frac{1}{1+mk_a}\right) = \ln\left(\frac{mk_a}{1+mk_a}\right) - \left(\frac{1+mk_a}{mk_a}\right) \beta_L . S_s t \tag{4}$$

where C_t and C_o (both in mg l^{-1}) are the concentration of sorbent at time t and zero, respectively, k_a (L g^{-1}) is a constant obtained by multiplying the Langmuir constants q_m and b, and m (g l^{-1}) and S_s (m^2 l^{-1}) are the adsorbent mass and surface area per unit of volume of solution, respectively. The mass transfer coefficient βL can be calculated from the slope of the regression line: $\ln[(C_t/C_o)-(1/1+mk_a)]$ vs. t.

The intra-particle diffusion was explored by using the following equation:

$$q_{(t)} = K_{dif} t^{1/2} + C \tag{5}$$

where C is the intercept and K_{dif} (mg g^{-1} min^{-2}) is the intra-particle diffusion rate constant. The value of $q_{(t)}$ correlated linearly with values of $t^{1/2}$ and the rate constant K_{dif} was directly evaluated from the slope of the regression line.

Biosorption isotherms describe the relationship between the biomass and the sorbate. Three isotherm models (Langmuir, Freundlich, and Dubinin-Radushkevich) were tested for modeling the biosorption isotherms.

Langmuir model is probably the most popular isotherm models due the simplicity and its usually good agreement with experimental data. The Langmuir isotherm is represented by the following equation:

$$q_{eq} = \frac{b q_m C_{eq}}{1 + b C_{eq}} \tag{6}$$

where q_{eq} is the uptake capacity at the equilibrium (mg g^{-1}); q_m is the maximum Langmuir uptake (mg g^{-1}); C_{eq} is the final concentration at the equilibrium (mg l^{-1}); b is the Langmuir affinity constant (L mg^{-1}). The parameters of the model were determined after linearization of Eq. (6) (i.e., C_{eq}/q_{eq} vs. C_{eq}).

The Freundlich isotherm is a nonlinear sorption model. The general form of this model is the following:

$$q_{eq} = K\, C_{eq}^{1/n} \tag{7}$$

where K is a constant related to the biosorption capacity and $1/n$ is an empirical parameter related to the sorption intensity of the adsorbent, which varies with the heterogeneity of the material. The parameters were obtained by linear regression after linearization ($\ln q_{eq}$ vs. $\ln C_{eq}$).

The equilibrium data were also analyzed with the Dubinin-Radushkevich (D-R) isotherm model to characterize the nature of the biosorption processes as physical or chemical. The linear representation of D-R isotherm equation is expressed by

$$\ln q_{eq} = \ln q_m - \beta\varepsilon^2 \quad (8)$$

where q_{eq} is the amount sorbed on per unit weight of biomass (mol g^{-1}), q_m is the maximum biosorption capacity (mol g^{-1}), β is the activity coefficient related to biosorption mean free energy (mol J^{-1}) and ε is the Polanyi potential ($\varepsilon = RT \ln(1+1/C_{eq})$). The sorption energy ($E$; Kj mol^{-1}) was calculated as follow:

$$E = \frac{1}{\sqrt{-2\beta}} \quad (9)$$

The desorbed dye was analyzed and desorption efficiency was calculated as follows:

$$Desorption(\%) = \frac{RB5(mg)r}{RB5(mg)s} x100 \quad (10)$$

where RB5 $(mg)_r$ is the amount of dye released from the loaded algal biomass and RB5 $(mg)_s$ is the dye sorbed onto the biomass.

Results and Discussion

Effect of the pretreatment on biomaterials

Biosorption efficiency of biomasses can be significantly enhanced by pre-treatment methods such as autoclaving, drying, and exposure to chemicals such as formaldehyde, HCl, NaOH, $NaHCO_3$, and $CaCl_2$ (Aksu 2005). However, in this case biosorption capacities of *M. pyrifera* and *U. pinnatifida* untreated biomasses were higher than those for treated biomasses (Fig. 2A). Comparing both treatments, $CaCl_2$-treated biomass showed higher dye removal than HCl-treated biomass.

Although the untreated biomass showed higher removal, biomass treated with $CaCl_2$ was used in the following studies due to the treatment allows conserving the structure of the biomass and reduces the leaching of many compounds from the cell wall.

Effect of the pH value

The effect of solution pH on the adsorption process is shown in Fig. 2B. The lower pH value the more RB5 uptake occurs. This effect was even more significant for *M. pyrifera* biomass. Since RB5 has sulfonic acids (functional group R-SO_2OH) which are stronger acids than most present on the biomass (mainly carboxylic groups), surely the dye is less positively charged at pH 1 than the biomass, allowing strong electrostatic interactions between them.

Effect of biosorbent mass

The effects of biosorbent dose on the removal of RB5 are shown in Fig. 2C. The behavior was similar for both biomasses. Percentage of dye removal increased from 20% to 99%

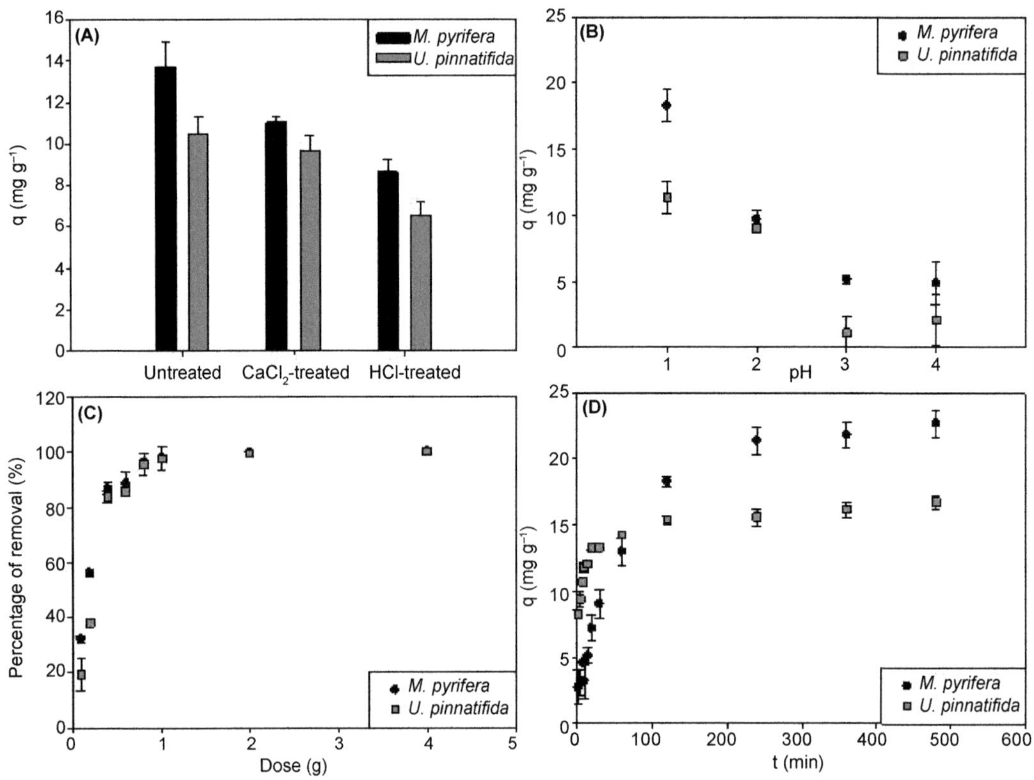

Figure 2. Influence of operational parameters on RB5 biosorption. (A) different treatment; (B) different pH value; (C) different biosorbent dose; (D) contact time (V: 100 ml; $[RB5]_i$: 50 mg l^{-1}; 24 hr; 180 rpm; 30°C). q is the sorption capacity.

for 0.1 g to 0.6 g of biosorbent, respectively. This increase is almost linear between both variables surely due to the increase of the surface exposed to the solution. After reaching the complete removal of RB5, new additions of biomass do not show any influence.

Effect of contact time

The influence of the contact time on the removal of the dye can be observed in Fig. 2D. Dye biosorption on *M. pyrifera* and *U. pinnatifida* biomasses showed the same kinetic profile. Fast uptake of dye occurs in the initial stages of the adsorption process, and then the adsorption rate becomes gradually slower until it reaches a constant value where no more dye could be removed (the equilibrium point). The equilibrium state was reached after 400 min (6 hr).

Kinetic data were tested using different models. Table 1 indicates the parameters for each kinetic model. The correlation coefficient values for pseudo-second-order model were slightly higher than those for pseudo-first-order model but in addition predicted sorption capacities at equilibrium were closer to the experimental values.

The kinetics of the overall biosorption process may be controlled by external or internal diffusion or even by a combination of both. Intraparticle diffusion and external mass transfer models were applied on the results of RB5 biosorption. According to the regression coefficients, external mass transfer controlled the RB5 dye biosorption process onto *M. pyrifera* and *U. pinnatifida* biomasses.

Table 1. Biosorption kinetic parameters.

Models	*M. pyrifera*	*U. pinnatifida*
Pseudo first order		
q_{exp} (mg g^{-1})	9.75	9.57
K_1 (min^{-1})	8.06 10^{-03}	8.06 10^{-03}
q_{eq} (mg g^{-1})	6.67	7.16
R^2	0.97	0.96
Pseudo second order		
K_2 [g (mg min)$^{-1}$]	2.0 10^{-03}	1.3 10^{-03}
q_{eq} (mg g^{-1})	10.5	10.8
R^2	0.99	0.98
Intraparticle diffusion		
K_{dif} (mg g^{-1} min^{-2})	0.32	0.34
C	2.77	2.11
R^2	0.88	0.82
External mass transfer		
β_L (m min^{-1})	4.23 10^{-4}	8.04 10^{-5}
R^2	0.97	0.99

Note: q_{exp}: is the experimental sorption capacity; q_{eq}: is the sorption capacities at equilibrium; K_1: is the pseudo-first order rate constant; K_2: is the pseudo-second order rate constant; K_{dif}: is the intra-particle diffusion rate constant; C: is the intercept; β_L: is the mass transfer coefficient.

Equilibrium studies

Results were expressed as plots of solid phase dye concentration against liquid phase concentration and they are shown in Fig. 3. Ca^{2+} released during the biosorption process was also included in Fig. 3. *M. pyrifera* seems to be much more efficient biosorbent than *U. pinnatifida*. The amount of Ca^{2+} released from two biosorbents did not suffer significant changes along the different RB5 concentrations and in addition was similar for both

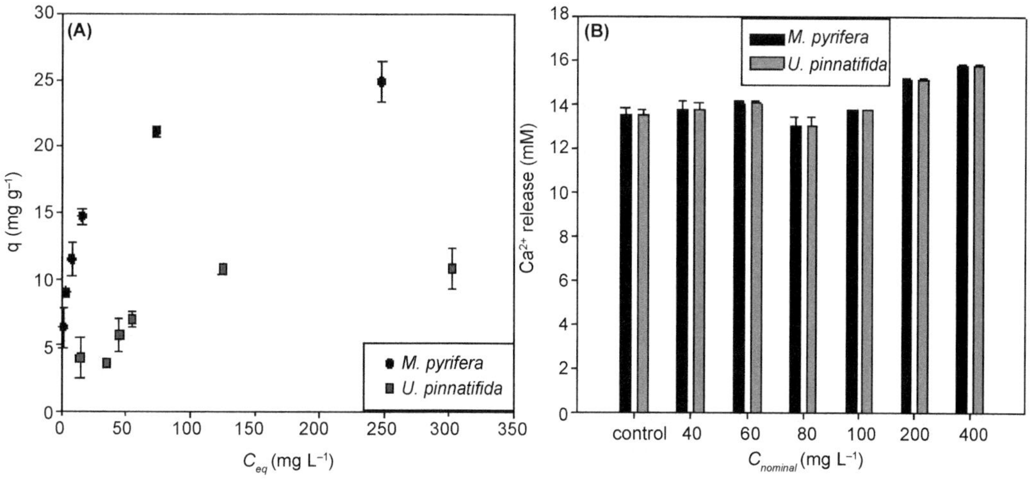

Figure 3. Biosorption on *M. pyrifera* and *U. pinnatifida* biomasses. (A) effect of initial RB5 dye concentration; (B) Ca^{2+} released during RB5 biosorption (V: 100 ml; $[RB5]_i$: 10–400 mg l^{-1}; 0.6 g; pH: 1.0; 24 hr; 180 rpm; 30°C). q is the sorption capacity, C_{eq} is the final concentration at equilibrium and $C_{nominal}$ is the initial concentration.

biomasses. These results would indicate the interchange did not have a main role in RB5 biosorption process.

Table 2 contains the main parameters from the application of Langmuir, Freundlich, and Dubinin-Radushkevich models to the RB5 dye biosorption.

Langmuir adsorption model fits better the experimental data than the other models suggesting that RB5 dye is adsorbed on active sites of both biosorbents forming a mono-layer coverage.

The experimental data were fitted to Dubinin-Radushkevich (D-R) model in order to determine if the adsorption occurred by a physical or chemical process (Table 2). However, there was not a good correlation.

The q_m value (26.3 mg g^{-1}) for the adsorption on *M. pyrifera* biomass was much higher than that for *U. pinnatifida* biomass (12.6 mg g^{-1}); also *M. pyrifera* biomass has a higher affinity (higher *b*) for RB5 dye than *U. pinnatifida* biomass. Both results confirm that *M. pyrifera* is a better RB5 biosorbent than *U. pinnatifida*. Since it was expected (Iscen et al. 2007) the percentage of dye removal for both biomasses decreased when the initial dye concentration was increased (Tables 3 and 4). For all concentrations dye removal

Table 2. Biosorption isotherm parameters.

Parameters	***M. pyrifera***	***U. pinnatifida***
Langmuir		
q_m (mg g^{-1})	26.3	12.6
b (L mg^{-1})	0.10	0.02
R^2	0.99	0.96
Freundlich		
K (L mg^{-1})	6.15	1.27
n	3.66	2.56
R^2	0.96	0.79
D-R		
E (Kj mol^{-1})	0.79	0.12
q_m	16.5	7.6
R^2	0.65	0.40

Note: q_m: is the maximum uptake capacity; b: is the Langmuir affinity constant; K: is the constant related to the biosorption capacity; n: is the sorption intensity of the adsorbent; E: is the sorption energy.

Table 3. Percentages of dye removal by *M. pyrifera* biomass.

[RB 5]$_i$, mg l^{-1}	**[RB 5]$_f$, mg l^{-1}**	**% R**
39.4 ± 1.5	1.6 ± 0.3	96
57.4 ± 0.2	3.7 ± 0.2	93
76.8 ± 0.1	8.1 ± 1.2	89
104.4 ± 0.2	16.3 ± 0.5	84
200 ± 0.1	73.7 ± 0.4	63
397.4 ± 1.2	248.0 ± 0.8	38

Note: [RB 5]$_i$: initial reactive black 5 dye concentration; [RB 5]$_f$: final reactive black 5 dye concentration; % R: removal percentage.

Table 4. Percentages of dye removal by *U. pinnatifida* biomass.

[RB 5]$_i$, mg l^{-1}	**[RB 5]$_f$, mg l^{-1}**	**% R**
38.5 ± 1.3	14.6 ± 1.5	62
57.0 ± 1.2	35.4 ± 1.8	38
79.0 ± 1.4	44.6 ± 1.2	43
96.8 ± 1.2	55.2 ± 0.5	43
189.9 ± 1.3	125.3 ± 0.8	34
398.4 ± 1.5	333.0 ± 1.3	16

Note: [RB 5]$_i$: initial reactive black 5 dye concentration; [RB 5]$_f$: final reactive black 5 dye concentration; % R: removal percentage.

by *M. pyrifera* was higher than that by *U. pinnatifida*; the maximum dye removals were 96% and 62% for *M. pyrifera* and *U. pinnatifida*, respectively, at the initial concentration of 40 mg l^{-1}.

Dye desorption studies

Since the RB5 adsorption onto algae biomasses increases at lower pH, desorption was attempted increasing the pH value. That is why 0.1 M NaOH was used as elutant. The results of RB5% desorption are illustrated in Fig. 4. The percentage of desorption reached almost 100% at pH about 12. Under such strong basic conditions, both RB5 and biomasses are negatively charged and the electrostatic repulsion contributes to separate the dye from the biosorbents. Similar results were reported by Vijayaraghavan and Yu (2008).

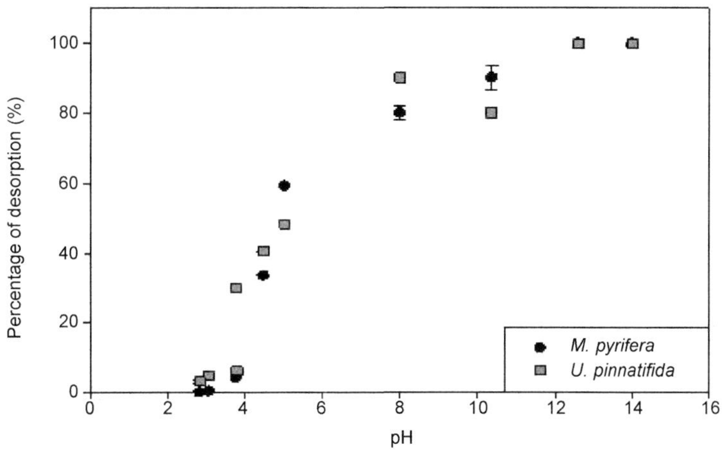

Figure 4. Efficiency of RB5 desorption (eluent: 0.1 M NaOH).

Analysis of biodegradation products using HPLC

Possible biodegradation products were analyzed using HPLC. Figure 5A shows the chromatogram for RB5 dye which presents a unique signal. The chromatograms shown in Fig. 5B and 5C correspond to the solution filtered from *M. pyrifera* biomass in contact with acidified water (pH: 1) without and with RB5 dye, respectively. The former chromatogram presents 5 signals corresponding to different compounds (III–VII) leached from the

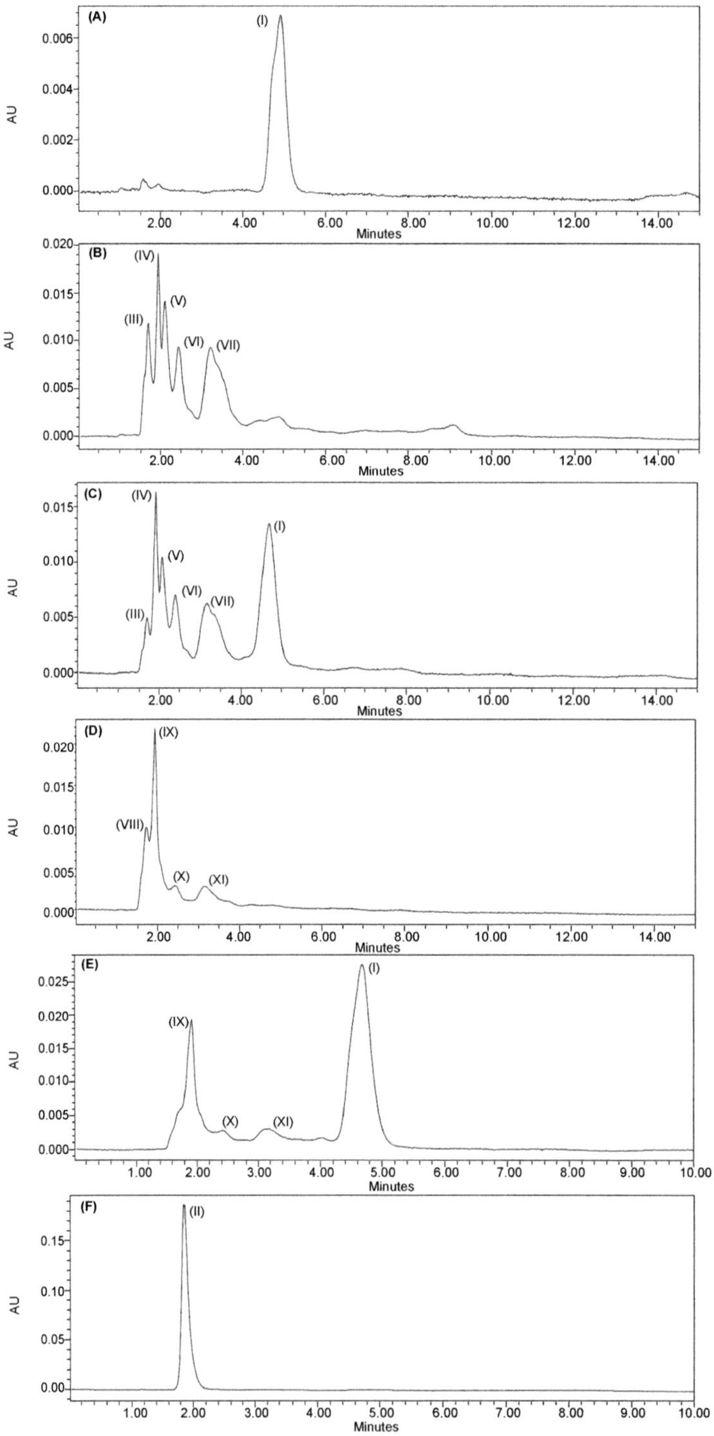

Figure 5. HPLC analysis. (A) RB5 dye (I); (B) *M. pyrifera* biomass (pH = 1, without dye); (C) *M. pyrifera* biomass (pH = 1; 400 mg l^{-1} RB5); (D) *U. pinnatifida* biomass (pH = 1, without dye); (E) *U. pinnatifida* biomass (pH = 1; 400 mg l^{-1} RB5); (F) Sulphanilic acid standard (II) (III)–(VII), and (VIII)–(XI) are organic leaching products released from *M. pyrifera* and *U. pinnatifida* biomasses, respectively.

biomass; these compounds have not been identified up to this moment. In the presence of RB5, the chromatogram shows the peak corresponding to free RB5 (there was an excess of dye due to the biomass has been in contact with 400 mg l^{-1} and at that concentration, just 38% was removed by biosorption) and some little modifications on the previous signals: the intensity of signal (III) significantly decreased while the signal (IV) increased. A similar situation occurred with the other biosorbent; Fig. 5D and 5E show the chromatograms for the solution filtered from *U. pinnatifida* biomass in contact with acidified water (pH: 1) without and with RB5 dye, respectively. According to the chromatogram 4 compounds have been released from the biomass. The retention times for III, IV, VI, and VIII (leached from *M. pyrifera* biomass) are similar to VIII, IX, X, and XI (leached from *U. pinnatifida* biomass); this would indicate that both biomasses release the same 4 compounds but more studies are needed to confirm that and to determine the identity of these compounds; on the other hand, signal V observed for *M. pyrifera* biomass does not have an equivalent for *U. pinnatifida* biomass. In the presence of RB5, in similar way to the first alga, the first peak (VIII) almost disappeared indicating that this compound is reacting with the dye although the second peak did not seem to be affected by RB5. The increase of peak II could indicate a little auto-oxidation of RB5 forming sulphanilic acid (Vyrides et al. 2014) whose peak (Fig. 5F) coincides with signals II and IX.

Concluding Remarks

M. pyrifera and *U. pinnatifida* biomasses are great adsorbents for RB5 dye mainly at very low pH value, reaching maximum adsorption capacity of 26.3 and 12.6 mg g^{-1}, respectively. The adsorption kinetics followed pseudo second order with combined influence of external mass transfer in the rate determining step. The equilibrium data fitted well with Langmuir adsorption model obeying monolayer adsorption. After using NaOH solution at pH higher than 12, not only RB5 could be recovered but also the biomasses could be repeatedly used.

Acknowledgements

Dr. E.R. Donati and Dr. J. Plaza Cazón are researchers from CONICET. This research was supported by ANPCyT (PICT 2013-630).

References Cited

Adnan, L.A., P. Sathishkumar, A.R. MohdYusoff and T. Hadibarata. 2015. Metabolites characterisation of laccase mediated Reactive Black 5 biodegradation by fast growing ascomycete fungus *Trichoderma atroviride* F03. Int. Biodeterior. Biodegr. 104: 274–282.

Aksu, Z. 2005. Application of biosorption for the removal of organic pollutants: a review. Process Biochem. 40: 997–1026.

Davis, T.A., B. Volesky and A. Mucci. 2003. A review of the biochemistry of heavy metal biosorption by brown algae. Water Res. 37: 4311–4330.

El-Zawahry, M.M., F. Abdelghaffar, R.A. Abdelghaffar and A.G. Hassabo. 2016. Equilibrium and kinetic models on the adsorption of Reactive Black 5 from aqueous solution using *Eichhornia crassipes*/chitosan composite. Carbohydr. Polym. 136: 507–515.

Gupta, V.K., M. Gupta and S. Sharma. 2001. Process development for the removal of lead and chromium from aqueous solutions using red mud-an aluminium industry waste. Water Res. 35: 1125–1134.

Hamzeh, Y., A. Ashori, E. Azadeh and A. Abdulkhani. 2012. Removal of Acid Orange 7 and Remazol Black 5 reactive dyes from aqueous solutions using a novel biosorbent. Mat. Sci. Eng. C. 32: 1394–1400.

Iscen, C.F., I. Kiran and S. Ilhan. 2007. Biosorption of Reactive Black 5 dye by *Penicillium restrictum*: The kinetic study. J. Hazard. Mater. 143: 335–340.

Jain, R.M., K.H. Mody, J. Keshri and B. Jha. 2014. Biological neutralization and biosorption of dyes of alkaline textile industry wastewater. Mar. Pollut. Bull. 84: 83–89.

Lim, C.K., H.H. Bay, T.C. Kee, Z.A. Majid and I. Zaharah. 2011. Decolourisation of Reactive Black 5 using *Paenibacillus* sp. immobilized onto macrocomposite. J. Bioremed. Biodegrad. S1: 004.
Pajot, H.P., J.I. Fariña and L.I. Castellanos de Figueroa. 2011. Evidence on manganese peroxidase and tyrosinase expression during decolorization of textile industry dyes by *Trichosporon akiyoshidainum*. Int. Biodeter. Biodegrad. 65: 1199–1207.
Pengthamkeerati, P., T. Satapanajaru and O. Singchan. 2008. Sorption of reactive dye from aqueous solution on biomass fly ash. J. Hazard. Mater. 153: 1149–1156.
Plaza Cazón, J., M. Viera, S. Sala and E. Donati. 2014. Biochemical characterization of *Macrocystis pyrifera* and *Undaria pinnatifida* (Phaeophyceae) in relation to their potentiality as biosorbents. Phycologia. 53: 100–108.
Romera, E., F. González, A. Ballester, M.L. Blázquez and J.A. Muñoz. 2007. Comparative study of biosorption of heavy metals using different types of algae. Bioresour. Technol. 98: 3344–3353.
Vijayaraghavan, K. and Y.-S. Yun. 2007. Utilization of fermentation waste (*Corynebacterium glutamicum*) for biosorption of Reactive Black 5 from aqueous solution. J. Hazard. Mater. 141: 45–52.
Vijayaraghavan, K. and Y.-S. Yun. 2008. Biosorption of C.I. Reactive Black 5 from aqueous solution using acid-treated biomass of brown seaweed *Laminaria* sp. Dyes Pigm. 76: 726–732.
Vyrides, I., B. Bonakdarpour and D.C. Stuckey. 2014. Salinity effects on biodegradation of Reactive Black 5 for one stage and two stages sequential anaerobic aerobic biological processes employing different anaerobic sludge. Bioresour. Technol. 99: 51–58.
Yagub, M.T., T.K. Sen, S. Afroze and H.M. Ang. 2014. Dye and its removal from aqueous solution by adsorption: A review. Adv. Colloid. Interface Sci. 209: 172–184.

6

Strategies for Biodegradation and Bioremediation of Pesticides in the Environment

Ma. Laura Ortiz-Hernández, María L. Castrejón-Godínez,*[a] *Elida C. Popoca-Ursino,*[b] *Fabricio R. Cervantes-Dacasa*[c] and *Maikel Fernández-López*[d]

Introduction

The increase in the world population has led to the accumulation of a wide variety of chemical compounds in the environment (Uqab et al. 2016). This problem has increased in recent years due to industrialization, intensive agriculture and anthropogenic activities (Biswas et al. 2015). There is also a need to produce food with greater yields and thus, in modern agriculture, the intensive use of pesticides is very frequent; millions of tons of pesticides are applied annually around the world, equivalent to billions of dollars in the market (Javaid et al. 2016). However, the application and mismanagement of pesticides has led to pest resistance, contamination of soil, air, and water, as well as the entry of these compounds into the food chain. Besides, they can cause adverse effects on different life forms, affecting populations, communities, and ecosystems in wildlife (Köhler and Triebskorn 2013). On the other hand, in several developing and transitional countries, there are more than half a million tons of obsolete, unused, forbidden or outdated pesticides, which endanger the environment and health of millions of people. In these countries, important quantities of obsolete pesticides have been accumulated since there is no strategy for their proper management (Dasgupta et al. 2010, Ortiz-Hernández et al. 2013).

In principle, only the use of pesticides that are not persistent in the environment are permitted; however, persistent pesticide residues are found worldwide (Fenner et al. 2013). Some authors report the pesticides presence in ecosystems even though their

Laboratorio de Investigaciones Ambientales, Centro de Investigación en Biotecnología, Universidad Autónoma del Estado de Morelos. Av. Universidad 1001, Col. Chamilpa, Cuernavaca Morelos, México.
[a] Email: mlcastrejon@uaem.mx
[b] Email: caro_popoca@hotmail.com
[c] Email: fabriciorcd@gmail.com
[d] Email: fmaikel44@yahoo.com
* Corresponding author: ortizhl@uaem.mx

use has been banned more than a decade ago (Muñoz-Arnanz and Jiménez 2011, Park et al. 2011). One important characteristic that a pesticide must have is its specificity; however, there are multiple reports about the adverse effects caused by pesticides on non-target organisms, including humans. Population-based studies have revealed a possible relationship between exposure to organophosphate pesticides and harmful health effects such as cardiovascular disease (Hung et al. 2015), negative effects on the reproductive and nervous systems (McKinlay et al. 2008, Jamal et al. 2015) and dementia (Lin et al. 2015). It has also been shown that there is a relationship between exposure to pesticides and the risk of suffering from some type of neurodegenerative disease such as Alzheimer's and Parkinson's diseases, as well as amyotrophic lateral sclerosis (Kamel 2013, Mostafalou and Abdollahi 2013).

All these problems make it necessary to develop strategies for the elimination of these compounds without causing additional damage to the environment. Different reports have shown that contaminated areas pose a potential risk to humans, leading to an international effort to remedy many of these contaminated sites (Kensa 2011).

There are different techniques for eliminating pesticides. The physicochemical methods are widely used; their main objective is to destroy a compound or to diminish its toxicity. Among the physical treatments, the most used is incineration, whose advantage is the complete elimination in a short time. However, its cost is high and atmospheric emissions can be generated (Ortiz-Hernández et al. 2011). Another method is the chemical decomposition by advanced oxidation, which uses UV and strong oxidizing agents such as H_2O_2, O_3, and TiO_2 to produce –OH radicals that can easily react with toxic compounds (Nandan et al. 2016). These methods can be very effective in reducing a range of pollutants but they also have disadvantages such as their complexity and high cost (Kensa 2011).

Concepts, Principles, and Relationship between Biodegradation and Bioremediation

Soil microorganisms play an important role in pesticides biodegradation. Because they are constantly exposed to these compounds, they have developed catabolic processes to degrade them (Jaiswal et al. 2017). Biodegradation is a process that eliminates pollutants or reduces their toxicity or concentration to levels permitted by environmental legislation (Cycoń et al. 2013). It involves the breakdown of pesticides by microorganisms, resulting in less complex substances, water, CO_2 and/or salts. Complete degradation of pesticides into inorganic compounds is known as mineralization. In some cases, the degradation produces less toxic and simpler compounds, which leads to partial biodegradation (Villaverde et al. 2017).

Studies on microbial pesticides degradation are useful for the development of bioremediation strategies. The use of microorganisms with the capacity to degrade xenobiotics is the basis for contaminated sites bioremediation and varies depending on the chemical properties of the contaminants (Prescott et al. 2008). Biodegradation and bioremediation are closely related processes and are based on the metabolism of pesticides by microorganisms. The difference between them is that biodegradation is a natural process and bioremediation is a technology (Singh 2008). Bioremediation uses the physiological potential of microorganisms and plants to degrade compounds, which enzymatically modifies contaminants to less harmful products (Villaverde et al. 2017). Bioremediation is effective under environmental conditions that allow growth and microbial activity, so it frequently requires the manipulation of environmental parameters in order to allow microbial growth and increase the rate of degradation (Sharma 2012).

Bioremediation is more economical than physicochemical methods such as incineration. Some contaminants can be treated *in situ*, reducing the risks for exposure or transportation accidents. This is why the public opinion considers that bioremediation is more acceptable than other technologies (Sharma 2012). Most bioremediation systems occur under aerobic conditions, but systems under anaerobic conditions can also be used to degrade xenobiotic compounds. Anaerobic microorganisms present in the subsoil can degrade a wide variety of organic contaminants (Coates and Anderson 2000); this process is of great importance since large volumes of contaminants found in aquifers can be degraded by anaerobic processes (Anderson and Lovley 1997). The control and optimization of biodegradation/bioremediation processes depend on many factors, including the physicochemical properties of pesticides (solubility, concentration, structure) and soil characteristics (type, humidity, temperature, pH, organic matter) (Jaiswal et al. 2017).

Bioremediation Technologies

Bioremediation can be classified into two categories (Das and Dash 2014, Juwarkar et al. 2014, Bhardwaj and Kapley 2015, Ajlan 2016):

1. ***In situ.*** The degradation or the removal of the contaminants is carried out in the impacted place. It involves techniques such as natural attenuation, bioventing, bioaugmentation, biostimulation, and phytoremediation.
2. ***Ex situ.*** These technologies require the mobilization of contaminated water or soil to a site under controlled conditions for its treatment.

In this chapter, some of these strategies are described.

In situ Bioremediation Technologies

Natural attenuation

Natural attenuation is a proactive approach that focuses on the verification and monitoring of natural remediation processes also known as passive remediation, *in situ* bioremediation, intrinsic remediation, bioattenuation, and intrinsic bioremediation. Natural attenuation is an *in situ* treatment method that uses natural processes to contain the spread of contamination and to reduce the concentration and amount of pollutants at contaminated sites (Juwarkar et al. 2014).

Bioventing

Bioventing is one of the most common *in situ* treatment techniques. This technology uses the organisms present in the contaminated site as well as the nutrients available to carry out the degradation process of the contaminants. This is an aerobic process involving air supply for the biodegradation while minimizing volatilization and release of contaminants to the atmosphere (Juwarkar et al. 2014).

Phytoremediation

The term phytoremediation comes from the Greek *phyto*: plant and from the Latin *remedium*: to restore the balance (Talukder et al. 2015, Banjoko and Eslamian 2016, Gomes et al. 2016, Ugya et al. 2016). It is defined as the use of plants and their associated microorganisms to absorb, accumulate, metabolize, volatilize or stabilize contaminants originated from human activities present in the environment, such as pesticides (herbicides, insecticides and fungicides), petroleum hydrocarbons, chlorinated solvents, explosives, heavy metal and radionuclides (Verma and Shukla 2016, Kovacs and Szemmelveisz 2017). This technology can be applied to extract organic and inorganic contaminants present in the soil, water, air, sludge, sediments and groundwater (Verma and Shukla 2016, Deng and Cao 2017).

The decontamination of polluted sites by phytoremediation is performed using at least one of the following mechanisms (López-Martínez et al. 2005, Zhang et al. 2010, Ijaz et al. 2016, Yadav et al. 2016):

- ***Phytoextraction or phytoaccumulation***: It is the capacity of some plants to accumulate contaminants in roots, stem or foliage.
- ***Rhizofiltration***: Ground plants with a well-developed root system are used to absorb, concentrate and precipitate contaminants.
- ***Phytostabilization***: Plant roots stimulate microorganisms present in the contaminated site to reduce the bioavailability and mobility of contaminants avoiding transport to underground layers or to the atmosphere.
- ***Phytodegradation***: The plants and microorganisms associated with them degrade the contaminants into simpler molecules. Sometimes these products serve as nutrients to the plant to accelerate its growth.
- ***Phytovolatilization***: During their growth, the plants absorb water with soluble organic and inorganic pollutants and some of them can reach the leaves and evaporate or volatilize to the atmosphere.
- ***Phytostimulation or rhizodegradation***: Plants generate root exudates that stimulate the growth of native microorganisms capable of degrading xenobiotic organic compounds.

Research in pesticide phytoremediation is currently focused on elucidating the plant mechanisms for pollutant transport across their structures, the mechanism by which some plants are capable of absorbing and tolerating high pesticides concentrations, and the genetic modification of plants for their use in the remediation of pesticide-contaminated sites. Understanding the interaction between microorganisms and plants at the rhizosphere level, and the possibility of using this knowledge for the remediation of sites contaminated with pesticides is also being investigated (Frazar 2000, Hussain et al. 2009).

Frazar (2000) suggests that the interaction between plant and microorganisms has led to increasing degradation of pesticides present in contaminated sites (Parween et al. 2016). The genus *Kochia* sp. is promising for the phytoremediation of soils contaminated with pesticides and groundwater.

Sorption of pesticides by plants is dependent on the physicochemical properties of the compounds, mode of application, soil type, climatic factors and plant species. Compounds absorbed through the plant roots can be translocated to other parts *via* xylem. The microbial activity in the rhizosphere also plays an important role in the transformation of chemical contaminants, which can help in the absorption through the roots and the degradation; so it should be considered an integral component of phytoremediation (Chaudhry et al. 2002, Mitton et al. 2016b).

Some plants have the ability to metabolize or accumulate organic compounds such as pesticides; for example 1,1,1-trichloro-2,2-bis (4-chlorophenyl) ethane (DDT), atrazine, dieldrin, endrin, alachlor, aldrin, polychlorinated biphenyls, and hexachlorobenzene (Green and Hoffnagle 2004, Matsumoto et al. 2009, Motoki et al. 2015). Table 1 shows examples of plants that are used in phytoremediation.

Table 1. Plants used in phytoremediation.

Plant	Pollutant	Reference
Kochia sp.	Atrazine, metolachlor, trifluralin	Frazar (2000)
Arabidopsis thaliana	Polychlorinated biphenyls (PCBs)	Asai et al. (2002)
Brassica napus	2,4-dichlorophenol (2,4-DCP)	Agostini et al. (2003)
Zea mays	Atrazine	Chang et al. (2005)
Populus spp.	Dieldrin	Skaates et al. (2005)
Cucurbita sp.	Dieldrin, endrin	Otani et al. (2007)
Populus spp.	PCBs	Campos del Pozo (2010)
Cichorium intybus	DDT	Delgadillo-López et al. (2011)
Zea mays	Atrazine	Ibrahim et al. (2013)
Salix spp.	Ethylene thiourea (ETU), atrazine	Lafleur et al. (2016)
Sunflower	Endosulfan	Mitton et al. (2016a)
Medicago sativa *Glycine max*	DDT	Mitton et al. (2016b)
Cucurbitaceae family	PCBs	Vergani et al. (2017)

Phytoremediation has disadvantages such as the plants' intolerance to pesticides or their inability to degrade or accumulate them; it also requires relatively large areas, is dependent on the weather or seasons, is a relatively slow process and requires long periods to effectively remediate a site. Besides, the plants containing contaminants may affect herbivores that feed from them (Verma and Shukla 2016, Thijs et al. 2017). However, phytoremediation also has numerous advantages in relation to the physicochemical methods currently used: it is a sustainable and low-cost alternative for the rehabilitation of sites affected by natural and anthropogenic contaminants (Delgadillo-López et al. 2011, Gomes et al. 2016, Malik et al. 2016, Verma and Shukla 2016). This technology is made more effective through genetic manipulation, which improves the remediation capacity of plants. Plant species have been designed with a greater capacity for degradation of organic pollutants or accumulation of heavy metals (Cherian and Oliveira 2005).

Bioaugmentation

Bioaugmentation is the supply of microorganisms for bioremediation (Díaz-Martínez et al. 2013, Azubuike et al. 2016). It is used to describe the addition of organisms or enzymes to a site for eliminating contaminants, in which allochthonous or genetically modified microorganisms capable of degrading pollutants are inoculated in the contaminated soil. This process can be carried out both *in situ* and *ex situ*, depending on economic factors, contaminant type and the facilities that exist for *ex situ* management (Windevoxhel et al. 2011, Ma et al. 2015, Ajlan 2016, Chaturvedi et al. 2016, Baćmaga et al. 2017, Nwankwegu and Onwosi 2017).

The bioaugmentation was first performed around 1968, inoculating *Cellumonas* sp. and nutrients under controlled conditions to degrade hydrocarbons, in which the degradation was more effective with the addition of bacteria than with nutrient application alone. From that moment, bioaugmentation is used in different environments to successfully degrade xenobiotic compounds (Dueñas and Santos 2006, Alvarez et al. 2017). This technology has become a new alternative to directly address many of the problems of soil contamination caused by different pollutants, including in particular the presence of fertilizers, pesticides, and herbicides that are commonly used in agriculture (Xu and Zhou 2017). However, for a successful bioaugmentation process, the following characteristics of microorganism should be considered: rapid growth, easily cultivable, withstanding high concentrations of pollutants and the ability to survive in a wide range of environmental conditions (Singer et al. 2005).

Bioaugmentation has been used to treat soils contaminated with herbicides (2,4-dichlorophenoxyacetic), insecticides (lindane, chlordane, parathion), chlorophenols (PCP) and nitrophenols (BCP, HTP, and HAP) (Sepúlveda and Trejo 2002, Xu and Zhou 2017). Gram-negative bacterial mixtures belonging to *Pseudomonas, Flavobacterium, Sphingobium,* and *Alcaligenes* genera, as well as Gram-positive bacteria such as *Arthrobacter, Nocardia, Rhodococus, Mycobacterium,* and *Bacillus* have been used in bioaugmentation. In addition to them, fungi also have the potential to be used in bioaugmentation, such as the genera *Achremonium, Aspergillus, Verticullium, Penicillium, Mucor,* and *Cupriavidus.* Individual strains or consortia of microorganisms can be used to treat different types of contaminants (Mrozik and Piotrowska-Seget 2010, Bhalerao 2012, Tiwari et al. 2017). Fantroussi and Agathos (2005) and Mrozik and Piotrowska-Seget (2010) reported that the bioaugmentation approach should be applied when biostimulation and bioattenuation have failed.

Table 2 shows some bacterial genera used in bioaugmentation processes for pesticide degradation.

Before carrying out a bioaugmentation process in a contaminated site, it is necessary to carry out three steps: to have enrichment bacterial cultures, to isolate the microorganisms capable of metabolizing or using the contaminant as a source of carbon, and cultivating microorganisms to obtain large amounts of biomass. Its use does not imply a high operating cost, but at the same time, it is a technology that can last several months or years

Table 2. Organisms used for pesticides biodegradation.

Organisms	Pesticide	Reference
Escherichia coli AtzA	Atrazine	Strong et al. (2000)
Pseudomonas sp. ADP	Atrazine and simazine	Morán et al. (2006)
Burkolderia sp. FDS-1	Fenitrothion	Hong et al. (2007)
Sphingobium chlorophenolicum ATCC 39723	Pentachlorophenol	Dams et al. (2007)
Mixed bacterial culture	Endosulfan	Kumar and Philip (2007)
Burkholderia cepacia PLC3	Carbofuran	Plangklang and Reungsang (2009)
Klebsiella sp.	Chlorpyrifos	Farhan et al. (2013)
Serratia marcescens	Deltamethrin	Cycoń et al. (2014)
Arthrobacter sp. AK-YN10	Atrazine	Sagarkar et al. (2015)
Novosphingobium DY4	2,4-Dichlorophenoxyacetic acid	Dai et al. (2015)
Bacillus	Insecticides	Parween et al. (2016)
Sphingomonas sp. PM2 *Sphingomonas* sp. ERG5 *Burkholderia* sp. TFD34 *Cupriavidus* sp. TFD38	Phenoxy acid herbicide 4-chloro-2-methyl-phenoxy-acetic acid (MCPA)	Samuelsen et al. (2017)
Arthrobacter sp.	Iprodione	Campos et al. (2017)
Trametes versicolor	Carbamates	Rodríguez-Rodríguez et al. (2017)

(Sepúlveda and Trejo 2002). According to Mrozik and Piotrowska-Seget (2010), one of the most difficult issues is the survival of the strain in the soil; it has been observed that the number of microorganisms is reduced after the inoculation. However, strain selection for bioaugmentation is clearly a key issue (Herrero and Stuckey 2015).

Biostimulation

Microbial populations have the potential to remove toxic compounds at contaminated sites, but some factors may inhibit or retard the bioremediation process. These factors include the lack of electron donors or acceptors, low availability of nitrogen and phosphorus compounds, and the absence of catalyst involved in the degradation pathways (Cosgrove et al. 2010). To address these problems biostimulation is used, which is based on the addition of electron acceptors or donors, as well as nutrients to stimulate the degradation of a compound by the endogenous microbial population (Scow and Hicks 2005, de Lorenzo 2008).

In situ biostimulation has been used to stimulate crude oil degradation (Mills et al. 2004), tetrachloroethane (Major et al. 2002), diesel (Namkoong et al. 2002, Rivera-Espinoza and Dendooven 2004) and aromatic hydrocarbons (Kalantary et al. 2014). However, for pesticide degradation, this effective bioremediation method has been limited to microcosm scales. Levi et al. (2014) investigated the biodegradation of the herbicides bentazone, mecoprop and dichloroprop, using a microcosm with the material of an anaerobic aquifer and with the addition of oxygen. They demonstrated that concentrations of oxygen between 4–11 mg l^{-1} increase the degradation of mecoprop (14–27%), dichlorprop (3–9%) and bentazone (15–20%) in a period of 200 d. In non-oxygenated controls, no degradation occurred.

In order to stimulate the toxic compounds' degradation, biostimulation can be complemented by a bioaugmentation process (Kanissery and Sims 2011). Silva et al. (2004) used both processes for the treatment of soil with high concentrations of atrazine. For bioaugmentation, they used *Pseudomonas* sp. ADP strain (10^7 cells g_{soil}^{-1}) and added citrate and succinate at 5.8–40 and 6.2–30.8 cells g_{soil}^{-1}, respectively. They achieved atrazine mineralization in soil up to 80% in 10 d. Although there is much evidence on the biostimulation benefits, when a bioremediation strategy needs to be implemented, there still lies a great challenge: to replicate the positive results obtained in the laboratory, in to the field (Bento et al. 2005).

Ex situ Bioremediation Technologies

Composting

Composting is an ancient method for obtaining organic fertilizer with high nutrients content and with the capacity to improve the soil properties. It is mainly made from agro-industrial waste, livestock and municipal organic waste, where a variety of organisms such as bacteria and ligninolytic fungi, decompose organic matter in aerobic conditions into a substrate known as "compost".

Composting can now be used not only for the recycling of organic matter but also for the removal of chemical contaminants such as synthetic organic compounds or other xenobiotics, including pesticides (Semple et al. 2001).

This technique is suitable for the degradation of pesticides because:

- The organisms have different characteristics and capacities. This diversity means that there is a greater chance that a pesticide will be degraded.
- The degradation of pesticides depends on their chemical structure that determines their solubility in water; therefore, those soluble in water will be easily degraded because they are bioavailable.
- Another feature that plays a role in the degradation of pesticides is the weak or labile bonds they present. The addition of water can break down many labile bonds by hydrolysis from hydrolytic enzymes such as esterases and phosphatases. Some other enzymes may oxidize the pesticide by increasing solubility through enzymes such as monooxygenases.

In the composting process four periods occur, which are observed according to the evolution of temperature:

1. Mesophilic: Mesophilic microorganisms grow rapidly, raising the temperature by the metabolic activity and causing the production of organic acids that lower the pH.
2. Thermophilic: When the temperature reaches 40°C, the thermophilic microorganisms, mainly fungi, transform the nitrogen into ammonia, and the pH becomes alkaline. At 60°C, these microorganisms disappear and spore bacteria and actinomycetes appear. These microorganisms are responsible for breaking down waxes, proteins, and hemicellulose.
3. Cooling: When the temperature is below 60°C, the thermophilic fungi reappear in the humus and degrade the cellulose. When lowering to 40°C, the mesophilic organisms also restart their activity and the pH of the medium drops slightly.
4. Maturation: It is a period that requires months at room temperature, during which secondary reactions of condensation and humus polymerization take place.

There are different works that report the composting process for pesticide degradation. Zbytniewski and Buszewski (2002) reported the compost use for the pesticides treatment, and they observed the sorption of the linuron herbicide in activated sludge of a wastewater treatment plant, obtaining better results compared with the sorption of linuron in soil. Jones and Huang (2003) observed the decrease in chlorpyrifos toxicity in margins from 100% to 4.4% by the addition of humic substances from compost. Said-Pullicino et al. (2004) studied the degradation effects of triasulfuron by compost from municipal and green waste. The results showed that the adsorption of triasulfuron to the soil increases with the presence of compost with respect to the soil without compost.

On the other hand, Mohee et al. (2009) evaluated the 2,4-dichlorophenoxyacetic (2,4-D) and atrazine degradation by composting with green waste for 110 d, and they found that 2,4-D degrades faster than atrazine. Karanasios et al. (2010) evaluated the degradation of a mixture of pesticides (dimethoate, indoxacarb, buprofezine, terbuthylazine, metribuzin, M-metalaxyl, iprodione and azoxystrobin) on an agricultural compost biomixture containing olive leaves, cotton and cotton seed. Spent mushroom substrate and commercial sea wrack were mixed with topsoil and straw (1:1:2). They found greater pesticides degradation in the biomixture than in compost at 42 d.

Purnomo et al. (2010) investigated the DDT degradation in manure for 28 d, and demonstrated that the greatest degradation of DDT was in the mesophilic stage; besides, they found 1,1-dichloro-2,2-bis (4-chlorophenyl) ethane (DDE) as one of the major degradation metabolites. Chin-Pampillo et al. (2016) measured the degradation of carbofuran and chlorpyrifos by means of a soil, coconut fiber and garden compost biomixture. They performed three different treatments: fresh biomixture, biomixture pre-exposed to chlorpyrifos and a mature biomixture exposed to chlorpyrifos. Carbofuran removal and two of its transformation products (3-hydroxycarbofuran and 3-ketocarbofuran) were greater in the pre-exposed biomixture compared to the mature biomixture. The application of chlorpyrifos did not affect the removal of carbofuran, suggesting that microorganisms are not inhibited by chlorpyrifos. The carbofuran mineralization was better in the mature biomixture treatment and the chlorpyrifos mineralized better in the fresh biomixture.

Huete-Soto et al. (2017) performed a bio-purification system using a biomixture to remove the active ingredients of a mixture of herbicide, insecticide, and fungicide. After 115 d, the biomixture was able to remove mainly the herbicides atrazine, linuron and ametryn. The total elimination of active ingredients varied from 40.9% to 61.2%, following the pattern: herbicides > fungicides > insecticides. Other bioremediation work with compost was done by Castro-Gutiérrez et al. (2017) where they evaluated the degradation of carbofuran by residues of coconut fiber and garden compost. They showed that during the first 48 hr, only 9.9% of initial carbofuran was eliminated; however, after 6 mon, it was possible to degrade up to 88.5%, due to the acclimation of the microorganisms after the first application. In experiments containing carbamates such as aldicarb, methomyl, and oxamyl, only aldicarb was degraded.

There are more studies on the pesticides degradation by composting; however, with the above mentioned works, it can be noted that there are differences in the compost composition, as well as the pesticides chemical characteristics. In order to define the appropriate treatment, the above mentioned variables must be taken into account.

Bioreactors use

A bioreactor (BR) is a container where chemical processes are performed by biochemically active organisms or substances derived from them. The processes can be aerobic or anaerobic and BRs commonly have a cylindrical shape; they are variable in size and manufacturing material, as acrylic for pilot designs and stainless steel for industrial scale (Ruíz-Leza et al. 2007). With the bioreactors, a biologically active environment is sought, maintaining the ideal environmental conditions (pH, temperature, oxygen concentration, etc.) for the organisms. The bioreactors design is different according to the treatment type, and it can be either in batch or continuous.

Other parameters to consider in a BR design are flow, temperature, pH, flow velocity or agitation, most being monitored for better performance. At present, there are bioreactors for water treatment as well as soil, which can be used to degrade different contaminants such as hydrocarbons, explosives, and pesticides (Robles-González et al. 2008).

Mansee et al. (2000) hydrolyzed up to 80% of coumaphos, an organophosphorus insecticide, by *E. coli* XL1-Blue cells immobilized on glass beads within a continuous flow reactor. Geetha and Fulekar (2008) conducted an experiment where they observed the hydrolysis of chlorpyrifos in concentrations up to 50 mg l^{-1} using *Pseudomonas aeruginosa*. Later Fulekar (2009) used the same strain to reduce concentrations of fenvalerate in a bioreactor, which completely mineralized the insecticide at a concentration of 10 mg l^{-1}. Also, Fulekar et al. (2009) were able to mineralize up to 25 mg l^{-1} of trichlorpyr ester butoxyethyl using the strain mentioned above.

Ghoshdastidar et al. (2012) performed an experiment using a bioreactor to remove five different organophosphorus insecticides (methyl azinphos, chlorpyrifos, diazinon, malate, and forate) at concentrations of 5 mg l^{-1}. They obtained removal efficiencies between 83 and 98% within the first five d. Marrón-Montiel et al. (2014) degraded linuron in a bioreactor packed with a microbial community capable of degrading it and obtained up to 100% degradation. However, intermediate products were found in the effluent.

Bioreactors use for pesticide degradation is efficient in most cases as mentioned above. Treatments are mainly in aqueous phase and laboratory scale; however, control of different variables to optimize the bioprocess is very important which must be taken into account for the construction of bioreactors on an industrial scale.

Other technologies

A tool that has had a great boom in recent years is bioinformatics that has allowed the identification and analysis of several cellular components such as the genes and proteins function, interactions and regulatory metabolic pathways associated with bioremediation. Another technique is electrobioremediation, which consists of the application of electric current directly to the soil, which increases the nutrients bioavailability and transforms the contaminants to simpler compounds, easier to degrade by the microorganisms, hence allowing the contaminants biodegradation, or the migration of pollutants towards the plants to be retained or eliminated (Martínez-Prado et al. 2014).

Application of Biosurfactants for Bioremediation

Pesticides are substances mainly used in agriculture (Nicolopoulou-Stamati et al. 2016). However, only a small part of the applied pesticides is bioactive, and the remainder is distributed dynamically in the environment. Once the pesticides enter the soil, they undergo processes such as transport, transformation, biotic/abiotic degradation and adsorption through a combination of physicochemical interactions (Lagaly 2001, Megharaj et al. 2011).

Adsorption is a complex process that depends on environmental factors (temperature, humidity, and rainfall), various soil properties (clay fraction composition, pH, cation exchange capacity, surface area, and organic matter content) and compound characteristics. Water-insoluble pesticides such as DDT, hexachlorocyclohexane (HCH) and endosulfan, persist in the soil due to their low bioavailability.

Moreover, poorly soluble in water pesticides such as atrazine, chlordane, and chlorpyrifos with a solubility of 0.030 g l^{-1}, 0.0001 g l^{-1} and 0.002 g l^{-1}, respectively, are difficult to degrade by the soil microorganisms (Bhardwaj and Kapley 2015), so that, at higher solubility, the amount of pesticide that may be bioavailable and susceptible to degradation will be high.

Processes such as adsorption, desorption, diffusion, and dissolution, control the bioavailability of a substance. Among the strategies to improve the bioavailability of the hydrophobic compounds, the use of biosurfactants (BS) is included, which increases their bioavailability through its solubilization.

BS are a broad group of structurally diverse compounds that are produced by different microorganisms including bacteria, fungi, and yeasts. It can be secreted extracellularly or be bound to the cell membrane. BS have a hydrophilic region formed by amino acids, peptides (anions or cations) or saccharides (mono- or poly-); besides a hydrophobic region, it consists of saturated and unsaturated fatty acids that facilitates the reduction of surface and interfacial tensions between two immiscible liquids, therefore increasing the solubility of hydrophobic compounds. Some of the microorganisms that produce BS are *Pseudomonas aeruginosa*, *Acinetobacter junii*, *Stenotrophomonas* sp., *Bacillus subtilis*, *Aneurinibacillus aneurinilyticus* and *Candida* sp. (Mnif et al. 2015, Dong et al. 2016, Gargouri et al. 2016, Balan et al. 2017).

Due to their physicochemical properties, there are several classifications for the BS. One of the most important is its molecular weight. Those of low molecular weight are generally glycolipids (rhamnolipids, trehalose lipids, sophorolipids and fructose lipids). Other compounds include lipopeptides, which are disaccharides that are bound to fatty acids or hydroxy acids and are effective in reducing interfacial and surface tension. BS with high molecular weight are polysaccharides, proteins, lipopolysaccharides, lipoproteins or biopolymers complex mixtures; they are less effective in reducing interfacial tension but are efficient in coating the oil droplets to prevent their binding. These compounds are efficient at low concentrations (0.01–0.001%) and have a substrate specificity (Ron and Rosenberg 2002).

The surfactant effectiveness is determined by its ability to reduce both surface tension and critical micellar concentration (CMC), which means that for the surface tension reduction; a low concentration of surfactant is required (Wittgens et al. 2016). After the micelles-pesticides interaction, the mass transfer to the microorganism is encouraged, and the BS action mode consists of three stages (De la Rosa Cruz et al. 2014):

i) The micelle with the substrate solubilized is transported.
ii) The BS exchange with the cell is favored; this stage can be interpreted as the degradation process.
iii) Transference of substrate to the bacterial cell is carried out.

Several studies report the use of BS to improve the pesticide biodegradation, mainly organochlorines due to their low solubility. The HCH isomers solubility was evaluated in presence of different BS, such as rhamnolipids, sophorolipids, and trehalose lipids, through the CMC. The most effective was the sophorolipid, which increased the HCH solubility by nine times and hence enhanced its biodegradation by *Sphingomonas* sp. NM05 (Manickam et al. 2012).

Another widely studied aspect is the source of C and N, as well as the addition of BS precursors since cells grown on different culture media exhibit different BS production. Wattanaphon et al. (2008) evaluated different sources of C and N as well as the effect of the addition of sunflower, soybean and olive oil on the growth of *Burkholderia cenocepacia* BSP3 and BS production. Glucose had the highest BS yield with 37 ± 3 g l^{-1}; the sodium nitrate addition increased 1.2-fold the cell growth and BS production; however, other N sources had an inhibitory effect on BS production. Finally, the addition of sunflower oil showed an increase of 1.49-fold yield. A BS, identified as a glycolipid, showed a marked improvement in the solubilization of methyl parathion (MP), ethyl parathion and trifluralin. Residues such as those from the soybean oil industry have started to be used as precursors to BS, which reduces production costs (Bagheri Lotfabad et al. 2017).

In order to have tools for a more effective bioremediation, the search for pesticides degrading strains that at the same time produce BS has been favored. Odukkathil and Vasudevan (2015) reported the isolation of two strains, *Bordetella petrii* I and *Bordetella petrii* II, which degrade α and β endosulfan and produce BS that reduce surface tension by 19.6 and 21.4%, respectively.

For all the characteristics presented, BS can be widely used in agricultural areas, mainly in farmland, improving bioavailability and biodegradation of pesticides. In addition, BS increase soil quality for the indirect promotion of plant growth, as they possess antimicrobial activity and increase the beneficial interaction plant-microorganisms (Sachdev and Cameotra 2013).

Use of Genetically Modified Microorganisms for Bioremediation Enhancing

Molecular biology and genetic engineering offer tools to optimize the xenobiotics biodegradability through the genetically modified organisms (GMO), opening up opportunities in the degradation routes manipulation (Joutey et al. 2013). The possibilities of using GMO in bioremediation are diverse. These include (I) enzyme modification to increase its specificity, (II) synthetic constructs for new metabolic pathway design, (III) introduction of a marker gene for the recombinant gene identification in the contaminated environment, and (IV) biosensor construction for detection of target chemical compounds (Wasilkowski et al. 2012).

The wide metabolic and physiological versatility of microorganisms to degrade various pesticides could be used. There are several reports describing the characterization of the enzymes involved in the degradation of these compounds. Methyl parathion hydrolase is a highly efficient metalloenzyme that hydrolyses MP with a kcat $Km^{-1} > 10^6$ s^1M^1 (Dong et al. 2005), but its activity towards other organophosphorus pesticides is lower. Therefore, Ng et al. (2015) performed a site-directed mutagenesis by site saturation and DNA shuffling to find mutants with improved activities for ethyl paraxone. They obtained the R2F3 mutant, which showed a kcat $Km^{-1} = 5.9 \times 10^5$ s^1M^1 that represents a nearly 100-fold increase in its activity with respect to the wild type.

Agricultural soils are often contaminated with a mixture of pesticides. Unfortunately, some microorganisms isolated from the natural environment do not have the ability to simultaneously degrade the different pesticide molecular structures. In *Pseudomonas putida* KT2440, the *mpd* and the *pyt*H genes were co-expressed. The modified strain could completely degrade MP, fenitrothion, chlorpyrifos, permethrin, fenpropathrin, and cypermethrin (0.2 mM each) in 48 hr. In suspension cultures and contaminated soils, the six pesticides could be completely degraded in 15 d (Zuo et al. 2015).

It is important to consider the characteristics of the microorganisms to be modified. *Pseudomonas putida* KT2440 was certified by the Recombinant DNA Advisory Committee as a biosafety strain for the manipulation of recombinant DNA (Nelson et al. 2002). In this strain, the *mpd* and *gfp* genes were integrated into the chromosome using a scarless chromosome modification strategy. The strain is capable of using MP and carbofuran as the only source of carbon. In addition to it, the introduced green fluorescent protein can be used as a biomarker to monitor the recombinant strain during bioremediation (Gong et al. 2016).

Likewise, Yang et al. (2013) used a green fluorescent protein to monitor the γ-HCH degrading *Sphingobium japonicum* UT26 strain, to which the *oph* gene was integrated to degrade parathion. The strain degraded 100 mg kg^{-1} and 10 mg kg^{-1} of parathion and γ-HCH, respectively, in 15 d when it was inoculated in soil. However, the instability of the plasmid reduces the efficacy of the recombinant UT26 strain for bioremediation.

Due to industrialization, intensive agriculture and anthropogenic activity, aquifers and soil contamination by xenobiotics and toxic metals has been increased in recent years (Biswas et al. 2015), for which the creation of GMO has focused on microorganisms design capable of degrading more than one type of contaminant. For example, Zhang et al. (2015) introduced the *mpd* gene in *Pseudomonas putida* X3, a bacterium capable of immobilizing cadmium (Cd). Using strain X3 in soils contaminated with 100 mg kg^{-1} of MP and/or 5 mg kg^{-1} of Cd, it was observed that MP degradation was delayed by the presence of Cd, but was completely eliminated after 40 hr. Application of strain X3 in soil contaminated with Cd strongly affected the Cd fractions distribution by reducing concentrations of bioavailable, soluble-interchangeable and bound to organic matter fractions.

In general, recombinant strains that possess degradation genes have higher degradation rates than natural microorganisms, since it is possible to achieve regular gene expression to promote a high level of expression (Gong et al. 2016); thus, GMO can be very effective in bioremediation, with lower cost. However, ecological and environmental implications, as well as regulatory constraints, are important factors for the GMO evaluation in the field (Joutey et al. 2013).

Hussain et al. (2009) reported that transgenic microorganisms and plants produce enzymes that can mineralize different pesticides groups and their metabolites with great efficiency. Genetically modified microorganisms can also be used in bioaugmentation; by genetic engineering, the degrading capacity of compounds is strengthened, thus implicating different kind of mutations and horizontal gene transfer, which is carried out constantly, but is relatively slow fashion in nature. Thus, there is a need to improve microbial degradation activity using genetic engineering (Dueñas and Santos 2006, Mrozik and Piotrowska-Seget 2010).

Advantages and Disadvantages of Bioremediation

The major bioremediation advantage is that it is a natural process because in most cases, the native microorganisms of the contaminated site are used. Normally, these organisms are exposed to low levels of pollutants over a period of time, which induces microorganisms to activate the metabolic pathways necessary to degrade them. This will initiate the degradation or absorption of the contaminant at an acceptable rate until we start getting less toxic products. Thus, the transfer of contaminants from one site to another is prevented.

In situ bioremediation can be carried out without disturbing the site activities. This also eliminates the need to transport off-site waste, as well as the potential health and environmental impact that may arise during transport. Another advantage is that they can be less costly and sustainable than other technologies used for the rehabilitation of environments affected by natural and anthropogenic pollutants.

Treatments can be more effective through genetic manipulation, which improves the remediation capacity of organisms. For example, plants species with a greater ability to degrade organic pollutants and heavy metals accumulation have been designed (Cherian and Oliveira 2005).

Laboratory bioremediation works are difficult to extrapolate on a pilot scale and subsequently on a large scale due to environmental factors influencing the removal efficiency, and biological processes are often highly specific. Different factors are important for the bioremediation process; these include microorganisms responsible for the pollutants degradation, adequate conditions for their growth, presence of enough nutrients and the pollutants' concentration. Thus, research is required to develop and engineer appropriate bioremediation technologies for a specific site, as they often contain complex mixtures of contaminants that are not uniformly dispersed in the environment. Bioremediation usually takes longer than other treatment options, such as excavation and soil removal or incineration.

Successful Cases of Bioremediation of Pesticides Contaminated Sites

Because organochlorine pesticides (OCPs) are highly persistent in the soil, efforts to remediate contaminated sites have been focused on this type of pesticide. Phillips et al. (2006) used the technology called DARAMEND® to treat 1,100 ton of contaminated soil with HCH in an old manufacture plant of this pesticide. The site was subjected to two different treatments: half of the site (zone A) was treated using anaerobic/aerobic conditions, starting with the addition of organic matter and moisture to near the saturation point of the soil, which improves the dechlorination rate by strongly reducing conditions. A change to aerobic conditions by aeration and reducing moisture content promotes the degradation of previously dechlorinated compounds. On the other hand, in zone B, a treatment only in aerobic conditions was applied, which involved the addition of organic matter, a lower moisture content, regular tillage, as well as aeration to promote and maintain aerobic processes. After 371 days of treatment, a reduction of up to 60 and 75% was obtained in zone A and B, respectively; however, environmental conditions such as rain had a negative impact on the biodegradation of HCH.

In another case of DARAMEND® technology in conjunction with EHC®, an injectable version of DARAMEND was used; the results were effective for the *in situ* treatment of OCPs in groundwater (Seech et al. 2008). In this case, 4,500 ton of contaminated soil was treated, achieving a removal of 99, 94, 85 and 87% of toxaphene, DDT, DDD, and DDE, respectively. The average cost was US$55 per ha^{-1}. In addition to this, an aquifer 110 feet wide, 190 feet long and 15–40 feet deep was treated; and although the OCPs concentrations were much lower because of their low solubility, a removal of 80% was obtained, at a cost of US $ 6.0 per m^3.

Soil contaminated with OCPs, located in an industrial site with a relatively high concentration of contaminants (> 500 mg g^{-1} soil), was treated by using a bioaugmentation process with strains such as *Pseudomonas* SF1, *Citrobacter freundii* G20, and *Acinetobacter calcoaceticus* A7. In addition to this, a biostimulation process was applied. After three months, the soil showed a reduction in the OCPs concentration of 90 ± 5% and 60 ± 5% when it was bioaugmented and biostimulated, respectively. The microbiological and molecular analyses used showed that the inoculated bacteria remained viable during the experimental time (Qureshi et al. 2009).

With respect to *in situ* bioremediation of pesticide-contaminated sites, there are few reports which may be attributed to environmental factors that cannot be controlled and affect the successful degradation of such compounds. These factors can be electron acceptor status, moisture, availability of nutrients, pH, temperature, rainfall, as well as a low population of microbiota with degradation capacities of pollutants (Sharma 2012, Juwarkar et al. 2014). Therefore, most of the work done for the pesticides biodegradation is still at the laboratory or pilot level. In spite of the long periods involved in *in situ* remediation, little or no disturbance to the soil structure is caused; additionally, these techniques tend to be more effective, eco-friendly and less expensive compared to other remediation techniques (Azubuike et al. 2016).

Concluding Remarks

There are many technologies for removing pesticides from the environment. They involve both physicochemical and biological approaches mentioned above. However, the first option is more expensive, has a high energy demand, and involves the use of chemical reagents. For this reason, biological methods through the use of microorganisms capable of degrading toxic compounds are more suitable; this type of technology is known as bioremediation (Mrozik and Piotrowska-Seget 2010, Chaturvedi et al. 2016). Several microorganisms with the ability to degrade different xenobiotics such as pesticides have been isolated and identified, for example, *Pseudomonas*, *Bacillus*, *Rhodococcus*, *Arthrobacter*, *Enterobacter*, *Sphingomonas*, among others (Bharadwaj et al. 2016, Dzionek et al. 2016, Sing and Singh 2016).

Biotechnological advances have improved the bioremediation processes allowing to overcome various limitations associated with the traditional bioremediation. Applications of these advances are shown in Table 3 (Ahmad and Ahmad 2014, Juwarkar et al. 2014).

Advances in bioremediation can be very attractive, but these technologies need to be proved on the field, since the pollutant toxicity and environmental factors can affect the site remediation. Therefore, it is necessary to develop hybrid technology for better results (Juwarkar et al. 2014). The introduction of new biotechnological tools in the bioremediation process adds new perspectives to research.

Table 3. Biotechnological tools applied to improve bioremediation processes.

Biotechnological tools	**Description**
Application of biosurfactants	Procedure for the bioavailability of hydrophobic pollutants.
Oxygenation	Oxygen is usually the limiting factor in aerobic bioremediation. The degradation of some pollutants is produced more quickly in aerobic conditions.
Phytoremediation with the association of microorganisms	Phytoremediation with microorganisms improves the process of remediation (Megharaj et al. 2011).
Tools and techniques of molecular biology (Genomics, transcriptomics, proteomics, metabolomics, interactomics)	The *omics* technologies has allowed the identification of genes or enzymes involved in degradation pathways, even in mineralization (Singh and Nagaraj 2006). The information generated through the use of these tools can be used to generate genetically modified microorganisms for their use in remediation processes of contaminated environments (Chaturvedi et al. 2016, Mahdi and Aziz 2017).

References Cited

Agostini, E., M.S. Coniglio, S.R. Milrad, H.A. Tigier and A.M. Giulietti. 2003. Phytoremediation of 2,4-dichlorophenol by *Brassica napus* hairy root cultures. Biotechnol. Appl. Biochem. 37: 139–144.

Ahmad, M. and I. Ahmad. 2014. Recent advances in the field of bioremediation. pp. 1–42. *In*: Ahmad, M. [ed.]. Biodegradation and Bioremediation. Studium Press LLC.

Ajlan, A. 2016. A review on bioremediation review. Microbiol. An. Int. J. 1(1): 1–7.

Alvarez, A., J.M. Saez, J.S. Davila Costa, V.L. Colin, M.S. Fuentes, S.A. Cuozzo et al. 2017. Actinobacteria: Current research and perspectives for bioremediation of pesticides and heavy metals. Chemosphere 166: 41–62.

Anderson, R.T. and D.R. Lovley. 1997. Ecology and biogeochemistry of *in situ* groundwater bioremediation. pp. 289–359. *In*: Jones, G. [ed.]. Advances in Microbial Ecology Vol. 15, Plenum Press, New York.

Asai, K., K. Takagi, M. Shimokawa, T. Sue, A. Hibi, T. Hiruta et al. 2002. Phytoaccumulation of coplanar PCBs by *Arabidopsis thaliana*. Environ. Pollut. 120: 509–511.

Azubuike, C.C., C.B. Chikere and G.C. Okpokwasili. 2016. Bioremediation techniques–classification based on site of application: principles, advantages, limitations and prospects. World J. Microbiol. Biotechnol. 32: 1–18.

Baćmaga, M., J. Wyszkowska and J. Kucharski. 2017. Bioaugmentation of soil contaminated with azoxystrobin. Water Air Soil Pollut. 228(19): 1–9.

Bagheri Lotfabad, T., N. Ebadipour, R. Roostaazad, M. Partovi and M. Bahmaei. 2017. Two schemes for production of biosurfactant from *Pseudomonas aeruginosa* MR01: Applying residues from soybean oil industry and silica sol-gel immobilized cells. Colloids Surfaces B Biointerfaces 152: 159–168.

Balan, S.S., C.G. Kumar and S. Jayalakshmi. 2017. Aneurinifactin, a new lipopeptide biosurfactant produced by a marine *Aneurinibacillus aneurinilyticus* SBP-11 isolated from Gulf of Mannar: Purification, characterization and its biological evaluation. Microbiol. Res. 194: 1–9.

Banjoko, B. and S. Eslamian. 2016. Phytoremediation. pp. 663–705. *In*: Eslamian, S. [ed.]. Urban Water Reuse Handbook. CRC Press Taylor & Francis Group.

Bento, F.M., F.A.O. Camargo, B.C. Okeke and W.T. Frankenberger. 2005. Comparative bioremediation of soils contaminated with diesel oil by natural attenuation, biostimulation and bioaugmentation. Bioresour. Technol. 96: 1049–1055.

Bhalerao, T.S. 2012. Bioremediation of endosulfan-contaminated soil by using bioaugmentation treatment of fungal inoculant *Aspergillus niger*. Turkish J. Biol. 36: 561–567.

Bharadwaj, A., N. Wahi, N. Nehra, M. Gupta, G. Pant and A.K. Bhatia. 2016. Bioremediation: An eco-friendly approach for treating pesticides. Adv. Biores. 7(3): 200–206.
Bhardwaj, P. and A. Kapley. 2015. Bioremediation of pesticide contaminated soil: Emerging options. pp. 293–313. *In*: Kalia, V.C. [ed.]. Microbial Factories: Biofuels, Waste treatment. Springer India.
Biswas, K., D. Paul and S.N. Sinha. 2015. Biological agents of bioremediation : A concise review. Front. Environ. Microbiol. 1(3): 39–43.
Campos, M., C. Perruchon, P.A. Karas, D. Karavasilis, M.C. Diez and D.G. Karpouzas. 2017. Bioaugmentation and rhizosphere-assisted biodegradation as strategies for optimization of the dissipation capacity of biobeds. J. Environ. Manage. 187: 103–110.
Campos del Pozo, V.M. 2010. Fitorremediación de contaminantes persistentes: una aproximación biotecnológica utilizando chopo (*Populus* spp.) como sistema modelo. Ph.D. Thesis, Universidad Politécnica de Madrid, España.
Castro-Gutiérrez, V., M. Masís-Mora, M.C. Diez, G.R. Tortella and C.E. Rodríguez-Rodríguez. 2017. Aging of biomixtures: Effects on carbofuran removal and microbial community structure. Chemosphere 168: 418–425.
Chang, S.W., S.J. Lee and C.H. Je. 2005. Phytoremediation of atrazine by *poplar* trees: toxicity, uptake, and transformation. J. Environ. Sci. Heal. Part. B 40(6): 801–811.
Chaturvedi, R., J. Prakash and G. Awasthi. 2016. Microbial bioremediation: An advanced approach for waste management. Int. J. Eng. Technol. Sci. Res. 3(5): 50–62.
Chaudhry, Q., P. Schröder, D. Werck-Reichhart, W. Grajek and R. Marecik. 2002. Prospects and limitations of phytoremediation for the removal of persistent pesticides in the environment. Environ. Sci. Pollut. Res. Int. 9(1): 4–17.
Cherian, S. and M.M. Oliveira. 2005. Transgenic plants in phytoremediation: Recent advances and new possibilities. Environ. Sci. Technol. 39(24): 9377–9390.
Chin-Pampillo, J.S., M. Masís-Mora, K. Ruiz-Hidalgo, E. Carazo-Rojas and C.E. Rodríguez-Rodríguez. 2016. Removal of carbofuran is not affected by co-application of chlorpyrifos in a coconut fiber/compost based biomixture after aging or pre-exposure. J. Environ. Sci. 46: 182–189.
Coates, J.D. and R.T. Anderson. 2000. Emerging techniques for anaerobic bioremediation of contaminated environments. Trends Biotechnol. 18: 408–412.
Cosgrove, L., P.L. McGeechan, P.S. Handley and G.D. Robson. 2010. Effect of biostimulation and bioaugmentation on degradation of polyurethane buried in soil. Appl. Environ. Microbiol. 76(3): 810–819.
Cycoń, M., A. Zmijowska, M. Wójcik and Z. Piotrowska-Seget. 2013. Biodegradation and bioremediation potential of diazinon-degrading *Serratia marcescens* to remove other organophosphorus pesticides from soils. J. Environ. Manage. 117: 7–16.
Cycoń, M., A. Zmijowska and Z. Piotrowska-Seget. 2014. Enhancement of deltamethrin degradation by soil bioaugmentation with two different strains of *Serratia marcescens*. Int. J. Environ. Sci. Technol. 11: 1305–1316.
Dai, Y., N. Li, Q. Zhao and S. Xie. 2015. Bioremediation using *Novosphingobium* strain DY4 for 2,4-dichlorophenoxyacetic acid-contaminated soil and impact on microbial community structure. Biodegradation 26: 161–170.
Dams, R.I., G. Paton and K. Killham. 2007. Bioaugmentation of pentachlorophenol in soil and hydroponic systems. Int. Biodeterior. Biodegrad. 60: 171–177.
Das, S. and H.R. Dash. 2014. Microbial bioremediation : A potential tool for restoration of contaminated areas. pp. 1–21. *In*: Das, S. [ed.]. Microbial Biodegradation and Bioremediation. Elsevier Inc.
Dasgupta, S., C. Meisner and D. Wheeler. 2010. Stockpiles of obsolete pesticides and cleanup priorities: A methodology and application for Tunisia. J. Environ. Manage. 91(4): 824–830.
Delgadillo-López, A.E., C.A. González-Ramírez, F. Prieto-García, J.R. Villagómez-Ibarra and O. Acevedo-Sandoval. 2011. Phytorremediation: an alternative to eliminate pollution. Trop. Subtrop. Agroecosystems 14: 597–612.
Deng, Z. and L. Cao. 2017. Fungal endophytes and their interactions with plants in phytoremediation: A review. Chemosphere 168: 1100–1106.
Díaz-Martínez, M.E., A. Alarcón, R. Ferrera-Cerrato, J.J. Almaraz-Suarez and O. García-Barradas. 2013. Crecimiento de *Casuarina equisetifolia* (Casuarinaceae) en suelo con diésel, y aplicación de bioestimulación y bioaumentación. Rev. Biol. Trop. J. Trop. Biol. Conserv. 61(3): 1039–1052.
Dong, H., W. Xia, H. Dong, Y. She, P. Zhu, K. Liang et al. 2016. Rhamnolipids produced by indigenous *Acinetobacter junii* from petroleum reservoir and its potential in enhanced oil recovery. Front. Microbiol. 7: 1–13.
Dong, Y.J., M. Bartlam, L. Sun, Y.F. Zhou, Z.P. Zhang, C.G. Zhang et al. 2005. Crystal structure of methyl parathion hydrolase from *Pseudomonas* sp. WBC-3. J. Mol. Biol. 353: 655–663.
Dueñas, C.M. and L.R.C. Santos. 2006. Evaluación de la bioestimulación (bacterias nativas y comerciales) en la biodegradación de hidrocarburos en suelos contaminados. Lic. Thesis, Universidad Industrial de Santander, Colombia.
Dzionek, A., D. Wojcieszyńska and U. Guzik. 2016. Natural carriers in bioremediation: A review. Electron. J. Biotechnol. 23: 28–36.

El Fantroussi, S. and S.N. Agathos. 2005. Is bioaugmentation a feasible strategy for pollutant removal and site remediation? Curr. Opin. Microbiol. 8: 268–275.
Farhan, M., A.U. Khan, A. Wahid, M. Ahmad, F. Ahmad, Z.A. Butt et al. 2013. Chlorpyrifos biodegradation in laboratory soil through bio-augmentation and its kinetics. Asian J. Chem. 25(17): 9994–9998.
Fenner, K., S. Canonica, L.P. Wackett and M. Elsner. 2013. Evaluating pesticide degradation in the environment: Blind spots and emerging opportunities. Science 341: 752–758.
Frazar, C. 2000. The bioremediation and phytoremediation of pesticide-contaminated sites. Washington, DC.
Fulekar, M.H. 2009. Bioremediation of fenvalerate by *Pseudomonas aeruginosa* in a scale up Bioreactor. Rom. Biotechnol. Lett. 14(6): 4900–4905.
Fulekar, M.H., M. Geetha and J. Sharma. 2009. Bioremediation of trichlorpyr butoxyethyl ester (TBEE) in bioreactor using adapted *Pseudomonas aeruginosa* in scale up process technique. Biol. Med. 1(3): 1–6.
Gargouri, B., M.D.M. Contreras, S. Ammar, A. Segura-Carretero and M. Bouaziz. 2016. Biosurfactant production by the crude oil degrading *Stenotrophomonas* sp. B-2: chemical characterization, biological activities and environmental applications. Environ. Sci. Pollut. Res. 1–11.
Geetha, M. and M.H. Fulekar. 2008. Bioremediation of pesticides in surface soil treatment unit using microbial consortia. African J. Environ. Sci. Technol. 2(2): 36–45.
Ghoshdastidar, A.J., J.E. Saunders, K.H. Brown and A.Z. Tong. 2012. Membrane bioreactor treatment of commonly used organophosphate pesticides. J. Environ. Sci. Health. B. 47(7): 742–50.
Gomes, M.A. da C., R.A. Hauser-Davis, A.N. de Souza and A.P. Vitória. 2016. Metal phytoremediation: General strategies, genetically modified plants and applications in metal nanoparticle contamination. Ecotoxicol. Environ. Saf. 134: 133–147.
Gong, T., R. Liu, Y. Che, X. Xu, F. Zhao, H. Yu et al. 2016. Engineering *Pseudomonas putida* KT2440 for simultaneous degradation of carbofuran and chlorpyrifos. Microb. Biotechnol. 9(6): 792–800.
Green, C. and A. Hoffnagle. 2004. Phytoremediation field studies for chorinated solvents, pesticides, explosives, and metals. Washington, D.C.
Herrero, M. and D.C. Stuckey. 2015. Bioaugmentation and its application in wastewater treatment: A review. Chemosphere 140: 119–128.
Hong, Q., Z. Zhang, Y. Hong and S. Li. 2007. A microcosm study on bioremediation of fenitrothion-contaminated soil using *Burkholderia* sp. FDS-1. Int. Biodeterior. Biodegrad. 59: 55–61.
Huete-Soto, A., M. Masís-Mora, V. Lizano-Fallas, J.S. Chin-Pampillo, E. Carazo-Rojas and C.E. Rodríguez-Rodríguez. 2017. Simultaneous removal of structurally different pesticides in a biomixture: Detoxification and effect of oxytetracycline. Chemosphere 169: 558–567.
Hung, D.Z., H.J. Yang, Y.F. Li, C.L. Lin, S.Y. Chang, F.C. Sung et al. 2015. The long-term effects of organophosphates poisoning as a risk factor of CVDs: A nationwide population-based cohort study. PLoS One 10(9): 1–15.
Hussain, S., T. Siddique, M. Arshad and M. Saleem. 2009. Bioremediation and phytoremediation of pesticides: Recent advances. Crit. Rev. Environ. Sci. Technol. 39(10): 843–907.
Ibrahim, S.I., M.F. Abdel Lateef, H.M.S. Khalifa and A.E. Abdel Monem. 2013. Phytoremediation of atrazine-contaminated soil using *Zea mays* (maize). Ann. Agric. Sci. 58(1): 69–75.
Ijaz, A., A. Imran, M. Anwar ul Haq, Q.M. Khan and M. Afzal. 2016. Phytoremediation: recent advances in plant-endophytic synergistic interactions. Plant Soil. 405: 179–195.
Jaiswal, D.K., J.P. Verma and J. Yadav. 2017. Microbe induced degradation of pesticides in agriculture soils. pp. 167–189. *In*: Singh, S.N. [ed.]. Microbe-induced Degradation of Pesticides. Springer International Publishing.
Jamal, F., Q.S. Haque, S. Singh and S. Rastogi. 2015. The influence of organophosphate and carbamate on sperm chromatin and reproductive hormones among pesticide sprayers. Toxicol. Ind. Health. 32(8): 1527–1536.
Javaid, M.K., M. Ashiq and M. Tahir. 2016. Potential of biological agents in decontamination of agricultural soil. Scientifica 2016: 1–9.
Jones, K.D. and W.H. Huang. 2003. Evaluation of toxicity of the pesticides, chlorpyrifos and arsenic, in the presence of compost humic substances in aqueous systems. J. Hazard. Mater. 103(1–2): 93–105.
Joutey, N.T., W. Bahafid, H. Sayel and N. El Ghachtouli. 2013. Biodegradation: Involved microorganisms and genetically engineered microorganisms. pp. 289–320. *In*: Chamy, R. and F. Rosenkranz [eds.]. Biodegradation—Life of Science. InTech.
Juwarkar, A.A., R.R. Misra and J.K. Sharma. 2014. Recent trends in bioremediation recent trends in bioremediation. pp. 81–100. *In*: Parmar, N. and A. Singh [eds.]. Geomicrobiology and Biogeochemistry. Springer-Verlag Berlin Heidelberg.
Kalantary, R.R., A. Mohseni-Bandpi, A. Esrafili, S. Nasseri, F.R. Ashmagh, S. Jorfi et al. 2014. Effectiveness of biostimulation through nutrient content on the bioremediation of phenanthrene contaminated soil. J. Environ. Heal. Sci. Eng. 12(143): 1–9.
Kamel, F. 2013. Paths from pesticides to Parkinson's. Science 341: 722–723.
Kanissery, R.G. and G.K. Sims. 2011. Biostimulation for the enhanced degradation of herbicides in soil. Appl. Environ. Soil Sci. 2011: 1–10.

Karanasios, E., N.G. Tsiropoulos, D.G. Karpouzas and C. Ehaliotis. 2010. Degradation and adsorption of pesticides in compost-based biomixtures as potential substrates for biobeds in southern Europe. J. Agric. Food Chem. 58(16): 9147–9156.
Kensa, V.M. 2011. Bioremediation—An overview. Jr. Ind. Pollut. Control. 27(2): 161–168.
Köhler, H.R. and R. Triebskorn. 2013. Wildlife ecotoxicology of pesticides: can we track efects to the popupation level and beyond? Science 341: 759–765.
Kovacs, H. and K. Szemmelveisz. 2017. Disposal options for polluted plants grown on heavy metal contaminated brownfield lands—A review. Chemosphere 166: 8–20.
Kumar, M. and L. Philip. 2007. Biodegradation of endosulfan-contaminated soil in a pilot-scale reactor-bioaugmented with mixed bacterial culture. J. Environ. Sci. Heal. Part B. 42(6): 707–715.
De la Rosa Cruz, N.L., E. Sánchez-Salinas and M.L. Ortiz-Hernández. 2014. Biosurfactantes y su papel en la biorremediación de suelos contaminados con plaguicidas. Rev. Latinoam. Biotecnol. Ambient. y Algal. 4(1): 47–67.
Lafleur, B., S. Sauvé, S.V. Duy and M. Labrecque. 2016. Phytoremediation of groundwater contaminated with pesticides using short-rotation willow crops: a case study of an apple orchard. Int. J. Phytoremediation 18(11): 1128–1135.
Lagaly, G. 2001. Pesticide-clay interactions and formulations. Appl. Clay Sci. 18: 205–209.
Levi, S., A.M. Hybel, P.L. Bjerg and H.J. Albrechtsen. 2014. Stimulation of aerobic degradation of bentazone, mecoprop and dichlorprop by oxygen addition to aquifer sediment. Sci. Total Environ. 473–474: 667–675.
Lin, J.N., C.L. Lin, M.C. Lin, C.H. Lai, H.H. Lin, C.H. Yang et al. 2015. Increased risk of dementia in patients with acute organophosphate and carbamate poisoning. Medicine 94(29): 1–8.
López-Martínez, S., M.E. Gallegos-Martínez, L.J. Pérez Flores and M.G. Rojas. 2005. Mecanismo de fitorremediación de suelos contaminados con moléculas orgánicas xenobióticas. Rev. Int. Contaminación Ambient. 21(2): 91–100.
de Lorenzo, V. 2008. Systems biology approaches to bioremediation. Curr. Opin. Biotechnol. 19: 579–589.
Ma, X.K., N. Ding and E.C. Peterson. 2015. Bioaugmentation of soil contaminated with high-level crude oil through inoculation with mixed cultures including *Acremonium* sp. Biodegradation 26: 259–269.
Mahdi, A.M. El and H.A. Aziz. 2017. Hydrocarbon biodegradation using agro-industrial wastes as co-substrates. pp. 155–185. *In*: Bhakta, J.N. [ed.]. Handbook of Research on Inventive Bioremediation Techniques. IGI Global, Hershey PA, USA.
Major, D.W., M.L. Mcmaster, E.E. Cox, E.A. Edwards, S.M. Dworatzek, E.R. Hendrickson et al. 2002. Field demonstration of successful bioaugmentation to achieve dechlorination of tetrachloroethene to ethene. Environ. Sci. Technol. 36: 5106–5116.
Malik, S., S.A.L. Andrade, M.H. Mirjalili, R.R.J. Arroo, M. Bonfill and P. Mazzafera. 2016. Biotechnological approaches for bioremediation: *In vitro* hairy root culture. pp. 1–23. *In*: Jha, S. [ed.]. Transgenesis and Secondary Metabolism. Springer International Publishing.
Manickam, N., A. Bajaj, H.S. Saini and R. Shanker. 2012. Surfactant mediated enhanced biodegradation of hexachlorocyclohexane (HCH) isomers by *Sphingomonas* sp. NM05. Biodegradation 23: 673–682.
Mansee, A.H., W. Chen and A. Mulchandani. 2000. Biodetoxification of coumaphos insecticide using immobilized *Escherichia coli* expressing organophosphorus hydrolase enzyme on cell surface. Biotechnol. Bioprocess Eng. 5: 436–440.
Marrón-Montiel, E., N. Ruiz-Ordaz, J. Galíndez-Mayer, S. Gonzalez-Cuna, F.S. Tepole and H. Poggi-Varaldo. 2014. Biodegradation of the herbicide linuron in a plug-flow packed-bed biofilm channel equipped with top aeration modules. Environ. Eng. Manag. J. 13(8): 1939–1944.
Martínez-Prado, M.A., J. Unzueta-Medina and M.E. Pérez-López. 2014. Electrobioremediation as a hybrid technology to treat soil contaminated whit total petroleum hydrocarbons. Rev. Mex. Ing. Química. 13(1): 113–127.
Matsumoto, E., Y. Kawanaka, S.J. Yun and H. Oyaizu. 2009. Bioremediation of the organochlorine pesticides, dieldrin and endrin, and their occurrence in the environment. Appl. Microbiol. Biotechnol. 84: 205–216.
McKinlay, R., J.A. Plant, J.N.B. Bell and N. Voulvoulis. 2008. Endocrine disrupting pesticides: Implications for risk assessment. Environ. Int. 34: 168–183.
Megharaj, M., B. Ramakrishnan, K. Venkateswarlu, N. Sethunathan and R. Naidu. 2011. Bioremediation approaches for organic pollutants: A critical perspective. Environ. Int. 37(8): 1362–1375.
Mills, M.A., J.S. Bonner, C.A. Page and R.L. Autenrieth. 2004. Evaluation of bioremediation strategies of a controlled oil release in a wetland. Mar. Pollut. Bull. 49: 425–435.
Mitton, F.M., M. Gonzalez, J.M. Monserrat and K.S.B. Miglioranza. 2016a. Potential use of edible crops in the phytoremediation of endosulfan residues in soil. Chemosphere 148: 300–306.
Mitton, F.M., J.L. Ribas Ferreira, M. Gonzalez, K.S.B. Miglioranza and J.M. Monserrat. 2016b. Antioxidant responses in soybean and alfalfa plants grown in DDTs contaminated soils: Useful variables for selecting plants for soil phytoremediation? Pestic. Biochem. Physiol. 130: 17–21.

Mnif, I., A. Grau-Campistany, J. Coronel-León, I. Hammami, M.A. Triki, A. Manresa et al. 2015. Purification and identification of *Bacillus subtilis* SPB1 lipopeptide biosurfactant exhibiting antifungal activity against *Rhizoctonia bataticola* and *Rhizoctonia solani*. Environ. Sci. Pollut. Res. 23(7): 6690–6699.

Mohee, R., V. Jumnoodoo, N. Sobratee, A. Mudhoo and G. Unmar. 2009. Assessing the suitability of the composting process in treating contaminating pesticides and pathogenic wastes. Dyn. Soil, Dyn. Plant. 3(1): 103–114.

Morán, A.C., A. Müller, M. Manzano and B. González. 2006. Simazine treatment history determines a significant herbicide degradation potential in soils that is not improved by bioaugmentation with *Pseudomonas* sp. ADP. J. Appl. Microbiol. 101: 26–35.

Mostafalou, S. and M. Abdollahi. 2013. Pesticides and human chronic diseases: Evidences, mechanisms, and perspectives. Toxicol. Appl. Pharmacol. 268: 157–177.

Motoki, Y., T. Iwafune, N. Seike, T. Otani and Y. Akiyama. 2015. Relationship between plant uptake of pesticides and water-extractable residue in Japanese soils. J. Pestic. Sci. 40(4): 175–183.

Mrozik, A. and Z. Piotrowska-Seget. 2010. Bioaugmentation as a strategy for cleaning up of soils contaminated with aromatic compounds. Microbiol. Res. 165: 363–375.

Muñoz-Arnanz, J. and B. Jiménez. 2011. New DDT inputs after 30 years of prohibition in Spain. A case study in agricultural soils from south-western Spain. Environ. Pollut. 159: 3640–3646.

Namkoong, W., E.Y. Hwang, J.S. Park and J.Y. Choi. 2002. Bioremediation of diesel-contaminated soil with composting. Environ. Pollut. 119: 23–31.

Nandan, S., D. Tailor and A. Yadav. 2016. Malathion pesticide degradation by advanced oxidation process (UV-irradiation). Int. Res. J. Adv. Eng. Sci. 1(4): 153–156.

Nelson, K.E., C. Weinel, I.T. Paulsen, R.J. Dodson, H. Hilbert, V.A.P. Martins dos Santos et al. 2002. Complete genome sequence and comparative analysis of the metabolically versatile *Pseudomonas putida* KT2440. Environ. Microbiol. 4(12): 799–808.

Ng, T.K., L.R. Gahan, G. Schenk and D.L. Ollis. 2015. Altering the substrate specificity of methyl parathion hydrolase with directed evolution. Arch. Biochem. Biophys. 573: 59–68.

Nicolopoulou-Stamati, P., S. Maipas, C. Kotampasi, P. Stamatis and L. Hens. 2016. Chemical pesticides and human health: The rrgent need for a new concept in agriculture. Front. Public Heal. 4(148): 1–8.

Nwankwegu, A.S. and C.O. Onwosi. 2017. Bioremediation of gasoline contaminated agricultural soil by bioaugmentation. Environ. Technol. Innov. 7: 1–11.

Odukkathil, G. and N. Vasudevan. 2015. Biodegradation of endosulfan isomers and its metabolite endosulfate by two biosurfactant producing bacterial strains of *Bordetella petrii*. J. Env. Sci. Heal. B. 50: 81–89.

Ortiz-Hernández, M.L., E. Sánchez-Salinas, A. Olvera-Velona and J.L. Folch-Mallol. 2011. Pesticides in the environment: Impacts and their biodegradation as a strategy for residues treatment. pp. 551–574. *In*: Stoytcheva, M. [ed.]. Pesticides-Formulations, Effects, Fate. InTech.

Ortiz-Hernández, M.L., E. Sanchez-Salinas, M.L. Castrejón-Godínez, E.D. González and E.C.P. Ursino. 2013. Mechanisms and strategies for pesticide biodegraation: opportunity for waste, soils and water cleaning. Rev. Int. Contam. Ambie. 29: 85–104.

Otani, T., N. Seike and Y. Sakata. 2007. Differential uptake of dieldrin and endrin from soil by several plant families and *Cucurbita* genera. Soil Sci. Plant Nutr. 53: 86–94.

Park, J.S., S.K. Shin, W. Il Kim and B.H. Kim. 2011. Residual levels and identify possible sources of organochlorine pesticides in Korea atmosphere. Atmos. Environ. 45: 7496–7502.

Parween, T., P. Bhandari, S. Jan and S.K. Raza. 2016. Interaction between pesticide and soil microorganisms and their degradation: A molecular approach. pp. 23–44. *In*: Hakeem, K.R. and M.S. Akhatar [eds.]. Plant, Soil and Microbes. Springer Nature.

Phillips, T.M., H. Lee, J.T. Trevors and A.G. Seech. 2006. Full-scale *in situ* bioremediation of hexachlorocyclohexane-contaminated soil. J. Chem. Technol. Biotechnol. 81: 289–298.

Plangklang, P. and A. Reungsang. 2009. Bioaugmentation of carbofuran residues in soil using *Burkholderia cepacia* PCL3 adsorbed on agricultural residues. Int. Biodeterior. Biodegrad. 63: 515–522.

Prescott, L.M., J.P. Harley and D.A. Klein. 2008. Microbiology. 7th ed. J.M. Willey, L.M. Sherwood and C.J. Woolverton [eds.]. McGraw-Hill, New York.

Purnomo, A.S., F. Koyama, T. Mori and R. Kondo. 2010. DDT degradation potential of cattle manure compost. Chemosphere 80: 619–624.

Qureshi, A., M. Mohan, G.S. Kanade, A. Kapley and H.J. Purohit. 2009. *In situ* bioremediation of organochlorine-pesticide-contaminated microcosm soil and evaluation by gene probe. Pest Manag. Sci. 65: 798–804.

Rivera-Espinoza, Y. and L. Dendooven. 2004. Dynamics of carbon, nitrogen and hydrocarbons in diesel-contaminated soil amended with biosolids and maize. Chemosphere 54: 379–386.

Robles-González, I.V., F. Fava and H.M. Poggi-Varaldo. 2008. A review on slurry bioreactors for bioremediation of soils and sediments. Microb. Cell Fact. 7(5): 1–16.

Rodríguez-Rodríguez, C.E., K. Madrigal-León, M. Masís-Mora, M. Pérez-Villanueva and J.S. Chin-Pampillo. 2017. Removal of carbamates and detoxification potential in a biomixture: Fungal bioaugmentation versus traditional use. Ecotoxicol. Environ. Saf. 135: 252–258.

Ron, E.Z. and E. Rosenberg. 2002. Biosurfactants and oil bioremediation. Curr. Opin. Biotechnol. 13: 249–252.
Ruíz-Leza, H.A., R.M. Rodríguez-Jasso, R. Rodríguez-Herrera, J.C. Contreras-Esquivel and C.N. Aguilar. 2007. Bio-reactors desing for solid state fermentation. Rev. Mex. Ing. Química. 6(1): 33–40.
Sachdev, D.P. and S.S. Cameotra. 2013. Biosurfactants in agriculture. Appl. Microbiol. Biotechnol. 97(3): 1005–1016.
Sagarkar, S., P. Bhardwaj, V. Storck, M. Devers-Lamrani, F. Martin-Laurent and A. Kapley. 2015. s-triazine degrading bacterial isolate *Arthrobacter* sp. AK-YN10, a candidate for bioaugmentation of atrazine contaminated soil. Appl. Microbiol. Biotechnol. 100(2): 903–913.
Said-Pullicino, D., G. Gigliotti and A.J. Vella. 2004. Environmental fate of triasulfuron in soils amended with municipal waste compost. J. Environ. Qual. 33: 1743–1751.
Samuelsen, E.D., N. Badawi, O. Nybroe, S.R. Sorensen and J. Aamand. 2017. Adhesion to sand and ability to mineralise low pesticide concentrations are required for efficient bioaugmentation of flow-through sand filters. Appl. Microbiol. Biotechnol. 101: 411–421.
Scow, K.M. and K.A. Hicks. 2005. Natural attenuation and enhanced bioremediation of organic contaminants in groundwater. Curr. Opin. Biotechnol. 16: 246–253.
Seech, A., K. Bolanos-Shaw, D. Hill and J. Molin. 2008. *In situ* bioremediation of pesticides in soil and groundwater. Remediat. J. 19(1): 87–98.
Semple, K.T., B.J. Reid and T.R. Fermor. 2001. Impact of composting strategies on the treatment of soil contaminated with organic pollutants. Environ. Pollut. 112: 269–283.
Sepúlveda, T.V. and J.A.V. Trejo. 2002. Tecnologías de remediación para suelos contaminados. INE-SEMARNAT, Mexico.
Sharma, S. 2012. Bioremediation: Features, strategies and applications. Asian J. Pharm. Life Sci. 2(2): 202–213.
Silva, E., A.M. Fialho, I. Sa-Correia, R.G. Burns and L.J. Shaw. 2004. Combined bioaugmentation and biostimulation to cleanup soil contaminated with high concentration of atrazine. Environ. Sci. Technol. 38: 632–637.
Sing, B. and K. Singh. 2016. Microbial degradation of herbicides. Crit. Rev. Microbiol. 42(2): 245–261.
Singer, A.C., C.J. Van Der Gast and I.P. Thompson. 2005. Perspectives and vision for strain selection in bioaugmentation. Trends Biotechnol. 23(2): 74–77.
Singh, D.K. 2008. Biodegradation and bioremediation of pesticide in soil: Concept, method and recent developments. Indian J. Microbiol. 48(1): 35–40.
Singh, O.V. and N.S. Nagaraj. 2006. Transcriptomics, proteomics and interactomics: Unique approaches to track the insights of bioremediation. Briefings Funct. Genomics Proteomics 4(4): 355–362.
Skaates, S.V., A. Ramaswami and L.G. Anderson. 2005. Transport and fate of dieldrin in poplar and willow trees analyzed by SPME. Chemosphere 61: 85–91.
Strong, L., H. McTavish, M. Sadowsky and L. Wackett. 2000. Field-scale remediation of atrazine-contaminated soil using recombinant *Escherichia coli* expressing atrazine chlorohydrolase. Environ. Microbiol. 2(1): 91–98.
Talukder, A.H., S. Mahmud, S.M. Shaon, R.Z. Tanvir, M.K. Saha, A. Al Imran et al. 2015. Arsenic detoxification by phytoremediation. Int. J. Basic Clin. Pharmacol. 4(5): 822–846.
Thijs, S., W. Sillen, N. Weyens and J. Vangronsveld. 2017. Phytoremediation: state-of-the-art and a key role for the plant microbiome in future trends and research prospects. Int. J. Phytoremediation 19(1): 23–38.
Tiwari, J., P. Naoghare, S. Sivanesan and A. Bafana. 2017. Biodegradation and detoxification of chloronitroaromatic pollutant by *Cupriavidus*. Bioresour. Technol. 223: 184–191.
Ugya, A.Y., T.S. Imam and S.M. Tahir. 2016. The role of phytoremediation in remediation of industrial waste. World J. Pharm. Res. 5(12): 1403–1430.
Uqab, B., S. Mudasir and R. Nazir. 2016. Review on bioremediation of pesticides. J. Bioremediation Biodegrad. 7(3): 1–5.
Vergani, L., F. Mapelli, E. Zanardini, E. Terzaghi, A. Di Guardo, C. Morosini et al. 2017. Phyto-rhizoremediation of polychlorinated biphenyl contaminated soils: An outlook on plant-microbe beneficial interactions. Sci. Total Environ. 575: 1395–1406.
Verma, A. and P.K. Shukla. 2016. A prospective study on emerging role of phytoremediation by endophytic microorganisms. pp. 236–265. *In*: Rathoure, A. and V. Dhatwalia [eds.]. Toxicity and Waste Management using Bioremediation. IGI Global, Hershey.
Villaverde, J., M. Rubio-Bellido, F. Merchán and E. Morillo. 2017. Bioremediation of diuron contaminated soils by a novel degrading microbial consortium. J. Environ. Manage. 188: 379–386.
Wasilkowski, D., Ż. Swędzioł and A. Mrozik. 2012. The applicability of genetically modified microorganisms in bioremediation of contaminated environments. Chemik. 66(8): 817–826.
Wattanaphon, H.T., A. Kerdsin, C. Thammacharoen, P. Sangvanich and A.S. Vangnai. 2008. A biosurfactant from *Burkholderia cenocepacia* BSP3 and its enhancement of pesticide solubilization. J. Appl. Microbiol. 105: 416–423.
Windevoxhel, R., S. Nereida and H. Bastardo. 2011. Bioaumentación y sustancias húmicas en la biodegradación de hidrocarburos del petróleo. Rev. Ing. UC. 18(1): 23–27.

Wittgens, A., F. Kovacic, M.M. Müller, M. Gerlitzki, B. Santiago-Schübel, D. Hofmann et al. 2016. Novel insights into biosynthesis and uptake of rhamnolipids and their precursors. Appl. Microbiol. Biotechnol. 1–14.
Xu, Y. and N.Y. Zhou. 2017. Microbial remediation of aromatics-contaminated soil. Front. Environ. Sci. Eng. 11(2): 1–9.
Yadav, A., N. Batra and A. Sharma. 2016. Phytoremediation and phytotechnologies. Int. J. Pure App. Biosci. 4(2): 327–331.
Yang, C., R. Liu, Y. Yuan, J. Liu, X. Cao, C. Qiao et al. 2013. Construction of a green fluorescent protein (GFP)-marked multifunctional pesticide-degrading bacterium for simultaneous degradation of organophosphates and γ-hexachlorocyclohexane. J. Agric. Food Chem. 61: 1328–1334.
Zbytniewski, R. and B. Buszewski. 2002. Sorption of pesticides in soil and compost. Polish J. Environ. Stud. 11(2): 179–184.
Zhang, B.Y., J.S. Zheng and R.G. Sharp. 2010. Phytoremediation in engineered wetlands: Mechanisms and applications. Procedia Environ. Sci. 2: 1315–1325.
Zhang, R., X. Xu, W. Chen and Q. Huang. 2015. Genetically engineered *Pseudomonas putida* X3 strain and its potential ability to bioremediate soil microcosms contaminated with methyl parathion and cadmium. Appl. Microbiol. Biotechnol. 100: 1987–1988.
Zuo, Z., T. Gong, Y. Che, R. Liu, P. Xu, H. Jiang et al. 2015. Engineering *Pseudomonas putida* KT2440 for simultaneous degradation of organophosphates and pyrethroids and its application in bioremediation of soil. Biodegradation 26: 223–233.

7

Pesticide Bioremediation

An Approach for Environmental Cleanup Using Microbial Consortia

Enzo E. Raimondo,[1,a] *Juliana M. Saez,*[1,b] *Gabriela E. Briceño,*[2] *María S. Fuentes*[1,c] and *Claudia S. Benimeli*[1,*]

Introduction

Agricultural ecosystems were created by humans during the invention of agriculture approximately 12000 years ago to generate a reliable food supply that could be easily stored (e.g., cereal grains), enabling the creation of cities and the rise of civilizations (McDonald and Stukenbrock 2016). For this reason, agricultural production is one of the most important economic activities in the world. The use of pesticides in agriculture for a long time represented an undeniable benefit, which allowed a greater agricultural production and, therefore, to have a better availability of food. However, although pesticides are necessary for agricultural development and pest control, other problems have occurred since these agrochemicals have been applied massively and indiscriminately, so that their residues remain in different environmental compartments, causing adverse effects on domestic animals, human health, and environment (Rouimi et al. 2012).

Pesticides were formulated to prevent, destroy or/and control several pests such as bacteria, fungi, plants, animals and/or insects, including species that affect agricultural production, food processing as well as vectors of human and animal diseases (FAO 2002). The chemical structure of these compounds determines their mechanisms of action, but they always interfere with vital cell pathways by acting on specific biological functions of pests.

[1] Planta Piloto de Procesos Industriales Microbiológicos (PROIMI-CONICET), Avenida Belgrano y Pasaje Caseros, 4000 Tucumán, Argentina.
[a] Email: enzo_er_25@hotmail.com
[b] Email: julianasaez@hotmail.com
[c] Email: soledadfs@gmail.com
[2] Departamento de Ingeniería Química, Universidad de La Frontera, Avenida Francisco Salazar 01145, 4780000 Temuco, Chile.
Email: gbriceno@ufro.cl
* Corresponding author: cbenimeli@yahoo.com.ar

Different pesticides are being used globally to control insects and to increase agricultural production. Although many developing countries have banned the use of several pesticides, studies carried out in recent years have reported increased presence of their residues in many areas of the environment because of the excessive application of these xenobiotics in different matrices. Indiscriminate and uncontrolled use of these chemicals in agricultural and urban areas has caused serious environmental problems and groundwater pollution. Agrochemicals residues can damage the ecosystem and cause risk to human, animal and plant health (Gomes et al. 2016, Rezende dos Santos et al. 2017, Zepeda-Arce et al. 2017).

Complex mechanisms such as volatilization, adsorption, desorption, and chemical and biological degradation determine the availability, movement, and fate of these complex organic compounds in the soil. However, the concentration of these xenobiotics in soils is principally regulated by adsorption (Gondar et al. 2013), which depends on the pesticide chemical properties (water solubility, polarity) and soil properties (pH, permeability, texture, organic matter content and clay) (Rama Krishna and Philip 2008).

There is substantial evidence that the uptake of organic pesticides is strongly dependent on soil organic matter and clay content (Đurović et al. 2009). It has been reported that if the content of organic matter and clay in the soil and hydrophobicity of the pesticide is higher, its adsorption and retention in the solid matrix will be greater too, whereas desorption may be insignificant. However, pesticides with low soil adsorption represent a high risk of contamination to the aquifers (Alfonso et al. 2017). Because the composition of pesticides in the environment is influenced by the factors mentioned above, the composition of pesticide residues in groundwater may be different from that of soils (Mishra et al. 2012). The lack of information on the composition and level of pesticide residues makes it difficult to plan for groundwater protection.

There are several criteria for the classification of pesticides. Based on their action towards target pests, the more widely used pesticides are insecticides, herbicides, fungicides, acaricides, nematicides, molluscicides, and rodenticides (Rani et al. 2017). Chemically, they can be mainly classified into four groups: organochlorine, organophosphorus, substituted urea and carbamides (Rama Krishna and Philip 2008). Organochlorine pesticides (OPs) are the more persistent compared to other groups of pesticides.

For many years, OPs were widely applied around the world as a wide spectrum insecticide on crops. In the 1940s, they were introduced as the first important synthetic organic pesticides applied to minimize pest attacks and benefit humans. Public health was another area where OPs were used for dengue, malaria, pediculosis and scabies controls. However, because of their environmental persistence, high toxicity on human health, chemical stability, long-range transport and resistance to microbial attack, a great number of OPs have been studied (Barakat et al. 2017).

Chemical formulas of representative OPs can be visualized in next Figure.

This chapter gives a general overview on the use of microbial consortia for the bioremediation of organochlorine pesticides, both individually and forming mixtures. The use of actinobacteria defined consortium in the bioremediation of OPs mixture in different systems is discussed in particular.

Organochlorine Pesticides: Characteristics, Effects and Presence in the Environment

The need to maximize agricultural crops in developing countries led to the wide application of organochlorine pesticides. For this reason, the contamination of many agricultural fields

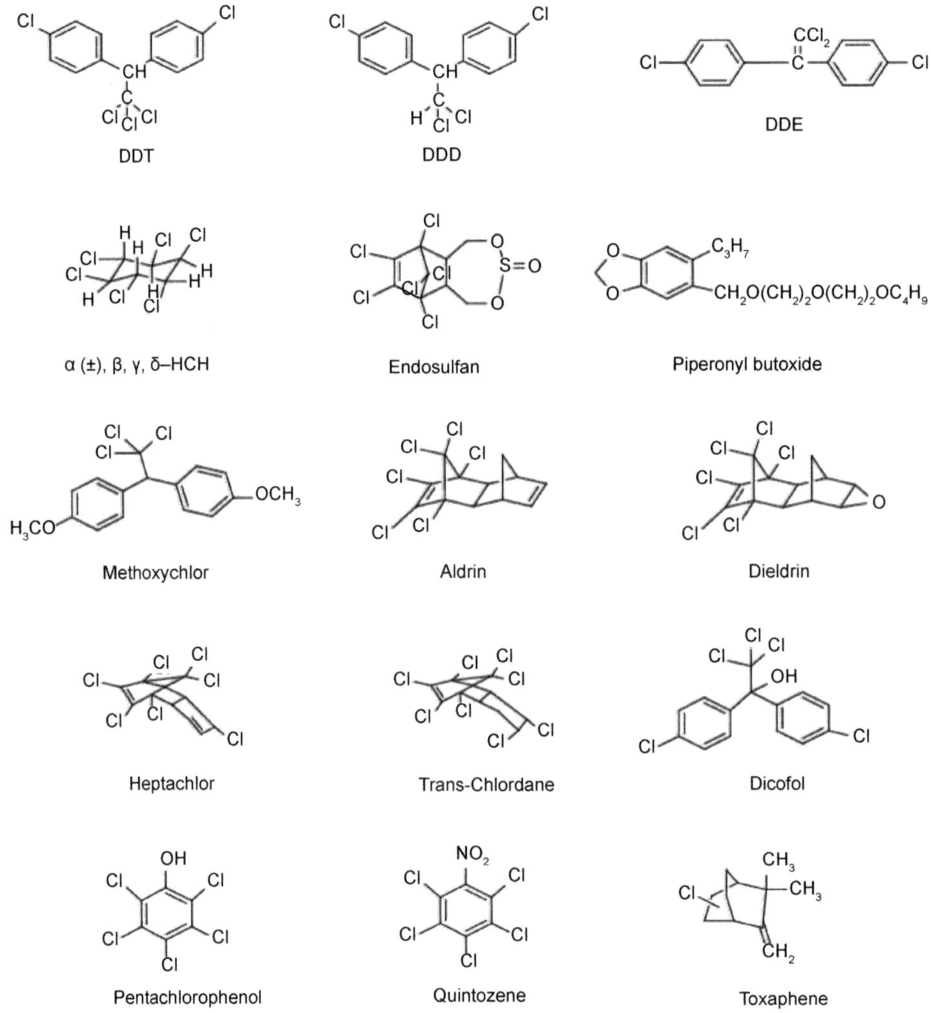

Figure 1. Structures of several organochlorine pesticides.

with their residues has resulted worldwide (Wang et al. 2016, Shruti et al. 2017). However, OPs are susceptible to be transported to other sites through the air and contribute to the pollution of different uncontaminated environmental areas. These compounds can reach aquatic systems by surface runoff, direct discharges or air travel, where they are strongly adsorbed by suspended particles and accumulated in the river and marine sediments due to their hydrophobic nature (Ballesteros et al. 2014, Jin et al. 2017). This represents a risk for aquatic biota, particularly fishes since they can bioaccumulate in organisms and biomagnify in food chains; then they are transferred to humans through ingestion of contaminated food and stored in adipose tissues, increasing cancer risk and others important public health problems (Shao et al. 2016).

It is well known that human exposure to these toxic compounds can induce mutagenesis, teratogenesis, reproductive disorders, neurotoxicity, abnormal functions of endocrine system, altered sex hormone levels and reduced sperm quality sufficiently to impair male fertility (Yaduvanshi et al. 2012, Chand et al. 2014, Costa 2015, Cremonese et al. 2017). In addition, a growing number of publications have linked OPs contact to liver, pancreas, breast and prostate cancers and testicular germ cell tumors (Chia et al. 2010, Eldakroory et al. 2016, VoPham et al. 2017). Therefore, the use and production of some organochlorine pesticides have been banned globally by the Stockholm Convention (UNEP 2016).

Although most developed countries established bans and restrictions on the use of OPs during the 1970s and 1980s, some developing countries continue using them because of their low cost and versatility in pest control (Kamel et al. 2015). For this reason, studies around the world continue reporting elevated levels of organochlorine pesticides in soils and aquatic ecosystems (Kamel et al. 2015, Pan et al. 2016, Sánchez-Osorio et al. 2017). Even, some pesticides were found in areas where they were never applied, such as Arctic and Antarctic (Dietz et al. 2004, Zhang et al. 2015). This poses a serious threat to humans, aquatic organisms and wildlife fauna living in agricultural areas, which may incorporate pesticides by ingestion, respiration, and skin contact. Recent data have demonstrated the presence of OPs in human blood, breast milk and fatty tissues (Elbashir et al. 2015, Tsygankov et al. 2015, Ploteau et al. 2016, Polanco Rodríguez et al. 2017).

In Argentina, several research works have shown the presence of organochlorine pesticide residues in different regions, because of their excessive use in the past, recalcitrance and transport processes. For instance, Chaile et al. (1999) revealed the presence of high concentrations of various OPs in the Salí River, the main hydrographical system of Tucumán province; the most abundant compounds were lindane, chlordane, and methoxychlor. Endosulfan, lindane, and dichlorodiphenyltrichloroethane (DDT), among others OPs, were found by Ballesteros et al. (2014) in water, sediments and suspended particulate material from Mar Chiquita (Buenos Aires province), in overpassed levels than allowed. Gonzalez et al. (2010) evaluated the distribution of OPs in agricultural soils from La Pampa and Patagonia, being endosulfan and p,p'-DDE the most predominant pollutants. Also, Lupi et al. (2016) detected high levels of OPs in different environmental matrices in the Quequén Grande River agricultural watershed. Besides, an extensive literature warns the presence of residues and metabolites in adipose tissues of aquatic animals such as fishes, whales, dolphins and mollusks, and even in feathers of avian scavengers from Argentinean Patagonia (Commendatore et al. 2015, Martínez-López et al. 2015, Torres et al. 2015, Silva Barni et al. 2016, Durante et al. 2016). For example, high OPs concentrations were detected in fishes from Negro River basin, exceeding the maximum levels for consumption, which is a potential risk to humans who consume these organisms (Ondarza et al. 2014).

The Role of Biotechnology in the Treatment of Contaminated Environments

For some decades, the society became aware of the real problem that environmental pollution means, and began to work to avoid and remedy such problems. Since soils affected by organochlorine pesticides represent potentially serious threats to the quality of surface and groundwater sources, several methodologies were developed to remove them: burial of contaminated soils in certified landfills, incineration and different chemical technologies for destruction of OPs, among others. However, most of these methods are costly, time-consuming and require certain treatments that increase the risk of worker exposure (Abhilash and Singh 2008).

Meanwhile, economic and eco-friendly methods that have no impact on the environment are being developed during the last decades, and biotechnology provides certain tools or mechanisms to remedy contaminated sites. Biotechnology is defined as "any technological application that uses biological systems and living organisms or their derivatives for the creation or modification of products or processes for specific uses" (ONU 1992). Therefore, environmental biotechnology refers to the application of biotechnology for the resolution of natural, agricultural and anthropic environmental problems and to achieve the conservation of environmental quality, taking advantage of properly qualified living organisms and genetic engineering for improving efficiency and cost of processes (Gómez Cruz 2010). Due to the urgent need to develop biotechnological strategies for the treatment of polluted environments, researchers have devised a technique called bioremediation.

Bioremediation consists in the use of the physiological potential of microorganisms and plants to eliminate environmental contaminants from the sites where these compounds have been thrown (Chishti et al. 2013). This technology involves the decomposition of pesticides or any other organic compound to substances less toxic, preferably to water and CO_2. In the last ten year, special emphasis has been placed on the bioremediation of several matrices contaminated with pesticides, that is, on the selection of microorganisms capable of degrading these compounds and/or their metabolites, which requires a complete study of the physiological, biochemical and microbiological aspects involved (Sardrood et al. 2013).

The processes used in bioremediation can be classified as: natural attenuation, a phenomenon by which autochthonous soil microorganisms can degrade contaminants without introducing modifications; biostimulation, which consists in the stimulation of microbial growth and biodegradation of contaminants by native microorganisms through the addition of oxygen, appropriate nutrients, surfactants, electron donors and/or moisture to the system; and bioaugmentation, which involves the inoculation of microbial strains or microbial consortia with desired metabolic pathways to degrade the xenobiotic of interest on the contaminated site and accelerate the process (Jiang et al. 2016, Zhang et al. 2016).

Microbial degradation is one of the most important factors in preventing the accumulation of pesticides in soils and groundwater sources. The ability of microorganisms to degrade pesticides and use them as carbon and energy source depends on the bioavailability of the chemical compounds as well as soil and contaminant characteristics (Rama Krishna and Philip 2011). These factors that can affect the effectiveness of the process must be previously investigated and understood. In this sense, previous works described the degradation of toxic compounds, especially organochlorines, by certain species of fungi, yeasts and Gram positive or Gram negative bacteria, isolated from aquatic soils and sediments (Kataoka et al. 2010, Satsuma and Masuda 2012, Abdul Salam et al. 2013, Teng et al. 2017).

Among the bacteria used to carry out a bioremediation process, actinobacteria can be highlighted, because they have a great catabolic diversity and relatively rapid capacity to grow in complex matrices, such as soils, and to colonize selective substrates. On the other hand, the main habitat of most actinobacteria is the aerobic zone of the soil, where they form a substantial component of the microbiota and they live in saprophyte form using a large variety of complex macromolecules (Shelton et al. 1996). An additional advantage is that the hyphae of actinobacteria can differentiate into spores that allow them to survive in the soil and resist unfavorable conditions such as lack of water and nutrients for long periods of time (Karagouni et al. 1993). For these reasons, actinobacteria are suitable agents for biotransformation and/or biodegradation of a wide range of contaminants, including pesticides with different chemical structures. There are several studies which demonstrate that actinobacteria have a great potential for the biotransformation of organic and inorganic toxic compounds (Polti et al. 2014, Bourguignon et al. 2014, Aparicio et al. 2015, Briceño et al. 2015, Sineli et al. 2016, Alvarez et al. 2017, Fuentes et al. 2017).

Microbial Consortia

In natural environments, many species of microorganisms are part of microbial consortia, formed by multiple populations that coexist and interact with each other by carrying out complex chemical processes and physiological functions in order to allow community survival (Polti et al. 2014). Natural microbial communities establish relationships of synergistic, antagonistic and/or neutral nature (Ho et al. 2016). These microbial interactions with other organisms and the environment are essential activities that serve to sustain life on Earth. Together, microorganisms exhibit great metabolic diversity and play essential roles in energy transformations and biogeochemical cycles, mainly carbon, sulfur, iron, and nitrogen, which serve to recycle the essential elements of living systems (Zhao et al. 2014).

A microbial consortium is an association of two or more microbial populations, which act together as a community in a complex system, where all members benefit from the activities of others. The association reflects synergistic or syntrophic lifestyles, where the range of growth and nutrient flow are conducted more effectively than in individual populations (Hero et al. 2017). Therefore, a microbial consortium, combining the catalytic specialties of different species, can perform complex functions that require multiple steps. These enzymatic activities could not be carried out by individual populations, whose metabolic pathways are limited. Also, they are very resistant to pollution. Due to these advantages, this microbial cooperation is observed mainly in the processes of industrial effluent treatments, bioremediation of multi-toxic contaminated matrices and degradation of lignocellulosic matter. For example, the degradation of lignocellulolytic materials requires the participation of different microbial populations able to carry out this complex process involving multiple stages since individual populations cannot carry out these functions alone. The degradation of plant biomass implies the cooperation of various cellulolytic, hemicellulolytic and ligninolytic enzymes, which are produced by several microorganisms that coexist in nature. To achieve hydrolysis efficiency, other non-cellulolytic species play an important role in the stabilization of the consortium (Hero et al. 2017).

On the other hand, microbial consortia are used in some parts of the world to produce fermented foods. A consortium of bacteria and yeasts is used in Europe to produce Kefir, e.g., a fermented milk beverage. This dairy beverage is manufactured by inoculating a kefir grain containing the symbiotic community into milk to initiate a natural fermentation. This consortium is not well defined because it varies substantially between different kefir grains (Walsh et al. 2016). Another complex microbial consortium containing several microorganisms is used to prepare many Korean traditional fermented soybean foods (Kim et al. 2011).

Microbial Consortia as a Bioremediation Tool

In natural environments, it has been observed that microbial communities frequently can metabolize a wide variety of persistent pesticides, using them as a source of energy and nutrients. In addition, several shreds of evidence have demonstrated that consortia are more efficient than pure microbial cultures for bioremediation process and for preventing the accumulation of toxic compounds derived from microbial degradation (Yang et al. 2010). The rationale for this approach is that microbial consortia can combine the catalytic activities of different species or strains, increasing the available metabolic pathways necessary to metabolize xenobiotic substrates, such as pesticides (Polti et al. 2014). Even more, metabolites produced by one strain could be consumed or transformed by other members of the consortium (Perruchon et al. 2017). Although it is known that there are some organisms which can completely degrade a specific organic pollutant, individual species generally do not contain the complete degradation pathway for xenobiotics (Gerhardt et al. 2009). On the other hand, their survival in the environment can be improved by the microbial biodiversity of the consortium (Siripattanakul et al. 2009). Since there is no competition between microbial populations or metabolites that inhibit them, they can attack the target compound more efficiently, which is an advantage over the total microbial population present in a contaminated environment (Pino et al. 2011). For the purpose of this chapter and in bioremediation studies in general, the terms "microbial consortia" and "mixed cultures" will be used as similar (Hamer 1997, Fuentes et al. 2011, Lee et al. 2013, Saez et al. 2015, Fuentes et al. 2016).

Microorganisms belonging to a microbial consortium are generally isolated from highly contaminated areas, where they can resist and degrade high concentrations of toxic compounds (Pino et al. 2011). This can be explained by the fact that prolonged exposure to microorganisms to certain extreme conditions causes selective pressure, which is responsible for allowing the development of the constituent microorganisms of a consortium, synthesizing enzymes capable of acting on a given contaminant (Carrillo-Pérez et al. 2004). For this reason, to obtain potent cultures and to achieve a successful degradation or removal of a certain compound, it is required the isolation of natural consortia from sites previously exposed to the interest xenobiotic through suitable enrichment techniques and their improvement in the laboratory by various techniques, such as acclimatization and stabilization (Elcey and Mohammad Kunhi 2009).

In many cases, when the pollutant toxicity is high for the autochthonous microorganisms of the soil or they are scarce or inappropriate in quantity and quality, soil inoculation with microbial consortia provides certain advantages over biostimulation of native microorganisms (Wu et al. 2013). The bioaugmentation with microbial cultures which have interesting catabolic capacities is a strategy that has been carried out since the 1970s. This technology is based on the fact that native microbial populations may not have the ability to degrade a wide range of toxic compounds present in the soil or they can do it but over a long period. In this way, the addition of microorganisms acting on the target contaminant complements the metabolic deficiencies of the indigenous microbial populations, achieving a greater rate of pollutant degradation (Adams et al. 2015). Therefore, there is a strong need to study microbial mixed cultures to evaluate their degradation activities for the treatment of contaminated environments (Manickam et al. 2008).

There are two types of mixed cultures used in a bioremediation process. One of them involves the use of exogenous degraders microorganisms, previously isolated, for the formulation of defined mixed cultures. The disadvantage of these consortia is that in some cases, previously isolated strains cannot be good degraders or have difficult to survive in different environments that are not their originals. The other type is based on the use of indigenous autochthonous cultures, which are a better alternative to achieve a successful bioremediation. Siripattanakul et al. (2009) demonstrated that two native mixed cultures isolated from atrazine-contaminated soils were more efficient in the removal of this compound than *Agrobacterium radiobacter* J14a, a strain whose ability to degrade atrazine had been reported previously. After 168 hr of incubation, the mixed cultures MC1 and MC2 removed 51.0% and 49.0% of atrazine respectively, while the strain J14a showed a removal of 38.0%. Other authors also efficiently used native microbial consortia to degrade pesticides such as methyl parathion. Pino et al. (2011) isolated three consortia from soil samples of Moravia, Czech Republic. They were able to degrade methyl-parathion in 120 hr in liquid culture, reaching the total removal when the three consortia were combined.

In soil, this compound was eliminated by the mixture of the three consortia, which was accompanied by a decrease of the toxicity.

Functional and structural stability of microbial communities is an important parameter that must be maintained in order to achieve applications in the future. However, the difficulty to define all the members included in a complex microflora impedes understanding the mechanisms responsible for achieving a stable coexistence between the different microbial species, resulting in uncertain biodegradation (Kato et al. 2005). In this sense, the use of defined mixed cultures is advantageous because it allows the study of the general characteristics of all its members and the monitoring of its dynamics. These data are important to explain the *in situ* role of the microorganisms involved in the degradation of complex substrates and its influence on the community (Hero et al. 2017). However, selecting the best defined consortium usually implies a large volume of work due to the numerous combinations of pure cultures to be performed in order to choose the best combination for pesticide degradation (Alvarez et al. 2012).

For all the mentioned above, the use of microbial consortia is presented as a potential alternative for the bioremediation of environments contaminated with pesticides.

Bioremediation of Organochlorine Pesticides Using Microbial Consortia

Organochlorine pesticides entered the market between the year 1940 and 1950 and they were very important for the control of pests and vectors of diseases (WHO 1990). The main OPs producing and selling countries are Japan, France, China, USA and Brazil. On the other hand, Europe and Asia, with India as the main example of this continent, have been the major pesticide consumers and there the role of these compounds was very large due to its strong agricultural activity (Rani et al. 2017). Unfortunately, these extremely stable and persistent substances have accumulated in the environment, and were even detected in remote regions where they were never produced and/or used (WHO 2003). Various physical, chemical, and biological methods were developed in order to achieve the elimination of OPs from contaminated sites. Among them, biodegradation is an eco-friendly process that allows the elimination of this type of xenobiotics, through the use of microorganisms, at a low cost and without damaging ecosystems (Adams et al. 2015). However, only specific bacterial and fungal species have been reported to be capable of degrading them, due to the recalcitrant structures of the OPs. It has been observed that in natural environments the degradation of several toxic compounds is carried out frequently by mixed microbial communities and the use of microbial consortia is successful in bioremediation processes (Nestler et al. 2001, Polti et al. 2014).

Defined microbial consortia have been reported in the literature as suitable tools for the degradation and removal of OPs. Among them, Awasthi et al. (2003) studied the degradation of α and β isomers of endosulfan by a defined co-culture of two *Bacillus* strains, in minimum medium supplemented with 20 $\mu g\ ml^{-1}$ of each isomer. In these conditions, the metabolisms of both pure isomers were comparable, detecting a degradation of the 92.0% of α-endosulfan and 86.0% of β-endosulfan at the end of the incubation period.

A research carried out in Tucumán (Argentina) evidenced the presence of a large amount of organochlorine pesticide residues in the Salí River, the main hydrographic system of that province. The presence of such residues is due to the large quantity of agricultural and industrial effluents, many of which have not an adequate final disposition and they are discharged throughout its extension. The most abundant OPs detected were lindane (LIN), chlordane (CLD), and methoxychlor (MTX) (Chaile et al. 1999). Lindane is the gamma isomer of hexachlorocyclohexane, the only isomer with insecticidal properties

(Cao et al. 2013). It is a highly chlorinated, saturated and cyclic insecticide used for crop protection and control of vector-borne diseases, such as malaria (Manickam et al. 2008). Its low water solubility and its chlorinated nature contribute to its persistence and resistance to microbial degradation (Phillips et al. 2001). Technical chlordane was used worldwide as a pesticide for croplands, lawns, gardens and as a germicide for houses. For this reason, this compound was detected in soils, atmosphere, vegetation, sediments and animal tissues (Li et al. 2007). Technical chlordane is a complex mixture of more than 140 compounds, being 120 of them already classified. The composition of this mixture varies with the manufacturing processes and the two most abundant components are γ-chlordane and α-chlordane (Dearth and Hites 1990). Methoxychlor is a potential carcinogen that acts as an endocrine disruptor and affects the activity of the central nervous system (Lafuente et al. 2007, Yim et al. 2008, Frye et al. 2012). This pesticide has a half-life of about one yr (Fogel et al. 1982).

Based on this background, Benimeli et al. (2003) and Fuentes et al. (2010) isolated indigenous actinobacteria strains from a wastewater sediment of a copper filter plant located in an agricultural area of Tucumán, Argentina, and from contaminated soil in Santiago del Estero, Argentina, where about 30 tons of organochlorine pesticides were found in 1994. Subsequently, Fuentes et al. (2011, 2013a, 2016) evaluated the ability of these actinobacteria to degrade OPs. The isolated strains were grown in liquid media artificially contaminated with LIN, CLD or MTX at a concentration of 1.66 mg l^{-1}. Then, defined consortia were formulated by using all possible combinations of two, three, four, five and six strains and the removal of LIN, CLD, and MTX by these consortia in liquid medium was evaluated. The authors demonstrated that LIN removal was substantially improved when defined consortia of two, three or four actinobacteria strains were used, as compared to the corresponding individual strains; the culture constituted by *Streptomyces* sp. A2-A5-M7-A11 was the most efficient for the degradation of this pesticide. It is important to highlight the presence of *Streptomyces* sp. A2 in this consortium, which presented the minimum removal capacity when it grew individually in minimum medium (MM). These results evidenced the potentiation between the microorganisms when acting together onto the pesticide. In the case of MTX, the mixed culture formed by *Streptomyces* sp. A6-A12-A14-M7 was selected, which almost completely removed MTX. For CLD, the mixed culture which presented the best capacity for removing the pesticide was *Streptomyces* sp. A2-A5-A13. Subsequently, Fuentes et al. (2011, 2013a, 2016) studied the removal of MXT, CLD, and LIN in soil microcosms using these defined consortia of actinobacteria, and observed a lower efficiency in the removal of three pesticides compared to values obtained in a liquid medium.

The ability of fungal consortia in the bioremediation of pesticides has also been previously reported. In this context, Hechmi et al. (2016) demonstrated the ability of *Byssochlamys nivea* and *Scopulariopsis brumptii* to remove pentachlorophenol (PCP) individually from soil microcosms. However, a synergistic effect on the degradation efficiency of PCP was observed when both strains were inoculated together, by comparison to fungi mono-cultures. High PCP degradation percentages were observed (95.0 and 80.0%) for the mixed culture when initial concentrations were 12.5 and 25.0 mg kg^{-1}, respectively.

On the other hand, Abraham and Silambarasan (2014) showed that a defined bacterial consortium and a defined fungal consortium were able to degrade completely the pesticide endosulfan in an aqueous medium as well as in soil amended with different nutrients, being the fungal consortium faster than the bacterial consortium.

Also, there are different studies on the biodegradation of OPs by native mixed cultures. Isolation of natural microbial communities through suitable enrichment techniques and their improvement in the laboratory by various techniques are necessary steps for obtaining potent cultures for bioremediation processes (Elcey and Mohammad Kunhi 2009). Moreover, assessments of biodegradation by indigenous microorganisms are helpful in understanding the dissipation of chemicals *in situ* (Chiu et al. 2004). In this sense, Hirano et al. (2007) measured 22.0 and 33.0% removal of cis- and trans-CLD, respectively, when they studied anaerobic biodegradation of this contaminant employing mixed cultures of indigenous bacteria isolated from a river sediment. Also, Elcey and Mohammad Kunhi (2009) isolated a consortium consisted of nine bacterial strains and a fungal strain, from a site having a long history of technical grade HCH contamination and evaluated its potential to degrade lindane. They demonstrated that this consortium was able to mineralize 300 $\mu g\ ml^{-1}$ of γ-HCH within 108 hr, but it degraded partially lindane when it was exposed to an amount of 400 $\mu g\ ml^{-1}$ of γ-HCH. Meanwhile, Rama Krishna and Philip (2008) studied individual biodegradation of different pesticides and they obtained 65.0% of lindane removal using enriched microbial cultures previously grown in this pesticide, under aerobic conditions. However, when they worked in anaerobic conditions, there was a slight increase in degradation efficiency by the same cultures. Also, Kumar and Philip (2006, 2007) reported a novel mixed bacterial culture isolated from an endosulfan contaminated soil, which degraded this insecticide in aerobic and facultative anaerobic systems, being the process faster in anaerobic conditions. Subsequently, this consortium was able to degrade endosulfan in soil samples by using a pilot scale soil reactor, achieving a more predominant degradation in the bottom portion of the reactor due to oxygen deficit condition. Chiu et al. (2004) isolated an anaerobic mixed culture from river sediments of Taiwan, and reported the anaerobic biodegradation of DDT and heptachlor by this consortium, evaluating the effects of temperature, chemical concentrations, carbon sources and electron acceptors. These factors affected the degradation rate of the pesticides under anaerobic conditions, but in all cases, the mixed culture was efficient in their degradation.

On the other hand, there are several reports demonstrating that the combination of different strains is not necessarily more efficient in the removal of OPs than the single cultures. Corona-Cruz et al. (1999) found no synergistic effect when a defined microbial consortium, including five species of *Pseudomonas*, one of *Klebsiella*, four of *Rhodococcus* and two strains of fungi, was inoculated into an aerobic reactor to remediate DDT-contaminated soils. This process was inefficient since 2.5% DDT removal was detected after four weeks of incubation. Similarly, Fuentes et al. (2011) detected a decrease in the efficiency of LIN removal respect to the high percentages obtained by some triple and quadruple actinobacteria consortia, when they used consortia of five and six strains, reaching removal percentages between 12.5 and 22.0%, respectively.

One of the difficulties to describe the characteristics of biological systems is due to their ability to adapt and evolve under different environmental conditions. Microorganisms are capable of adapting to various non-native carbon compounds by using a number of strategies to meet their growth requirements, achieving an optimal overall performance in the new conditions (Lee and Palsson 2010). The mineralization of many organic compounds that are introduced into treatment systems or natural environments is often preceded by an acclimation period. It has been postulated that a gradual increase in the concentration of a determined substrate during the acclimation may result in the induction of enzymes necessary for the degradation of such substrate. Further, in a microbial consortium, the net result of this process may be a synergism through complementation of the enzymes present in the different strains that compose such consortium (Saez et al. 2015). Currently, some of the most common biotechnological applications of microbial acclimation processes include the activation of latent pathways for using non-natural substrates or for producing non-natural products, increasing the speed of substrate utilization, as well as improving growth velocity to reach detectable bioconversion rates (Fong et al. 2005). In this sense, Bidlan and Manonmani (2002) demonstrated that a microbial consortium, isolated from DDT-contaminated fields and enriched in flasks containing DDT as sole carbon source, was able to degrade 25 ppm of DDT in 144 hr, after being acclimated by repeated passages through a mineral salt medium containing increasing concentrations of DDT. Also, Elcey and Mohammad Khuni (2009) demonstrated that a microbial consortium, isolated from a soil having a long history of technical grade HCH application, was able to degrade high concentrations of lindane after an acclimation period with gradual increasing in the concentration of the pesticide. However, none of the individual strains members of the consortium could degrade high concentrations of HCH. These results suggest that a proper acclimation could be necessary for the strains to be able to mineralize higher pesticide concentrations. On the other hand, Saez et al. (2014) reported that the acclimation of a *Streptomyces* defined consortium stimulated growth and improved lindane removal in liquid systems. However, the authors observed that the acclimation was not necessary for a slurry system since it did not increase significantly the pesticide removal.

The immobilization of microbial cells is increasingly applied in biotechnological processes for chemical conversion and bioremediation processes, obtaining greater biodegradation activity, higher dilution rate without elimination by washing in continuous processes and greater tolerance to high concentrations of toxic compounds (Wang et al. 2004). Immobilized microbial cells have been used for the degradation of organochlorine pesticides. Therefore, one of the most promising areas of research is the use of this technology to reduce environmental pollution. In this sense, Saez et al. (2014) reported that a *Streptomyces* consortium immobilized in cloth sachets was efficient to remove lindane from a concentrated slurry system, contaminated with an initial pesticide concentration of 50 mg kg^{-1}. This immobilized consortium was able to remove 35.3 mg kg^{-1} of lindane after 7 d of incubation.

It is well known that removal of the halogen atoms from halogenated compounds is the key reaction during its microbial degradation because the halogen atom is responsible for the toxic and recalcitrant character of the organochlorine compounds. Generally, hydrogen atoms or hydroxyl groups replace the halogen, and its elimination reduces the risk of forming intermediates compounds during the following metabolic steps (Camacho-Pérez et al. 2012). The degradation pathway of HCH and MTX has been intensively studied. Nagata et al. (1999, 2007) identified two types of dehalogenase enzymes involved in the early stages of lindane degradation pathway in *Sphingobium japonicum* UT26. Lee et al. (2006) observed that the fungus *Stereum hirsutum* was able to degrade high concentrations of MTX and that the rapid degradation by the fungus occurred due to a mono-dechlorination reaction at initial stages, which was confirmed by GC/MS spectrometry analysis. Dechlorination is a very important step in organochlorine pesticide degradation, being dehalogenases the enzymes involved in this step (Nagata et al. 2007). In the environment, dechlorination of organochlorine compounds usually involves the cooperation of several kinds of microorganisms (Chiu et al. 2004). In this regard, Awasthi et al. (1997) studied the biodegradation of endosulfan by a bacterial co-culture, obtaining a stoichiometric ratio between the disappearance of endosulfan and the release of chloride ions, which indicated that dehalogenation is a part of the degradative activity of this consortium. In addition, Elcey and Mohammad Kunhi (2009) reported a 100% Cl^- release when they studied the degradation of different concentrations of LIN by a native mixed culture. Further, Fuentes et al. (2011, 2013a, 2016) determined dechlorinase activity in cell-free extracts of 171 mixed cultures of *Streptomyces* sp., grown in MM in the presence of LIN, CLD and/or MTX as the sole carbon source. In most cases, it was demonstrated that mixed cultures had an observed specific dechlorinase activity higher than the expected activity, suggesting that the degradation of the pesticide could be facilitated by the use of microbial consortia. Similar results were found by Saez et al. (2015) who determined the presence of chloride ions, during a bioremediation of lindane by using a consortium of immobilized actinobacteria in slurry systems, reaching a maximum value of chloride ions release and lindane removal at 21 d, which confirmed lindane degradation in slurries. During a bioremediation study by using an anaerobic mixed culture, Chiu et al. (2004) postulated that DDT degradation occurred by a reductive dechlorination mechanism, where chlorine is substituted by hydrogen, being DDD the major product of this process. Moreover, they observed that for degradation the heptachlor, only reductive dechlorination happened, with a removal of chlorine on C_1 and formation of chlordane. Rama Krishna and Philip (2008) also detected chlorine released during the degradation of LIN by different enriched cultures, in liquid medium under aerobic and anaerobic conditions, which reduce its toxicity and enhanced the biodegradation.

Bioremediation of Organochlorine Pesticides Mixtures Using Microbial Consortia

The increased use of pesticides in agriculture and domestic activities are polluting the environment day by day. Soil, groundwater and surface water contamination by pesticides is currently a significant concern worldwide because these compounds are toxic to human life and environment. Different types of pesticides are normally used to control specific pests of various crops. As a result, agricultural soils are contaminated with a mixture of pesticides. Among these compounds the organochlorine pesticides are of global concern. The application of them for a long time and their high persistence allowed their entry and accumulation in the environment. Several studies have shown an increase in the presence of residues of OPs in different environmental because of their excessive application, generating contaminated agricultural soils even with mixtures of them (Rama Krishna and Phillip 2011). For all the above, it is important to develop systems capable of biodegrading several groups of pesticides, studying the behavior of mixtures of OPs in different matrices and microorganisms with the capacity to degrade them. However, most of the previous researches focused on the degradation of individual pesticides or a series of pesticides belonging to the same group. In the next section, bioremediation of OPs mixtures in liquid medium, soil and slurry systems is considered.

Bioremediation of Organochlorine Pesticide Mixtures in Liquid Medium

Studies carried out by several researchers showed the ability of mixed cultures to degrade OPs forming a mixture with other pesticides. Kulshrestha and Kumari (2010) observed that a fungal mixed culture isolated from sewage sludge was able to degrade co-metabolically a DDT-chlorpyrifos mixture. In addition, they reported that the degradation of both insecticides was highest at lowest concentration and declined with increasing pesticides level, due to an inhibition of microbial growth and pesticide biodegradation by the toxicity exerted by the high concentrations of the contaminants. Rajashekara Murthy and Manonmani (2007) formulated a microbial consortium, which was able to degrade technical HCH (a mixture of four major isomers of HCH) up to 25 ppm, under shaking conditions, ambient temperature and neutral pH, being the gamma isomer the most removed compound in the mixture. On the other hand, Fuentes et al. (2017)

evaluated the capacity of an actinobacteria consortium constituted by *Streptomyces* sp. A2-A5-A11-M7 to grow in the presence of a mixture of LIN, CLD, and MTX and to remove them. They cultivated this consortium in MM supplemented with the three pesticides (1.66 mg l^{-1} each one) as the only carbon source for 16 d at 30°C. In these conditions, a biomass increase through the incubation time was observed, reaching the stationary phase at 16 d of incubation, with a final biomass of 9.63 × 10^4 CFU ml^{-1} (Table 1). In contrast, no microbial growth was observed in the controls carried out in MM without the addition of OPs. The fact that growth inhibition was not observed in the presence of pesticides could imply that they are not toxic to the microorganisms at the concentration evaluated and that the pesticides are being used to support its growth. The four actinobacteria could have adequate enzymes to carry out the degradation of these compounds because they were isolated from an environment contaminated with OPs (Benimeli et al. 2003, Fuentes et al. 2010). In addition, the mixed culture of *Streptomyces* sp. was able to remove the three OPs of the mixture, detecting the maximum removal percentages at 16 d of incubation (Table 1). It is important to remark that the abiotic controls (contaminated MM, in the absence of the mixed culture), did no show significant pesticides removal (Table 1).

In this sense, Fuentes et al. (2013b) had reported that the actinobacteria consortium was able to remove a mixture of chlorpyrifos (40.2 and 71.0%) and pentachlorophenol (5.2 and 14.7%) from MM without another additional carbon source when this consortium was assayed either as free or immobilized cells, respectively.

Table 1. Microbial growth and removal of the pesticides mixture by a *Streptomyces* sp. consortium inoculated in MM (Adapted from Fuentes et al. 2017).

Incubation Time (d)	Microbial growth (CFU ml^{-1}) × 10^4	LIN removal (%)	MTX removal (%)	CLD removal (%)
0	0.48 ± 0.02	ND	ND	ND
8	3.55 ± 0.07	14.8 ± 6.4	42.3 ± 0.2	99.5 ± 0.1
16	9.63 ± 0.04	40.4 ± 2.0	99.5 ± 0.1	99.8 ± 0.2

LIN: lindane; MTX: methoxychlor; CLD: chlordane; ND: not detected.

Bioremediation of Organochlorine Pesticide Mixtures in Soil and Slurry Systems

Based on the above, it can be inferred that the combined metabolic activity of actinobacteria in mixed cultures stimulates the degradation of OPs mixtures in liquid systems since they would increase the possible metabolic pathways for the use of these contaminants as a carbon source. Therefore, it is necessary to evaluate the degradation of OPs mixtures in other environmental matrices, specifically soils and slurries. However, there are very few reports on this topic. For example, Baczynski and Pleissner (2010) studied anaerobic bioremediation of a non-sterile soil contaminated with a mixture of OPs, including LIN, DDT, and MTX, using methanogenic granular and wastewater fermented sludges, rich in microorganisms, as inoculum. These authors detected a similar performance in both processes: all contaminants were removed considerably (80–90% and over). In addition, they observed a lack of degradative capacity of autochthonous soil microflora in uninoculated controls.

Among microbial communities, actinobacteria have a great potential for the biodegradation of OPs (Alvarez et al. 2017). Fuentes et al. (2017) evaluated the bioremediation of an OPs mixture (LIN + MTX + CLD) by using the consortium *Streptomyces* sp. A2-A5-A11-M7 in soils of different texture. The soil samples were taken from different experimental sites of the province of Tucumán (Argentina), free of contamination with organochlorine pesticides. Soil microcosms were sterilized and artificially contaminated with the three OPs (1.66 mg kg^{-1} each one). Contaminated and uncontaminated (control) microcosms were inoculated with the mentioned consortia and incubated for 16 d at 30°C. The authors informed that the actinobacteria consortium was able to grow in all the soil types (clay silt loam, sandy, and loam), showing similar growth profiles in all of them (Table 2). The cell population increased rapidly in 4 d, followed by a stationary phase from the 4th to the 16th d. In addition, there were no differences in microbial growth between contaminated soils and pesticide-free controls during the 16 d of incubation. This reinforce the hypothesis that actinobacteria are well adapted to proliferate in natural soils contaminated with xenobiotics and that they are able to colonize these matrices.

Regarding the bioremediation process, it was demonstrated that not all the OPs of the mixture were removed from the soils in the same way. The pesticides removal percentages from each soil type are shown in Table 2. In the clay silt loam soil (CSLS), without inoculate, an abiotic removal of 18.0%, 10.0%, and 5.0% was reported for MTX, LIN, and CLD, respectively. This finding could be due to pesticides adsorption to soil particles such as clay and organic matter (Cao et al. 2013). In sandy soil (SS) sterile microcosms, the percentages of pesticide dissipation were similar to those detected in uninoculated contaminated controls, not detecting significant differences between them. The authors suggest that these results are due to the characteristics of the sandy soils, with low content of organic matter and large pore size, which allowed the percolation of the pesticides through them. Similarly, Rama Krishna and Philip (2008) found that sandy soils with small amounts of organic matter retained pesticides in a lesser extent than clayey and composted soils. Finally, Table 2 shows the removal percentages in loam soil (LS) microcosms, not detecting CDL removal (Fuentes et al. 2017). In this case, there were also no significant differences between the microcosms inoculated with actinobacteria and abiotic controls.

It is important to note that CSLS presented high content of small particles. This soil characteristic could provide more sites for the bacteria to use the OPs as a source of carbon and energy. In this way, several studies showed that the greater specific surface area of silts and clays in fine texture soils probably increased the microbial availability of the contaminant adsorbed on the soil particles (Cui et al. 2011). In the case of LS microcosms, the low percentages of removal obtained suggest that the actinobacteria consortium could not have adapted to this type of soil during the time of the incubation, which was also observed by Cuozzo et al. (2012) in a bioremediation study in this type of soil, during the first 15 d. All these results demonstrated that the removal of OPs mixture from soil by microbial consortia does not occur at the same rate for each toxic compound. This finding did not coincide with that observed by Rama Krishna and Philip (2011). These authors studied the degradation of a mixture of LIN, methyl parathion and carbofuran in sterile soils of different textures by using a mixed pesticide-enriched culture, showing

Table 2. Comparison of the simultaneous removal of LIN, CLD, and MTX in different texture soils inoculated with a *Streptomyces* consortium (Adapted from Fuentes et al. 2017).

Type of soil	Microbial growth ($CFU\ g^{-1}$) × 10^7	LIN removal (%)	MTX removal (%)	CLD removal (%)
CSLS	3.93 ± 0.48	30.0 ± 0.1	39.0 ± 0.3	23.0 ± 0.1
SS	3.75 ± 0.62	48.0 ± 0.1	27.0 ± 0.1	23.0 ± 0.1
LS	3.18 ± 0.60	12.0 ± 0.1	4.0 ± 0.1	ND

CSLS: clay silty loam soil; SS: sandy soil; LS: loam soil; LIN: lindane; MTX: methoxychlor; CLD: chlordane; ND: not detected.

different percentages of removal in all cases. Their results demonstrated a higher removal percentage in sandy soils than in clay soils and the lowest pesticide removal was observed in compost soil. In addition, minimum degradation was detected for lindane in all soil types, due to its organochlorine nature that provides a high resistance to microbial attack.

Since the highest microbial removal percentages for all pesticides were obtained in CSLS, Fuentes et al. (2017) selected this type of soil to evaluate the biodegradation of the pesticide mixture in sterile and non-sterile soil microcosms and also in slurry systems (Table 3). In the first case, no significant removal of CLD was detected in both controls without inoculum and in the contaminated and inoculated samples. With regard to MTX, 21.0% of removal was obtained after 16 d of treatment, while for LIN a depletion of 25.0% was detected at the end of incubation time. In the uninoculated controls, no removal of CLD or MTX was observed, but LIN removal was 14.0%. For this reason, the authors assumed that the native soil microflora was involved in the removal of lindane, or that the concentration decrease in uninoculated controls could be due to the adsorption of LIN to clay particles.

Therefore, the defined actinobacteria consortium presented the ability to grow in a non-sterile soil system and to remove an OPs mixture in the presence of native microflora. However, their effectiveness was lower in comparison to the obtained in sterile conditions. The researchers suggest that changes in soil properties during the sterilization process can affect the microbial growth and thus the degradation of pesticides. It has been reported that all sterilization methods alter the physical and chemical properties of soils (Berns et al. 2008, Zamani et al. 2015).

In contrast, Cycoń et al. (2013) observed that the native microflora of different types of soil was involved in the degradation of a mixture of organophosphorus pesticides when they carried out a bioremediation study of these compounds by using a *Serratia marcescens* strain.

Slurry bioreactors are useful technologies that have been used for the bioremediation of soils contaminated with pesticides, explosives, and aromatic hydrocarbons, among others. They have the advantage of facilitating the treatment of contaminated soils with high clay content and organic matter in a short period (Robles-González et al. 2012, Saez et al. 2014). For this reason, Fuentes et al. (2017) studied the ability of the mixed culture to remove the OPs mixture in this kind of system (Table 3). The microbial growth was exponential until the 4th d of incubation, and then it remained constant, reaching biomass values of $4.94 \pm 0.74 \times 10^7$ and $4.83 \pm 0.12 \times 10^7$ CFU g^{-1} soil for assay and uncontaminated control, respectively, at the end of the incubation. As in the soil microcosms, not all OPs were eliminated in the same way. MTX dissipation was the most important, reaching a removal of 35.8% at the end of the 16 d of incubation, but in uninoculated controls, a 9.8% removal was detected. Regarding to LIN and CLD, the decreases in their concentrations were lower during the same period, reaching removal values of 13.5% and 15.0%, respectively. The concentration variation in the abiotic controls was 0.9% for LIN and 5.0% for CLD (Table 3). It is important

Table 3. Comparison of the simultaneous removal of LIN, CLD, and MTX in three soil systems inoculated with an actinobacteria consortium (Adapted from Fuentes et al. 2017).

Type of systems	LIN removal (%)	MTX removal (%)	CLD removal (%)
CSLS sterile microcosm	30.0 ± 0.1	39.0 ± 0.3	23.0 ± 0.1
CSLS non-sterile microcosm	25.0 ± 0.1	21.0 ± 0.2	5.0 ± 0.1
CSLS sterile slurry	13.5 ± 0.1	35.8 ± 0.1	15.0 ± 0.1

CSLS: clay silty loam soil; LIN: lindane; MTX: methoxychlor; CLD: chlordane.

to note that the use of reactors allowed an increase in the percentage of elimination of MTX by the bacterial consortium, in relation to that obtained in microcosms with this type of soil (26.0% in slurry systems and 21.0% in soil microcosms), which could be due to the fact that this system allows to increase the contact microorganisms-contaminants and microorganisms-nutrients (Robles-González et al. 2008). Fuentes et al. (2013a) obtained similar results when they studied bioremediation of MTX in slurry systems and sterile soil microcosms by using a defined microbial consortium. In contrast, this was not the behavior observed with LIN and CLD. Perhaps, an increase in LIN and CLD degradation efficiency for the soil slurry system tested would require longer incubation periods to achieve higher values (Fuentes et al. 2017).

These reports found in the literature show the efficiency of defined and native mixed cultures to degrade organochlorine pesticides and mixtures of them in different kind of systems since the use of microbial consortia is more adequate to achieve the complete mineralization of toxic and dangerous chemical compounds, or their transformation to nontoxic products in a bioremediation process (Castillo et al. 2006).

Concluding Remarks

The development of efficient remediation strategies requires the knowledge of environmental parameters and metabolic performance that govern a microbial consortium. The use of mixed cultures to remediate pesticide contaminated soils shows great potential for their application in bioremediation. Prolonged exposure of consortia to these xenobiotics makes them more efficient to degrade the contaminants. Native consortia are formed naturally and, therefore, they have the greatest potential for degradation; this is the main advantage of using a microbial consortium for bioremediation processes. However, defined consortia prepared in the laboratory can present more efficient degradation kinetics and tolerate higher concentrations of contaminants. Even so, formulation of harmonious consortia that do not compete with each other is difficult. Nevertheless, several researchers have used defined and native microbial consortia for the bioremediation of organochlorine pesticides in different matrices, consisting of bacteria, fungi or mixtures of both, in a free or immobilized way, evaluating the release of chloride ions as fundamental step for the degradation of these compounds. Among them, microbial consortia, especially those formed by actinobacteria, have a great potential to bioremediate matrices contaminated with OPs.

Acknowledgments

The authors gratefully acknowledge the financial assistance of Secretaría de Ciencia, Arte e Innovación Tecnológica (SCAIT), Consejo Nacional de Investigaciones Científicas y Técnicas (CONICET, PIO CONICET-YPF 13320130100022CO) and Agencia Nacional de Promoción Científica y Tecnológica (ANPCyT, PICT 2013 N° 0141, PICT 2014 N° 2893), Argentina. We also thank Mr. Borchia and Mrs. Colombres for their technical assistance.

References Cited

Abdul Salam, J., V. Lakshmi, D. Das and N. Das. 2013. Biodegradation of lindane using a novel yeast strain, *Rhodotorula* sp. VITJzN03 isolated from agricultural soil. World J. Microbiol. Biotechnol. 29(3): 475–487.

Abhilash, P.C. and N. Singh. 2008. Influence of the application of sugarcane bagasse on lindane (γ-HCH) mobility through soil column: Implication for biotreatment. Bioresour. Technol. 99(18): 8961–8966.

Abraham, J. and S. Silambarasan. 2014. Biomineralization and formulation of endosulfan degrading bacterial and fungal consortiums. Pestic. Biochem. Physiol. 116: 24–31.

Adams, G.O., P.T. Fufeyin, S.E. Okoro and I. Ehinomen. 2015. Bioremediation, biostimulation and bioaugmetation: a review. Int. J. Environ. Bior. Biod. 3(1): 28–39.

Alfonso, L.F., G.V. Germán, P.C.M. del Carmen and G. Hossein. 2017. Adsorption of organophosphorus pesticides in tropical soils: The case of karst landscape of northwestern Yucatan. Chemosphere 166: 292–299.

Alvarez, A., C.S. Benimeli, J.M. Saez, M.S. Fuentes, S.A. Cuozzo, M.A. Polti et al. 2012. Bacterial bio-resources for remediation of hexachlorocyclohexane. Int. J. Mol. Sci. 13(11): 15086–15106.

Alvarez, A., J.M. Saez, J.S. Dávila Costa, V.L. Colin, M.S. Fuentes, S.A. Cuozzo et al. 2017. Actinobacteria: Current research and perspectives for bioremediation of pesticides and heavy metals. Chemosphere 166: 41–62.

Awasthi, N., N. Manickam and A. Kumar. 1997. Biodegradation of endosulfan by a bacterial coculture. Bull. Environ. Contam. Toxicol. 59(6): 928–934.

Awasthi, N., A.K. Singh, R.K. Jain, B.S. Khangarot and A. Kumar. 2003. Degradation and detoxification of endosulfan isomers by a defined co-culture of two *Bacillus* strains. Appl. Microbiol. Biotechnol. 62(2-3): 279–283.

Aparicio, J.D., M.Z. Simón Solá, C.S. Benimeli, M.J. Amoroso and M.A. Polti. 2015. Versatility of *Streptomyces* sp. M7 to bioremediate soils co-contaminated with Cr (VI) and lindane. Ecotoxicol. Environ. Saf. 116: 34–39.

Baczynski, T.P. and D. Pleissner. 2010. Bioremediation of chlorinated pesticide-contaminated soil using anaerobic sludges and surfactant addition. J. Environ. Sci. Health B. 45(1): 82–88.

Ballesteros, M.L., K.S.B. Miglioranza, M. Gonzalez, G. Fillmann, D.A. Wunderlin and M.A. Bistoni. 2014. Multimatrix measurement of persistent organic pollutants in Mar Chiquita, a continental saline shallow lake. Sci. Total. Environ. 490: 73–80.

Barakat, A.O., M.A. Khairy and M.R. Mahmoud. 2017. Organochlorine pesticides and polychlorinated biphenyls in sewage sludge from Egypt. J. Environ. Sci. Health, Part A 1–7.

Benimeli, C.S., M.J. Amoroso, A.P. Chaile and G.R. Castro. 2003. Isolation of four aquatic streptomycetes strains capable of growth on organochlorine pesticides. Bioresour. Technol. 89(2): 133–138.

Berns, A.E., H. Philipp, H.D. Narres, P. Burauel, H. Vereecken and W. Tappe. 2008. Effect of gamma-sterilization and autoclaving on soil organic matter structure as studied by solid state NMR, UV and fluorescence spectroscopy. Eur. J. Soil Sci. 59(3): 540–550.

Bidlan, R. and H.K. Manonmani. 2002. Aerobic degradation of dichlorodiphenyltrichloroethane (DDT) by *Serratia marcescens* DT-1P. Process Biochem. 38(1): 49–56.

Bourguignon, N., P. Isaac, H. Alvarez, M.J. Amoroso and M.A. Ferrero. 2014. Enhanced polyaromatic hydrocarbon degradation by adapted cultures of actinomycete strains. J. Basic Microbiol. 54(12): 1288–1294.

Briceño, G., M.S. Fuentes, O. Rubilar, M. Jorquera, G. Tortella, G. Palma et al. 2015. Removal of the insecticide diazinon from liquid media by free and immobilized *Streptomyces* sp. isolated from agricultural soil. J. Basic Microbiol. 55(3): 293–302.

Camacho-Pérez, B., E. Ríos-Leal, N. Rinderknecht-Seijas and H.M. Poggi-Varaldo. 2012. Enzymes involved in the biodegradation of hexachlorocyclohexane: a mini review. J. Environ. Manage. 95: S306–S318.

Cao, X., C. Yang, R. Liu, Q. Li, W. Zhang, J. Liu et al. 2013. Simultaneous degradation of organophosphate and organochlorine pesticides by *Sphingobium japonicum* UT26 with surface-displayed organophosphorus hydrolase. Biodegradation 24(2): 295–303.

Carrillo-Pérez, E., R.M. Arturo and H. Yeomans-Reina. 2004. Aislamiento, identificación y evaluación de un cultivo mixto de microorganismos con capacidad para degradar DDT. Revista Internacional de Contaminación Ambiental. 20(2): 69–75.

Castillo, M.A., N. Felis, P. Aragon, G. Cuesta and C. Sabater. 2006. Biodegradation of the herbicide diuron by streptomycetes isolated from soil. Int. Biodeter. Biodegr. 58(3): 196–202.

Chaile, A.P., N. Romero, M.J. Amoroso, M.D.V. Hidalgo and M.C. Apella. 1999. Organochlorine pesticides in Sali River. Tucumán-Argentina. Rev. Boliv. Ecol. 6: 203–209.

Chand, S., M.D. Mustafa, B.D. Banerjee and K. Guleria. 2014. CYP17A1 gene polymorphisms and environmental exposure to organochlorine pesticides contribute to the risk of small for gestational age. Eur. J. Obstet. Gynecol. Reprod. Biol. 180: 100–105.

Chia, V.M., Y. Li, S.M. Quraishi, B.I. Graubard, J.D. Figueroa, J.P. Weber et al. 2010. Effect modification of endocrine disruptors and testicular germ cell tumour risk by hormone-metabolizing genes. Int. J. Androl. 33(4): 588–596.

Chishti, Z., S. Hussain, K.R. Arshad, A. Khalid and M. Arshad. 2013. Microbial degradation of chlorpyrifos in liquid media and soil. J. Environ. Manage. 114: 372–380.

Chiu, T.C., J.H. Yen, T.L. Liu and Y.S. Wang. 2004. Anaerobic degradation of the organochlorine pesticides DDT and heptachlor in river sediment of Taiwan. Bull. Environ. Contam. Toxicol. 72(4): 821–828.

Commendatore, M.G., M.A. Franco, P. Gomes Costa, I.B. Castro, G. Fillmann, G. Bigatti et al. 2015. Butyltins, polyaromatic hydrocarbons, organochlorine pesticides, and polychlorinated biphenyls in sediments and bivalve mollusks in a mid-latitude environment from the Patagonian coastal zone. Environ. Toxicol. Chem. 34(12): 2750–2763.

Costa, L.G. 2015. The neurotoxicity of organochlorine and pyrethroid pesticides. Occup. Neurol. 131: 135.

Corona-Cruz, A., G. Gold-Bouchot, M. Gutiérrez-Rojas, O. Monroy-Hermosillo and E. Favela. 1999. Anaerobic-aerobic biodegradation of DDT (dichlorodiphenyl trichloroethane) in soils. Bull. Environ. Contam. Toxicol. 63(2): 219–225.

Cremonese, C., C. Piccoli, F. Pasqualotto, R. Clapauch, R.J. Koifman, S. Koifman et al. 2017. Occupational exposure to pesticides, reproductive hormone levels and sperm quality in young Brazilian men. Reprod. Toxicol. DOI: 10.1016/j.reprotox.2017.01.001.

Cui, X., W. Hunter, Y. Yang, Y. Chen and J. Gan. 2011. Biodegradation of pyrene in sand, silt and clay fractions of sediment. Biodegradation 22(2): 297–307.

Cuozzo, S.A., M.S. Fuentes, N. Bourguignon, C.S. Benimeli and M.J. Amoroso. 2012. Chlordane biodegradation under aerobic conditions by indigenous *Streptomyces* strains. Int. Biodeter. Biodegr. 66(1): 19–24.

Cycoń, M., A. Żmijowska, M. Wójcik and Z. Piotrowska-Seget. 2013. Biodegradation and bioremediation potential of diazinon-degrading *Serratia marcescens* to remove other organophosphorus pesticides from soils. J. Environ. Manage. 117: 7–16.

Dearth, M.A. and R.A. Hites. 1990. Highly chlordane dimethano fluorenes in technical chlordane and in human adipose tissue. J. Am. Soc. Mass Spectrom. 1(1): 99–103.

Dietz, R., F.F. Riget, C. Sonne, R. Letcher, E.W. Born and D.C.G. Muir. 2004. Seasonal and temporal trends in polychlorinated biphenyls and organochlorine pesticides in East Greenland polar bears (*Ursus maritimus*), 1990–2001. Sci. Total Environ. 331(1): 107–124.

Durante, C.A., E.B. Santos-Neto, A. Azevedo, E.A. Crespo and J. Lailson-Brito. 2016. POPs in the South Latin America: Bioaccumulation of DDT, PCB, HCB, HCH and Mirex in blubber of common dolphin (*Delphinus delphis*) and Fraser's dolphin (*Lagenodelphis hosei*) from Argentina. Sci. Total. Environ. 572: 352–360.

Đurović, R., J. Gajić-Umiljendić and T. Đorđević. 2009. Effects of organic matter and clay content in soil on pesticide adsorption processes. Pestic. Fitomed. 24(1): 51–57.

Elbashir, A.B., A.O. Abdelbagi, A.M. Hammad, G.A. Elzorgani and M.D. Laing. 2015. Levels of organochlorine pesticides in the blood of people living in areas of intensive pesticide use in Sudan. Environ. Monit. Assess. 187(3): 68.

Elcey, C.D. and A.A. Mohammad Kunhi. 2009. Substantially enhanced degradation of hexachlorocyclohexane isomers by a microbial consortium on acclimation. J. Agric. Food Chem. 58(2): 1046–1054.

Eldakroory, S.A., D.E. Morsi, R.H. Abdel-Rahman, S. Roshdy, M.S. Gouida and E.O. Khashaba. 2016. Correlation between toxic organochlorine pesticides and breast cancer. Hum. Exp. Toxicol. 1–9.

FAO. 2002. Food and Agriculture Organization of the United Nations. International code of conduct on the distribution and use of pesticides. Retrieved on 2007-10-25.

Fogel, S., R.L. Lancione and A.E. Sewall. 1982. Enhanced biodegradation of methoxychlor in soil under sequential environmental conditions. Appl. Environ. Microbiol. 44(1): 113–120.

Fong, S.S., A.R. Joyce and B.Ø. Palsson. 2005. Parallel adaptive evolution cultures of *Escherichia coli* lead to convergent growth phenotypes with different gene expression states. Genome Res. 15(10): 1365–1372.

Frye, C.A., E. Bo, G. Calamandrei, L. Calzà, F. Dessì-Fulgheri, M. Fernández et al. 2012. Endocrine disrupters: a review of some sources, effects, and mechanisms of actions on behaviour and neuroendocrine systems. J. Neuroendocrinol. 24(1): 144–159.

Fuentes, M.S., C.S. Benimeli, S.A. Cuozzo and M.J. Amoroso. 2010. Isolation of pesticide-degrading actinomycetes from a contaminated site: bacterial growth, removal and dechlorination of organochlorine pesticides. Int. Biodeter. Biodegr. 64(6): 434–441.

Fuentes, M.S., J.M. Sáez, C.S. Benimeli and M.J. Amoroso. 2011. Lindane biodegradation by defined consortia of indigenous *Streptomyces* strains. Water Air Soil Pollut. 222(1-4): 217–231.

Fuentes, M.S., A. Alvarez, J.M. Saez, C.S. Benimeli and M.J. Amoroso. 2013a. Methoxychlor bioremediation by defined consortium of environmental *Streptomyces* strains. Int. J. Environ. Sci. Technol. 11(4): 1147–1156.

Fuentes, M.S., G.E. Briceño, J.M. Saez, C.S. Benimeli, M.C. Diez and M.J. Amoroso. 2013b. Enhanced removal of a pesticides mixture by single cultures and consortia of free and immobilized *Streptomyces* strains. Biomed. Res. Int. 2013: 1–9.

Fuentes, M.S., V.L. Colin, M.J. Amoroso and C.S. Benimeli. 2016. Selection of an actinobacteria mixed culture for chlordane remediation. Pesticide effects on microbial morphology and bioemulsifier production. J. Basic. Microbiol. 56(2): 127–137.

Fuentes, M.S., E.E. Raimondo, M.J. Amoroso and C.S. Benimeli. 2017. Removal of a mixture of pesticides by a *Streptomyces* consortium: Influence of different soil systems. Chemosphere 173: 359–367.

Gerhardt, K.E., X.D. Huang, B.R. Glick and B.M. Greenberg. 2009. Phytoremediation and rhizoremediation of organic soil contaminants: potential and challenges. Plant Sci. 176(1): 20–30.

Gomes, M.P., S.G. Le Manac'h, S. Maccario, M. Labrecque, M. Lucotte and P. Juneau. 2016. Differential effects of glyphosate and aminomethylphosphonic acid (AMPA) on photosynthesis and chlorophyll metabolism in willow plants. Pest. Biochem. Physiol. 130: 65–70.

Gómez Cruz, R. 2010. Biotecnología ambiental: Un acercamiento a la química y a los compuestos xenobióticos. Kuxulkab. 30: 77–79.

Gondar, D., R. López, J. Antelo, S. Fiol and F. Arce. 2013. Effect of organic matter and pH on the adsorption of metalaxyl and penconazole by soils. J. Hazard. Mater. 260: 627–633.

Gonzalez, M., K.S. Miglioranza, J.E. Aizpún, F.I. Isla and A. Peña. 2010. Assessing pesticide leaching and desorption in soils with different agricultural activities from Argentina (Pampa and Patagonia). Chemosphere 81(3): 351–358.

Hamer, G. 1997. Microbial consortia for multiple pollutant biodegradation. Pure Appl. Chem. 69(11): 2343–2356.

Hechmi, N., L. Bosso, L. El-Bassi, R. Scelza, A. Testa, N. Jedidi et al. 2016. Depletion of pentachlorophenol in soil microcosms with *Byssochlamys nivea* and *Scopulariopsis brumptii* as detoxification agents. Chemosphere 165: 547–554.

Hero, J.S., J.H. Pisa, N.I. Perotti, C.M. Romero and M.A. Martínez. 2017. Endoglucanase and xylanase production by *Bacillus* sp. AR03 in co-culture. Prep. Biochem. Biotechnol. 1–8.

Hirano, T., T. Ishida, K. Oh and R. Sudo. 2007. Biodegradation of chlordane and hexachlorobenzenes in river sediment. Chemosphere 67(3): 428–434.

Ho, A., R. Angel, A.J. Veraart, A. Daebeler, Z. Jia, S.Y. Kim et al. 2016. Biotic interactions in microbial communities as modulators of biogeochemical processes: Methanotrophy as a Model System. Front. Microbiol. 7: 1–11.

Jiang, Y., K.J. Brassington, G. Prpich, G.I. Paton, K.T. Semple, S.J. Pollard et al. 2016. Insights into the biodegradation of weathered hydrocarbons in contaminated soils by bioaugmentation and nutrient stimulation. Chemosphere 161: 300–307.

Jin, M., J. Fu, B. Xue, S. Zhou, L. Zhang and A. Li. 2017. Distribution and enantiomeric profiles of organochlorine pesticides in surface sediments from the Bering Sea, Chukchi Sea and adjacent Arctic areas. Environ. Pollut. DOI: 10.1016/j.envpol.2016.12.075.

Kamel, E., S. Moussa, M.A. Abonorag and M. Konuk. 2015. Occurrence and possible fate of organochlorine pesticide residues at Manzala Lake in Egypt as a model study. Environ. Monit. Assess. 187(1): 4161.

Karagouni, A.D., A.P. Vionis, P.W. Baker and E.M.H. Wellington. 1993. The effect of soil moisture content on spore germination, mycelium development and survival of a seeded streptomycete in soil. Microb. Releases. 2: 47–51.

Kataoka, R., K. Takagi, I. Kamei, H. Kiyota and Y. Sato. 2010. Biodegradation of dieldrin by a soil fungus isolated from a soil with annual endosulfan applications. Environ. Sci. Technol. 44(16): 6343–6349.

Kato, S., S. Haruta, Z.J. Cui, M. Ishii and Y. Igarashi. 2005. Stable coexistence of five bacterial strains as a cellulose-degrading community. Appl. Environ. Microbiol. 71(11): 7099–7106.

Kim, Y.S., M.C. Kim, S.W. Kwon, S.J. Kim, I.C. Park, J.O. Ka et al. 2011. Analyses of bacterial communities in meju, a Korean traditional fermented soybean bricks, by cultivation-based and pyrosequencing methods. J. Microbiol. 49(3): 340–348.

Kulshrestha, G. and A. Kumari. 2010. Simultaneous degradation of mixed insecticides by mixed fungal culture isolated from sewage sludge. J. Agric. Food Chem. 58(22): 11852–11856.

Kumar, M. and L. Philip. 2006. Enrichment and isolation of a mixed bacterial culture for complete mineralization of endosulfan. J. Environ. Sci. Health B. 41(1): 81–96.

Kumar, M. and L. Philip. 2007. Biodegradation of endosulfan-contaminated soil in a pilot-scale reactor-bioaugmented with mixed bacterial culture. J. Environ. Sci. Health B. 42(6): 707–715.

Lafuente, A., T. Cabaleiro, A. Caride, A. Gutiérrez and A.I. Esquifino. 2007. Toxic effects of methoxychlor in rat striatum: modifications in several neurotransmitters. J. Physiol. Biochem. 63(2): 171–178.

Lee, S.M., J.W. Lee, K.R. Park, E.J. Hong, E.B. Jeung, M.K. Kim et al. 2006. Biodegradation of methoxychlor and its metabolites by the white rot fungus *Stereum hirsutum* related to the inactivation of estrogenic activity. J. Environ. Sci. Health B. 41(4): 385–397.

Lee, D.H. and B.Ø. Palsson. 2010. Adaptive evolution of *Escherichia coli* K-12 MG1655 during growth on a nonnative carbon source, L-1, 2-propanediol. Appl. Environ. Microbiol. 76(13): 4158–4168.

Lee, D.J., K.Y. Show and A. Wang. 2013. Unconventional approaches to isolation and enrichment of functional microbial consortium—a review. Bioresour. Technol. 136: 697–706.

Li, X., L. Yang, U. Jans, M.E. Melcer and P. Zhang. 2007. Lack of enantioselective microbial degradation of chlordane in Long Island Sound sediment. Environ. Sci. Technol. 41(5): 1635–1640.

Lupi, L., F. Bedmar, D.A. Wunderlin and K.S. Miglioranza. 2016. Organochlorine pesticides in agricultural soils and associated biota. Environ. Earth Sci. 75(6): 1–11.

Manickam, N., M.K. Reddy, H.S. Saini and R. Shanker. 2008. Isolation of hexachlorocyclohexane-degrading *Sphingomonas* sp. by dehalogenase assay and characterization of genes involved in γ-HCH degradation. J. Appl. Microbiol. 104(4): 952–960.

Martínez-López, E., S. Espín, F. Barbar, S.A. Lambertucci, P. Gómez-Ramírez and A.J. García-Fernández. 2015. Contaminants in the southern tip of South America: Analysis of organochlorine compounds in feathers of avian scavengers from Argentinean Patagonia. Ecotoxicol. Environ. Saf. 115: 83–92.

McDonald, B.A. and E.H. Stukenbrock. 2016. Rapid emergence of pathogens in agro-ecosystems: global threats to agricultural sustainability and food security. Phil. Trans. R. Soc. B. 371 (1709).

Mishra, K., R.C. Sharma and S. Kumar. 2012. Contamination levels and spatial distribution of organochlorine pesticides in soils from India. Ecotoxicol. Environ. Saf. 76: 215–225.

Nagata, Y., A. Futamura, K. Miyauchi and M. Takagi. 1999. Two different types of dehalogenases, LinA and LinB, involved in γ-Hexachlorocyclohexane degradation in *Sphingomonas paucimobilis* UT26 are localized in the periplasmic space without molecular processing. J. Bacteriol. 181(17): 5409–5413.

Nagata, Y., R. Endo, M. Ito, Y. Ohtsubo and M. Tsuda. 2007. Aerobic degradation of lindane (γ-hexachlorocyclohexane) in bacteria and its biochemical and molecular basis. Appl. Microbiol. Biotechnol. 76(4): 741.

Nestler, C., L.D. Hansen, D. Ringelberg and J.W. Talley. 2001. Remediation of soil PAH: comparison of biostimulation and bioaugmentation. pp. 43–50. *In*: Sixth International *In Situ* and On Site Bioremediation Symposium. San Diego, California.

Ondarza, P.M., M. Gonzalez, G. Fillmann and K.S.B. Miglioranza. 2014. PBDEs, PCBs and organochlorine pesticides distribution in edible fish from Negro River basin, Argentinean Patagonia. Chemosphere 94: 135–142.

ONU. 1992. Organización de las Naciones Unidas. Convenio sobre diversidad biológica. Artículo 2. pp 3. www.cbd.int/.

Pan, H.W., H.J. Lei, X.S. He, B.D. Xi, Y.P. Han and Q.G. Xu. 2016. Levels and distributions of organochlorine pesticides in the soil–groundwater system of vegetable planting area in Tianjin City, Northern China. Environ. Geochem. Health. 1–13.

Perruchon, C., A. Chatzinotas, M. Omirou, S. Vasileiadis, U. Menkissoglou-Spiroudi and D.G. Karpouzas. 2017. Isolation of a bacterial consortium able to degrade the fungicide thiabendazole: the key role of a *Sphingomonas phylotype*. Appl. Microbiol. Biotechnol. 1–13.

Phillips, T.M., A.G. Seech, H. Lee and J.T. Trevors. 2001. Colorimetric assay for lindane dechlorination by bacteria. J. Microbiol. Methods 47(2): 181–188.

Pino, N.J., M.C. Domínguez and G.A. Penuela. 2011. Isolation of a selected microbial consortium capable of degrading methyl parathion and p-nitrophenol from a contaminated soil site. J. Environ. Sci. Health, Part B. 46(2): 173–180.

Ploteau, S., J.P. Antignac, C. Volteau, P. Marchand, A. Vénisseau, V. Vacher et al. 2016. Distribution of persistent organic pollutants in serum, omental, and parietal adipose tissue of French women with deep infiltrating endometriosis and circulating versus stored ratio as new marker of exposure. Environ. Int. 97: 125–136.

Polanco Rodríguez, Á.G., M.I. Riba López, T.A. Del Valls Casillas, J.A. Araujo León, B.A. Kumar Prusty and F.J.A. Cervera. 2017. Levels of persistent organic pollutants in breast milk of Maya women in Yucatan, Mexico. Environ. Monit. Assess. 189(2): 59.

Polti, M.A., J.D. Aparicio, C.S. Benimeli and M.J. Amoroso. 2014. Simultaneous bioremediation of Cr (VI) and lindane in soil by actinobacteria. Int. Biodeter. Biodegr. 88: 48–55.

Rajashekara Murthy, H. and H.K. Manonmani. 2007. Aerobic degradation of technical hexachlorocyclohexane by a defined microbial consortium. J. Hazard. Mater. 149(1): 18–25.

Rama Krishna, K. and L. Philip. 2008. Adsorption and desorption characteristics of lindane, carbofuran and methyl parathion on various Indian soils. J. Hazard. Mater. 160(2): 559–567.

Rama Krishna, K. and L. Philip. 2011. Bioremediation of single and mixture of pesticide-contaminated soils by mixed pesticide-enriched cultures. Appl. Biochem. Biotechnol. 164(8): 1257–1277.

Rani, M., U. Shanker and V. Jassal. 2017. Recent strategies for removal and degradation of persistent & toxic organochlorine pesticides using nanoparticles: A review. J. Environ. Manage. 190: 208–222.

Rezende dos Santos, A.P., T.L. Rocha, C.L. Borges, A.M. Bailão, C.M. de Almeida Soares and S.M.T. de Sabóia-Morais. 2017. A glyphosate-based herbicide induces histomorphological and protein expression changes in the liver of the female guppy *Poecilia reticulata*. Chemosphere 168: 933–943.

Robles-González, I.V., F. Fava and H.M. Poggi-Varaldo. 2008. A review on slurry bioreactors for bioremediation of soils and sediments. Microb. Cell Fact. 7(1): 5.

Robles-González, I.V., E. Ríos-Leal, I. Sastre-Conde, F. Fava, N. Rinderknecht-Seijas and H.M. Poggi-Varaldo. 2012. Slurry bioreactors with simultaneous electron acceptors for bioremediation of an agricultural soil polluted with lindane. Process Biochem. 47(11): 1640–1648.

Rouimi, P., N. Zucchini-Pascal, G. Dupont, A. Razpotnik, E. Fouché, G. De Sousa et al. 2012. Impacts of low doses of pesticide mixtures on liver cell defence systems. Toxicol. *In Vitro*. 26(5): 718–726.

Saez, J.M., A. Álvarez, C.S. Benimeli and M.J. Amoroso. 2014. Enhanced lindane removal from soil slurry by immobilized *Streptomyces* consortium. Int. Biodeter. Biodegr. 93: 63–69.

Saez, J.M., J.D. Aparicio, M.J. Amoroso and C.S. Benimeli. 2015. Effect of the acclimation of a *Streptomyces* consortium on lindane biodegradation by free and immobilized cells. Process Biochem. 50(11): 1923–1933.

Sánchez-Osorio, J.L., J.V. Macías-Zamora, N. Ramírez-Álvarez and T.F. Bidleman. 2017. Organochlorine pesticides in residential soils and sediments within two main agricultural areas of northwest Mexico: Concentrations, enantiomer compositions and potential sources. Chemosphere DOI: 10.1016/j.chemosphere.2017.01.010.

Sardrood, B.P., E.M. Goltapeh and A. Varma. 2013. An introduction to bioremediation. pp. 3–27. *In*: Goltapeh, E.M., Y.R. Danesh and A. Varma [eds.]. Fungi as Bioremediators. Springer, Berlin, Heidelberg.

Satsuma, K. and M. Masuda. 2012. Reductive dechlorination of methoxychlor by bacterial species of environmental origin: evidence for primary biodegradation of methoxychlor in submerged environments. J. Agric. Food Chem. 60(8): 2018–2023.

Shao, W.T. and A.H. Gu. 2016. Effects of organochloride pesticides on dyslipidemias. Chin. J. Prev. Med. 50(11): 1011.

Shelton, D.R., S. Khader, J.S. Karns and B.M. Pogell. 1996. Metabolism of twelve herbicides by *Streptomyces*. Biodegradation 7(2): 129–136.

Silva Barni, M.F., P.M. Ondarza, M. Gonzalez, R. Da Cuña, F. Meijide, F. Grosman et al. 2016. Persistent organic pollutants (POPs) in fish with different feeding habits inhabiting a shallow lake ecosystem. Sci. Total Environ. 550: 900–909.

Sineli, P.E., G. Tortella, J.S. Costa, C.S. Benimeli and S.A. Cuozzo. 2016. Evidence of α-, β- and γ-HCH mixture aerobic degradation by the native actinobacteria *Streptomyces* sp. M7. World J. Microbiol. Biotechnol. 32(5): 1–9.

Siripattanakul, S., W. Wirojanagud, J. McEvoy, T. Limpiyakorn and E. Khan. 2009. Atrazine degradation by stable mixed cultures enriched from agricultural soil and their characterization. J. Appl. Microbiol. 106(3): 986–992.

Sruthi, S.N., M.S. Shyleshchandran, S.P. Mathew and E.V. Ramasamy. 2017. Contamination from organochlorine pesticides (OCPs) in agricultural soils of Kuttanad agroecosystem in India and related potential health risk. Environ. Sci. Pollut. Res. 24(1): 969–978.

Teng, Y., X. Wang, Y. Zhu, W. Chen, P. Christie, Z. Li et al. 2017. Biodegradation of pentachloronitrobenzene by *Cupriavidus* sp. YNS-85 and its potential for remediation of contaminated soils. Environ. Sci. Pollut. Res. Int. 1–10.

Torres, P., K.S. Miglioranza, M.M. Uhart, M. Gonzalez and M. Commendatore. 2015. Organochlorine pesticides and PCBs in southern right whales (*Eubalaena australis*) breeding at Península Valdés, Argentina. Sci. Total. Environ. 518: 605–615.

Tsygankov, V.Y., M.D. Boyarova, P.F. Kiku and M.V. Yarygina. 2015. Hexachlorocyclohexane (HCH) in human blood in the south of the Russian Far East. Environ. Sci. Pollut. Res. Int. 22(18): 14379–14382.

UNEP. 2016. Listing of original and added persistent organic pollutants (POPs) in the Stockholm Convention. http://chm.pops.int/TheConvention/ThePOPs/tabid/673/Default.aspx (accessed July 17, 2016).

VoPham, T., K.A. Bertrand, J.E. Hart, F. Laden, M.M. Brooks, J.M. Yuan et al. 2017. Pesticide exposure and liver cancer: a review. Cancer Causes Control. 1–14.

Walsh, A.M., F. Crispie, K. Kilcawley, O. O'Sullivan, M.G. O'Sullivan, M.J. Claesson et al. 2016. Microbial succession and flavor production in the fermented dairy beverage Kefir. mSystems. 1(5): 1–16.

Wang, Y., Y. Fan and J.D. Gu. 2004. Dimethyl phthalate ester degradation by two planktonic and immobilized bacterial consortia. Int. Biodeter. Biodegr. 53(2): 93–101.

Wang, B., C. Wu, W. Liu, Y. Teng, Y. Luo, P. Christie et al. 2016. Levels and patterns of organochlorine pesticides in agricultural soils in an area of extensive historical cotton cultivation in Henan province, China. Environ. Sci. Pollut. Res. 23(7): 6680–6689.

WHO. 1990. World Health Organization. Environmental program on chemical safety. Environmental health criteria. 124, Lindane, World Health Organization, Geneva, Switzerland. http://www.inchem.org/documents/ehc/ehc/ehc124.htm.

WHO. 2003. World Health Organization. Hexachlorocyclohexanes. pp. 61–85. In Health risks of persistent organic pollutants from long-range trans-boundary air pollution. Joint WHO/Convention Task Force on the Health Aspects of Air Pollution. Geneva, Switzerland.

Wu, M., L. Chen, Y. Tian, Y. Ding and W.A. Dick. 2013. Degradation of polycyclic aromatic hydrocarbons by microbial consortia enriched from three soils using two different culture media. Environ. Pollut. 178: 152–158.

Yaduvanshi, S.K., N. Srivastava, F. Marotta, S. Jain and H. Yadav. 2012. Evaluation of micronuclei induction capacity and mutagenicity of organochlorine and organophosphate pesticides. Drug. Metab. Pharmacokinet. 6(3): 187–197.

Yang, C., Y. Li, K. Zhang, X. Wang, C. Ma, H. Tang et al. 2010. Atrazine degradation by a simple consortium of *Klebsiella* sp. A1 and *Comamonas* sp. A2 in nitrogen enriched medium. Biodegradation 21(1): 97–105.

Yim, Y., J. Seo, S. Kang, J. Ahn and H. Hur. 2008. Reductive dechlorination of methoxychlor and DDT by human intestinal bacterium *Eubacterium limosum* under anaerobic conditions. Arch. Environ. Contam. Toxicol. 54(3): 406–411.

Zamani, J., M.A. Hajabbasi and E. Alaie. 2015. The effect of steam sterilization of a petroleum-contaminated soil on PAH concentration and maize (*Zea mays* L.) growth. Int. J. Curr. Microbiol. App. Sci. 4(8): 93–104.

Zhang, Q., Z. Chen, Y. Li, P. Wang, C. Zhu, G. Gao et al. 2015. Occurrence of organochlorine pesticides in the environmental matrices from King George Island, west Antarctica. Environ. Pollut. 206: 142–149.

Zhang, S., Z. Hou, X.M. Du, D.M. Li and X.X. Lu. 2016. Assessment of biostimulation and bioaugmentation for removing chlorinated volatile organic compounds from groundwater at a former manufacture plant. Biodegradation 27(4–6): 223–236.

Zhao, M., K. Xue, F. Wang, S. Liu, S. Bai, B. Sun et al. 2014. Microbial mediation of biogeochemical cycles revealed by simulation of global changes with soil transplant and cropping. ISME J. 8(10): 2045–2055.

Zepeda-Arce, R., A.E. Rojas-García, A. Benitez-Trinidad, J.F. Herrera-Moreno, I.M. Medina-Díaz, B.S. Barrón-Vivanco et al. 2017. Oxidative stress and genetic damage among workers exposed primarily to organophosphate and pyrethroid pesticides. Environ. Toxicol. DOI: 10.1002/tox.22398.

8

Mycoremediation

Fungal Mediated Processes for the Elimination of Organic Pollutants

Carlos E. Rodríguez-Rodríguez,[1,*] *Víctor Castro-Gutiérrez*[1] and *Gonzalo Tortella*[2,3]

Introduction

This chapter focuses on solid-phase and liquid-phase fungal mediated processes applied in the bioremediation of real polluted matrices. Emphasis is given to two groups of organic pollutants: pesticides as traditional contaminants, and pharmaceutical compounds as a group of emerging contaminants. The first part of the chapter presents general considerations on the potential of fungi to be used in bioremediation processes; the second part includes a brief summary of the characteristics of solid-phase and liquid-phase strategies, including enzymatic approaches. The third part describes the most relevant findings regarding the application of such approaches in the elimination of pesticides and pharmaceutical compounds; in particular, the use of biopurification systems for pesticide degradation in agricultural wastewaters and slurry reactors and biopiles for the removal of pharmaceuticals from wastewater and sludge are explored in more detail.

Saprophytic fungi produce a wide array of extracellular enzymes, essential for the degradation of plant material, and which often permit the fungal degradation of organic contaminants (Kjøller and Struwe 2002). Because of their low substrate specificity, these enzymes allow these organisms to cometabolize structurally diverse compounds, belonging to different classes of contaminants (Harms et al. 2011). These characteristics, along with the ability of fungi to form extended mycelial networks, enable these organisms to be efficient for different xenobiotic bioremediation purposes (Harms et al. 2011). In an

[1] Centro de Investigación en Contaminación Ambiental (CICA), Universidad de Costa Rica, 2060 San José, Costa Rica.
Email: victormanuel.castro@ucr.ac.cr
[2] Departamento de Ingeniería Química, Universidad de La Frontera, Casilla 54-D, Temuco, Chile.
[3] Centro de Excelencia en Investigación Biotecnológica Aplicada al Medio Ambiente (CIBAMA), Universidad de La Frontera, Casilla 54-D, Temuco, Chile.
Email: gonzalo.tortella@ufrontera.cl
* Corresponding author: carlos.rodriguezrodriguez@ucr.ac.cr

alternate approach of microbial collaboration for pollutant removal, some authors even propose fungi as vectors for the dispersion of pollutant-degrading bacteria (Kohlmeier et al. 2005).

Several ecophysiological groups of fungi have been associated with bioremediation potential, including white-rot fungi, brown-rot fungi, litter fungi, soil fungi and mycorrhizal fungi (Anastasi et al. 2013). In particular, in the last few decades, white-rot fungi (WRF) have been regarded as important organisms for xenobiotic bioremediation due to their powerful enzymatic machinery (Pointing 2001, Reddy and Mathew 2001). WRF constitute a physiological group of fungi capable of lignin biodegradation; their name is due to the white appearance of wood when it is colonized by these organisms (Pointing 2001). This group is mainly composed of Basidiomycetes, nonetheless some Ascomycetes are also capable of carrying out this type of degradation (Eaton and Hale 1993).

Lignin degradation is in charge of lignin-modifying enzymes (LMEs), a group of non-specific extracellular enzymes (Martínez et al. 2005). The mode of action and high inespecificity permit LMEs to degrade a diverse range of organic environmental contaminants, including dyes, polycyclic aromatic hydrocarbons (PAH), chlorophenols, explosives and pesticides, among others (Korcan et al. 2013, Rodríguez-Rodríguez et al. 2013). Moreover, certain low molecular weight metabolites act as enzymatic mediators and improve contaminant degradation (Asgher et al. 2008). The ability of WRF to stand a wide range of pH values further improves its contaminant-degrading potential (Verma and Madamwar 2002). The list of WRF with proven capabilities of degrading organic contaminants includes *Phanerochaete chrysosporium*, *Pleurotus ostreatus*, *Bjerkandera adusta*, *Irpex lacteus* and *Trametes versicolor*, among others (Asgher et al. 2008).

LMEs are essential in the process of lignin degradation; the most important LMEs include the phenoloxydase laccase (Lac; E.C. 1.10.3.2) and the peroxidases lignin peroxidase (LiP; E.C. 1.11.1.14), manganese dependent peroxidase (MnP; E.C. 1.11.1.13) and versatile peroxidase (VP; E.C. 1.11.1.16). The oxidative mechanism of these enzymes has been comprehensively reviewed by Hofrichter et al. (2010), Martínez (2002) and Morozova et al. (2007). LME production is dependent on the fungal species, and enzymatic profiles vary according to different experimental conditions evaluated; Table 1 shows LMEs (Lac, MnP or LiP) commonly produced by different species of WRF. Typically, LMEs are produced in the secondary metabolism of WRF, and production requires nutrient limitations of C or N. Depending on growth conditions, several isoforms of the enzymes are expressed and some are considered constitutive while others have been determined to be inducible.

Even though the physiology of LMEs production and its relation with the degradation of pollutants have been extensively studied, removal of a contaminant does not necessarily imply a correlation with LMEs production; also, there are other mechanisms involved in the transformation of organic compounds. In this respect, the participation of the intracellular cytochrome P450 enzymatic complex in the oxidation of pollutants has been demonstrated through experiments that applied cytochrome P450 inhibitors or the overexpression of the cytochrome genes in the presence of pollutants (Bezalel et al. 1997, Doddapaneni and Yadav 2004).

The capacity of WRF to transform organic pollutants has been demonstrated for diverse chemicals. Their degrading scope includes compounds such as hydrocarbons, among which transformation of PAH has been widely described (Valentín et al. 2006), even at the level of field applications (Winquist et al. 2014); polychlorinated biphenyls (PCBs) (Čvančarová et al. 2012, Kamei et al. 2006); agrochemicals from several chemical structure groups such as triazines (Bastos and Magan 2009), organophosphorus pesticides (Karas et al. 2011), organochlorine pesticides (Kennedy et al. 1990), carbamates (Ruiz-Hidalgo et al.

Table 1. Selected white-rot fungi used in bioremediation, and their profile on the production of lignin modifying enzymes (Based on Hatakka 1994 and Novotný et al. 2000).

White-rot fungi	Lignin-modifying enzymes		
	Laccase	Manganese dependent peroxidase	Lignin peroxidase
Bjerkandera adusta	√	√	√
Dichomitus squalens	√	√	
Ganoderma australe	√	√	
Irpex lacteus	√	√	√
Lentinus tigrinus	√	√	
Phanerochaete chrysosporium		√	√
Phlebia radiata	√	√	√
Pleurotus ostreatus	√	√	
Pleurotus sajor-caju	√	√	
Stereum hirsutum	√	√	
Trametes versicolor	√	√	√

2014); and textile dyes (Kaushik and Malik 2009) among others. In recent decades, research on WRF has expanded to include the degradation of emerging pollutants, defined as compounds that are not currently covered by existing water-quality regulations, that have been scarcely studied before, and that could exert adverse effects on ecosystems and human health (Farré et al. 2008). Among these pollutants, the transformation of pharmaceutical and personal care products (PPCPs) (Rodarte-Morales et al. 2011, Rodríguez-Rodríguez et al. 2012a), brominated flame retardants (Uhnáková et al. 2009), ultraviolet filters (Gago-Ferrero et al. 2012) and endocrine disrupting compounds (Cajthalm et al. 2009) has been explored.

Bioremediation processes mediated by WRF have been applied for the treatment of several contaminated matrices, including soil, sewage sludge, and wastewaters from different origins (e.g., paper, olive oil or textile processing industries). Table 2 depicts examples of the use of WRF for the degradation of diverse organic contaminants from several polluted matrices.

Solid-Phase versus Liquid-Phase Approaches

Liquid-Phase Approaches

Liquid phase approaches may be used to degrade pollutants in matrices that originally have a mostly liquid composition or in solid matrices which are subsequently suspended in a liquid phase. In this type of systems there is an increased contact between contaminants/nutrients and the degrading microorganisms and also a higher mass transfer rate; these characteristics frequently lead to shorter treatment times when compared to solid phase strategies (Robles-González et al. 2008). The WRF *T. versicolor* has been employed for the successful removal of pharmaceuticals in different kinds of wastewater effluents, including hospital wastewaters (Cruz-Morató et al. 2014, Ferrando-Climent et al. 2015), urban wastewaters (Cruz-Morató et al. 2013), and sludge from the anaerobic digester of a wastewater treatment plant (Rodríguez-Rodríguez et al. 2012a).

Table 2. Diversity of organic pollutants removed by white-rot fungi on different contaminated matrices.

Group of organic pollutants	Organic pollutant	Fungi	Contaminated matrix	References
Textile dyes	Azo dyes (not specified)	*Thelephora* sp.	Industrial textile dye effluent	Selvam et al. 2003
	Two antraquinone dyes (not specified)	*Trametes versicolor*	Carpet dye effluent	Ramsay and Goode 2004
Polychlorinated biphenyls (PCBs)	Commercial mixture Delor 103	*Pleurotus ostreatus*	Soil	Kubátová et al. 2001
	Mixture of PCBs (not specified)	*Trametes versicolor, Phanerochaete chrysosporium, Lentinus edodes*	Extract from contaminated soil	Ruiz-Aguilar et al. 2002
Hydrocarbons	Complex mixture of substituted mono- and polycyclic aromatic hydrocarbons (PAHs)	*Phanerochaete chrysosporium, Pleurotus pulmonarius*	Aged contaminated soil	D'Annibale et al. 2005
	14 PAHs (components of creosote)	*Phanerochaete sórdida*	Soil contaminated with creosote	Davis et al. 1993
Components of munitions	2,4,6-trinitrotoluene (TNT), 1,3,5,7-tetranitroperhydro-1,3,5,7-tetrazocine (HMX), 1,3,5-trinitroperhydro-1,3,5-triazine (RDX)	*Pleurotus ostreatus*	Soil	Axtell et al. 2000
	TNT, HMX, RDX	*Phanerochaete chrysosporium*	Extract from contaminated soil	Spiker et al. 1992
Pesticides	Lindane (γ-hexachlorocyclohexane)	*Bjerkandera adusta*	Soil (treated in a slurry system)	Quintero et al. 2007
	Simazine, trifluralin, dieldrin	*Phanerochaete chrysosporium, Trametes versicolor*	Soil	Fragoerio and Magan 2008
Pharmaceuticals	Complex mixture of analgesics, antibiotics and psychiatric drugs	*Trametes versicolor*	Hospital wastewater	Cruz-Morató et al. 2014
	Fenofibrate, atorvastatin, hydrochlorothiazide, diclofenac, clarithromycin, ranitidine, ibuprofen, furosemide	*Trametes versicolor*	Sewage sludge	Rodríguez-Rodríguez et al. 2012c
Endocrine disrupting compounds	17β-estradiol, 17α-ethynylestradiol	*Trametes versicolor*	Synthetic wastewater	Blánquez and Guieysse 2008
	Nonylphenol, octylphenol, bisphenol A, ethynylestradiol	*Trametes* sp. (enzymatic treatment with laccase)	Sand	Tanaka et al. 2001
Brominated flame retardants	Several polybromodiphenylethers and decabromodiphenylethane	*Trametes versicolor*	Sewage sludge	Rodríguez-Rodríguez et al. 2012a

Popular configurations employed for this purpose include the slurry reactors (for the sludge) in the form of traditional stirred-tank reactors, and the air fluidized-bed bioreactors which are fluidized thanks to air pulses that provide oxygen for fungal metabolism and favor the homogenization of the system (Blánquez et al. 2008). Such configurations employ fungal biomass in the form of pellets (Blánquez et al. 2004), given among other reasons, to their potential reutilization in several cycles of degradation (Yesilada et al. 2003). These configurations have usually been applied for the reactor-scale treatment of other organic pollutants such as textile dyes (Borchert and Libra 2001) or polycyclic aromatic hydrocarbons (Valentín et al. 2007) in liquid phase systems.

Enzymatic reactors

Given that the ability of LMEs to non-specifically oxidize organic pollutants has been widely studied, the enzymatic reactors are included here as a particular case of liquid-phase approaches. The use of LMEs has led to the application of enzymatic reactors; nonetheless, as stated above, the degradation of pollutants by fungi may be mediated by enzyme complexes other than LMEs. In particular, the use of enzymatic reactors has been successfully demonstrated in the removal of textile dyes. Several systems have employed mediators, which are low molecular weight, diffusible redox molecules that provide high redox potentials to enhance the production of reactive radical species that attack xenobiotics (Wesenberg et al. 2003). Mendoza et al. (2011) used a membrane reactor system for the decolorization of several azo dyes with laccase from *T. versicolor* and syringaldehyde as mediator; similarly, the continuous decolorization of the Acid Violet 17 was achieved in an enzyme membrane reactor using laccase from *Cyathus bulleri* and 2,2′-azino-bis(3-ethylbenzothiazoline-6-sulfonic acid) (ABTS) as the mediator (Chhabra et al. 2009). López et al. (2004) employed MnP from *Bjerkandera* and Mn^{2+} as mediator in the elimination of Orange II in an enzymatic membrane reactor. In addition, immobilization of the enzymes is also a common practice, in order to increase the enzyme stability and the reusability of the catalyst, which can be easily recovered if immobilized (Sheldon 2007). The decolorization of Reactive Black 5 was achieved with laccase from *Trametes pubescens* immobilized on alumina pellets coated with polyelectrolytes in fluidized-bed and stirred tanks reactors (Osma et al. 2010). Similarly, Champagne and Ramsay (2010) explored the removal of anthraquinone and azo dyes in packed and fluidized-bed reactors using laccase from *T. versicolor* immobilized in silica beads, and Katuri et al. (2009) removed Acid Black 10BX in a membrane reactor containing laccase from *Pleurotus ostreatus* immobilized in chitosan.

Solid-Phase Approaches

Solid-phase systems are of particular interest in the case of fungal bioremediation processes since they mimic the natural conditions of growth and colonization of most filamentous fungi (Hölker and Lenz 2005). Compared to liquid-phase reactors, solid-phase systems are inexpensive and the energy requirements for aeration and mixing are considerably lower (Nano et al. 2003), making them cost-effective for long treatment periods, usually required for the removal of highly recalcitrant pollutants. Moreover, these systems frequently take advantage of the use of agricultural wastes, which can be used as carbon and energy sources for microbial growth (Hölker and Lenz 2005). In this respect, lignocellulosic residues have been used as carriers or co-substrate materials for fungal biomass in bioremediation strategies (Covino et al. 2010, Steffen and Tuomela 2011) and for the production of LMEs

(Rodríguez-Couto and Sanromán 2005, Rodríguez-Couto and Toca-Herrera 2006), as they favor the colonization and enzymatic activity of WRF (Sánchez 2009). Such lignocellulosic wastes include wheat grains or straw, wood chips, maize stalks, grape stalks, olive leaves, and rice husk among others (Acevedo et al. 2011, D'Annibale et al. 2005, Karanasios et al. 2010b, Rubilar et al. 2011, Ruiz-Hidalgo et al. 2016b). These substrates can be applied following a pre-colonization process by the fungus or may be added to contaminated soil (Covino et al. 2016) or sludge (Rodríguez-Rodríguez et al. 2010a, 2011) at the same time as the fungal inoculum; some works even report the development of an exploratory mycelium originated in chopped straw, to remove PAHs in the adjoining contaminated soil (Novotný et al. 1999, 2001). To improve the performance of these strategies, the addition of surfactants such as Triton X-100, sodium dodecyl sulfate, Tween-20 and Tween-80 to increase the solubility and bioavailability of some pollutants has been practiced for the bioremediation of contaminated soil (Garon et al. 2002, Leonardi et al. 2007, 2008).

Both, solid and liquid phase approaches have been assayed for the fungal degradation of pesticides and pharmaceutical compounds.

Case I: Removal of Pesticides by Fungal Approaches

Bioreactors for the Degradation of Pesticides

The transformation of pesticides by the action of fungi has been largely studied, and several reports describe the production of metabolites from insecticides (Kullman and Matsumara 1996, Mougin et al. 1996), herbicides (Mougin et al. 1994), and fungicides (Pothuluri et al. 2000), from liquid media in flask-scale laboratory assays. Nonetheless, the removal of pesticides from aqueous matrices at higher bioreactor scale has been scarcely explored. Rubilar et al. (2007) employed a 5l-stirred tank reactor to treat soil contaminated with the highly persistent pentachlorophenol, using the fungus *Anthracophyllum discolor*. The system was operated as a slurry containing 5% soil (w/v), and achieved removal of 62% and 81% after 14 d and 28 d (from an initial concentration of 250 mg kg^{-1}), representing a similar performance to that of the flask-scale assays. A similar reactor configuration was used for the removal of hexachlorocyclohexane (HCH) with *Bjerkandera adusta* (Quintero et al. 2007); different slurry loads and HCH concentrations were assayed, and at optimal conditions (slurry 10% w/v and 100 mg l^{-1} HCH), elimination of 94.5%, 78.5% and 66.1% were achieved after 30 d for γ-, α-, and δ-HCH isomers respectively. The performance of this reactor surpassed the efficiency observed at flask-scale. Tekere et al. (2002) described the elimination of the organochlorine pesticide lindane (γ-HCH) (up to 81%) in a packed-bed bioreactor using a non-identified WRF; likewise, *T. versicolor* was able to remove carbofuran in a fluidized-bed bioreactor (100 mg l^{-1}) during continuous operation at a hydraulic residence time of 23 hr (Castro-Gutiérrez et al. 2016).

Enzymatic Systems for the Elimination of Pesticides

The transformation of pesticides by fungal enzymes has been described for compounds such as methoxychlor (using MnP from *P. chrysosporium* and laccase from *T. versicolor*; Hirai et al. 2004), dichlorophen, bromoxynil and pentachlorophenol (using VP from *Bjerkandera adusta*; Dávila-Vázquez et al. 2005). However, as sometimes fungal degradation of pesticides is not dependent on LMEs, other enzymatic systems may also play important roles in the removal, as demonstrated for the case of carbofuran (Mir-Tutusaus et al. 2014). In this respect, the use of complete fungal cells provides a wider application spectrum, as intracellular complexes such as the cytochrome P450 may play important roles in the detoxification process.

Most studies employed laccase as catalyst in the elimination of pesticides by fungal enzymes. A work by Zhang et al. (2009) reported the use of laccase from *T. versicolor* immobilized on chitosan to remove 2,4-dichlorophenol, a compound employed in the synthesis of the herbicide 2,4-dichlorophenoxyacetic acid (2,4-D). Córdova Juárez et al. (2011) proposed the use of enzymatic extracts derived from spent substrates used for the cultivation of *Pleurotus pulmonaris*, to remove the fungicide chlorothalonil; the uncharacterized extracts proved to be highly efficient (100% removal) when used fresh. Enzymatic studies at reactor scale are scarce; moreover, application to real polluted matrices was not found in scientific literature. At reactor scale, Jolivalt et al. (2000) employed immobilized laccase (from *T. versicolor*) onto a hydrophilic polyvinylidene fluoride (PVDF) microfiltration membrane which operated as a membrane reactor in the elimination of the herbicide derivative *N'*,*N'*-(dimethyl)-*N*-(2-hydroxyphenyl)urea; the oxidation of the contaminant resulted in the formation of an insoluble metabolite that was retained by the membrane.

Biopurification Systems for the Treatment of Pesticide-Containing Wastewaters

Biopurification systems (BPS) were developed specifically for the treatment of pesticide-containing wastewaters produced in agricultural labors of pest control. BPS are focused on reducing point source contamination with pesticides, which is considered as the main route of environmental contamination with these agrochemicals; point source contamination is linked to pesticide leftovers and the production of wastewaters during filling and washing of pesticide application equipment (de Wilde et al. 2007). The biological treatment of pesticides in BPS takes place in their active core or biomixture, where a joint activity by microbial populations promotes an improved degradation of these compounds (Castillo et al. 2008). The degradation of numerous pesticides has been reported in BPS, including different chemical families of herbicides (Huete-Soto et al. 2017), fungicides (Tortella et al. 2013) and insecticides (Rodríguez-Rodríguez et al. 2017).

Biomixtures are made up by three main components: soil, a substrate rich in humic compounds, and a lignocellulosic residue. Soil is regarded as the main provider of microbial degrading communities, including bacteria and fungi. Amounts of soil as low as 0.5% (v/v), have been sufficient to establish proper degrading communities in biomixtures (Sniegowski et al. 2012). To improve their performance, it is desirable that soil used in biomixture preparation had been previously exposed to the target pesticides, in order to employ acclimated microbial populations to favor the degrading capacity, which is particularly important for pesticides that exhibit accelerated degradation. This phenomenon refers to the significant increase in the rate of degradation of a pesticide in soil following previous applications of the same or a structurally similar pesticide (Arbeli and Fuentes 2007). In this respect, enhanced removal and mineralization of linuron was achieved with the use of pesticide-primed soil (Sniegowski et al. 2011). Similarly, this process has been shown to occur after repeated applications of carbendazim (Tortella et al. 2013).

The aim of adding the humic-rich component is to increase the retention of pesticides in the biomixture, to avoid their leaching or their accumulation at the bottom layers of BPS and to enhance their homogeneous distribution; this substrate also provides moisture control and may play a role in pesticide degradation (Castillo et al. 2008, Castillo and Torstensson 2007). In the original biomixture configurations, the most common humic-rich component was peat (Karanasios et al. 2012), however, this material is not easily available (at low prices) everywhere, reason why its substitution by garden or urban composts has attracted interest (Fogg et al. 2003). Moreover, compost is capable of enhancing the removal capacity of biomixtures, as it has been demonstrated in the cases of bentazone, dimethoate, isoproturon and terbuthylazine, among others (Coppola et al. 2011, Karanasios et al. 2010a, Omirou et al. 2012).

The third component, the lignocellulosic waste, is intended to act as a substrate to promote the colonization and enzymatic activity of WRF (Castillo et al. 2008). Furthermore, fungal metabolism of this substrate releases carbon sources that can be more easily assimilated by other microbial communities than the original materials (de Wilde et al. 2007).

The most commonly employed composition of biomixtures is a volumetric ratio of 25:25:50 (soil:humic-rich component:lignocellulosic residue) (Castillo et al. 2008); nonetheless other compositions have been assessed including the use of mixtures containing only compost and lignocellulosic wastes (Coppola et al. 2007, 2011). Some authors suggest that the optimization of the composition according to the target pesticides provides better results, as demonstrated for the insecticide/nematicide carbofuran (Chin-Pampillo et al. 2015b).

Although BPS rely on the simultaneous biological activity of fungi and bacteria, the high content of lignocellulosic residues used has led to fungal bioaugmentation approaches being tested in order to enhance the performance of biomixtures. In this respect, their composition can be modified to enhance the fungal role in pesticide degradation. For instance, the use of peat instead of compost results in more acidic biomixtures, thus providing more favorable conditions for the activity of LMEs (Tavares et al. 2006). Biomixtures with pH values closer to neutrality such as those containing compost, promote higher metabolic activity of bacterial communities, in detriment of fungal communities. However, efficient removal of emerging organic pollutants has been achieved with *T. versicolor* in solid matrices of neutral pH (Rodríguez-Rodríguez et al. 2012a, 2014), and this parameter should not represent a complete limitation for fungal activity. Similarly, the lower content of C in compost with respect to peat, results in high C/N ratios in peat-based biomixtures and low C/N ratios in compost-based biomixtures; the N-limitation conditions in the former are known to enhance the production of LMEs (Eggert et al. 1996) and in general the cometabolic transformation of pesticides by fungal-mediated reactions (Karanasios et al. 2012). On the contrary, compost-based biomixtures are likely to favor bacterial transformation of the pesticides.

Bioaugmentation of biomixtures comprises an interesting approach to improve the efficiency of the matrix, and some successful bacterial-mediated processes have been described (Karas et al. 2016). However, the high amount of lignocellulosic wastes contained in biomixtures, which usually reaches up to 50% (Castillo et al. 2008), has made the option of fungal bioaugmentation with ligninolytic fungi an attractive, though seldom explored approach to enhance the performance of BPS (Rodríguez-Rodríguez et al. 2013).

The degradation of the insecticide carbofuran has been largely studied in biomixtures bioaugmented with *T. versicolor*. The use of bioaugmented peat-based biomixtures using rice husk as lignocellulosic substrate produced a 5-fold increase in the ^{14}C-carbofuran mineralization rate with respect to the non-bioaugmented biomixture; however, when peat was substituted by compost, the mineralization rate increased only by 18%. Similarly, bioaugmentation resulted in a decrease in carbofuran half-life in the peat-based matrix from 10.3 d to 3.4 d; while in the compost-based biomixture bioaugmentation increased this value from 3.8 d to 8.1 d. Also, bioaugmentation of the peat-based matrices promoted a reduced accumulation of the transformation products 3-ketocarbofuran and 3-hydroxycarbofuran (Chin-Pampillo et al. 2015a, Madrigal-Zúñiga et al. 2016).

The optimization of the biomixture to remove carbofuran in the bioaugmented system revealed an optimal volumetric composition of 27:43:30 (soil:humic-rich component:lignocellulosic substrate), which maximized degradation and minimized residual toxicity and accumulation of transformation products (Ruiz-Hidalgo et al. 2016a); this composition widely differs from the traditional volumetric proportion (25:25:50).

During pesticide treatment, these bioaugmented matrices were able to decrease the residual acute toxicity, though at a similar extent as the non-bioaugmented systems (Chin-Pampillo et al. 2015a, Madrigal-Zúñiga et al. 2016). Despite achieving high degradation of carbofuran, chronic toxic effects were still observed in tests on *Daphnia magna* and the fish *Oreochromis aureus* (Ruiz-Hidalgo et al. 2016b).

The similarity in chemical structure with carbofuran was exploited in a biomixture (optimized for carbofuran degradation), to remove other carbamates (aldicarb, methomyl and methiocarb); however, in this case the fungal bioaugmentation did not improve pesticide elimination (Rodríguez-Rodríguez et al. 2017).

The removal of other pesticides has been also improved in biomixtures after bioaugmentation with WRF. Pinto et al. (2016) described accelerated degradation of terbuthylazine, difenoconazole, diflufenican and pendimethalin when biomixtures lacking a humic-rich component were bioaugmented with *Lentinula edodes*; the authors identified the transformation product 2-hydroxyterbuthylazine as the dominant metabolite in the bioaugmented system. Similarly, the addition of immobilized *Stereum hirsutum*, *T. versicolor* and *Inonotus* sp. increased the elimination of atrazine in peat-based biomixtures; *S. hirsutum* yielded the best results, which were linked to higher respiration activity (Elgueta et al. 2016). Likewise, the removal of isoproturon improved from 78% to > 99% by bioaugmentation with *Phanerochaete chrysosporium* in peat-based biomixtures (von Wirén-Lehr et al. 2001). Bending et al. (2002) evaluated the efficiency of three WRF to remove several pesticides (metalaxyl, terbuthylazine, atrazine, diuron, iprodione, and chlorpyrifos) in a barley straw soil-compost biomixture; the bioagumentation with *S. hirsutum* resulted in the highest elimination (average 55.4% after 42 d) with respect to *T. versicolor* (average 46.2%) and *Hypholoma fasciculare* (average 38.1%) for every pesticide except atrazine and chlorpyrifos. In other cases, bioaugmentation has proven efficient only at specific conditions; in this respect, Bastos and Magan (2009) observed a significant contribution of *T. versicolor* in atrazine removal only at dry operation conditions in a clay soil-sawdust biomixture.

However, not all the fungal bioaugmentation studies have resulted in successful processes. During the bioaugmentation of a coconut fiber-compost-soil biomixture, *T. versicolor* did not improve the elimination of fungicides including several triazoles, metalaxyl, carbendazim and edifenphos, even after re-inoculation with fungal biomass (Murillo-Zamora et al. 2017). On the other hand, a similar approach resulted in a clear decrease in the removal capacity of the biomixture for the treatment of atrazine, terbuthylazine, terbutryn and chlorpyrifos when bioaugmentation was applied (Lizano-Fallas et al. 2017). Therefore, the feasibility of fungal bioaugmentation should be assayed for every biomixture composition and the target pesticides to be treated in the matrix.

Case II: Removal of Pharmaceuticals by Fungal-Mediated Processes

Removal of Pharmaceuticals from Wastewater and Sewage Sludge in Bioslurry Reactors

The few studies that report the removal of pharmaceuticals from wastewater using fungi usually make use of *T. versicolor*. Rodríguez-Rodríguez et al. (2012a) assayed the removal of pharmaceuticals at pre-existent concentrations using a stirred tank reactor (working volume: 6 l) containing a bioslurry made of sewage sludge obtained from the outlet of an anaerobic digester; the system was operated at 25°C in batch mode for 26 d, a period that mimicked the real residence time of the sludge in the anaerobic reactor of the wastewater treatment plant. On the other hand, Cruz-Morató et al. (2013, 2014) and Gros et al. (2014) employed 10 l air-pulsed fluidized-bed bioreactors to treat sterile and non-sterile urban and hospital wastewaters. The system was operated in batch conditions for 8 d; however, continuous additions of glucose and ammonium tartrate were applied in some cases. Table 3 shows some examples of selected pharmaceuticals removed from real matrices by different approaches using WRF or their enzymes.

The success on the removal of pharmaceutical compounds from polluted matrices largely varies depending on the chemical group and the removal conditions. Table 4 summarizes the degree of elimination of pharmaceuticals in liquid phase systems with real matrices.

In general terms, most of the pharmaceuticals were highly removed (> 75%), particularly antihypertensives, β-blockers, histamine H1 and H2 receptor antagonists and several hormones. Some analgesics/anti-inflammatories such as acetaminophen, codeine and ibuprofen were usually highly removed, while others such as diclofenac exhibited high or medium removal (35–75%) depending on the treatment. On the contrary, dexamethasone, piroxicam and salicylic acid were poorly removed (< 35%). In the case of ibuprofen, previous flask-scale assays demonstrated that transformation by *T. versicolor* produces 1,2-dihydroxyibuprofen, which is not further degraded and presents higher toxicity than the parental compound (Marco-Urrea et al. 2009).

A similar panorama was observed for the antibiotics: while many of them including the groups of sulfonamides and macrolides were efficiently eliminated, others such as cefalexine and dimetridazole were not removed at all. For the lipid regulators and diuretics, removals were over 41% and 50%, respectively in every case. The group of psychiatric drugs comprises several compounds of poor removal, such as carbamazepine and the benzodiazepines diazepam and lorazepam. The removal of the transformation products from carbamazepine (some previously demonstrated to be produced by fungal metabolism, Jelić et al. 2012), 10,11-epoxy carbamazepine and 2-hydroxy carbamazepine was highly successful, however, for acridone contrasting results were obtained depending on the treatment conditions. On the contrary, antidepressant drugs of the selective serotonin reuptake inhibitors class were mostly removed (100%), except citalopram in the specific condition of using sterile hospital wastewaters.

In an interesting approach, Accinelli et al. (2010) employed propagules of *P. chrysosporium* entrapped in a granular bioplastic formulation; this immobilized fungal biomass was able to increase the removal of the antiviral drug oseltamivir and the antibiotics erythromycin, sulfamethoxazole and ciprofloxacin from wastewater and sludge at flask-scale.

Plenty of reports for the removal of pharmaceuticals at reactor scale are available from synthetic wastewater, including the simultaneous elimination of several sulfonamides with *T. versicolor* in a continuous fluidized-bed reactor (Rodríguez-Rodríguez et al. 2012b) and

Table 3. Removal of selected pharmaceuticals (analgesic/anti-inflammatories and antibiotics) by ligninolytic fungi or their enzymes from real matrices using liquid-phase processes.

Pharmaceutical classification	Compound	Removed from		
		UWW	HWW	Sludge
Analgesics/anti-inflammatories	Acetaminophen	STBR/Flask (1), MBR (3)	STBR (2)	-
	Codeine	STBR/Flask (1)	STBR (2)	-
	Dexamethasone	-	STBR (2)	-
	Diclofenac	-	STBR(2)	Bioslurry (4)
	Ibuprofen	STBR/Flask (1)	STBR(2)	Bioslurry (4)
	Indomethacine	-	-	Bioslurry (4)
	Ketoprofen	STBR/Flask (1)	STBR(2)	Bioslurry (4)
	Mefenamic acid	MBR(3)	-	Bioslurry (4)
	Naproxen	STBR/Flask (1)	STBR(2)	Bioslurry (4,5)
	Phenazone	-	STBR(2)	Bioslurry (4)
	Piroxicam	-	STBR(2)	-
	Salicylic acid	STBR/Flask (1)	STBR(2)	-
Antibiotics	Azithromycin	STBR/Flask (1)	STBR(2)	-
	Cefalexine	STBR/Flask (1)	-	-
	Ciprofloxacin	STBR/Flask (1), Flask (6)	STBR (2)	Flask (6)
	Clarithromycin	-	STBR (2)	-
	Dimetridazole	-	STBR (2)	-
	Erythromycin	STBR/Flask (1), Flask (6)	-	Flask (6)
	Metronidazole	STBR/Flask (1)	STBR (2)	-
	MetronidazoleOH	-	STBR (2)	-
	Ofloxacin	-	STBR (2,7)	-
	Ronidazole	-	STBR (2)	-
	Sulfamethazine	-	-	Bioslurry (4)
	Sulfamethoxazole	Flask (6)	STBR (2)	Flask (6)
	Sulfapyridine	-	-	Bioslurry (4)
	Sulfathiaziole	-	-	Bioslurry (4)
	Trimethroprim	-	STBR (2)	-

UWW: urban wastewater; HWW: hospital wastewater
STBR: stirred tank bioreactor; MBR: membrane bioreactor

1. Cruz-Morató et al. (2013)
2. Cruz-Morató et al. (2014)
3. Ba et al. (2014)
4. Rodríguez-Rodríguez et al. (2012a)
5. Rodríguez-Rodríguez et al. (2010a)
6. Accinelli et al. (2010)
7. Gros et al. (2014)

diclofenac, ibuprofen and naproxen with *P. chrysosporium* in a stirred tank reactor (Rodarte-Morales et al. 2011); nonetheless, as they employ optimal conditions for fungal growth, they do not properly represent the complex interactions taking place in a real matrix.

Table 4. Pharmaceutical compounds removed from real liquid polluted matrices (urban and hospital wastewater, sludge, leachates from landfill) using WRF (*P. chrysosporium* or *T. versicolor*). Based on works by Accinelli et al. (2010), Castellana and Loffredo (2014), Cruz-Morató et al. (2013, 2014); Gros et al. (2014) and Rodríguez-Rodríguez et al. (2010a, 2012a).

Group of pharmaceuticals	**Degree of removal**		
	High removal (> 75%)	**Medium removal (35%–75%)**	**Low removal (< 35%)**
Analgesic/antiinflammatories	Acetaminophen	**Diclofenac**	Dexamethasone
	Codeine	Indomethacin	Piroxicam
	Diclofenac	Ketoprofen	Salicylic acid
	Ibuprofen	Mefenamic acid	
	Naproxen	**Naproxen**	
		Phenazone	
Antibiotics	Azithromycin	**Ciprofloxacin**	Cefalexine
	Ciprofloxacin	Metronidazole	Dimetridazole
	Clarithromycin	Sulfamethoxazole	
	Erythromycin	Trimethoprim	
	Metronidazole		
	Ofloxacin		
	Sulfamethazine		
	Sulfapyridine		
	Sulfathiazole		
Anticoagulants	Clopidrogel		
Antidiabetics	Glibenclamide		
Antihypertensives	Amlodipine		Losartan
	Diltiazem		
	Irbesartan		
	Valsartan		
	Verapamil		
Antiparasitics	Albendazole		
Antivirals		Oseltamivir	
β2 adrenergic receptor agonists	Salbutamol		
Barbiturics	Butalbital		
β-blocker	**Atenolol**	**Atenolol**	
	Corazolol		
	Metoprolol		
	Nodalol		
	Propanolol		
	Sotalol		
Endocrine disruptors and hormones	Bisphenol A		
	Estrone		
	Estradiol		
	Estriol		

Table 4 contd. ...

...Table 4 contd.

Group of pharmaceuticals	Degree of removal		
	High removal (> 75%)	**Medium removal (35%–75%)**	**Low removal (< 35%)**
Histamine H1 and H2 receptor antagonists	Cimetidine		
	Desloratadine		
	Famotidine		
	Ranitidine		
Lipid regulators	Fenofibrate	Atorvastatin	
	Pravastatin	Bezafibrate	
		Gemfibrozil	
Diuretics	**Furosemide**	**Furosemide**	
	Hydrochlorothiazide	**Hydrochlorothiazide**	
	Torasemide		
Psychiatric drugs	**Citalopram**	**Carbamazepine**	**Carbamazepine**
	Fluoxetine	**Venlafaxine**	**Citalopram**
	Olanzapine		Diazepam
	Paroxetine		Lorazepam
	Sertraline		
	Venlafaxine		
X-ray contrast agent		Iopromide	

Bolded names represent pharmaceutical compounds for which different degrees of removal have been reported.

Enzymatic Systems for the Elimination of Pharmaceutical Compounds

Among pharmaceuticals, the use of fungal enzymes on real matrices has seen more development on the removal of endocrine disrupting compounds. The complete elimination of the natural estrogens estrone, 17β-estradiol and estriol was achieved in wastewater with non-immobilized laccase derived from *T. versicolor* (Auriol et al. 2007); moreover, the authors found a negligible matrix effect by comparison with synthetic wastewater. Furthermore, the removal of estrogenic activity was also demonstrated (Auriol et al. 2008). Several organic contaminants including naproxen, ketoprofen, diclofenac and salicylic acid were removed (70–80%) from wastewater by an enzymatic treatment with laccase from *Trametes pubescens,* which also partially detoxified the matrix and decreased its estrogenic capacity (Spina et al. 2015). An alternative approach by Tanaka et al. (2001) employed a rotating reactor with laccase from *Trametes* sp. to efficiently remove bisphenol A and 17α-ethinylestradiol (as well as endocrine disrupting alkylphenols) from sea sand.

Besides these successful attempts with free enzymes, the immobilization of the catalyst has been employed in enzyme-mediated systems. Ba et al. (2014) described a hybrid reactor of cross-linked enzyme aggregates of laccase from *T. versicolor* and a polysulfone hollow fiber microfiltration membrane used for the elimination of acetaminophen, mefenamic acid and carbamazepine; the hybrid system exhibited higher efficiency than the membrane alone. Similarly, Nair et al. (2013) achieved more than 85% removal of bisphenol A and 17α-ethinylestradiol and 30% removal of diclofenac from wastewater using laccase from *Coriolopsis gallica*, immobilized on mesoporus silica spheres by a two-step adsorption-crosslinking process. Recently, the concept of applying laccase immobilized onto nanomaterials for the treatment of polluted effluents in reactors with retention of nanoparticles (combination of cascade fixed-bed reactor with membrane technology and perfused based bioreactor with membrane technology), began to be explored (Corvini and Shahgaldian 2010).

Biopiles for the Treatment of Pharmaceuticals

An approach that resembles the biomixtures from biopurification systems has been used in the fungal treatment of pharmaceuticals and other emerging pollutants contained in sludge from wastewater treatment plants. The system is called "biopile", and consists of a matrix made up by mixing the polluted material, in this case the sludge, with a fungal pre-colonized lignocellulosic residue; as with other previously described approaches, *T. versicolor* is the fungus traditionally reported in this treatment.

The use of biopiles required the previous determination of the capacity of *T. versicolor* to colonize the sludge; in this respect, the sludge was amended with different concentrations of wheat straw pellets (WSP) pre-colonized by the fungus (0%, 24% and 38% w/w). Over a 60 d period data of ergosterol and laccase indicated an efficient colonization of the sludge in the WSP-containing systems, while colonization failed in the systems lacking WSP. Considering the overall results, the system containing 38% WSP was employed for the evaluation of degrading capacity and showed a complete removal of naproxen and around 48% for carbamazepine after 72 hr (Rodríguez-Rodríguez et al. 2010a).

Taking into account the results of naproxen elimination, a degrading test called ND24 (naproxen degradation in 24 hr) was developed, intended to monitor the degrading ability of *T. versicolor* in solid-state processes (Rodríguez-Rodríguez et al. 2010b). The test consisted of measuring the removal (%) of a determined concentration of naproxen spiked on a solid-phase system after 24 hr. The test can be applied as a complement to other analyses such as ergosterol, and laccase, as ND24 does not necessarily correlate those parameters. For samples collected over 45 d in a sludge-containing biopile, high ND24 values supported the feasibility of applying *T. versicolor* as a bioremediation agent, as previously suggested in terms of colonization and removal of model drugs.

For the elimination of pharmaceuticals at pre-existent concentrations, the treatment time was set in 42 d, similar to that employed in typical sludge-composting processes. Overall biological parameters indicated that *T. versicolor* had been active throughout the treatment period. The treatment was conducted both in conditions of sterilized (Rodríguez-Rodríguez et al. 2011, García-Galán et al. 2011) and non-sterilized sludge (Rodríguez-Rodríguez et al. 2012c, 2014). The abundant anti-inflammatory compounds ibuprofen and diclofenac were eliminated at 75% and 64%, respectively, using sterilized sludge, and around 54% for both compounds when non-sterilized sludge was employed; the elimination of mefenamic acid (72%) and phenazone (100%) was also reported. Several compounds were completely removed when sterile sludge was used prior fungal inoculation, including the lipid regulators bezafibrate and fenofibrate; the antibiotics clarithromycin, sulfamethazine, sulfapyridine and sulfathiazole; the beta-blocker atenolol and the histamine H2-receptor antagonist cimetidine. The lipid regulator atorvastatin and the diuretic hydrochlorothiazide exhibited removals of 80% and 52%, respectively. In conditions of non-sterilized sludge, the removal of atorvastatin and clarithromycin was slightly reduced to 65% and 82%, respectively, while it increased for hydrochlorothiazide (65%); for fenofibrate, furosemide and ranitidine complete elimination was achieved. The psychiatric drugs carbamazepine and diazepam exhibited the lowest removal, both in sterile (43% for both compounds) and non-sterile (9%, only carbamazepine) sludge. In general, the fact that in some cases the use of non-sterile sludge increases the pollutant elimination, suggest that removal is favored by the joint effect of the fungus and the indigenous microbiota of the effluent.

In the case of non-sterile sludge, the fungal bioaugmentation accelerated the removal capacity of several compounds with respect to the non-bioaugmented biopile: diclofenac, atorvastatin and fenofibrate; nonetheless DGGE analysis revealed that *T. versicolor* remained as the main fungal taxon in the system only for half the treatment period (Rodríguez-Rodríguez et al. 2012c). Therefore, re-inoculation strategies were suggested. In this respect, two approaches were considered: the addition of fungal blended mycelium or the addition of the fungal pre-colonized lignocellulosic substrate (wheat straw), after 22 d of treatment (Rodríguez-Rodríguez et al. 2014). Both strategies improved the efficiency of the biopiles, however the former was the most successful, achieving 86% removal of the total pharmaceuticals concentration (versus 69%) by the end of the process after 42 d; in contrast, the removal at the time of re-inoculation was only 35%. The authors ascribed the better performance of the re-inoculation with blended mycelium alone, to a more direct re-colonization of the matrix. Moreover, they highlighted that this strategy represents an advantage from the operational perspective, as the additional use of substrate would result in the increase of costs and the dilution of the pollutants.

Concluding Remarks

Ligninolytic fungi have high potential for the degradation of diverse groups of organic pollutants. This potential has been translated into the use of different approaches for pollutant removal. Liquid phase approaches include the use of bioreactor configurations including slurry stirred tank reactors and fluidized-bed reactors, which usually employ fungi in the form of pellets, or even cell-free enzymatic reactors. On the other hand, solid-phase approaches are sometimes preferred as they mimic the natural conditions of fungal growth, and in many cases are considered more cost-effective. In the particular case of pesticide removal, scarce research has been performed in liquid-phase or enzymatic systems; conversely, their treatment has been relatively successful in the so-called biopurification systems, in which bacteria and fungi cooperate in the transformation; specific fungal bioaugmentations have improved the removal process depending on the matrix composition. In the case of pharmaceuticals, fast removal has been demonstrated for some therapeutic groups contained in wastewater and sludge in fungal liquid-phase reactors and solid biopiles; variations in the degradation efficiency depend on the sterilization/non-sterilization of the contaminated matrix previous to fungal inoculation, which reveals a combined effect of the fungi and the indigenous microbial communities contained in the matrix. Future research should be focused on the scale-up of the successful configurations to determine the potential applicability of fungal systems at large scale for the treatment of polluted matrices.

Acknowledgements

This work was supported by Vicerrectoría de Investigación, Universidad de Costa Rica (802-B4-503 and 802-B6-137), and the Costa Rican Ministry of Science, Technology and Telecommunications, MICITT (project FI-093-13).

References Cited

Accinelli, C., M.L. Saccà, I. Batisson, J. Fick, M. Mencarelli and R. Grabic. 2010. Removal of oseltamivir (Tamiflu) and other selected pharmaceuticals from wastewater using a granular bioplastic formulation entrapping propagules of *Phanerochaete chrysosporium*. Chemosphere 81: 436–443.

Acevedo, F., L. Pizzul, M.d.P. Castillo, R. Cuevas and M.C. Diez. 2011. Degradation of polycyclic aromatic hydrocarbons by the Chilean white-rot fungus *Antracophyllum discolor*. J. Hazard. Mater. 185: 212–219.

Anastasi, A., V. Tigini and G.C. Varese. 2013. The bioremediation potential of different ecophysiological groups of fungi. pp. 29–49. *In*: Goltapeh, E.M., Y.R. Danesh and A. Varma [eds.]. Soil Biology. Fungi as Bioremediators. Springer-Verlag, Berlin Heidelberg.

Arbeli, Z. and C.L. Fuentes. 2007. Accelerated biodegradation of pesticides: An overview of the phenomenon, its basis and possible solutions; and a discussion on the tropical dimension. Crop Prot. 26: 1733–1746.

Asgher, M., H.N. Bhatti, M. Ashraf and R.L. Legge. 2008. Recent developments in biodegradation of industrial pollutants by white rot fungi and their enzyme system. Biodegradation 19: 771–783.

Auriol, M., Y. Filali-Meknassi, R.D. Tyagi and C.D. Adams. 2007. Laccase-catalyzed conversion of natural and synthetic hormones from a municipal wastewater. Water Res. 41: 3281–3288.

Auriol, M., Y. Filali-Meknassi, C.D. Adams, R.D. Tyagi, T.N. Noguerol and B. Pina. 2008. Removal of estrogenic activity of natural and synthetic hormones from a municipal wastewater: efficiency of horseradish peroxidase and laccase from *Trametes versicolor*. Chemosphere 70: 445–452.

Axtell, C., C.G. Johnston and J.A. Bumpus. 2000. Bioremediation of soil contaminated with explosives at the Naval Weapons Station Yorktown. Soil Sediment Contam. 9: 537–548.

Ba, S., J.P. Jones and H. Cabana. 2014. Hybrid bioreactor (HBR) of hollow fiber microfilter membrane and cross-linked laccase aggregates eliminate aromatic pharmaceuticals in wastewaters. J. Hazard. Mater. 280: 662–670.

Bastos, A.C. and N. Magan. 2009. *Trametes versicolor*: potential for atrazine bioremediation in calcareous clay soil, under low water availability conditions. Int. Biodeter. Biodegr. 63: 389–394.

Bending, G.D., M. Friloux and A. Walker. 2002. Degradation of contrasting pesticides by white rot fungi and its relationship with ligninolytic potential. FEMS Microbiol. Lett. 212: 59–63.

Bezalel, L., Y. Hadar and C.E. Cerniglia. 1997. Enzymatic mechanisms involved in phenanthrene degradation by the white rot fungus *Pleurotus ostreatus*. Appl. Environ. Microbiol. 63: 2495–2501.

Blánquez, P., N. Casas, X. Font, X. Gabarrell, M. Sarrà, G. Caminal et al. 2004. Mechanism of textile metal dye biotransformation by *Trametes versicolor*. Water Res. 38: 2166–2172.

Blánquez, P. and B. Guieysse. 2008. Continuous biodegradation of 17β-estradiol and 17α-ethynylestradiol by *Trametes versicolor*. J. Hazard. Mater. 150: 459–462.

Blánquez, P., M. Sarrà and T. Vicent. 2008. Development of a continuous process to adapt the textile wastewater treatment by fungi to industrial conditions. Process Biochem. 43: 1–7.

Borchert, M. and J.A. Libra. 2001. Decolorization of reactive dyes by the white rot fungus *Trametes versicolor* in sequencing batch reactors. Biotechnol. Bioeng. 75: 313–321.

Cajthaml, T., Z. Křesinová, K. Svobodová and M. Möder. 2009. Biodegradation of endocrine-disrupting compounds and suppression of estrogenic activity by ligninolytic fungi. Chemosphere 75: 745–750.

Castellana, G. and E. Loffredo. 2014. Simultaneous removal of endocrine disruptors from a wastewater using white rot fungi and various adsorbents. Water Air Soil Pollut. 225: 1872.

Castillo, M.d.P. and L. Torstensson. 2007. Effect of biobed composition, moisture and temperature on the degradation of pesticides. J. Agric. Food Chem. 55: 5725–5733.

Castillo, M.d.P., L. Torstensson and J. Stenström. 2008. Biobeds for environmental protection from pesticide use. A review. J. Agric. Food Chem. 56: 6206–6219.

Castro-Gutiérrez, V., M. Masís-Mora, G. Caminal, T. Vicent, E. Carazo-Rojas, M. Mora-López et al. 2016. A microbial consortium from a biomixture swiftly degrades high concentrations of carbofuran in fluidized-bed reactors. Process Biochem. 51: 1585–1593.

Champagne, P.P. and J.A. Ramsay. 2010. Dye decolorization and detoxification by laccase immobilized on porous glass beads. Bioresour. Technol. 101: 2230–2235.

Chhabra, M., S. Mishra and T.R. Sreekrishnan. 2009. Laccase/mediator assisted degradation of triarylmethane dyes in a continuous membrane reactor. J. Biotechnol. 143: 69–78.

Chin-Pampillo, J.S., K. Ruiz-Hidalgo, M. Masís-Mora, E. Carazo-Rojas and C.E. Rodríguez-Rodríguez. 2015a. Adaptation of biomixtures for carbofuran degradation in on-farm biopurification systems in tropical regions. Environ. Sci. Pollut. Res. 22: 9839–9848.

Chin-Pampillo, J.S., K. Ruiz-Hidalgo, M. Masís-Mora, E. Carazo-Rojas and C.E. Rodríguez-Rodríguez. 2015b. Design of an optimized biomixture for the degradation of carbofuran based on pesticide removal and toxicity reduction of the matrix. Env. Sci. Pollut. Res. 22: 19184–19193.

Coppola, L., M.d.P. Castillo, E. Monaci and C. Vischetti. 2007. Adaptation of the biobed composition for chlorpyrifos degradation to southern Europe conditions. J. Agric. Food Chem. 55: 396–401.

Coppola, L., P. Castillo and C. Vischetti. 2011. Degradation of isoproturon and bentazone in peat- and compost-based biomixtures. Pest Manage. Sci. 67: 107–113.

Córdova Juárez, R.A., L.L. Gordillo Dorry, R. Bello-Mendoza and J.E. Sánchez. 2011. Use of spent substrate after *Pleurotus pulmonarius* cultivation for the treatment of chlorothalonil containing wastewater. J. Environ. Manage. 92: 948–952.

Corvini, P.F. and P. Shahgaldian. 2010. LANCE: laccase-nanoparticle conjugates for the elimination of micropollutants (endocrine disrupting chemicals) from wastewater in bioreactors. Rev. Environ. Sci. Bio/Technol. 9: 23–27.

Covino, S., K. Svobodová, M. Čvančarová, A. D'Annibale, M. Petruccioli, F. Federici et al. 2010. Inoculum carrier and contaminant bioavailability affect fungal degradation performances of PAH-contaminated solid matrices from a wood preservation plant. 79: 855–864.

Covino, S., T. Stella, A. D'Annibale, S. Lladó, P. Baldrian, M. Čvančarová et al. 2016. Comparative assessment of fungal augmentation treatments of a fine-textured and historically oil-contaminated soil. Sci. Total Environ. 566-567: 250–259.

Cruz-Morató, C., L. Ferrando-Climent, S. Rodríguez-Mozaz, D. Barceló, E. Marco-Urrea, T. Vicent et al. 2013. Degradation of pharmaceuticals in non-sterile urban wastewater by *Trametes versicolor* in a fluidized bed bioreactor. Water Res. 47: 5200–5210.

Cruz-Morató, C., D. Lucas, M. Llorca, S. Rodríguez-Mozaz, M. Gorga, M. Petrovic et al. 2014. Hospital wastewater treatment by fungal bioreactor: Removal efficiency for pharmaceuticals and endocrine disruptor compounds. Sci. Total Environ. 493: 365–376.

Čvančarová, M., Z. Křesinová, A. Filipová, S. Covino and T. Cajthaml. 2012. Biodegradation of PCBs by ligninolytic fungi and characterization of the degradation products. Chemosphere 88: 1317–1323.

D'Annibale, A., M. Ricci, V. Leonardi, D. Quarantino, E. Mincione and M. Petruccioli. 2005. Degradation of aromatic hydrocarbons by white-rot fungi in a historically contaminated soil. Biotechnol. Bioeng. 90: 723–731.

Dávila-Vázquez, G., R. Tinoco, M.A. Pickard and R. Vazquez-Duhalt. 2005. Transformation of halogenated pesticides by versatile peroxidase from *Bjerkandera adusta*. Enzyme Microb. Technol. 36: 223–231.

Davis, M.W., J.A. Glaser, J.W. Evans and R.T. Lamar. 1993. Field evaluation of the lignin-degrading fungus *Phanerochaete sordida* to treat creosote-contaminated soil. Environ. Sci. Technol. 27: 2572–2576.

De Wilde, T., P. Spanoghe, C. Debaer, J. Ryckeboer, D. Springael and P. Jaeken. 2007. Overview of on-farm bioremediation systems to reduce the occurrence of point source contamination. Pest Manage. Sci. 63: 111–128.

Doddapaneni, H. and J.S. Yadav. 2004. Differential regulation and xenobiotic induction of tandem P450 monooxygenase genes pc-1 (CYP63A1) and pc-2 (CYP63A2) in the white-rot fungus *Phanerochaete chrysosporium*. Appl. Microbial. Biotechnol. 65: 559–565.

Eaton, R.A. and M.D.C. Hale. 1993. Wood: Decay, Pests and Protection. Chapman and Hall, London.

Eggert, C., U. Temp and K.E. Eriksson. 1996. The ligninolytic system of the white rot fungus *Pycnoporus cinnabarinus*: purification and characterization of the laccase. Appl. Environ. Microbiol. 62: 1151–1158.

Elgueta, S., C. Santos, N. Lima and M.C. Diez. 2016. Atrazine dissipation in a biobed system inoculated with immobilized white-rot fungi. Arch. Agron. Soil Sci. 62: 1451–1461.

Farré, M., S. Pérez, L. Kantiani and D. Barceló. 2008. Fate and toxicity of emerging pollutants, their metabolites and transformation products in the aquatic environment. TRAC Trend Anal. Chem. 27: 991–1007.

Ferrando-Climent, L., C. Cruz-Morató, E. Marco-Urrea, T. Vicent, M. Sarrà, S. Rodríguez-Mozaz et al. 2015. Non conventional biological treatment based on *Trametes versicolor* for the elimination of recalcitrant anticancer drugs in hospital wastewater. Chemosphere 136: 9–19.

Fogg, P., A.B. Boxall, A. Walker and A.A. Jukes. 2003. Pesticide degradation in a 'biobed' composting substrate. Pest Manage. Sci. 59: 527–537.

Fragoeiro, S. and N. Magan. 2008. Impact of *Trametes versicolor* and *Phanerochaete chrysosporium* on differential breakdown of pesticide mixtures in soil microcosms at two water potentials and associated respiration and enzyme activity. Int. Biodeter. Biodegr. 62: 376–383.

Gago-Ferrero, P., M. Badia-Fabregat, A. Olivares, B. Piña, P. Blánquez, T. Vicent et al. 2012. Evaluation of fungal- and photo-degradation as potential treatments for the removal of sunscreens BP3 and BP1. Sci. Total Environ. 427: 355–363.

García-Galán, M.J., C.E. Rodríguez-Rodríguez, T. Vicent, G. Caminal, M.S. Díaz-Cruz and D. Barceló. 2011. Biodegradation of sulfamethazine by *Trametes versicolor*: Removal from sewage sludge and identification of intermediate products by UPLC-QqTOF-MS. Sci. Total Environ. 409: 5505–5512.

Garon, D., S. Krivobok, D. Woeessidjewe and F. Seigle-Murandi. 2002. Influence of surfactants on solubilization and fungal degradation of fluorene. Chemosphere 47: 303–309.

Gros, M., C. Cruz-Morató, E. Marco-Urrea, P. Longrée, H. Singer, M. Sarrà et al. 2014. Biodegradation of the X-ray contrast agent iopromide and the fluoroquinolone antibiotic ofloxacin by the white rot fungus *Trametes versicolor* in hospital wastewaters and identification of degradation products. Water Res. 60: 228–241.

Harms, H., D. Schlosser and L.Y. Wick. 2011. Untapped potential: exploiting fungi in bioremediation of hazardous chemicals. Nat. Rev. Microbiol. 9: 177–192.

Hatakka, A. 1994. Lignin-modifying enzymes from selected white-rot fungi: production and role from in lignin degradation. FEMS Microbiol. Rev. 13: 125–135.

Hirai, H., S. Nakanishi and T. Nishida. 2004. Oxidative dechlorination of methoxychlor by ligninolytic enzymes from white-rot fungi. Chemosphere 55: 641–645.

Hofrichter, M., R. Ullrich, M.J. Pecyna, C. Liers and T. Lundell. 2010. New and classic families of secreted fungal heme peroxidases. Appl. Microbiol. Biotechnol. 87: 871–897.

Hölker, U. and J. Lenz. 2005. Solid-state fermentation—are there any biotechnological advantages? Curr. Opin. Microbiol. 8: 301–306.

Huete-Soto, A., M. Masís-Mora, V. Lizano-Fallas, J.S. Chin-Pampillo, E. Carazo-Rojas, C.E. Rodríguez-Rodríguez. 2017. Simultaneous removal of structurally different pesticides in a biomixture: Detoxification and effect of oxytetracycline. Chemosphere 169: 558–567.

Jelić, A., C. Cruz-Morató, E. Marco-Urrea, M. Sarrà, S. Perez, T. Vicent et al. 2012. Degradation of carbamazepine by *Trametes versicolor* in an air pulsed fluidized bed bioreactor and identification of intermediates. Water Res. 46: 955–964.

Jolivalt, C., S. Brenon, E. Caminade, C. Mougin and M. Pontié. 2000. Immobilization of laccase from *Trametes versicolor* on a modified PVDF microfiltration membrane: characterization of the grafted support and application in removing a phenylurea pesticide in wastewater. J. Membrane Sci. 180: 103–113.

Kamei, I., R. Kogura and R. Kondo. 2006. Metabolism of 4,4′-dichlorobiphenyl by white-rot fungi *Phanerochaete chrysosporium* and *Phanerochaete* sp. MZ142. Appl. Microbiol. Biotechnol. 72: 566–575.

Karanasios, E., N.G. Tsiropoulos, D.G. Karpouzas and C. Ehaliotis. 2010a. Degradation and adsorption of pesticides in compost-based biomixtures as potential substrates for biobeds in Southern Europe. J. Agric. Food Chem. 58: 9147–9156.

Karanasios, E., N.G. Tsiropoulos, D.G. Karpouzas and U. Menkissoglu-Spiroudi. 2010b. Novel biomixtures based on local Mediterranean lignocellulosic materials: evaluation for use in biobed systems. Chemosphere 80: 914–921.

Karanasios, E., N.G. Tsiropoulos and D.G. Karpouzas. 2012. On-farm biopurification systems for the depuration of pesticide wastewaters: recent biotechnological advances and future perspectives. Biodegradation 23: 787–802.

Karas, P.A., C. Perruchon, K. Exarhou, C. Ehaliotis and D.G. Karpouzas. 2011. Potential for bioremediation of agro-industrial effluents with high loads of pesticides by selected fungi. Biodegradation 22: 215–228.

Karas, P.A., C. Perruchon, E. Karanasios, E.S. Papadopoulou, E. Manthou, S. Sitra et al. 2016. Integrated biodepuration of pesticide-contaminated wastewaters from the fruit-packaging industry using biobeds: Bioaugmentation, risk assessment and optimized management. J. Hazard. Mater. 320: 635–644.

Katuri, K.P., S.V. Mohan, S. Sridhar, B.R. Pati and P.N. Sarma. 2009. Laccase-membrane reactors for decolorization of an acid azo dye in aqueous phase: process optimization. Water Res. 43: 3647–3658.

Kaushik, P. and A. Malik. 2009. Fungal dye decolourization: recent advances and future potential. Environ. Int. 35: 127–141.

Kennedy, D.W., S.D. Aust and J.A. Bumpus. 1990. Comparative biodegradation of alkyl halide insecticides by the white rot fungus, *Phanerochaete chrysosporium* (BKM-F-1767). Appl. Environ. Microbiol. 56: 2347–2353.

Kjøller, A.H. and S. Struwe. 2002. Fungal communities, succession, enzymes, and decomposition. Enzymes in the environment: activity, ecology and applications. Marcel Dekker, New York.

Kohlmeier, S., T.H.M. Smits, R.M. Ford, C. Keel, H. Harms and L.Y. Wick. 2005. Taking the fungal highway: Mobilization of pollutant-degrading bacteria by fungi. Env. Sci. Technol. 39: 4640–4646.

Korcan, S.E., İ.H. Ciğerci and M. Konuk. 2013. White-rot fungi in bioremediation. pp. 371–390. *In*: Goltapeh, E.M., Y.R. Danesh and A. Varma [eds.]. Soil Biology. Fungi as Bioremediators. Springer-Verlag, Berlin Heidelberg.

Kubátová, A., P. Erbanová, I. Eichlerová, L. Homolka, F. Nerud and V. Šašek. 2001. PCB congener selective biodegradation by the white rot fungus *Pleurotus ostreatus* in contaminated soil. Chemosphere 43: 207–215.

Kullman, S.W. and F. Matsumura. 1996. Metabolic pathways utilized by *Phanerochaete chrysosporium* for degradation of the cyclodiene pesticide endosulfan. Appl. Environ. Microbiol. 62: 593–600.

Leonardi, V., V. Šašek, M. Petruccioli, A. D'Annibale, P. Erbanová and T. Cajthaml. 2007. Bioavailability modification and fungal biodegradation of PAHs in aged industrial soils. Int. Biodeter. Biodegr. 60: 165–170.

Leonardi, V., M.A. Giubilei, E. Federici, R. Spaccapelo, V. Šašek, Č. Novotný et al. 2008. Mobilizing agents enhance fungal degradation of polycyclic aromatic hydrocarbons and affect diversity of indigenous bacteria in soil. Biotechnol. Bioeng. 101: 273–285.

Lizano-Fallas, V., M. Masís-Mora, D. Espinoza-Villalobos, M. Lizano-Brenes and C.E. Rodríguez-Rodríguez. 2017. Removal of pesticides and ecotoxicological changes during the simultaneous treatment of triazines and chlorpyrifos in biomixtures. Chemosphere 182: 106–113.

López, C., M.T. Moreira, G. Feijoo and J.M. Lema. 2004. Dye decolorization by manganese peroxidase in an enzymatic membrane bioreactor. Biotechnol. Prog. 20: 74–81.

Madrigal-Zúñiga, K., K. Ruiz-Hidalgo, J.S. Chin-Pampillo, M. Masís-Mora, V. Castro-Gutiérrez and C.E. Rodríguez-Rodríguez. 2016. Fungal bioaugmentation of two rice husk-based biomixtures for the removal of carbofuran in on-farm biopurification systems. Biol. Fertil. Soils. 52: 243–250.

Marco-Urrea, E., M. Pérez-Trujillo, T. Vicent and G. Caminal. 2009. Ability of white-rot fungi to remove selected pharmaceuticals and identification of degradation products of ibuprofen by *Trametes versicolor*. Chemosphere 74: 765–772.

Martínez, A.T. 2002. Molecular biology and structure-function of lignin-degrading heme peroxidases. Enzyme Microb. Technol. 30: 425–444.

Martínez, A.T., M. Speranza, F.J. Ruiz-Dueñas, P. Ferreira, S. Camarero, F. Guillén et al. 2005. Biodegradation of lignocellulosics: microbial, chemical and enzymatic aspects of the fungal attack of lignin. Int. Microbiol. 8: 195–204.

Mendoza, L., M. Jonstrup, R. Hatti-Kaul and B. Mattiasson. 2011. Azo dye decolorization by a laccase/mediator system in a membrane reactor: enzyme and mediator reusability. Enzyme Microb. Technol. 49: 478–484.

Mir-Tutusaus, J.A., M. Masís-Mora, C. Corcellas, E. Eljarrat, D. Barceló, M. Sarrà et al. 2014. Degradation of selected agrochemicals by the white rot fungus *Trametes versicolor*. Sci. Total Environ. 500-501: 235–242.

Morozova, O.V., G.P. Shumakovich, S.V. Shleev and Y.I. Yaropolov. 2007. Laccase-mediator systems and their applications: a review. Appl. Biochem. Microbiol. 43: 523–535.

Mougin, C., C. Laugero, M. Asther, J. Dubroca, P. Frasse and M. Asther. 1994. Biotransformation of the herbicide atrazine by the white rot fungus *Phanerochaete chrysosporium*. Appl. Environ. Microbiol. 60: 705–708.

Mougin, C., C. Pericaud, C. Malosse, C. Laugerob and M. Astherb. 1996. Biotransformation of the insecticide lindane by-the white rot basidiomycete *Phanerochaete chrysosporium*. Pestic. Sci. 47: 51–59.

Murillo-Zamora, S., V. Castro-Gutiérrez, M. Masís-Mora, V. Lizano-Fallas and C.E. Rodríguez-Rodríguez. 2017. Elimination of fungicides in biopurification systems: effect of fungal bioaugmentation on removal performance and microbial community structure. Chemosphere 186: 625–634.

Nair, R.R., P. Demarche and S.N. Agathos. 2013. Formulation and characterization of an immobilized laccase biocatalyst and its application to eliminate organic micropollutants in wastewater. New Biotechnol. 30: 814–823.

Nano, G., A. Borroni and R. Rota. 2003. Combined slurry and solid-phase bioremediation of diesel contaminated soils. J. Hazard. Mater. 100: 79–94.

Novotný, Č., P. Erbanová, V. Šašek, A. Kubátová, T. Cajthaml, E. Lang et al. 1999. Extracellular oxidative enzyme production and PAH removal in soil by exploratory mycelium of white rot fungi. Biodegradation 10: 159–168.

Novotný, Č., P. Erbanova, T. Cajthaml, N. Rothschild, C. Dosoretz and V. Šašek. 2000. *Irpex lacteus*, a white rot fungus applicable to water and soil bioremediation. Appl. Microbiol. Biotechnol. 54: 850–853.

Novotný, Č., B. Rawal, M. Bhatt, M. Patel, V. Šašek and H.P. Molitoris. 2001. Capacity of *Irpex lacteus* and *Pleurotus ostreatus* for decolorization of chemically different dyes. J. Biotechnol. 89: 113–122.

Omirou, M., P. Dalias, C. Costa, C. Papastefanou, A. Dados, C. Ehaliotis et al. 2012. Exploring the potential of biobeds for the depuration of pesticide-contaminated wastewaters from the citrus production chain: Laboratory, column and field studies. Environ. Poll. 166: 31–39.

Osma, J.F., J.L. Toca-Herrera and S. Rodríguez-Couto. 2010. Biodegradation of a simulated textile effluent by immobilised-coated laccase in laboratory-scale reactors. Appl. Catal. A. 373: 147–153.

Pinto, A.P., S.C. Rodrigues, A.T. Caldeira and D.M. Teixeira. 2016. Exploring the potential of novel biomixtures and *Lentinula edodes* fungus for the degradation of selected pesticides. Evaluation for use in biobed systems. Sci. Total Environ. 541: 1372–1381.

Pointing, S. 2001. Feasibility of bioremediation by white-rot fungi. Appl. Microbiol. Biotechnol. 57: 20–33.

Pothuluri, J.V., J.P. Freeman, T.M. Heinze, R.D. Beger and C.E. Cerniglia. 2000. Biotransformation of vinclozolin by the fungus *Cunninghamella elegans*. J. Agric. Food Chem. 48: 6138–6148.

Quintero, J.C., T.A. Lu-Chau, M.T. Moreira, G. Feijoo and J.M. Lema. 2007. Bioremediation of HCH present in soil by the white-rot fungus *Bjerkandera adusta* in a slurry batch bioreactor. Int. Biodeter. Biodegr. 60: 319–326.

Ramsay, J.A. and C. Goode. 2004. Decoloration of a carpet dye effluent using *Trametes versicolor*. 26: 197–201.

Reddy, C.A. and Z. Mathew. 2001. Bioremediation potential of white rot fungi. British Mycological Society Symposium Series. pp. 52–78.

Robles-González, I.V., F. Fava and H.M. Poggi-Varaldo. 2008. A review on slurry bioreactors for bioremediation of soils and sediments. Microb. Cell Fact. 7: 5–20.

Rodarte-Morales, A.I., G. Feijoo, M.T. Moreira and J.M. Lema. 2011. Degradation of selected pharmaceutical and personal care products (PPCPs) by white-rot fungi. World J. Microbiol. Biotechnol. 27: 1839–1846.

Rodríguez-Couto, S. and M.A. Sanromán. 2005. Application of solid-state fermentation to ligninolytic enzyme production. Biochem. Eng. J. 22: 211–219.

Rodríguez-Couto, S. and J.L. Toca-Herrera. 2006. Industrial and biotechnological applications of laccases: A review. Biotechnol. Adv. 24: 500–513.

Rodríguez-Rodríguez, C.E., E. Marco-Urrea and G. Caminal. 2010a. Degradation of naproxen and carbamazepine in spiked sludge by slurry and solid-phase *Trametes versicolor* systems. Bioresour. Technol. 101: 2259–2266.
Rodríguez-Rodríguez, C.E., E. Marco-Urrea and G. Caminal. 2010b. Naproxen degradation test to monitor *Trametes versicolor* activity in solid-state bioremediation processes. J. Hazard. Mater. 179: 1152–1155.
Rodríguez-Rodríguez, C.E., A. Jelić, M. Llorca, M. Farré, G. Caminal, M. Petrović et al. 2011. Solid-phase treatment with the fungus *Trametes versicolor* substantially reduces pharmaceutical concentrations and toxicity from sewage sludge. Bioresour. Technol. 102: 5602–5608.
Rodríguez-Rodríguez, C.E., E. Barón, P. Gago-Ferrero, A. Jelić, M. Llorca, M. Farré et al. 2012a. Removal of pharmaceuticals, polybrominated flame retardants and UV filters from sludge by the fungus *Trametes versicolor* in bioslurry reactor. J. Hazard. Mater. 233-234: 235–243.
Rodríguez-Rodríguez, C.E., M.J. García-Galán, P. Blánquez, M.S. Díaz-Cruz, D. Barceló, G. Caminal et al. 2012b. Continuous degradation of a mixture of sulfonamides by *Trametes versicolor* and identification of metabolites from sulfapyridine and sulfathiazole. J. Hazard. Mater. 213-214: 347–354.
Rodríguez-Rodríguez, C.E., A. Jelić, M.A. Pereira, D.Z. Sousa, M. Petrović, M.M. Alves et al. 2012c. Bioaugmentation of sewage sludge with *Trametes versicolor* in solid-phase biopiles produces degradation of pharmaceuticals and affects microbial communities. Environ. Sci. Technol. 46: 12012–12020.
Rodríguez-Rodríguez, C.E., V. Castro-Gutiérrez, J.S. Chin-Pampillo and K. Ruiz-Hidalgo. 2013. On-farm biopurification systems: role of white rot fungi in depuration of pesticide-containing wastewaters. FEMS Microbiol. Lett. 345: 1–12.
Rodríguez-Rodríguez, C.E., D. Lucas, E. Barón, P. Gago-Ferrero, D. Molins-Delgado, S. Rodríguez-Mozaz et al. 2014. Re-inoculation strategies enhance the degradation of emerging pollutants by fungal bioaugmentation in sewage sludge. Bioresour. Technol. 168: 180–189.
Rodríguez-Rodríguez, C.E., K. Madrigal-León, M. Masís-Mora, M. Pérez-Villanueva and J.S. Chin-Pampillo. 2017. Removal of carbamates and detoxification potential in a biomixture: Fungal bioaugmentation versus traditional use. Ecotox. Environ. Safe. 135: 252–258.
Rubilar, O., G. Feijoo, C. Diez, T.A. Lu-Chau, M.T. Moreira and J.M. Lema. 2007. Biodegradation of pentachlorophenol in soil slurry cultures by *Bjerkandera adusta* and *Anthracophyllum discolor*. Ind. Eng. Chem. Res. 46: 6744–6751.
Rubilar, O., G. Tortella, M. Cea, F. Acevedo, M. Bustamante, L. Gianfreda et al. 2011. Bioremediation of a Chilean Andisol contaminated with pentachlorophenol (PCP) by solid substrate cultures of white-rot fungi. Biodegradation 22: 31–41.
Ruiz-Aguilar, G.M., J.M. Fernández-Sánchez and H. Poggi-Varaldo. 2002. Degradation by white-rot fungi of high concentrations of PCB extracted from a contaminated soil. Adv. Environ. Res. 6: 559–568.
Ruiz-Hidalgo, K., J.S. Chin-Pampillo, M. Masís-Mora, E. Carazo R. and C.E. Rodríguez-Rodríguez. 2014. Degradation of carbofuran by *Trametes versicolor* in rice husk as a potential lignocellulosic substrate for biomixtures: from mineralization to toxicity reduction. Process Biochem. 49: 2266–2271.
Ruiz-Hidalgo, K., J.S. Chin-Pampillo, M. Masís-Mora, E. Carazo R. and C.E. Rodríguez-Rodríguez. 2016a. Optimization of a fungally bioaugmented biomixture for carbofuran removal in on-farm biopurification systems. Water Air Soil Pollut. 227: 3.
Ruiz-Hidalgo, K., M. Masís-Mora, E. Barbieri, E. Carazo-Rojas and C.E. Rodríguez-Rodríguez. 2016b. Ecotoxicological analysis during the removal of carbofuran in fungal bioaugmented matrices. Chemosphere 144: 864–871.
Sánchez, C. 2009. Lignocellulosic residues: Biodegradation and bioconversion by fungi. Biotechnol. Adv. 27: 185–194.
Selvam, K., K. Swaminathan and K.S. Chae. 2003. Decolourization of azo dyes and a dye industry effluent by a white rot fungus *Thelephora* sp. Bioresour. Technol. 88: 115–119.
Sheldon, R.A. 2007. Enzyme immobilization: the quest for optimum performance. Adv. Synth. Catal. 349: 1289–1307.
Sniegowski, K., K. Bers, K. Van Goetem, J. Ryckeboer, P. Jaeken, P. Spanoghe et al. 2011. Improvement of pesticide mineralization in on-farm biopurification systems by bioaugmentation with pesticide-primed soil. FEMS Microb. Ecol. 76: 64–73.
Sniegowski, K., K. Bers, J. Ryckeboer, P. Jaeken, P. Spanoghe and D. Springael. 2012. Minimal pesticide-primed soil inoculum density to secure maximum pesticide degradation efficiency in on-farm biopurification systems. Chemosphere 88: 1114–1118.
Spiker, J.K., D.L. Crawford and R.L. Crawford. 1992. Influence of 2, 4, 6-trinitrotoluene (TNT) concentration on the degradation of TNT in explosive-contaminated soils by the white rot fungus *Phanerochaete chrysosporium*. Appl. Environ. Microbiol. 58: 3199–3202.
Spina, F., C. Cordero, T. Schilirò, B. Sgorbini, C. Pignata, G. Gilli et al. 2015. Removal of micropollutants by fungal laccases in model solution and municipal wastewater: evaluation of estrogenic activity and ecotoxicity. J. Cleaner Prod. 100: 185–194.

Steffen, K. and M. Tuomela. 2011. Fungal soil bioremediation: Developments towards large-scale applications. pp. 451–467. *In*: Hofrichter, M. [ed.]. The Mycota. Industrial Applications. Springer-Verlag, Berlin Heidelberg.
Tanaka, T., T. Tonosaki, M. Nose, N. Tomidokoro, N. Kadomura, T Fujii et al. 2001. Treatment of model soils contaminated with phenolic endocrine-disrupting chemicals with laccase from *Trametes* sp. in a rotating reactor. J. Biosci. Bioeng. 92: 312–316.
Tavares, A.P.M., M.A.Z. Coelho, M.S.M. Agapito, J.A.P. Coutinho and A.M.R.B. Xavier. 2006. Optimization and modeling of laccase production by *Trametes versicolor* in a bioreactor using statistical experimental design. Appl. Biochem. Biotechnol. 134: 233–248.
Tekere, M., I. Ncube, J.S. Read and R. Zvauya. 2002. Biodegradation of the organochlorine pesticide, lindane by a sub-tropical white rot fungus in batch and packed bed bioreactor systems. Environ. Technol. 23: 199–206.
Tortella, G.R., R.A. Mella-Herrera, D.Z. Sousa, O. Rubilar, G. Briceño, L. Parra et al. 2013. Carbendazim dissipation in the biomixture of on-farm biopurification systems and its effect on microbial communities. Chemosphere 93: 1084–1093.
Uhnáková, B., A. Petříčková, D. Biedermann, L. Homolka, V. Vejvoda, P. Bednář et al. 2009. Biodegradation of brominated aromatics by cultures and laccase of *Trametes versicolor*. Chemosphere 76: 826–832.
Valentín, L., G. Feijoo, M.T. Moreira and J.M. Lema. 2006. Biodegradation of polycyclic aromatic hydrocarbons in forest and salt marsh soils by white-rot fungi. Int. Biodeter. Biodegr. 58: 15–21.
Valentín, L., T.A. Lu-Chau, C. López, G. Feijoo, M.T. Moreira and J.M. Lema. 2007. Biodegradation of dibenzothiophene, fluoranthene, pyrene and chrysene in a soil slurry reactor by the white-rot fungus *Bjerkandera* sp. BOS55. Process Biochem. 42: 641–648.
Verma, P. and D. Madamwar. 2002. Production of lignolytic enzymes for dye decolorization by co-cultivation of white rot fungi *Pleurotus ostreatus* and *Phanerochaete chrysosporium* under solid state fermentation. Appl. Biochem. Biotechnol. 102/103: 109–118.
von Wirén-Lehr, S., M.d.P. Castillo, L. Torstensson and I. Scheunert. 2001. Degradation of isoproturon in biobeds. Biol. Fertil. Soils. 33: 535–540.
Wesenberg, D., I. Kyriakides and S.N. Agathos. 2003. White-rot fungi and their enzymes for the treatment of industrial dye effluents. Biotechnol. Adv. 22: 161–187.
Winquist, E., K. Björklöf, E. Schultz, M. Räsänen, K. Salonen, F. Anasonye et al. 2014. Bioremediation of PAH-contaminated soil with fungi—From laboratory to field scale. Int. Biodeter. Biodegr. 86: 238–247.
Yesilada, O., D. Asma and S. Cing. 2003. Decolorization of textile dyes by fungal pellets. Process Biochem. 38: 933–938.
Zhang, J., Z. Xu, H. Chen and Y. Zong. 2009. Removal of 2, 4-dichlorophenol by chitosan-immobilized laccase from *Coriolus versicolor*. Biochem. Eng. J. 45: 54–59.

9

Pesticides in the Environment

Biobed Systems as an Innovative Biotechnological Tool to Minimize Pollution

Gonzalo Tortella Fuentes,[1,2,*] *Gabriela Briceño,*[1,a] *Carlos E. Rodríguez-Rodríguez,*[3] *Sergio Cuozzo*[4,5] and *Olga Rubilar*[1,2,b]

Introduction

There are several chemical compounds widely used throughout the world to control a great variety of pests, principally in agriculture. These compounds are commonly known as insecticides, herbicides, nematicides, fungicides, among others. Pesticides have been in use since the 1960s to date (pyrethroids, organophosphate, carbamates, etc.) and have contributed significantly to pest control and food production. However, while benefits of pesticide application are evident, negative environmental impacts must also be considered. In this sense, pesticides were formulated to act on target microorganisms, but only a small proportion of the pesticide is effectively applied, and subsequently, a great proportion is localized in other environment compartments. The implications of pesticides in soil and water, especially regarding their persistence, biodegradation, impact on microbial communities, and effects on soil fertility, have become a matter of environmental concern in recent years.

Structure is an important condition for pesticide transference from soil to groundwater or surface water. The structure of each pesticide is closely tied to physical and chemical

[1] Departamento de Ingeniería Química, Universidad de La Frontera, Casilla 54-D, Temuco, Chile.
[a] Email: gabriela.briceno@ufrontera.cl
[b] Email: olga.rubilar@ufrontera.cl
[2] Centro de Excelencia en Investigación Biotecnológica Aplicada al Medio Ambiente (CIBAMA), Universidad de La Frontera, Casilla 54-D, Temuco, Chile.
[3] Centro de Investigación en Contaminación Ambiental (CICA), Universidad de Costa Rica, 2060 San José, Costa Rica.
Email: carlos.rodriguezrodriguez@ucr.ac.cr
[4] Planta Piloto de Procesos Industriales Microbiológicos (PROIMI-CONICET), Avenida Belgrano y Pasaje Caseros, 4000 Tucumán, Argentina.
Email: sergio.cuozzo@gmail.com
[5] Facultad de Ciencias Naturales e Instituto Miguel Lillo, Universidad Nacional de Tucumán, Miguel Lillo 205, 4000 Tucumán, Argentina.
* Corresponding author: gonzalo.tortella@ufrontera.cl

properties and therefore, inherent biodegradability. Moreover, other important factors must also be considered, such as pesticide concentration, soil characteristics (physical and chemical properties), among others. Figure 1 shows the principal movement of pesticides into the environment following applications.

It is important to note that there are two principal ways that pesticides enter the environment, known as point source and non-point source (diffuse) contamination. Non-point source is known as the contamination associated when pesticides are applied in the field and a great proportion of these are dispersed principally by the air and residues are localized distantly from the application point (Ravier et al. 2005). In this sense, pesticide residues are frequently found in surface and groundwater (Gilliom 2007, Hildebrandt et al. 2008). The possible damage from these pesticide residues or their metabolites is mainly related to the amount of pesticide and the duration of exposure (Onneby et al. 2010). However, although pollution can occur through diffusion, point source contamination has been indicated as the more prevalent cause of water pollution (Carter 2000, Mason et al. 1999). Therefore, in this chapter is described the pollution by pesticides in soil and water environment, effects on microbial populations and application of biobeds system to treat the point source contamination by pesticides.

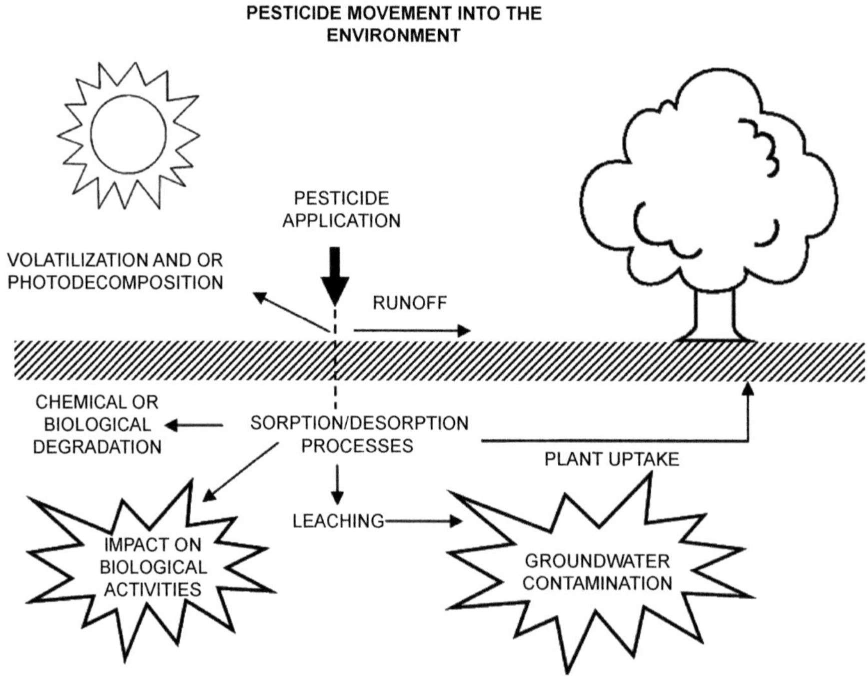

Figure 1. Schematic representation of the movement of pesticides in the environment.

Pesticide Point Source Contamination in the Agriculture

The use of chemical compounds as pesticides is necessary in modern farming and large-scale food production. These compounds allow control of plant pathogens and ensure high production levels. As pesticides are dangerous by nature, they can cause damage to aquatic and soil ecosystems. Therefore, the general public concern is to evaluate the possible effects of pesticides on ecosystems and human health (Tarazona et al. 2000, Karpouzas and Capri 2006).

The exposition of pesticides to the environment occurs, as was previously mentioned, generally by diffuse (non-point source) or localized (point source) contamination. Point source can occur through spillages, tank filling, cleaning the spraying equipment, among other ways (Fischer et al. 1996, Mason et al. 1999, Spliid et al. 1999). In countries including Germany and Belgium, it is estimated that between 40% and 90% of surface water contaminations are by point source (Huber et al. 2000, Müller et al. 2002, Jaeken and Debaer 2005). In another work, Neumann et al. (2002) evaluated water samples from five entry routes for 2 insecticides, 5 fungicides and 13 herbicides. Interestingly, they found that point source contamination was relevant for all the water samples evaluated and consequently non-point contamination was insignificant. Similarly, Helweg et al. (2002) evaluated secondary groundwater in two Danish counties below loading and mixing sites of pesticides. Relatively high concentrations of 24 pesticides and their metabolites were found and about 10% of samples contained more than 50 μg l^{-1}. In some investigation at the catchment scale, losses of 1–2% of applied pesticides have been found from diffuse contamination (non-point source), while most contamination derives from bad agronomic practices, for example, regarding preservation and spilling zones (Isensse and Sadeghi 1996, Torstensson and Castillo 1997). Studies conducted by Hildebrandt et al. (2007) reported that several pesticides in level between 0.57 and 5.37 μg l^{-1} were detected in water samples in agricultural areas of the Ebro River. Studies in Tanzania demonstrated that other types of plants have been affected by point source contamination (Mahugija and Kishimba 2005, Mahugija and Kishimba 2007). They found high concentrations of pesticides and metabolites in grasses, sedges and young leaves of mango tree, near the point source contamination. In Chile, studies of water and soil contamination are scarce. However, Báez et al. (1996) detected methabenzothiazuron (MBT) and other pesticides in the surface water in Aconcagua Valley. Barra et al. (2001) reported that organochlorine pesticides were found in four lakes in the Bío-Bío region of Chile. Further, Focardi et al. (1996) discovered pesticides pp′-DDT and lindane in the mouth of the Bío-Bío river.

Therefore, the study of new sustainable technologies or methods for reducing point-source pollution is critical to avoid difficult soil and water decontamination processes later on. In this sense, some studies have shown that certain materials, such as wood residues, can prevent soil and water pollution when used in biotechnology and biobeds (Rodriguez-Cruz et al. 2007, De Wilde et al. 2009). Biobed systems in agriculture have demonstrated the ability to retain and degrade pesticides from point source contamination.

However, before going into detail about the use of biobeds and their interesting characteristics, it is important to know the environmental impacts of pesticides. The main objective of pesticides is to maintain high food production and control plant pathogens, given the demand for food grows annually due to an increasing world population. Moreover, it is important to consider that pesticides and their residues affect not only target microorganisms, but also non-target microorganisms.

Effects of Pesticides on Soil Microbial Communities and Biological Activities

Several pesticides (herbicides, fungicides, acaricides, and insecticides) with a wide range of physicochemical properties are being applied in agriculture to protect crops against plant pathogens. These pesticides were formulated to inhibit the growth of target organisms; however, their residual effects have often been extended to non-target soil microorganisms, thereby causing changes in the turnover rate of nutrients, microbial community structure and soil quality, as has been reviewed by several authors (Wainwright 1978, Subhani et al. 2000, Johnsen et al. 2001, Chowdhury et al. 2008).

Digrak and Özcelik (1998) evaluated the effects of several insecticides and fungicides on actinomycetes, total aerobic bacteria, anaerobic bacteria and cellulolytic bacteria in agricultural soils contaminated with 1 mg kg^{-1} of pesticide. They found that the pesticides did not inhibit several bacterial and fungal groups, with the fungi being the least inhibited. Gigliotti et al. (1998) reported the effects produced by the application of two doses (16 and 160 μg kg^{-1}) of bensolfuron-methyl (BSM) on microbial quantity and activities. These authors reported that a concentration greater than normally applied in agriculture resulted in altered aspects of structure and activity in the microbial soil community. Moreover, only cellulolytic microorganisms in soil without previous pesticide applications were affected significantly in number, and with a higher dose (160 μg kg^{-1}) in soil pre-treated with BSM, substantial inhibition in nitrification was observed. Soil respiration was not influenced by herbicide application in any of the samples. In another work, Saeki and Toyota (2004) also evaluated the effects of BSM on microbial community and nitrification potential. They found that nitrification potential was significantly suppressed, although total bacterial number was not remarkably affected. Moreover, assays with denaturing gradient gel electrophoresis (DGGE) showed that bacterial community structure was not affected by BSM after 8 weeks of exposure. Others works with herbicides have also been evaluated.

Prado and Airoldi (2001) studied the toxicity on soil microflora caused by application of picloram in concentrations between 0 and 20 μg g^{-1} through microcalorimetric experiments. They reported that total thermal effects provoked in soil microorganisms were affected by increasing doses of picloram between 0 and 10 μg g^{-1} per g of soil. When 20 μg g^{-1} were added, the thermal was markedly affected, reaching a null value. Baxter and Cummings (2008) evaluated degradation of the herbicide bromoxynil and its impact on bacterial diversity by DGGE and quantitative PCR of five bacterial taxa. The results obtained showed that three applications of 10 μg kg^{-1} of bromoxynil resulted in rapid degradation (between 7 and 9 d) of the pesticide, and bacterial population profiles showed a significant similarity throughout time (28 d). However, when bromoxynil was added at 50 μg kg^{-1}, the degradation was preceded by a lag phase and only 50% of the concentration was degraded in 28 d. In another work, Ratcliff et al. (2006) reported that the herbicide glyphosate applied at the recommended field doses resulted in minimal changes to the microbial community structure. Similar results were found for total culturable bacteria, fungal hyphal length, biomass, carbon utilization profiles (using BIOLOG Ecoplate) and fungal and bacterial phospholipid fatty acid. On the other hand, the same authors reported that a high concentration of glyphosate (100 fold the field rate), simulating a spill of this compound, altered the bacterial community in all soil evaluated; however, an increment was observed in total bacteria, culturable bacteria, in relation at fungal:bacteria biomass after the application, indicating an enrichment of generalist bacteria. Changes in fungal properties were few and transient. Conversely, Ayansina and Oso (2006) reported that the herbicide atrazine and atrazine + metolachlor caused a decrease in microbial population, both at recommended and high doses. Herbicide application resulted in the elimination of some microbial species.

Several studies have also evaluated the effect of insecticides on soil microbiological communities. Digrak and Kazanici (2001) evaluated some organophosphorous insecticides and their effect on soil microbiota. They found that total viable bacteria, actinomycetes, anaerobic bacteria, aerobic endospores, cellulotytic microorganism and yeast-mold were not inhibited by pesticide presence. Moreover, pesticide application caused stimulation effects and consequently, total viable bacteria were found to be higher than the control. Another organophosphorous insecticide (chlorpyrifos) showed no significant effect on microbial biomass or microbial respiration of soil when applied at recommended doses (Adesodun et al. 2005). Similar results were found by Eisenhauer et al. (2009) where chlorpyrifos increased soil microbial biomass parameters. In the same way, Fang et al. (2008) studied the effects and persistence of repeated applications of chlorpyrifos. They reported that degradation rate for the insecticide increased whereas its inhibitory effects on soil microbial communities gradually decreased with the various applications. Different concentrations of chlorpyrifos in soil (4, 8, and 12 mg kg^{-1}) and their impact on soil microbial diversity was investigated by Fang et al. (2009). They reported a half-life of 14, 16 and 18 d. Further, the average well color development (AWCD, indicator of overall microbial activity) obtained with BIOLOG Microplate was inhibited by chlorpyrifos, but only within the first two weeks and thereafter showed recovery levels similar to the control. Still, when high concentrations of chlorpyrifos were used, inhibitory effects on soil microorganisms were aggravated.

Chu et al. (2008) evaluated the effects of chlorpyrifos alone or in combination with chlorothalonil on soil microbial populations. They reported that degradation of chlorpyrifos was not altered by combination with chlorothalonil, but inhibitory effects were observed on soil microorganisms when chlorpyrifos was applied in combination with this pesticide. When chlorpyrifos was applied alone, population of bacteria, fungi and actinomycetes were significantly reduced by 44, 61 and 72% respectively. With the combination of chlorothalonil, the inhibitory effects were incremented to 55, 79 and 85%, respectively. Therefore, the authors suggested that combined effects of pesticides should be accounted for to assess the actual impacts of pesticide application. The insecticide cypermethrin has been also assessed. Zhang et al. (2009) evaluated the effect of this insecticide on microbial community in the plant phyllosphere. DGGE and fatty acid analysis revealed that an increase in bacterial abundance and a shift in community composition were obtained after pesticide application. Similar results were reported by Zhang et al. (2008). Application of cypermethrin caused a significant increase in both total bacteria and bacterial biomass, but a decrease in fungal biomass. On the other hand, a significant change in bacterial community structure was observed.

Other studies have evaluated the effects of fungicides on microbial diversity and enzymes in soil. Yu et al. (2006) investigated the effect of repeated applications of chlorothalonil on bacterial, fungal, actinomycetes population, and several enzymes. Following the first application, both bacterial and actinomycetes populations were significantly reduced; however, population of fungi was unchanged. Soil microorganisms were affected strongly by the second application, but after initial variations, all microorganisms showed a gradual adaptation to chlorothalonil. On the other hand, these authors reported that the enzymes evaluated (acid phosphatase, alkaline phosphatase, urease, catalase and invertase) followed a similar trend to the inhibitory effects of soil microorganisms caused by chlorothalonil.

In more recent works, Gu et al. (2010) reported that through DGGE analysis, significant changes were observed in bacterial community following the application of enostroburin in wheat phyllosphere. In another investigation, Cycón et al. (2010) studied the response of indigenous microorganisms to fungicidal mixture application at three doses (15, 75, and 1500 mg kg^{-1}). They reported that the combination of mancozeb and dimethomorph in all concentrations evaluated, increased the ammonification rate and total number of culturable bacteria; however, it did not change the number of nitrate-reducing bacteria. White et al. (2010) evaluated the effects of 4 fungicides (chlorothalonil, 1.26 kg ha^{-1}; tebuconazole, 0.23 kg ha^{-1}; flutriafol, 0.13 kg ha^{-1} and cyproconazole, 0.084 kg ha^{-1}) on soil microbial dynamics. The results of soil microbial phospholipid fatty acids (PLFA) and neutral lipid fatty acids (NLFA) showed that the evaluated fungicides had little impact on microbial community dynamics.

As priorly exposed, bacterial community and biological activity in soil can be affected by a pesticide in a single dose or repeated applications; however, successive exposures to these agrochemicals (soil with historical applications) can invoke adaptation of these microorganisms, both in biomass or metabolic activities, which often causes a rapid degradation of pesticides.

Effects of Pesticides in Aquatic Ecosystems

After each application, the pesticide can be incorporated into water ecosystems through processes such as surface runoff, lixiviation, spray drifts, among others. An important issue is that pesticides rarely enter the environment as a single compound, and these can be found as a mixture of pesticides (Scholz et al. 2003). Depending on the chemical and physical properties of the pesticides, soil or water, the produced effects could vary and be dangerous or harmless. Pesticides are known to have negative impacts for target and non-target microorganisms. In this sense, Ahmed et al. (1998) evaluated pesticide residues such as chlorinated hydrocarbon in rainwater, soil and groundwater. They reported that residues detected in groundwater were higher than residues reported in studies conducted in Europe and that the concentration of pesticides evaluated caused a significant reduction in both heterotrophic bacteria and fungi. Rousis et al. (2017) evaluated pesticide residues and their metabolites in wastewater and surface water from Spain and Italy. They reported that seventeen pesticides and metabolites were found, with a higher concentration in surface waters than in wastewaters. Pinto et al. (2016) evaluated priority pesticides found in sediments from Obidos Lagoon in Portugal. However, based on sediment quality guidelines, the pesticides found showed no negative biological effects, except DDT, heptachlor and heptachlor epoxide. Qiu et al. (2017) evaluated persistent organic pollutants in seawater, sediments, phytoplankton and macroalgae (*Ulva lactuca*). They reported that all collected samples indicated polybrominated diphenyl ethers, and organochlorine pesticides were more prevalent in phytoplankton. Therefore, bioconcentration is an important route for pesticides and other pollutants towards higher trophic levels. Mauffret et al. (2017) reported that triazines at 1 $\mu g\ l^{-1}$ caused major effects in groundwater bacterial communities with no prior exposure to pesticides. However, in previously contaminated water, the effects are caused with concentration > 10 $\mu g\ l^{-1}$, demonstrating adaptation of the microorganism to triazines.

Adaptation of microorganisms to pesticides can be seen from two different points of view. On one hand, microbial adaptation to pesticides could signify an increase in the pesticide concentration required to obtain the desired effects. On the other hand, to obtain microorganisms adapted to pesticides, that could be used in bioremediation processes to allow a rapid and efficient degradation of the pollutants (accelerated degradation of pesticides) (Racke 1990, Arbeli and Fuentes 2007). This process has been evidenced in water, soil and bioprophylaxis systems (biobeds) as shown below.

Pesticide Biodegradation in Cultures Enriched and in Soils with Historical Application

Microbial communities, including fungi and bacteria, often have or develop the capacity to degrade xenobiotics in soil (Rabinovich et al. 2004, Ryan et al. 2008). These microorganisms can adapt in biomass or in metabolic activities due to availability and response to the presence of, for example, pesticides (Arbeli and Fuentes 2007). Application of pesticides in soils, with one or more historical applications of the same pesticide or different pesticides with similar chemical characteristics, can increase the rate of microbial degradation. This process is known as accelerated or enhanced degradation or cross acclimation (Racke 1990, Arbeli and Fuentes 2007).

Several pesticides have been satisfactorily degraded through accelerated biodegradation, as reviewed by Arbeli and Fuentes (2007). Early, Kirkland and Fryer (1972) observed that several herbicides, including 2-methyl-4-chlorophenoxyacetic acid (MCPA) and 4-(4-chloro-o-tolyloxy)butyric acid (MCPB), degraded more rapidly in soil treated with MCPA than untreated control soil. However, the rates of phytotoxic residues of dichlorprop, mecoprop, fenoprop and dicamba were unaffected by prior applications of MCPA. Later, Chapman and Harris (1990) reported that enhanced degradation of insecticides was observed when the soil was treated with 10 ppm of different pesticides. They reported an increase in the degradation rates from 2 to 100 folds depending on type of soil and that cross enhancement occurred among aryl methylcarbamates and a few similar organophosphorous insecticides.

Turco and Konopka (1990) studied the enhanced degradation of the insecticide carbofuran for soils with and without history of carbofuran application. The results showed a rapid breakdown of this insecticide (10 or 100 $\mu g\ g^{-1}$) in previously exposed soil, corresponding with a decrease in size of microbial biomass. They concluded that enhanced biodegradation of carbofuran occurs in a two-step process; first, hydrolysis and adsorption to soil, and later, following removal of carbofuran, biomass returned to nearly the control size. Cross-enhancement biodegradation of carbofuran in two soils (Montardon and Dijon) treated with carbamate pesticides were evaluated by Morel-Chevillet et al. (1996). For Montardon soil, degradation of carbofuran increased along with the number of carbofuran-degrading microorganisms. On the contrary, no degradation enhancement was observed for Dijon soil. Smelt et al. (1996) evaluated the accelerated degradation of nematicides in different soils treated annually during 3–10 yr with aldicarb, oxamyl, ethoprophos, fenamiphos or 1,3-dichloropropene. The results showed accelerated degradation for aldicarb, oxamyl, ethoprophos. Following five years without application, high degradation rates were also obtained for aldicarb and oxamyl. Cullington and Walker (1999) reported that repeated applications of diuron to the soil caused a rapid degradation of this pesticide (half-life time (DT_{50}) < 24 h).

In another work, Hole et al. (2001) reported that rapid degradation of the herbicide carbetamide was observed in soils pre-treated with this pesticide. DT_{50} of carbetamide in soil was reduced from 54 to 9 d. The authors indicate that rapid carbetamide degradation was probably linked to adaptation of soil bacteria. This observation was supported, given a rapid degradation of this herbicide was found in untreated soil amended with 1% of the adapted soil, when compared to the respective control. El-Sebai et al. (2005) evaluated mineralization of the herbicide isoproturon in soils previously exposed to this herbicide. They found 30–51% mineralization of isoproturon after one treatment and about 45–67% after two treatments. Moreover, after the second application of isoproturon, the rate of mineralization was enhanced in several soils. Walker and Welch (2006) reported no effects on simazine degradation in laboratory assays when the soil was pre-treated with four doses over 12 mon. However, propyzamide, linuron and alachlor degraded more quickly in treated soils than in control soils.

Sirisha et al. (2006) evaluated effects of repeated applications of the insecticide chlorpyrifos in surface (0–15 cm) and subsurface (40–50 cm) soil. They reported that degradation rate of chlorpyrifos was enhanced for both soils; however, the effects varied according to soil, mainly attributed to higher nutrient availability and microbial biomass in the surface soil. On the other hand, chlorpyrifos degradation was inhibited when soil was spiked with 1000 μg g^{-1}. Zabletewicz et al. (2006) evaluated atrazine degradation in soil and observed that mineralization of atrazine was extensive in all soils with prior applications. Cumulative mineralization was observed after 30 d of 45–72% in pre-treated soils, and only 5 to 10% in soil control without acclimation. Enhanced degradation of dichlobenil (2,6-dichlorobenzonitrile) and three metabolites: 2,6-dichlorobenzamide (BAM), 2,6-dichlorobenzoic acid (2,6-DCBA) and *ortho*-chlorobenzamide (OBAM) was evaluated by Holtze et al. (2007a). They found that dichlobenil and OBAM were degraded in soils with and without previous exposure to the herbicide; however, BAM and 2,6-DCBA were degraded 100% and 85–100% respectively in soils pre-treated with dichlobenil. Several studies with atrazine have been also realized.

Shaner and Henry (2007) evaluated the degradation of atrazine and metolachlor in soil pre-treated with atrazine. They found that half-life for atrazine and metolachlor ranged between 3.5 and 7.2 d and 17.9 and 18.8 d, respectively, in soil pre-treated with atrazine. Krutz et al. (2007) studied the enhanced degradation of atrazine in soils with continuous application of atrazine each yr, once every two yr, and soil with no history of atrazine. They reported that the persistence of atrazine in soil without atrazine history applications was 2-fold greater than in soil pre-treated with this pesticide, and that atrazine persistence may be reduced by at least 50% if this pesticide is applied to soil with historical applications. Krutz et al. (2008a) reported that cross adaptation occurred in soil pre-treated with s-triazines. Both atrazine and simazine were mineralized 3.5-fold higher in atrazine adapted soil than in soil that had not been pre-treated with the herbicide. Krutz et al. (2008b) evaluated atrazine dissipation in s-triazine adapted and non-adapted soil from Colorado and Mississippi. The degradation assays realized with atrazine showed that the average of half-life in adapted soil was 10-fold lower than in non-adapted soil. Moreover, degradation of atrazine metabolites (deisopropylatrazine and desethylatrazine) were also degraded in adapted soil. A contrary situation was observed in non-adapted soil where an accumulation of these metabolites was found. On the other hand, these authors reported that soil genes that code for enzymes able to rapidly catabolize atrazine (atzABC and trzN) were only detected in adapted soil. Tryky-Dotan et al. (2010) studied the effects in the accelerate degradation of methyl isothiocyanate (MITC), that is used in agriculture as a soil pesticide, mainly for protection against fungi and nematodes, following repeated application of metam sodium, a parent compound of MITC. They reported that MITC degradation was reduced in sterilized soil pre-treated with metam sodium, thereby suggesting a key role of soil microorganisms in accelerated degradation of MITC. On the other hand, rapid degradation of MITC was observed when the soil that had not been pre-treated with the pesticide was inoculated with soil with a history of metam sodium application. Moreover, the authors observed that an extensive change in total bacterial community occurred in concentrated soil extracts after a single metam sodium application.

Some studies in liquid cultures supplemented with soils pre-treated with pesticides have also been evaluated. Parekh et al. (1994) evaluated the degradation of carbofuran (10 mg l^{-1}) supplemented with pre-treated soils with the pesticide. They reported that all pre-treated soils degraded the pesticide more rapidly than untreated soils. In liquid enrichment culture, Cullington and Walker (1999) evaluated a bacterial isolate from an enhanced soil pre-treated with diuron. They observed the degradation of diuron occurred without the requirement of supplementary carbon and nitrogen sources. Furthermore, an addition of this isolate at 9.3×10^6 cfu g^{-1} to soil containing "aged" diuron residues resulted in rapid diuron degradation in both pre-fumigated and non-fumigated soil. Sørensen and Aamand (2003) studied the mineralization of [phenyl–U–^{14}C] labeled isoproturon (IPU), monodesmethyl-isoproturon (MDIPU) and 4-isopropylaniline (4-IA) (10 mg l^{-1}) in cultures supplemented with 5 g of soil pre-treated with isoproturon. They found an extensive mineralization for all three evaluated compounds in soils previously exposed to isoproturon. ^{14}C-MDIPU was mineralized between 50.9 to 57.8% after 30 d, and ^{14}C-IPU and ^{14}C-4IA were mineralized between 33.9–44.8% and 32.9–36%, respectively. Holtze et al. (2007b) evaluated the degradation of the herbicide 2,6-dichlorobenzamide (BAM) (50 mg l^{-1}) in eleven bacteria cultures from a soil exposed to dichlobenil. They reported that ten cultures mineralized the pesticide only in presence of supplementary carbon, and one was able to utilize BAM as a sole source of both carbon and nitrogen. Hussain et al. (2009) reported that isoproturon (50 mg l^{-1}) was rapidly mineralized by a bacteria culture isolated from an agricultural soil pre-exposed to isoproturon. After 48 hr, about 95% of the initial added isoproturon disappeared from the liquid culture, but 36% was mineralized to $^{14}CO_2$. Additionally, other structurally related phenylurea herbicides were evaluated; however no degradation of these compounds was obtained.

Biobed System: Organic Matrix, Pesticide Degradation and Microbial Communities

Biobeds are a simple and low-cost biological system used in farms that apply high quantities of pesticides. Initially, this bioprophylaxis system was developed in Sweden and is used to reduce point source water and soil contamination by pesticides (Fig. 2) (Torstensson and Castillo 1997). In Europe, the biobed system is widely used in agricultural activities and several studies have demonstrated that this biological system can effectively retain and degrade a wide range of pesticides alone or in combinations (Torstensson 2000, Coppola et al. 2007, Vischetti et al. 2008). In its original conformation, biobeds are built with simple and cheap components, composed mainly of a biomixture obtained through a mixture of straw (50%), peat (25%) and top soil (25%) (Torstensson and Castillo 1997). The straw is considered the main substrate because it stimulates the development of the white-rot fungi (lignin degrading fungi). The soil provides sorption sites and promotes microbial activity. The peat contributes to sorption capacity and moisture control. Biobed system has been studied in several countries and has been adapted to the local conditions of each region

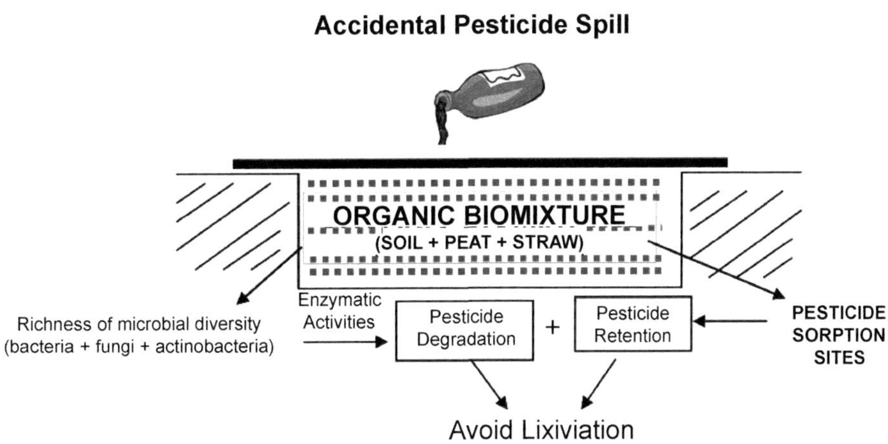

Figure 2. Schematic representation of biobed system.

(Henriksen et al. 2003, Fogg et al. 2003a,b, 2004a,b, Vischetti et al. 2004, Coppola et al. 2007, Chin-Pampillo et al. 2015). On the other hand, several factors can affect the performance and operation of biobeds, as has been reported by De Wilde et al. (2007) and Castillo et al. (2008).

Several pesticides and their metabolites have been evaluated in the context of the biobed system. In this sense, Coppola et al. (2007) evaluated chlorpyrifos degradation at 100 mg g^{-1} and its principal metabolite 3,5,6-trichloro-2-pyridinol (TCP), using a typical Swedish biomix and other biomixture composed of urban and garden composts. They reported that a higher ^{14}C-chlorpyrifos mineralization and lower TCP levels were obtained in the biomixture with garden and urban compost, alone or amended with straw. In another work, Castillo and Torstensson (2007) studied the effects of the composition, moisture and temperature on degradation of isoproturon, linuron, metamitron, methabenzthizuron, metribuzin, chloridazon and terbuthylazine in biobeds in a concentration of 700 µg g^{-1} of pesticide biomixture. They reported that most pesticides dissipated completely and their dissipation was correlated with respiration and/or phenoloxidase content. Moreover, more pesticides were dissipated at biobed moisture levels of 60% water holding capacity and 20°C. Vischetti et al. (2004) also evaluated the use of citrus peel, urban waste and public green compost in a modified biobeds system to the degradation of chlorpyrifos (insecticide, at 0.75 g l^{-1}), metalaxyl (fungicide, at 0.96 g l^{-1}) and imazamox (herbicide, at 0.075 g l^{-1}). They reported that chlorpyrifos was retained more than metalaxyl and imazamox probably due to physico-chemical characteristics of chlorpyrifos. Moreover, a rapid degradation of all pesticides occurred in the modified biobed, and the half-life was less than 14 d, compared with the values for soil alone (60–70 d).

Degradation of mecoprop and isorpoturon at concentrations from 0.0005 to 25.000 mg kg^{-1} were evaluated in biobed by Henriksen et al. (2003). They reported that mecoprop was degraded quickly in concentration between 0.0005 to 500 mg kg^{-1} and in elevated concentrations, the degradation was much delayed. For isoproturon, the degradation was rapid in low concentration and decreasing in high concentration. After 120 d, between 5 and 8% of ^{14}C was involved at concentrations between 0.0005 and 25.000 mg kg^{-1}. Fogg et al. (2003a) studied the degradation of a mixture of pesticides (isoproturon, pendimethalin, chlorpyrifos, chlorothalonil, epoxiconazole and dimethoate) in the biobed system at concentrations between 13 and 94 mg kg^{-1} and when these are applied in repeated applications. They found some interaction between pesticides, but were less significant in the biomix more than in top soil. On the other hand, when the pesticides were added in three occasions at 30 d intervals, the degradation was significantly quicker in biomix more than in top soil; however, the degradation was affected with the successive application. The authors reported that negative effects in the degradation were possibly due to the toxic effect of the pesticide on the microbial community. However, the authors did not evaluate biological activity. Fogg et al. (2004a) evaluated degradation of several pesticides in the biobed system and found that all applied pesticides were effectively degraded and retained within the 0–5 cm. The most mobile pesticides leached in the biobeds, but 99% of these were removed. In another work, Fogg et al. (2004b) evaluated the degradation of isoproturon, dimethioate and mecoprop-P in biobeds system with different soil textures. They reported that > 98% of the pesticides retained were degraded in all evaluated soils. Values of DT_{50} determined for all pesticides evaluated in this study, and applied 4-fold more than of the recommended doses, were similar across the biomix types and less than or equal to reported DT_{50} values for soil. Vischetti et al. (2008) studied the degradation of chlorpyrifos (10 and 50 mg kg^{-1}) and metalaxyl (100 mg kg^{-1}) in biobeds system. These authors evaluated the effects of initial concentration, co-application and repeated application. In all assays, results showed degradation of chlorpyrifos decreased with an increasing concentration of chlorpyrifos. On the other hand, degradation of metalaxyl increased when this pesticide was applied together with chlorpyrifos. Spliid et al. (2006) evaluated the degradation of 21 pesticides in a full-scale model biobed. They found that after 169 d, all pesticides in the biobed profile were degraded more than 50% of the initial dose and the pesticides with soil organic carbon-water partitioning coefficient (K_{oc}) values above 100 were degraded more slowly due to low bioavailability.

Recently, Urrutia et al. (2013) evaluated different lignocellulosics wastes including barley husk, sawdust and oat husk, as an alternative to wheat straw, given this lignocellulosic substrate is not readily available everywhere. They reported that atrazine, chlorpyrifos and isoproturon were efficiently degraded when oat husk was added to the biomixture. Half-life ($t_{1/2}$) values were 28, 58 and 26 d respectively. Moreover, when the biomixture was amended with barley husk and sawdust as partial replacement (25%) of wheat straw, similar degradation values were obtained. Interestingly, when the biomixture was amended with barley husk and sawdust as total replacement of wheat straw, half-life values increased to over 100 d. Diez et al. (2013) evaluated the degradation of six pesticides applied repeatedly and a pesticide mixture in a biomixture where straw was partially replaced by barley husk (25%), and pine sawdust. These authors reported a quick degradation (between 50 to 90%) of all pesticides when barley husk was added to the biomixture.

Traditional biomixture, composed of straw, peat and top soil, has been also evaluated to degrade pesticides applied in high concentrations. Tortella et al. (2013a) evaluated atrazine degradation at 40 mg kg^{-1} applied at 0, 30 and 60 d. They reported that the pesticide was degraded quickly in the biomixture and low metabolite formation was detected. Similar results were reported by Tortella et al. (2013b, 2014), wherein repeated applications of carbendazime and diazinon were quickly degraded, even observing an accelerated degradation after each application of pesticides, which resulted in a decrease of $t_{1/2}$ of the pesticides. More recently, Castro-Gutiérrez et al. (2017) evaluated the capacity of a biomixture composed of straw, compost and soil to degrade carbofuran for one continual yr under successive applications. They reported pesticide degradation was low (9.9%) 48 hr after the first application, but a possible microbial adaptation was observed and carbofuran degradation reached > 88.5% after 6 mon of pesticide application.

Microbiological Performance of Biomixture of Biobeds

Although the degradation of pesticides in biobeds has been reported in the literature, as shown above, few studies have correlated pesticide degradation with microbial activity or evaluated the effect of pesticides on microbial community and enzymatic activities in the biobed systems (Coppola et al. 2007, Castillo and Torstensson 2007, Vischetti et al. 2008, Karanasios et al. 2012). This knowledge is important because an understanding of these processes in biobeds has the potential to guide the performance of microbiological parameters when the biomixture is exposed to higher pesticide concentrations.

A good biomixture is based in their efficiency to retain and degrade pesticides alone or in mixture and, therefore, the biomixture must have good adsorption capacity and high biological catalytic activity for pesticide degradation (Castillo et al. 2008). Several factors have been reported to affect efficiency of the biomixture in relation to the pesticide degradation, such as temperature, composition, moisture, maturity (age) of the biomixture, and components of the substrates used, such as terpenes, among others (Fogg et al. 2004b, Castillo and Torstensson 2007, Castillo et al. 2008, Tortella et al. 2012, 2013c, Castro-Gutierrez et al. 2017). However, microbial robustness is an important parameter, given that pesticide degradation is only dependent on physico-chemical characteristics of the biomixture and pesticides. Favorable pesticide degradation in biobeds system can only be achieved by maintaining the environmental (moisture, temperature, C/N relation, nutrients, among others) conditions necessary to the development of a robust microbial community with capacity to degrade pesticides (Castillo et al. 2008, Vischetti et al. 2007, Coppola et al. 2011). Knowledge of microbial parameters is important because pesticide exposure can alter enzymatic activities, microbial community structure and biomixture quality, affecting efficient pesticide degradation, as has been reported in soil and reviewed by several authors (Subhani et al. 2000, Johnsen et al. 2001, Chowdhury et al. 2008). Microbiological information will provide an understanding of the long-term sustainability of the biological system, in order for application in agricultural processes and as a complement to good agricultural practices.

Microorganism responses have varied in soil or similar substrates, according to the biomixture as compost or the addition of pesticides, and either stimulation or inhibition has been reported (Johnsen et al. 2001, Fernández-Gómez et al. 2011). Some studies have evaluated the effect of pesticides on microbial communities in the biomixture of biobeds. In comparison to the traditional biomixture, some change in the composition were made (soil:peat:straw). In general, it has been reported that pesticide persistence increased with higher concentrations and repeated applications, which could cause an inhibitory effect on microbial activity (Fogg et al. 2003a, Castillo et al. 2008, Vischetti et al. 2008). Coppola et al. (2011) evaluated the effect of different fungicides on microbial diversity of a biomixture composed of ligneous pruning residues and straw (60:40 vv^{-1}), and observed some changes with DGGE analysis after the addition of pesticide. Vischetti et al. (2008) evaluated the impact of chlorpyrifos, metalaxyl and glyphosate on biochemical parameters in a biomixture composed of urban waste compost. They reported that chlorpyrifos initially had a negative effect on microbial biomass content, and also observed a decrease in the fluorescein diacetate hydrolysis, while metalaxyl and glyphosate increased the microbial biomass content during assays. In another work, Sniegowski et al. (2011) evaluated degradation of the pesticide linuron in a biomixture of biobed system. Biomixture was evaluated with and without addition of linuron-primed soil. They showed that addition of linuron-primed soil allowed a rapid establishment of the microbial community with a higher capacity to degrade linuron. Moreover, no important changes were observed in microbial communities when evaluated by DGGE. The authors observed some variations, generally associated to environmental parameters such as low temperatures or periods of drought. Marinozzi et al. (2013) evaluated dissipation of fungicides used commonly in vineyards, such as azoxystrobin, fludioxonil and peconazole and their effects on microbial communities of biomixture of biobeds system. The authors reported that the effect of pesticides on microbial communities is dependent on the type of pesticide used. Peconazole caused adverse effects on size and microbial activity, and changes in structure were observed for peconazole and fludioxonil. However, the authors reported that adverse effects were only transitory and a complete recovery of the activity was noted after 60 d of the pesticide application. Tortella et al. (2013a,b) evaluated degradation of atrazine and carbendazime when the pesticides were applied successively (three times) at higher concentrations, and its effects on microbial communities and biological activities in a biomixture composed by straw-peat-soil (50–25–25% vv^{-1}). The authors reported that biological activities such as alkaline phosphatase, dehydrogenase, phenoloxidase and fluorescein diacetate hydrolysis were negatively affected after the first addition of pesticide. However, all activities were recovered within 60 d following application. In these same works, the authors evaluated community level physiological profiles by BiologEcoplate™ system and microbial communities by DGGE system. The authors reported that physiological profiles showed no important changes, and the only minimal variations found were not attributed to the pesticide. More recently, Castro-Gutierrez et al. (2017) evaluated the effects of carbofuran on microbial communities of a biomixture composed of straw-compost-soil, and exposed to pesticides during one yr. The authors reported rapid pesticide degradation and observed only minimal changes in bacterial and fungal communities.

In general, studies related with microbial communities in biobed systems have demonstrated that the composition of biomixture straw-peat-soil, or modifications, create conditions for the development of a robust microbial community that allows pesticide degradation, even when pesticides are applied as mixtures and in high concentrations. However, it is necessary to research microbiological parameters in the long term. The biobed system is built to operate for a long period of time, and thus far, no works have evaluated microbiological quality after intensive use. Moreover, effects of other chemicals applied in agriculture, in combination with pesticides such as sulfates, copper derivates, and organics oils, have not been evaluated.

Concluding Remarks and Future Perspectives

It is well known that pesticides can be detrimental to the environment. Surface water, groundwater and soil are the principal sink of these pollutants and their effects are well documented. Agricultural activities can be seen as the principal point where pesticides are released, and point source and non-point source contamination are well identified. It is clear that pesticides came into the world to stay for a long time. Higher demand for food, climate change, development of new pests, economic loss, among others conditions, make pesticides a fundamental tool, principally in extensive farm activities. In this context, good practices in agriculture reach a higher demand every day, especially to reduce pesticide pollution. Point source contamination has been indicated as the principal cause of the presence of pesticides in soil and groundwater. Accidental spills and cleaning equipment can cause a great accumulation of pesticides in the environment. Therefore, biotechnological applications such as the biobed system have been demonstrated to be an interesting alternative to decrease environmental pesticide levels. Biobeds are a bioprophylaxis system which take advantage of the capacity of the microorganisms (fungi, bacteria and actinobacteria) to degrade a wide variety of compounds and furthermore, is an easy and cheap technology that can be installed in farms. Biobeds have been implemented in several countries in Europe and Latin America, and several studies have demonstrated that this simple biotechnological tool can efficiently degrade pesticides.

Although the effectiveness of the biobeds system has been established and evidenced in several scientific reports, its installation in a specific locality requires availability of substrates previously evaluated to ensure optimum performance. As mentioned above, it is also necessary to understand the effects caused by the presence of other components such as oils, sulfur compound or pesticides, derivates of copper as copper oxychloride, given that these could alter chemical properties or damage microbiological communities, thereby affecting the degradation capacity of biobed systems. On the other hand, it is important to consider a future change in environmental conditions, such as temperature increase, decreasing pluviometry, among others, that could have significant effects on the behavior of microbial communities in the biomixture or on the behavior of contaminants. Finally, and equally important, is the necessity of taking into account the new pollutants that are being produced and proposed as possible substitutes of traditional pesticides, such as metal nanoparticles. Therefore, future studies should evaluate their incorporation into the biomixture, and subsequent impacts in pesticide degradation and microbial community in order to contribute a better understanding of the interaction of metal nanoparticles and pesticides in biobed systems, providing relevant ecotoxicological information as a basis for a comprehensive risk assessment.

Acknowledgements

The authors gratefully acknowledge the support of the Centro de Excelencia en Investigación Biotecnológica Aplicada al Medio Ambiente (CIBAMA), Centro de Investigación en Contaminación Ambiental (CICA), Universidad de Costa Rica, Planta Piloto de Procesos Industriales Microbiológicos (PROIMI-CONICET) and Fondecyt Project 1161713.

References Cited

Adesodun, J.K., D.A. Davidson and D.W. Hopkins. 2005. Micro-morphological evidence for changes in soil faunal activity following application of sewage sludge and biocide. Appl. Soil Ecol. 29: 39–45.
Ahmed, M.T., S.M.M. Ismail and S.S. Mabrouk. 1998. Residues of some chlorinated hydrocarbon pesticides in rain water, soil and ground water, and their influence on some soil microorganisms. Environ. Int. 24: 665–670.
Arbeli, Z. and C.L. Fuentes. 2007. Accelerated biodegradation of pesticides: An overview of the phenomenon, its basis and possible solutions; and a discussion on the tropical dimension. Crop Pro. 26: 1733–1746.
Ayansina, A.D.V. and B.A. Oso. 2006. Effect of two commonly used herbicides on soil microflora at two different concentrations. Afr. J. Biotechnol. 5: 129–132.
Baéz, M., M. Rodriguez, O. Lastra, A. Peña, C. De La Colina and F. Sanchez-Rasero. 1996. Residuos de plaguicidas en aguas superficiales de la V región de Chile. Estudio prospectivo. Bol. Soc. Chil. Quim. 41: 271–276.
Barra, R., M. Cisternas, R. Urrutia, K. Pozo, P. Pacheco, O. Parra et al. 2001. First report on chlorinated pesticide deposition in a sediment core from a small lake in central Chile. Chemosphere 45: 749–757.
Baxter, J. and S.P. Cummings. 2008. The degradation of the herbicide bromoxynil and its impact on bacterial diversity in a top soil. J. Appl. Microbiol. 104: 1605–1616.
Carter, A.D. 2000. How pesticides get into water and proposed reduction measures. Pesticide Outlook. 11: 149–157.
Castillo, M.D. and L. Torstensson. 2007. Effect of biobed composition moisture, and temperature on the degradation of pesticides. J. Agric. Food. Chem. 55: 5725–5733.
Castillo, M. del P., L. Torstenson and J. Stentröm. 2008. Biobeds for environmental protection from pesticide Use—A review. J. Agric. Food Chem. 56: 6206–6219.
Castro-Gutiérrez, V., M. Masís-Mora, M.C. Diez, G.R. Tortella and C.E. Rodríguez-Rodríguez. 2017. Aging of biomixtures: Effects on carbofuran removal and microbial community structure. Chemosphere 168: 418–425.
Chapman, R.A. and C.R. Harris. 1990. Enhanced degradation of insecticides in soil: Factors influencing the development and effects of enhanced microbial activity. pp. 82–96. *In*: Racke, K.D. and J.R. Coats [eds.]. Enhanced Biodegradation of Pesticides in the Environment. American Chemical Society, Washington, DC.
Chin-Pampillo, J.S., K. Ruiz-Hidalgo, M. Masís-Mora, E. Carazo-Rojas and C. Rodríguez-Rodríguez. 2015. Adaptation of biomixtures for carbofuran degradation in on-farm biopurification systems in tropical regions. Environ. Sci. Pollut. Res. 22: 9839–9848.
Chowdhury, A., S. Pradhan, M. Saha and N. Sanyal. 2008. Impact of pesticides on soil microbiological parameters and possible bioremediation strategies. Indian J. Microbiol. 48: 114–127.
Chu, X., H. Fang, X. Pan, X. Wang, M. Shan, B. Feng et al. 2008. Degradation of chlorpyrifos alone and in combination with chlorothalonil and their effects on soil microbial populations. J. Environ. Sci. 20: 464–469.
Coppola, L., M.d.P. Castillo, E. Monaci and C. Vischetti. 2007. Adaptation of the biobed composition for chlorpyrifos degradation to southern Europe conditions. J. Agric. Food Chem. 55: 396–401.
Coppola, L., F. Comitini, C. Casucci, V. Milanovic, E. Monaci, M. Marinozzi et al. 2011. Fungicides degradation in an organic biomixture: Impact on microbial diversity. New Biotech. 29: 99–106.
Cycón, M., Z. Piotrwska-Seget and J. Kozdrój. 2010. Responses of indigenous microorganisms to a fungicidal mixture of mancozed and dimethomorph added to sandy soils. Int. Biodeter. Biodegr. 64: 316–323.
Cullington, J.E. and A. Walker. 1999. Rapid biodegradation of diuron and other phenylurea herbicides by a soil bacterium. Soil Biol. Biochem. 31: 677–686.
De Wilde, T., P. Spanoghe, R.C. Debae, J. Ryckeboer, D. Springael and P. Jaeken. 2007. Overview of on-farm bioremedation systems to reduce the occurrence of point source contamination. Pest. Manag. Sci. 63: 111–128.
De Wilde, T., P. Spanoghe, J. Ryckeboer, P. Jaeken and D. Springael. 2009. Sorption characteristics of pesticides on matrix substrates used in biopurification systems. Chemosphere 75: 100–108.

Diez, M.C., G.R. Tortella, G. Briceño, M. del P. Castillo, J. Díaz, G. Palma et al. 2013. Influence of novel lignocellulosic residues in a biobed biopurification system on the degradation of pesticides applied in repeatedly high doses. Electron J. Biotechnol. 16: 1–11.
Digrak, M. and S. Özçelik. 1998. Effect of some pesticides on soil microorganisms. Bull. Environ. Contam. Toxicol. 60: 916–922.
Digrak, M. and F. Kazanici. 2001. Effect of some organophosphorous insectiside on soil microorganisms. Turk. J. Biol. 25: 51–58.
Eisenhauer, N., M. Klier, S. Partsch, A.C.W. Sabais, Ch. Scherber, W.W. Weisser et al. 2009. No interactive effects of pesticide and plant diversity on soil microbial biomass and respiration. Appl. Soil Ecol. 42: 31–36.
El-Sebai, T., B. Lagacherie, J.F. Cooper, G. Soulas and F. Martin-Laurent. 2005. Enhanced isoproturon mineralisation in a clay silt loam agricultural soil. Agron. Sustain. Dev. 25: 271–277.
Fang, H., Y.L. Yu, X.G. Wang, X.Q. Chu, X.D. Pan and X.E. Yang. 2008. Effects of repeated applications of chlorpyrifos on its persistence and soil microbial functional diversity and development of its degradation capability. Bull. Environ. Contam. Toxicol. 81: 397–400.
Fang, H., Y. Yu, X. Chu, X. Wang, X. Yang and J. Yu. 2009. Degradation of chlorpyrifos in laboratory soil and its impact on soil microbial functional diversity. J. Environ. Sci. 21: 380–386.
Fernández-Gómez, M.J., R. Nogales, H. Insam, E. Romero and M. Goberna. 2011. Role of vermicompost chemical composition, microbial functional diversity, and fungal community structure in their microbial respiratory response to three pesticides. Bioresource Technol. 102: 9638–9645.
Fischer, P., M. Bach, J. Burhenne, M. Spiteller and H.G. Frede. 1996. Pesticides in streams. Part 3: non-point and point sources in small streams. DGM 40: 168–173.
Focardi, S., C. Fossi, C. Leonzio, S. Corsolini and O. Parra. 1996. Persistent organochlorine residues in fish and water birds from the Biobio River, Chile. Environ. Monitor. Assess. 43: 73–92.
Fogg, P., A.B. Boxall and A. Walker. 2003a. Degradation of pesticides in biobeds: the effect of concentration and pesticide mixtures. J. Agric. Food Chem. 51: 5344–5349.
Fogg, P., A.B. Boxall, A. Walker and A.A. Jukes. 2003b. Pesticide degradation in a "biobed" composting substrate. Pest. Manag. Sci. 59: 527–537.
Fogg, P., A.B. Boxall, A. Walker and A. Jukes. 2004a. Effect of different soil textures on leaching potential and degradation of pesticides in biobeds. J. Agric. Food Chem. 52: 5643–5652.
Fogg, P., A.B. Boxall, A. Walker and A. Jukes. 2004b. Effect of different soil textures on leaching potential and degradation of pesticides in biobeds. J. Agric. Food Chem. 52: 5643–5652.
Gilliom, R.J. 2007. Pesticides in U.S. streams and groundwater. Environ. Sci. Technol. 41: 3408–3414.
Gigliotti, C., L. Allievi, C. Salardi, F. Ferrari and A. Farini. 1998. Microbial ecotoxicity and persistence in soil of the herbicide bensulfuron-methyl. J. Environ. Sci. Heal. B. 33: 399–409.
Gu, L., Z. Bai, B. Jin, Q. Hu, H. Wang, G. Zhuang et al. 2010. Assessing the impact of fungicide enostroburin application on bacterial community in wheat phyllosphere. J. Environ. Sci. 22: 134–141.
Helweg, A., H. Bay, H.P.B. Hansen, M. Rabolle, A. Sonnenborg and L. Stenvang. 2002. Pollution at and below sites used for mixing and loading of pesticides. Int. J. Environ. An. Ch. 82: 583–590.
Henriksen, V.V., A. Helweg, N.H. Spliid, G. Felding and L. Stenvang. 2003. Capacity of model biobeds to retain and degrade mecoprop and isoproturon. Pest. Manag. Sci. 59: 1076–1082.
Hildebrandt, A., S. Lacorte and D. Barceló. 2007. Assessment of priority pesticides, degradation products and pesticide adjuvant in groundwater and top soils from agricultural areas of the Ebro river basin. Anal. Bioanal. Chem. 387: 1459–1468.
Hildebrandt, A., M. Guillamón, S. Lacorte, R. Tauler and D. Barceló. 2008. Impact of pesticides used in agriculture and vineyards to surface and groundwater quality (North Spain). Water Res. 42: 3315–3326.
Holtze, M.S., H.C. Hansen, R.K. Juhler, J. Sørensen and J. Aamand. 2007a. Microbial degradation pathways of the herbicide dichlobenil in soils with different history of dichlobenil-exposure. Environ. Pollut. 148: 343–351.
Holtze, M.S., S.R. Sørensen, J. Sørensen, H.C.B. Hansen and J. Aamand. 2007b. Biostimulation and enrichment of 2,6-dichlorobenzamide-mineralising soil bacterial communities from dichlobenil-exposed soil. Soil Biol. Bioch. 39: 216–223.
Hole, S.J., N.C. McClure and S.B. Powles. 2001. Rapid degradation of carbetamide upon repeated application to Australian soils. Soil Biol. Biochem. 33: 739–745.
Huber, A., M. Bach and H.G. Frede. 2000. Pollution of surface waters with pesticides in Germany: modeling non-point source inputs. Agric. Ecosys. Environ. 80: 191–204.
Hussain, S., S.R. Sorensen, M. Devers-Lamrani, T. El-Sebai and F. Martin-Laurent. 2009. Characterization of an isoproturon mineralizing bacterial culture enriched from a French agricultural soil. Chemosphere 77: 1052–1059.
Isensee, A.R. and A.M. Sadeghi. 1996. Effect of tillage reversal on herbicide leaching to groundwater. Soil Sci. 161: 382–389.
Jaeken, P. and C. Debaer. 2005. Risk of water contamination by plant protection products (PPP) during pre- and post treatment operations. Annu. Rev. Agric. Eng. 4: 93–114.

Johnsen, K., C. Jacobsen, V. Torsvik and J. Sorensen. 2001. Pesticide effect on bacterial diversity in agricultural soils—a review. Biol. Fert. Soils. 33: 443–453.
Karanasios, E., D.G. Karpouzas and N.G. Tsiropoulos. 2012. Key parameters and pesticide practices controlling pesticide degradation efficiency of biobed substrates. J. Environ. Sci. Health B. 47: 589–598.
Karpouzas, D. and E. Capri. 2006. Risk analysis of pesticides applied to rice paddies using RICEWQ 1.6.2v and RIVWQ 2.02. Paddy Water Environ. 4: 29–38.
Kirkland, K. and J.D. Fryer. 1972. Degradation of several herbicides in a soil previously treated with MCPA. Weed Res. 12: 90–95.
Krutz, L.J., R.M. Zablotowicz, K.N. Reddy, C.H. Koger and M.A. Weaver. 2007. Enhanced degradation of atrazine under field conditions correlates with a loss of weed control in the glasshouse. Pest. Manag. Sci. 63: 23–31.
Krutz, L.J., I.C. Burke, K.N. Reddy and R. Zablotowicz. 2008a. Evidence for cross-adaptation between s-triazine herbicides resulting in reduced efficacy under field conditions. Pest. Manag. Sci. 64: 1024–1030.
Krutz, L.J., D.L. Shaner, C. Accinelli, R.M. Zablotowicz and W.B. Henry. 2008b. Atrazine dissipation in s-triazine-adapted and nonadapted soil from Colorado and Mississippi: implications of enhanced degradation on atrazine fate and transport parameters. J. Environ. Qual. 37: 848–857.
Mahugija, J. and M. Kishimba. 2005. Concentrations of pesticide residues in grasses and sedges due to point source contamination and the indications for public health risks, Vikuge, Tanzania. Chemosphere 61: 1293–1298.
Mahugija, J. and M. Kishimba. 2007. Organochlorine pesticides and metabolites in young leaves of Mangifera indica from sites near a point source in Coast region, Tanzania. Chemosphere 68: 832–837.
Marinozzi, M., L. Coppola, E. Monaci, D.G. Karpouzas, E. Papadopoulou, U. Menkissoglu-Spiroudi et al. 2013. The dissipation of three fungicides in a biobed organic substrate and their impact on the structure and activity of the microbial community. Environ. Sci. Pollut. Res. Int. 20: 2546–2555.
Mason, P.J., I.D.L. Foster, A.D. Carter, A. Walker, S. Higginbotham, R.L. Jones et al. 1999. Relative importance of point source contamination of surface waters: River Cherwell catchment monitoring study. pp. 405–412. *In*: Del Re, A.A.M., C. Brown, E. Capri, G. Errera, S.P. Evans and M. Trevisan [eds.]. Human and Environmental Exposure to Xenobiotics. La Goliardica Pavese: Pavia, Italy.
Maufrfet, A., N. Baran and C. Joulian. 2017. Effects of pesticides and metabolites on groundwater bacterial communitiy. Sci. Total Environ. 576: 879–887.
Morel-Chevillet, C., N.R. Parekh, D. Pautrel and J.C. Fournier. 1996. Cross-enhancement of carbofuran biodegradation in soil previously treated with carbamate pesticides. Soil Biol. Biochem. 28: 1767–1776.
Müller, K., M. Bach, H. Hartmann, M. Spiteller and H. Frede. 2002. Point- and nonpoint-source pesticide contamination in the Zwester Ohm Catchment, Germany. J. Environ. Qual. 31: 309–318.
Neumann, M., R. Schulz, K. Schäfer, W. Müller, W. Mannheller and M. Liess. 2002. The significance of entry routes as point and non-point sources of pesticides in small streams. Water Res. 36: 835–842.
Onneby, K., A. Jonsson and J. Stenstrom. 2010. A new concept for reduction of diffuse contamination by simultaneous application of pesticide-degrading microorganisms. Biodegradation 21: 21–29.
Parekh, N.R., D.L. Suett, S.J. Roberts, T. McKeown, E.D. Shaw and A.A. Jukes. 1994. Carbofuran-degrading bacteria from previously treated field soils. J. Appl. Bacteriol. 76: 559–567.
Pinto, M.I., C. Vale, G. Sontang and J.P. Noronha. 2016. Pathways of priority pesticides in sediments of coastal lagoons: The case study of Óbidos Lagoon, Portugal. Mar. Pollut. Bull. 106: 335–340.
Prado, A. and C. Airoldi. 2001. Toxic effect caused on microflora of soil by pesticide picloram application. J. Environ. Monitor. 3: 394–397.
Qiu, Y.W., E.Y. Zeng, H. Qiu, K. Yu and S. Cai. 2017. Bioconcentration of polybrominated diphenyl ethers and organochlorine pesticides in algae is an important contaminant route to higher trophic levels. Sci. Total Environ. 579: 1885–1893.
Racke, K.D. 1990. Pesticides in the soil microbial ecosystem. Enhanced biodegradation of pesticides in the environment. pp. 1–12. *In*: Racke, K.D. and J.R. Coats [eds.]. Enhanced Biodegradation of Pesticides in the Environment. American Chemical Society, Washington, D.C.
Ratcliff, A., M. Busse and C. Shestak. 2006. Changes in microbial community structure following herbicide (glyphosate) additions to forest soil. Appl. Soil Ecol. 34: 114–124.
Ravier, I., E. Haouisee, M. Clement, R. Seux and O. Briand. 2005. Field experiments for the evaluation of pesticide spray drift on arable crops. Pest. Manag. Sci. 61: 728–736.
Rabinovich, M.L., A.V. Bolobova and L.G. Vasil'chenko. 2004. Fungal decomposition of natural aromatic structures and xenobiotics: A review. Appl. Biochem. Microbiol. 401: 1–17.
Ryan, R.P., K. Germaine, A. Franks, D.J. Ryan and D.N. Dowling. 2008. Bacterial endophytes: recent developments and applications. FEMS Microbiol. Lett. 278: 1–9.
Rodriguez-Cruz, S., M. Andrade, M. Sanchez-Camazano and M. Sanchez-Martin. 2007. Relation between the adsorption capacity of pesticides by wood residues and the properties of wood and pesticides. Environ. Sci. Technol. 42: 3613–3619.

Rousis, N.I., R. Bade, L. Bijlsma, E. Zuccato, J.V. Sancho, F. Hernandez et al. 2017. Monitoring a large number of pesticides and transformation products in water samples from Spain and Italy. Environ. Res. 156: 31–38.
Scholz, N.L., J.P. Incardona, D.H. Baldwin, B.A. Berejikan, A.H. Dittman, B.E. Feist et al. 2003. Evaluating the sublethal impacts of current use pesticides on the environmental health of salmonids in Columbia River Basin. Bonneville Power Administration FY 2003 Provincial Project Review, 1–41.
Saeki, M. and K. Toyota. 2004. Effect of bensolfuron-methyl (a sulfonylurea herbicide) on the soil bacterial community of a paddy soil microcosm. Biol. Fert. Soils. 40: 110–118.
Shaner, D. and B. Henry. 2007. Field history and dissipation of atrazine and metolachlor in Colorado. J. Environ. Qual. 36: 128–134.
Sniegowski, K., K. Bers, K. Van Goetem, J. Ryckeboer, P. Jaeken, P. Spanoghe et al. 2011. Improvement of pesticide mineralization in on-farm biopurication systems by bioaugmentation with pesticide-primed soil. Fems Microbiol. Ecol. 76: 64–73.
Sirisha, K., S. Venkata and S. Jayarama. 2006. Effect of repeated application of chlorpyrifos on its degradation in surface and subsurface soil. Toxicol. Environ. Chem. 88: 373–384.
Smelt, J.H., A.E. Van De Peppel-Groen, L.J.T. Van Der Pas and A. Dijksterhuis. 1996. Development and duration of accelerated degradation of nematicides in different soils. Soil Biol. Biochem. 28: 1757–1765.
Sørensen, S.R. and J. Aamand. 2003. Rapid mineralisation of the herbicide isoproturon in soil from a previously treated Danish agricultural field. Pest Manag. Sci. 59: 1118–1124.
Spliid, N.H., W. Brusch, O.S. Jacobsen and S.U. Hansen. 1999. Pesticide point sources and dispersion of pesticides from a site previously used for handling of pesticides. 16th Danish Plant Protection Conference, Side Effects of Pesticides. Weeds. 33–46.
Spliid, N.H., A. Helweg and K. Heinrichson. 2006. Leaching and degradation of 21 pesticides in a full-scale model biobed. Chemosphere 65: 2223–2232.
Subhani, A., A.M. El-ghamry, H. Changyong and X. Jianming. 2000. Effects of pesticides (herbicides) on soil microbial biomass—A review. Pakistan J. Biol. Sci. 3: 705–709.
Tarazona, J.V., A. Fresno, S. Aycard, C. Ramos, M.M. Vega and G. Carbonell. 2000. Assessing the potential hazard of chemical substances for the terrestrial environment. Development of hazard classification criteria and quantitative environmental indicators. Sci. Total Environ. 247: 151–164.
Torstensson, L. and M. de P. Castillo. 1997. Use of biobeds in Sweden to minimize environmental spillages from agricultural spraying equipment. Pesticide. Outlook. 8: 24–27.
Tortensson, L. 2000. Experiences of biobeds in practical use in Sweden. Pesticide Outlook. 11: 206–212.
Tortella, G.R., O. Rubilar, M.P. Castillo, M. Cea, R. Mella-Herrera and M.C. Diez. 2012. Chlorpyrifos degradation in a biomixture of biobed at different maturity stages. Chemosphere 88: 224–228.
Tortella, G.R., R.A. Mella-Herrera, D.Z. Sousa, O. Rubilar, J.J. Acuna, G. Briceño et al. 2013a. Atrazine dissipation and its impact on the microbial communities and community level physiological profiles in a microcosm simulating the biomixture of on-farm biopurification system. J. Hazard. Mater. 260: 459–467.
Tortella, G.R., R.A. Mella-Herrera, D.Z. Sousa, O. Rubilar, G. Briceño, L. Parra et al. 2013b. Carbendazim dissipation in the biomixture of on-farm biopurification systems and its effect on microbial communities. Chemosphere 93: 1084–1093.
Tortella, G.R., O. Rubilar, J. Stenstrom, M. Cea, G. Briceño, A. Quiroz et al. 2013c. Using volatile organic compounds to enhance atrazine biodegradation in a biobed system. Biodegradation 24: 711–720.
Tortella, G.R., O. Rubilar, M. Cea, G. Briceño, A. Quiroz, M.C. Diez et al. 2013d. Natural wastes rich in terpenes and their relevance in the matrix of an on-farm biopurification system for the biodegradation of atrazine. Int. Biodeter. Biodegr. 85: 8–15.
Tortella, G.R., E. Salgado, S.A. Cuozzo, R.A. Mella-Herrera, L. Parra, M.C. Diez et al. 2014. Combined microbiological test to assess changes in an organic matrix used to avoid agricultural soil contamination, exposed to an insecticide. J. Soil Sci. Plant Nut. 14: 869–880.
Triky-Dotan, S., M. Ofek, M. Austerweil, B. Steiner, D. Minz, J. Katan et al. 2010. Microbial aspects of accelerated degradation of metam sodium in soil. Phytopatology. 100: 367–375.
Turco, R. and A. Konopka. 1990. Biodegradation of carbofuran in enhanced and non-enhanced soils. Soil Biol. Biochem. 22: 195–201.
Urrutia, C., O. Rubilar, G.R. Tortella and M.C. Diez. 2013. Degradation of pesticide mixture on modified matrix of a biopurification system with alternatives lignocellulosic wastes. Chemosphere 92: 1361–1366.
Vischetti, C., E. Capri, M. Trevisan, C. Casucci and P. Perucci. 2004. Biomassbed: a biological system to reduce pesticide point contamination at farm level. Chemosphere 55: 823–828.
Vischetti, C., L. Coppola, E. Monaci, A. Cardinali and M.d.P. Castillo. 2007. Microbial impact of the pesticide chlorpyrifos in Swedish and Italian biobeds. Agron. Sustain. Dev. 27: 267–272.
Vischetti, C., E. Monaci, A. Cardinali, C. Casucci and P. Perucci. 2008. The effect of initial concentration, co-application and repeated applications on pesticide degradation in a biobed mixture. Chemosphere 72: 1739–1743.
Wainwright, M. 1978. A review of the effects of pesticides on microbial activity in soils. Eur. J. Soil Sci. 29: 287–298.

Walker, A. and S. Welch. 2006. Enhanced degradation of some soil-applied herbicides. Weed Res. 31: 49–57.
White, P.M., T.L. Potter and A.K. Culbreath. 2010. Fungicide dissipation and impact on metolachlor aerobic soil degradation and soil microbial dynamics. Sci. Total Environ. 408: 1393–1402.
Yu, L., M. Shan, H. Fang, X. Wang and X. Chu. 2006. Responses of soil microorganisms and enzymes to repeated applications of chlorothalonil. J. Agr. Food Chem. 54: 10070–10075.
Zabletewicz, R., M. Weaver and M. Locke. 2006. Microbial adaptation for accelerated atrazine mineralization/degradation in Mississippi Delta soil. Weed Sci. 54: 538–547.
Zhang, B., H. Zhang, B. Jin, L. Tang, J. Yang, B. Li et al. 2008. Effect of cypermethrin insecticide on the microbial community in cucumber phyllosphere. J. Environ. Sci. 20: 1356–1362.
Zhang, B., Z. Bai, D. Hoefel, L. Tang, X. Wang, B. Li et al. 2009. The impacts of cypermethrin pesticide application on the non-target microbial community of the pepper plant phyllosphere. Sci. Total Environ. 407: 1915–1922.

10

Hexachlorocyclohexane

Sources, Harmful Effects and Promising Bacteria for its Bioremediation

María Z. Simón Solá,[1,a] *Daiana E. Pérez Visñuk,*[1,b] *Paula Paterlini,*[2] *Marta A. Polti*[1,3] and *Analía Alvarez*[1,3,*]

Introduction

Highly toxic organic wastes have been released into the environment over a long time. Persistent organic pollutants (POPs) such as pesticides, fuels, polycyclic aromatic hydrocarbons (PAHs), polychlorinated biphenyls (PCBs), and dyes are some of these compounds (Wania et al. 1993). POPs are resistant to biodegradation by native flora (Diez et al. 2010) compared to the naturally occurring organic compounds that are readily degraded. Therefore, hazardous compounds and chemicals have become one of the worldwide problems today. Among POPs, organochlorine pesticides (OPs) constitute a major environmental problem, mainly because of their high persistence and ability to bioaccumulate in the food chain, as well as for their toxicity (Xue et al. 2006, Fuentes et al. 2010).

The OP 1,2,3,4,5,6 hexachlorocyclohexane (HCH) is a broad-spectrum insecticide that was used on a large scale since the 1940s, and is available in two formulations: technical-grade HCH [a mixture of different isomers (in percentages): α-(60–70), β-(5–12), γ-(10–15) and δ-HCH (6–10)] and lindane (almost pure γ-HCH). This mixture was marketed as an inexpensive insecticide, but since γ-HCH is the only isomer that exhibits insecticidal properties, it has been common to refine it from technical-grade HCH and

[1] Planta Piloto de Procesos Industriales Microbiológicos (PROIMI-CONICET), Avenida Belgrano y Pasaje Caseros, Tucumán 4000, Argentina.
[a] Email: zoleicas@hotmail.com
[b] Email: dayp.visnuk90@gmail.com
[2] Facultad de Bioquímica, Química y Farmacia, Universidad Nacional de Tucumán (UNT), Ayacucho 471, Tucumán, 4000, Argentina.
Email: paula_paterlini@hotmail.com
[3] Facultad de Ciencias Naturales e Instituto Miguel Lillo, Universidad Nacional de Tucumán (UNT), Miguel Lillo 205, Tucumán 4000, Argentina.
Email: marpolti@hotmail.com
* Corresponding author: alvanalia@gmail.com

marketed as "lindane". Therefore, large quantities of byproducts were discarded leading to the contamination of many areas worldwide. There are very few officially recognized OP-contaminated sites in South America, and these are located mainly in heavily populated industrial sites as Sao Paulo (Brazil), Buenos Aires (Argentina), and Santiago and Concepción (Chile). However, these recognized OP-contaminated sites grossly underestimate the real situation because of the existence of illegally polluted sites (Fuentes et al. 2010).

All HCH isomers are acutely toxic to mammals, due to their mutagenic, teratogenic and carcinogenic properties (Willet et al. 1998). Although nowadays its use is restricted or completely banned, it continues posing serious environmental and health concerns (Lal et al. 2010). This is because γ-HCH was widely used as insecticide, which added to its high persistence, ensures lindane residues are found all over the world reaching aquatic environments through effluent release, atmospheric deposition, and runoff, among other ways (Willet et al. 1998, Yang et al. 2005). The inclusion of the α, β and γ-HCH isomers in the Stockholm Convention list of POPs renewed the interest in these contaminants and the sanitation of polluted sites.

Currently, ecofriendly techniques using different microbial species have become the focus of interest for cleaning up contaminated environments. This approach, known as bioremediation, is considered to be less invasive and more restorative of soil functions compared to conventional physicochemical methods (Kidd et al. 2009). Microbial degradation is regarded as an important process of OPs-dissipation, and to date, many researchers have studied this issue (Diez et al. 2010). Particularly, bacteria are the most used agent for bioremediation (De Lorenzo 2008), playing a significant role in the transformation and degradation of xenobiotics. Even the most POPs can be metabolized by microbial cultures, either by utilization of the compounds as a source of energy or nutrients or by cometabolism with other substances supporting microbial growth.

As a way of not altering the ecosystems, the actual trend is the utilization of microorganisms native to the polluted sites. Many Gram-negative bacteria have been reported to have metabolic properties to degrade HCHs. Such is the case of several *Sphingomonas* strains isolated from polluted soils, which were capable of degrading HCH isomers (Böltner et al. 2005, Phillips et al. 2005, Mohn et al. 2006, Kidd et al. 2008). Meanwhile, among Gram-positive microorganisms, actinobacteria have received great global interest for several biotechnological applications, including bioremediation. In fact, many actinobacteria strains have a great potential for HCHs removal (Benimeli et al. 2007, Fuentes et al. 2010, 2011, Saez et al. 2014, Alvarez et al. 2015, 2017).

In this chapter, it is intended to compile and update the information available on the bacterial aerobic degradation of HCHs. Besides, an account on the sources and harmful effects of OPs is also given.

Sources and Distribution of Organochlorine Pesticides in the Environment

Agricultural production is one of the largest and most important economic activities in the world. Therefore, since the 1950s the use of pesticides to control pests has continuously grown.

Multiple wastes of chemicals discharged by factories or as result of their extensive use in agriculture have been monitored. These residues contaminate rivers including sediments and aquatic biota, causing harmful effects in humans through food and drinking water (Chopra et al. 2011). At present, several freshwater bodies are contaminated with OPs since only about 10 percent of the wastewater produced by human activities is treated;

the rest is discharged into the water bodies. Besides, subsurface run-off from agricultural fields contains several fertilizers and pesticides that generally flow into local rivers. In the last five years, a great number of researchers have reported OPs residues in groundwater and drinking water worldwide. Moreira et al. (2012) revealed the presence of OPs in groundwater samples collected in two areas of Mato Grosso (Brazil), localized among the major soybean, corn, and cotton producers of Brazil.

The water scarcity and over application of OPs, has nowhere been more complex than in China. Li et al. (2015) demonstrated that both shallow and deep aquifers in the Shanxi Province (China) were polluted with OPs such as HCHs, 2,4′-dichlorodiphenyl dichloroethane (o,p′-DDD), aldrin, and endosulfan-sulfate. Meanwhile, several researchers have reported pesticides residues in ground water and drinking water in India. Yadav et al. (2015) reviewed the presence of POPs in multi-component environmental samples, finding that different freshwater bodies were contaminated with dicloro difenil tricloroetano (DDT), endosulfan, HCHs, parathion-methyl, among others, many of them prohibited in the 12th Stockholm Convention.

Pesticides movement in soils depends on their solubility in water, their adsorption by soil particles, and their persistence. The organic matter content is a factor which defines the pesticide retention in soils and sediments because it provides binding sites for organic contaminants, especially hydrophobic xenobiotic. For instance, the retention of HCHs in different soil types is mainly governed by organic matter. In general, the pollutant retention by soil components leads to a decrease in bioavailability and obstruct degradation (Becerra-Castro et al. 2013).

Many pesticides can volatilize and move far away from the application site. This explains traces of pesticides detected in pristine regions, suggesting that their occurrence is due to the atmospheric redistribution rather than a result of direct application (Shegunova et al. 2007). Despite the importance of this world environmental problem, the information published on pesticide wastes in the air is limited. Yadav et al. (2015) reviewed the information available on POPs wastes found in the Indian air and all the studies found high levels of these compounds. Zhang et al. (2008) monitored POPs along the coastal region of India, and they found DDT and its metabolites, HCHs, and Chlordane.

In Africa, atmospheric measurement of pesticides is also very limited. However, recently, Arinaitwe et al. (2016) reported that the air of Lake Victoria basin, a system that has a high level of commercial agriculture activity, was contaminated with OPs.

In 2008, Spain began monitoring POPs in air, since the air was recognized as the major route of long-range transport of pesticides through the world, by the Global Monitoring Plan of the Stockholm Convention. Passive air samplers were placed in seven remote points and four urban Spanish locations to assess levels of DDTs and HCHs. Results revealed that HCHs were the major pollutants, followed by DDTs when urban and remote locations were evaluated together. Urban areas presented statistically significant higher levels for all families of pesticides studied, compared to remote locations, revealing anthropogenic activities as sources for HCHs and DDTs (Torre et al. 2016). In opposite, Estellano et al. (2015) did not find any difference between urban and rural contents of OPs in air samples taken of ten areas in the Tuscan region (Italy).

Harmful Effects of Pesticides on Human Health

Ideally, a pesticide must be lethal to the target pests, but not to non-target species. However, this is not the case and inadequate management of pesticides constitutes occupational and environmental hazards (Lake et al. 2012). Besides, pesticide wastes retained in crops may influence public health via food consumption. In 1990, the World Health Organization (WHO) reported that occupational poisoning by pesticides resulted in a million cases worldwide and provided evidence that pesticides were responsible for pathologies affecting several aspects of human health. In fact, pesticide contamination has been implicated in the appearance of 'cancer villages'. This term refers to a site in which the mortality ratio of cancer is significantly higher than the average, probably because of widespread pesticide pollution, mainly in water (Lu et al. 2015).

According to their acute toxicity, the most hazardous pesticides are (in decreasing order) the organophosphorus, carbamate, and OPs pesticides. These compounds are powerful chemicals that act primarily by disrupting nervous system function (Ridolfi et al. 2014). Organophosphorus and carbamates pesticides act as acetyl cholinesterase inhibitors, affecting several organs such as peripheral and central nervous systems, muscles, liver, pancreas, and brain; whereas OPs are neurotoxic involved in the alteration of ion channels. As a shared mechanism, organophosphorus, carbamates, and OPs produce cellular oxidative stress and therefore disrupt neuronal and hormonal status of the body (Karami-Mohajeri and Abdollahi 2010).

Organochlorine pesticides constitute a serious environmental problem and considerable risks for human health. This is because its biological degradation is difficult, they are highly soluble in lipids (and consequently biomagnified in the food chain), and they are persistent in the environment because of its chlorine groups (Wang et al. 2013). Consequently, OPs exert several toxic effects on human health (mainly as endocrine disruptors) producing infertility, cancer of reproductive systems, neurotoxicity, and immunotoxicity.

Bacterial Degradation of Hexachlorocyclohexane

For a long time, it was assumed that HCHs degradation was an anaerobic process, but it was observed that it could be carried out in both aerobic and anaerobic conditions (Phillips et al. 2005, Lal et al. 2010). However, in this chapter, we will refer to aerobic degradation of HCHs. Most of the HCH-degrading aerobes known worldwide are the Gram-negative members of the family Sphingomonadaceae (Lal et al. 2006, 2008). Genes encoding the lindane degradation enzymes have been cloned, sequenced and characterized (Manickam et al. 2007). These genes (*lin* genes) were initially studied in *Sphingobium japonicum* UT26 (Nagata et al. 1999). Since plasmids were found to be associated with the horizontal gene transfer of *lin* genes among the degraders (Ceremonie et al. 2006), very similar *lin* genes have also been identified for all the other HCHs degrading *Sphingomonas* (Böltner et al. 2005, Ceremonie et al. 2006, Ito et al. 2007, Lal et al. 2006, 2008, Yamamoto et al. 2009). Thus, all the strains isolated from different geographical locations have a very similar set of genes (Lal et al. 2006). In *Sphingobium japonicum* UT26, the pathway is comprised as follows: *linA*, encoding a dehydrochlorinase (Imai et al. 1991); *linB*, encoding a haloalkane dehalogenase (Nagata et al. 1993); *linC*, encoding a dehydrogenase (Nagata et al. 1994); *linD*, encoding a reductive dechlorinase (Miyauchi et al. 1998); *linE/linEb*, encoding a ring cleavage oxygenase (Endo et al. 2005); *linF*, encoding a maleylacetate reductase (Endo et al. 2005); *linGH*, encoding an acyl-CoA transferase, and *linJ*, encoding a thiolase (Nagata et al. 2007), plus *linR/linI*, which are regulatory genes (Miyauchi et al. 2002, Nagata et al. 2007). There is evidence that LinA dehydrochlorinates the γ, α and δ-HCH isomers, while LinB hydrolytically dechlorinates β and δ-HCH, in all the Gram-negative strains examined.

Other Gram-negative genera were also recognized as lindane degrading bacteria, such as *Pandoraea* sp. LIN-3, isolated from an enrichment culture, which could simultaneously degrade α and γ isomers of HCH in both liquid and soil systems (Okeke et al. 2002), and

Xanthomonas sp., strain isolated from contaminated soil from an industrial site in India, was also capable of degrading lindane (Manickam et al. 2007).

In the last decade, there has been a great increase in the number of studies about HCHs degradation by other bacteria than the Gram-negative *Sphingomonas* strains. Such is the case of several Gram-positive or Gram-variable strains belonging to the Actinobacteria phylum. Even though the metabolic pathway for HCHs degradation by actinobacteria is not fully known, it is known that these microorganisms produce enzymes that degrade a wide range of complex organic compounds, including HCHs.

There are differences with respect to the hydrolytic dechlorination pathway for β- and δ-HCH comparing among Gram-negative and some actinobacteria. For instance, Datta et al. (2000) found that γ-HCH aerobic degradation by *Arthrobacter citreus* BI-100 does not present the formation of tetrachloro-cyclohexadiene (1,3,4,6-TCDN) as a transient product by dehydrochlorination of γ-pentachloro-cyclohexene (γ-PCCH) that occurs during the metabolism of γ-HCH by *S. japonicum* UT26 (Nagata et al. 1994). It is interesting to note the formation of trichlorocyclohexa-diene (TCCD) by *A. citreus* BI-100, in contrast to the production of 1,2,4-trichlorobenzene (1,2,4-TCB) as a dead-end product by *S. japonicum* UT26 during the metabolism of γ-HCH (Nagata et al. 1994). Furthermore, *S. japonicum* UT26 pathway does not show the appearance of 2-chlorophenol, catechol, and phenol which are present in *A. citreus* BI-100 lindane degradation pathway. The study of Datta et al. (2000) concludes that some γ-HCH metabolites produced by *A. citreus* BI-100 are quite different from those produced from γ-HCH by any other single microorganism reported in the literature.

Meanwhile, others researchers found similarities among Gram-negative and actinobacteria dechlorination pathways. Such is the case of the findings of Manickam et al. (2006) which isolated the actinobacteria *Microbacterium* sp. ITRC1 that degraded all four major isomers of HCH. DNA fragments corresponding to the two initial genes involved in γ-HCH degradative pathway encoding enzymes LinB and LinC were amplified and sequenced showing that the two genes present in *Microbacterium* sp. ITRC1 were homologous to those present in *S. japonicum* UT26. Similarly, Isaza et al. (2012) identify a haloalkane dehalogenase (LinB) belonging to the actinobacteria *Mycobacterium tuberculosis* CDC1551. This protein of 300 aminoacids, showed 100 percent homology to *M. tuberculosis* KZN 605, *M. bobis* BCG STR Pasteur 117 and 98 percent homology to *Sphingobiun francensis* and *Sphingobiun* sp. SSD4-1, which would indicate the stability of the gene in Gram-positive as well as in Gram-negative bacteria.

Cuozzo et al. (2009) informed a specific dechlorinase activity in *Streptomyces* M7 isolated from wastewater sediments of a copper filter plant. Using lindane as a specific substrate, the authors demonstrated that synthesis of the dechlorinase enzyme was induced by the presence of lindane in the culture medium. In addition, the first two degradation products γ-pentachlorocyclohexene and 1,3,4,6-tetrachloro-1,4 cyclohexadiene produced by the action of dechlorinase over lindane were detected in the cell-free extract of *Streptomyces* M7. When this strain was cultured in minimal medium with γ-HCH, a differential protein band corresponding to polynucleotide phosphorylase appeared in SDS-PAGE. The authors hypothesized this enzyme could be involved in the regulation of the dechlorinase gene. Later, Sineli et al. (2016) worked also with *Streptomyces* M7 and reported this strain has the ability to degrade α- and β-HCHs (besides lindane) since pentachlorocyclohexenes and tetrachlorocyclohexenes were detected as degradation products. In addition, the formation of possible persistent compounds such as chlorobenzenes and chlorophenols were informed, while no phenolic compounds were detected. The formation of possible persistent compounds as products of HCHs degradation had previously been studied by De Paolis et al. (2013) who reported the presence of pentachlorocyclohexenes and tetrachlorocyclohexenes, while phenolic compounds were not found.

Indirect Detection of Hexachlorocyclohexane Degradation

Biodegradation addresses the biological bases of the metabolism of recalcitrant compounds, which implies the determination of the parent compound as well as their intermediates. In this context, gas chromatography (GC) is the specific tool for study OPs degradation (Manickam et al. 2006, 2007, 2008, Quintero et al. 2007, Poggi-Varaldo et al. 2010, Camacho-Pérez et al. 2012). However, this technique requires extraction and cleaning up of the samples which imply expensive instrumentation and a considerable time.

Another possibility for the preliminary study of HCHs-degrading activities involves a colorimetric assay that detects chloride released from HCHs by dehalogenase enzymes (Phillips et al. 2001). Since the molecule of HCH has six chlorine atoms, dechlorination is a significant step for its degradation. The colorimetric assay is sensitive, rapid, inexpensive and highly useful for the screening of many samples (Phillips et al. 2001). For instance, Fuentes et al. (2010) used this methodology to determine *Streptomyces* and *Micromonospora* strains able to dechlorinate γ-HCH. Those strains were isolated from polluted soil samples collected from an illegal storage of OPs found in Santiago del Estero (Argentina). The authors determine the pesticide removal by GC with electron microcapture detector (GC-μECD) and indirectly probe its degradation by the detection of chloride ions. Thus, 12 out of 18 studied actinobacteria released chloride to the culture medium.

In addition, to providing a rapid screening tool for the detection of OPs degrading microorganisms, the colorimetric assay that detects chloride released was used also to select microbial consortia for γ-HCH degradation. Mixed cultures are considered to be potential agents for biodegradation of recalcitrant compounds because, in some cases, they are more efficient than pure cultures (Fuentes et al. 2011). However, selecting the best consortium usually implies a large volume of work due to the numerous assays that should be made to choose the cultures combination for the best degradation of a given pollutant. Fuentes et al. (2011) measured dechlorinase activity in cell-free extracts of 57 consortia of actinobacteria strains using the colorimetric technique. The authors found that the combination of four specific strains had the highest specific dechlorinase activity, compared to the pure cultures. Similarly, Saez et al. (2014) demonstrated the ability of this defined *Streptomyces* consortium to degrade lindane by the colorimetric determination of chloride ions released into a slurry system. The authors confirmed the validation of this technique by the identification of three intermediate metabolites of lindane degradation.

Plant-associated Bacteria for Hexachlorocyclohexane Dissipation

Nowadays there is a growing interest in the influence of microorganisms on plant growth and contaminant bioavailability and degradation. Consequently, more and more studies are focusing on the role of plant-associated microorganisms in improving phytoremediation efficiency (Kidd et al. 2009, Weyens et al. 2009). Phytoremediation techniques, based on the interactions between plants and microorganisms, have been proposed as cost-effective and ecofriendly methods for cleaning up polluted soils (San Miguel et al. 2013).

Microbe amended phytoremediation appears to be especially effective for organic contaminants, including OPs (Gerhardt et al. 2009). Several studies have demonstrated enhanced dissipation of OPs at the root-soil interface (Gerhardt et al. 2009, Kidd et al. 2009, Becerra-Castro et al. 2013). The rhizosphere effect is attributed to an increase in microbial activity caused by the release of plant root exudates (REs) containing amino acids, enzymes, carbohydrates, carboxylic acids, and phenolic compounds, among others (Curl and Truelove 1986). Regarding this approach, Alvarez et al. (2015) found that four *Streptomyces* strains were able to grow with maize plants as the sole carbon source, confirming that the plants, and/or its REs, represent a viable carbon source for the *Streptomyces* strains microorganisms. In fact, in a previous experiment, Alvarez et al. (2012) had assayed the effect of isolated maize REs on growth and γ-HCH removal by three *Streptomyces* strains. The authors found that *Streptomyces* sp. A5 showed maximum biomass and the highest pesticide dissipation (55 percent) in presence of REs. As part of this study, carbohydrates, proteins, and phenolic compounds were detected in the collected REs. Alvarez et al. (2015) also cultivated a mixed culture of four *Streptomyces* strains with maize plants on soils artificially polluted with γ-HCH. Similar levels of γ-HCH dissipation were registered in both inoculated and non-inoculated planted systems, suggesting that pesticide removal was not significantly affected by the bacterial inoculant. However, the inoculation of the actinobacteria consortium led to an increase in the vigor index of the maize plants and protected them against the existing toxicity. In a phytoremediation context, consortia can provide multiple benefits to plants, including the synthesis of protective compounds, chelators for delivering key plant nutrients, and degradation of contaminants before they can negatively impact to the plants (Gerhardt et al. 2009). Similar observations were made by Becerra-Castro et al. (2013), who inoculated substrates seeded with *Cytisus striatus*, which growing spontaneously on HCH-polluted sites, with the endophytic actinobacteria *Rhodococcus erythropolis* ET54b and *Sphingomonas* sp. D4. The authors found that systems planted showed a higher removal of the HCH isomers (including γ-HCH) and that the bacteria protected the plants against the negative effects of the pollutant.

Concluding Remarks

Hexachlorocyclohexane was one of the most extensively used OPs for both agriculture and medical purposes. Currently, its use is banned because of its toxicity, environmental persistence, and accumulation in the food chain. However, many developing countries continue using γ-HCH (lindane) for economic reasons. Thus, new sites are continuously being contaminated by γ-HCH and the other HCH-isomers. A possible way for remediation of HCH-polluted soils is the use of bacteria that inhabit these soils. There have been many reports regarding aerobic degradation of HCH by Gram-negative bacteria, especially those belonging to the *Sphingomonas* genus. The *Sphingomonas* strains are clearly the group of the first choice for bioremediation purposes, mainly due to the extensive knowledge reached about its metabolic pathways of HCH-degradation. Meanwhile, Gram-positive bacteria like actinobacteria, are highly promising for HCH-bioremediation purposes. Especially strains of the *Streptomyces* genus have great potential for remediation of toxic organic and inorganic pollutants. Several researches informed about *Streptomyces* sp. able to dechlorinate HCHs and some strains are proposed to be used for soil decontamination. Furthermore, phytostimulation of γ-HCH-degrading *Streptomyces* sp. appears to be a successful strategy for the remediation of γ-HCH-contaminated environments.

Acknowledgements

This work was supported by Consejo Nacional de Investigaciones Científicas y Técnicas (CONICET, PIP 0372) and Agencia Nacional de Promoción Científica y Tecnológica (ANPCyT, PICT 0480).

References Cited

Alvarez, A., C.S. Benimeli, J.M. Saez, M.S. Fuentes, S.A. Cuozzo, M.A. Polti et al. 2012. Bacterial bio-resources for remediation of hexachlorocyclohexane. Int. J. Mol. Sci. 13: 15086–15106.

Alvarez, A., C.S. Benimeli, J.M. Saez, A. Giuliano and M.J. Amoroso. 2015. Lindane removal using *Streptomyces* strains and maize plants: a biological system for reducing pesticides in soils. Plant Soil 395: 401–413.

Alvarez, A., J.M. Saez, J.S. Davila Costa, V.L. Colin, M.S. Fuentes, S.A. Cuozzo et al. 2017. Actinobacteria: current research and perspectives for bioremediation of pesticides and heavy metals. Chemosphere 166: 41–62.

Antizar-Ladislao, B. 2010. Bioremediation-Bacterial Alchemists. Elements 6: 389–394.

Arinaitwe, K., B.T. Kiremire, D.C.G. Muir, P. Fellin, H. Li, C. Teixeira et al. 2016. Legacy and currently used pesticides in the atmospheric environment of Lake Victoria, East Africa. Sci. Total Environ. 543: 9–18.

Becerra-Castro, C., A. Prieto-Fernández, P.S. Kidd, N. Weyens, B. Rodríguez-Garrido, M. Touceda-Gonzalez et al. 2013. Improving performance of *Cytisus striatus* on substrates contaminated with hexachlorocyclohexane (HCH) isomers using bacterial inoculants: developing a phytoremediation strategy. Plant Soil 362: 247–260.

Benimeli, C.S., M.J. Amoroso, A. Chaile and G. Castro. 2003. Isolation of four aquatic streptomycetes strains capable of growth on organochlorine pesticides. Bioresour. Technol. 89: 133–138.

Benimeli, C.S., G. Castro, A. Chaile and M.J. Amoroso. 2007. Lindane uptake and degradation by aquatic *Streptomyces* sp. strain M7. Int. Biodeterior. Biodegrad. 59: 148–155.

Benimeli, C.S., M.S. Fuentes, C.M. Abate and M.J. Amoroso. 2008. Bioremediation of lindane contaminated soil by *Streptomyces* sp. M7 and its effects on *Zea mays* growth. Int. Biodeterior. Biodegrad. 61: 233–239.

Böltner, D., S. Moreno-Morillas and J.L. Ramos. 2005. 16S rDNA phylogeny and distribution of *lin* genes in novel hexachlorocyclohexane-degrading *Sphingomonas* strains. Environ. Microbiol. 7: 1329–1338.

Camacho-Pérez, B., E. Ríos-Leal, N. Rinderknecht-Seijas and H.M. Poggi-Varaldo. 2012. Enzymes involved in the biodegradation of hexachlorocyclohexane: A mini review. J. Environ. Manage 95: S306–S318.

Cérémonie, H., H. Boubakri, P. Mavingui, P. Simonet and T.M. Vogel. 2006. Plasmid-encoded γ-hexachlorocyclohexane degradation genes and insertion sequences in *Sphingobium francense* (ex-*Sphingomonas paucimobilis* sp+). FEMS Microbiol. Lett. 257: 243–252.

Chopra, A.K., M.K. Sharma and S. Chamoli. 2011. Bioaccumulation of organochlorine pesticides in aquatic system—an overview. Environ. Monit. Assess. 73: 905–16.

Cuozzo, S.A., G.C. Rollán, C.M. Abate and M.J. Amoroso. 2009. Specific dechlorinase activity in lindane degradation by *Streptomyces* sp. M7. World J. Microbiol. Biotechnol. 25: 1539–1546.

Curl, E.A. and B. Truelove. 1986. The Rhizosphere. Springer-Verlag, Berlin.

Datta, J., A.K. Maiti, D.P. Modak, P.K. Chakrabartty, P. Bhattacharyya and P.K. Ray. 2000. Metabolism of γ-hexachlorocyclohexane by *Arthrobacter citreus* strain BI-100: Identification of metabolites. J. Gen. App. Microbiol. 46: 59–67.

De Lorenzo, V. 2008. Systems biology approaches to bioremediation. Curr. Opin. Biotechnol. 19: 579–589.

De Paolis, M.R., D. Lippi, E. Guerriero, C.M. Polcaro and E. Donati. 2013. Biodegradation of α-, β-, and γ-hexachlorocyclohexane by *Arthrobacter fluorescens* and *Arthrobacter giacomelloi*. Appl. Biochem. Biotechnol. 170: 514–524.

De Schrijver, A. and R. de Mot. 1999. Degradation of pesticides by actinomycetes. Crit. Rev. Microbiol. 25: 85–119.

Diez, M.C. 2010. Biological aspects involved in the degradation of organic pollutants. J. Soil Sci. Plant. Nut. 10: 244–267.

Endo, R., M. Kamakura, K. Miyauchi, M. Fukuda, Y. Ohtsubo, M. Tsuda et al. 2005. Identification and characterization of genes involved in the downstream degradation pathway of γ hexachlorocyclohexane in *Sphingomonas paucimobilis* UT26. J. Bacteriol. 187: 847–853.

Estellano, V.H., K. Pozo, C. Efstathiou, K. Pozo, S. Corsolini and S. Focardi. 2015. Assessing levels and seasonal variations of current-use pesticides (CUPs) in the Tuscan atmosphere, Italy, using polyurethane foam disks (PUF) passive air samplers. Environ. Pollut. 205: 52–59.

Fuentes, M.S., C.S. Benimeli, S.A. Cuozzo and M.J. Amoroso. 2010. Isolation of pesticide-degrading actinomycetes from a contaminated site: bacterial growth, removal and dechlorination of organochlorine pesticides. Int. Biodeterior. Biodegrad. 64: 434–441.

Fuentes, M.S., J.M. Sáez, C.S. Benimeli and M.J. Amoroso. 2011. Lindane biodegradation by defined consortia of indigenous *Streptomyces* strains. Water Air. Soil. Pollut. 222: 217–231.

Gerhardt, K.E., X.D. Huang, B.R. Glick and B.M. Greenberg. 2009. Phytoremediation and rhizoremediation of organic soil contaminants: Potential and challenges. Plant. Sci. 176: 20–30.

Imai, R., Y. Nagata, M. Fukuda, M. Takagi and K. Yano. 1991. Molecular cloning of a *Pseudomonas paucimobilis* gene encoding a 17-kilodalton polypeptide that eliminates HCl molecules from γ-hexachlorocyclohexane. J. Bacteriol. 173: 6811–6819.

Isaza, J.P., C. Duque, V. Gomez, J. Robledo, L.F. Barrera and J.F. Alzate. 2012. Whole genome shotgun sequencing of one Colombian clinical isolate of *Mycobacterium tuberculosis* reveals DosR regulon gene deletions. FEMS Microbiol. Lett. 330: 113–120.

Ito, M., Z. Prokop, M. Klvaňa, Y. Otsubo, M. Tsuda, J. Damborský et al. 2007. Degradation of β-hexachlorocyclohexane by haloalkane dehalogenase LinB from γ-hexachlorocyclohexane-utilizing bacterium *Sphingobium* sp. MI1205. Arch. Microbiol. 188: 313–325.

Karami-Mohajeri, S. and M. Abdollahi. 2010. Toxic influence of organophosphate, carbamate, and organochlorine pesticides on cellular metabolism of lipids, proteins, and carbohydrates: a systematic review. Hum. Exp. Toxicol. 30: 1119–1140.

Kidd, P., A. Prieto-Fernández, C. Monterroso and M.J. Acea. 2008. Rhizosphere microbial community and hexachlorocyclohexane degradative potential in contrasting plant species. Plant Soil 302: 233–247.

Kidd, P.S., J. Barceló, M.P. Bernal, F. Navari-Izzo, C. Poschenrieder, E. Shilev et al. 2009. Trace element behaviour at the root–soil interface: Implications in phytoremediation. Environ. Exp. Bot. 67: 243–259.

Lake, I.R., L. Hooper, A. Abdelhamid, G. Bentham, A.B.A. Boxall, A. Draper et al. 2012. Climate change and food security: health impacts in developed countries. Environ. Health Perspect. 120: 1520–1526.

Lal, R., C. Dogra, S. Malhotra, P. Sharma and R. Pal. 2006. Diversity, distribution and divergence of *lin* genes in hexachlorocyclohexane-degrading sphingomonads. Trends Biotechnol. 24: 121–130.

Lal, R., M. Dadhwal, K. Kumari, P. Sharma, A. Singh, H. Kumari et al. 2008. *Pseudomonas* sp. to *Sphingobium indicum*: A journey of microbial degradation and bioremediation of hexachlorocyclohexane. Indian J. Microbiol. 48: 3–18.

Lal, R., G. Pandey, P. Sharma, K. Kumari, S. Malhotra, R. Pandey et al. 2010. Biochemistry of microbial degradation of hexachlorocyclohexane and prospects for bioremediation. Microbiol. Mol. Biol. Rev. 74: 58–80.

Li, J.L., C.X. Zhang, Y.X. Wang, X.P. Liao, L.L. Yao, M. Liu et al. 2015. Pollution characteristics and distribution of polycyclic aromatic hydrocarbons and organochlorine pesticides in groundwater at Xiaodian Sewage Irrigation Area, Taiyuan City. Huanjing Kexue 36: 172–178.

Lu, Y., S. Song, R. Wang, Z. Liu, J. Meng, A.J. Sweetman et al. 2015. Impacts of soil and water pollution on food safety and health risks in China. Environ. Int. 77: 5–15.

Manickam, N., M. Mau and M. Schloemann. 2006. Characterization of the novel HCH-degrading strain, *Microbacterium* sp. ITRC1. Appl. Microbiol. Biotechnol. 69: 580–588.

Manickam, N., R. Misra and S. Mayilraj. 2007. A novel pathway for the biodegradation of γ-hexachlorocyclohexane by a *Xanthomonas* sp. strain ICH12. J. Appl. Microbiol. 102: 1468–1478.

Manickam, N., M.K. Reddy, H.S. Saini and R. Shanker. 2008. Isolation of hexachlorocyclohexane-degrading *Sphingomonas* sp. by dehalogenase assay and characterization of genes involved in γ-HCH degradation. J. Appl. Microbiol. 104: 952–960.

Manonmani, H.K., D.H. Chandrashekariah, N. Sreedhar Reddy, C.D. Elecy and A.A.M. Kunhi. 2000. Isolation and acclimation of a microbial consortium for improved aerobic degradation of α-hexachlorocyclohexane. J. Agric. Food Chem. 48: 4341–4351.

Miyauchi, K., S.K. Suh, Y. Nagata and M. Takagi. 1998. Cloning and sequencing of a 2,5-dichlorohydroquinone reductive dehalogenase gene whose product is involved in degradation of hexachlorocyclohexane by *Sphingomonas paucimobilis*. J. Bacteriol. 180: 1354–1359.

Miyauchi, K., H.S. Lee, M. Fukuda, M. Takagi and Y. Nagata. 2002. Cloning and characterization of *linR*, involved in regulation of the downstream pathway for γ-hexachlorocyclohexane degradation in *Sphingomonas paucimobilis* UT26. Appl. Environ. Microbiol. 68: 1803–1807.

Mohn, W.W., B. Mertens, J.D. Neufeld, W. Verstraete and V. de Lorenzo. 2006. Distribution and phylogeny of hexachlorocyclohexane degrading bacteria in soils from Spain. Environ. Microbiol. 8: 60–68.

Moreira, J.C., F. Peres, A.C. Simões, W.A. Pignati, E. de C. Dores, S.N. Vieira et al. 2012. Ground water and rainwater contamination by pesticides in an agricultural region of Mato Grosso state in central Brazil. Cien. Saude. Colet. 17: 1557–68.

Nagata, Y., T. Nariya, R. Ohtomo, M. Fukuda, K. Yano and M. Takagi. 1993. Cloning and sequencing of a dehalogenase gene encoding an enzyme with hydrolase activity involved in the degradation of hexachlorocyclohexane in *Pseudomonas paucimobilis*. J. Bacteriol. 175: 6403–6410.

Nagata, Y., R. Ohtomo, K. Miyauchi, M. Fukuda, K. Yano and M. Takagi. 1994. Cloning and sequencing of a 2,5-dichloro-2,5-cyclohexadiene-1,4-diol dehydrogenase gene involved in the degradation of hexachlorocyclohexane in *Pseudomonas paucimobilis*. J. Bacteriol. 176: 3117–3125.

Nagata, Y., K. Miyauchi and M. Takagi. 1999. Complete analysis of genes and enzymes for γ-hexachlorocyclohexane degradation in *Sphingomonas paucimobilis* UT26. J. Ind. Microbiol. Biotechnol. 23: 380–390.

Nagata, Y., R. Endo, M. Ito, Y. Ohtsubo and M. Tsuda. 2007. Aerobic degradation of lindane (gamma-hexachlorocyclohexane) in bacteria and its biochemical and molecular basis. Appl. Microbiol. Biotechnol. 76: 741–752.

Okeke, B.C., T. Siddique, M. Camps Arbestain and W.T. Frankenberger. 2002. Biodegradation of γ-Hexachlorocyclohexane (Lindane) and α-Hexachlorocyclohexane in Water and a Soil Slurry by a *Pandoraea* Species. J. Agric. Food. Chem. 50: 2548–2555.

Phillips, T.M., A.G. Seech, H. Lee and J.T. Trevors. 2001. Colorimetric assay for lindane dechlorination by bacteria. J. Microbiol. Methods 47: 181–188.

Phillips, T.M., A.G. Seech, H. Lee and J.T. Trevors. 2005. Biodegradation of hexachlorocyclohexane (HCH) by microorganisms. Biodegradation 16: 363–392.
Poggi-Varaldo, H.M. 2010. Bioremediation of an agricultural soil polluted with lindane in triphasic, sequential methanogenic-sulfate reducing slurry bioreactors. J. Biotechnol. 150: 561–562.
Quintero, J.C., T.A. Lú-Chau, M.T. Moreira, G. Feijoo and J.M. Lema. 2007. Bioremediation of HCH present in soil by the white-rot fungus *Bjerkandera adusta* in a slurry bacth bioreactor. Int. Biodeterior. Biodegradation 60: 319–326.
Ridolfi, A.S., G.B. Álvarez and M.E. Rodríguez Giraul. 2014. Organochlorinated contaminants in general population of Argentina and other Latin American Countries. pp. 17–40. *In*: Alvarez, A. and M.A. Polti [eds.]. Bioremediation in Latin America. Current Research and Perspectives. Springer, New York, NY; USA.
Rodriguez, R.A. and G.A. Toranzos. 2003. Stability of bacterial populations in tropical soil upon exposure to Lindane. Int. Microbiol. 6: 253–258.
Saez, J.M., A. Alvarez, C.S. Benimeli and M.J. Amoroso. 2014. Enhanced lindane removal from soil slurry by immobilized *Streptomyces* consortium. Int. Biodeterior. Biodegrad. 93: 63–69.
San Miguel, A., P. Ravanel and M. Raveton. 2013. A comparative study on the uptake and translocation of organochlorines by *Phragmites australis*. J. Hazar. Mater. 244: 60–69.
Shegunova, P., J. Klánová and I. Holoubek. 2007. Residues of organochlorinated pesticides in soils from the Czech Republic. Environ. Pollut. 146: 257–261.
Siddique, T., B.C. Okeke, M. Arshad and W.T. Jr. Frankerberger. 2002. Temperature and pH effects on biodegradation of hexachlorocyclohexane isomers in water and soil slurry. J. Agric. Food Chem. 50: 5070–5076.
Sineli, P.E., G. Tortella, J.S. Dávila Costa, C.S. Benimeli and S.A. Cuozzo. 2016. Evidence of α-, β- and γ-HCH mixture aerobic degradation by the native actinobacteria *Streptomyces* sp. M7. World J. Microbiol. Biotechnol. 32: 81. doi: 10.1007/s11274-016-2037-0.
Siripattanakul, S., W. Wirojanagud, J. McEvoy, T. Limpiyakorn and E. Khan. 2009. Atrazine degradation by stable mixed cultures enriched from agricultural soil and their characterization. J. Appl. Microbiol. 106: 986–999.
Torre, A., P. Sanz, I. Navarro and M.A. Martínez. 2016. Time trends of persistent organic pollutants in spanish air. Environ. Pollut. doi: 10.1016/j.envpol.2016.01.040.
Wang, L., H. Jia, X. Liu, Y. Sun, M. Yang, W. Hong et al. 2013. Historical contamination and ecological risk of organochlorine pesticides in sediment core in northeastern Chinese river. Ecotox. Environ. Safe 93: 112–120.
Wania, F. and D. Mackay. 1993. Global fractionation and cold condensation of low volatility organochlorine compounds in polar regions. Ambio 22: 10–18.
Weyens, N., D. van der Lelie, S. Taghavi and J. Vangronsveld. 2009. Phytoremediation: plant–endophyte partnerships take the challenge. Curr. Opin. Biotechnol. 20: 248–254.
Willet, K.L., E.M. Ulrich and R.A. Hites. 1998. Differential toxicity and environmental fates of hexachlorocyclohexane isomers. Environ. Sci. Technol. 32: 2197–2207.
Xue, N., D. Zhang and X. Xu. 2006. Organochlorinated pesticide multiresidues in surface sediments from Beijing Guanting reservoir. Water Res. 40: 183–194.
Yadav, I.C., N.L. Devi, J.H. Syed, Z. Cheng, J. Li, G. Zhang et al. 2015. Current status of persistent organic pesticides residues in air, water, and soil, and their possible effect on neighboring countries: a comprehensive review of India. Sci. Total Environ. 511: 123–137.
Yamamoto, S., S. Otsuka, Y. Murakami, M. Nishiyama and K. Senoo. 2009. Genetic diversity of gamma-hexachlorocyclohexane-degrading sphingomonads isolated from a single experimental field. Lett. Appl. Microbiol. 49: 472–477.
Yang, R., G. Ji, Q. Zhoe, C. Yaun and J. Shi. 2005. Occurrence and distribution of organochlorine pesticides (HCH and DDT) in sediments collected from East China Sea. Environ. Int. 31: 799–804.
Zhang, G., P. Chakraborty, J. Li, P. Sampathkumar, T. Balasubramanian and K. Kathiresan. 2008. Passive atmospheric sampling of organochlorine pesticides, polychlorinated biphenyls, and polybrominated diphenyl ethers in urban, rural, and wetland sites along the coastal length on India. Environ. Sci. Technol. 42: 8218–8223.
Zheng, G., A. Selvam and J.W.C. Wong. 2011. Rapid degradation of lindane (γ-hexachlorocyclohexane) at low temperature by *Sphingobium* strains. Int. Biodeterior. Biodegradation 65: 612–618.

11

Actinobacteria as Bio-Tools for Removing and Degrading α-, β- and γ-Hexachlorocyclohexane

Pedro E. Sineli,[1] *Gonzalo R. Tortella*[2,3] and *Sergio A. Cuozzo*[1,4,*]

Introduction

A wide variety of pesticides is used on agricultural crops to control various insects. In spite of their benefits, pesticides are often considered a serious threat because of their long environmental persistence. Therefore, the removal of pesticides from the environment, or their transformation into less toxic compounds, is a topic of research interest worldwide.

For a long time, organochlorine pesticides (OPs) were widely used in agriculture and medicine; however, the use of these compounds is now prohibited on account of their toxicity, environmental persistence, and bioaccumulation throughout the food chain (Abhilash 2009). Thus in 2009, OPs were added to the UNEP Stockholm Convention list as new persistent organic pollutants (Vijgen et al. 2011).

Hexachlorocyclohexane (HCH) belongs to the organochlorine pesticides and was commercially available in two formulations, technical HCH and lindane. Technical HCH consists of a stereoisomers mixture where the isomers α-, β- and δ-HCH represent up to 85% of the mixture. The γ-isomer is the only isomer having insecticidal properties and represents around 10–15% of the technical HCH. Lindane is the γ-isomer of HCH at 99% (γ-HCH). During lindane production, the other stereoisomers are separated out and frequently dumped as waste at production sites worldwide, causing a negative environmental impact, as determined by Heeb et al. (2014). In addition, Vijgen (2006) estimated that for each ton of lindane, between 8 and 12 tons of the other isomers are produced, indicating

[1] Planta Piloto de Procesos Industriales Microbiológicos (PROIMI-CONICET), Avenida Belgrano y Pasaje Caseros, 4000 Tucumán, Argentina.
Email: sinelip@gmail.com
[2] Departamento de Ingeniería Química, Universidad de La Frontera, Casilla 54-D, Temuco, Chile.
Email: gonzalo.tortella@ufrontera.cl
[3] Centro de Excelencia en Investigación Biotecnológica Aplicada al Medio Ambiente (CIBAMA), Universidad de La Frontera, Temuco-Chile.
[4] Facultad de Ciencias Naturales e Instituto Miguel Lillo, Universidad Nacional de Tucumán, Miguel Lillo 205, 4000 Tucumán, Argentina.
* Corresponding author: scuozzo@proimi.org.ar

the large amount that is deposited in dumps. Although the use of lindane has been strictly forbidden in most countries since the '90s, γ-HCH and its non-insecticidal isomers (α-, β-, and δ-HCH) continue to pose a real risk to the environment and to human health (Pavlíková et al. 2012). These isomers are highly hydrophobic, persistent, and are widespread in the environment. They also accumulate in the food chain (Manickam et al. 2006) on account of their lipophilic properties, which leads to toxicity (ATSDR 2011), as determined by Liu et al. (2010). Lindane was primarily used as a domestic and agriculture fumigant based on its insecticidal properties against a wide range of insects. Nowadays its use has been banned in most developed countries; however, due to economic factors, some developing countries still use lindane (Johri et al. 1998). On account of its prevalence, lindane-contaminated sites are prominent worldwide. Once HCH enters the environment, it can be distributed globally (Simonich and Hitéis 1995) and can also persist in the environment (Abhilash et al. 2008, Abhilash 2009). Lindane contamination, as well as HCH isomers-contaminated sites, has been reported in different continents, viz., Europe (Concha-Graña et al. 2006), America (Österreicher-Cunha et al. 2003, Phillips et al. 2006), and Asia (Prakash et al. 2004, Zhu et al. 2005). The half-life period reported for lindane in soil and water is 708 and 2292 days, respectively (Beyer and Matthies 2001).

Based on toxicological studies, HCH isomers damage the central nervous, renal, and reproductive systems, as well as the liver and kidney (Quintero et al. 2005, Colt et al. 2009, Chia et al. 2010, Manna et al. 2015). Also, Manickam et al. (2006) and Quintero et al. (2007) detected that, during their presence in the environment, HCH isomers can volatilize and be transported through the atmosphere, producing greater pollution. This was then confirmed by findings of HCH residues in regions without human activity (Abhilash 2009).

The α-HCH isomer is considered a potential carcinogen and it is listed as a priority pollutant by the US EPA. Additionally, β-HCH isomer has been classified within group 2B, as possibly carcinogenic to humans according to the International Agency Research and Cancer, as reported by Ingelido et al. (2009). Furthermore, abiotic and biotic degradation of HCH isomers in nature leads to accumulation of the β-isomer, which is the most persistent and resistant to bacterial attack (Phillips et al. 2005).

Non-insecticidal HCH-isomers cause serious environmental impacts and persist in the soil for long periods. Therefore, toxicity and threats of environmental contamination are of great concern, and this problem can be solved through biodegradation-based approaches. In this context, it is essential to develop procedures to remove these toxic compounds or transform them to non-toxic forms (Kumar et al. 2016). Although several techniques for HCH decontamination are available, such as chemical treatment, incineration, and landfilling, they lack widespread application due to their cost factor and toxicity concerns to the living systems. Aversely, bioremediation technology has been proposed as a promising tool for *in situ* detoxification of pesticide-contaminated sites. Bioremediation-based approaches possess high efficiency, sustainability, and their eco-friendly nature provides a solution to traditional physicochemical remediation (Mani and Kumar 2013). Diverse soil microorganisms capable of degrading and utilizing the organochlorine HCH isomers as a source of carbon have been reported in various places over the last decades (Okeke et al. 2002, Nawab et al. 2003, Abhilash et al. 2011).

Sun et al. (2015) conducted four studies of microcosms' bioaugmentation for the remediation of organochlorine pesticides-contaminated soil. They evaluated the effects of the bioaugmentation on the degradation of HCHs and dichlorodiphenyltrichloroethanes (DDTs) and found that nutrients/plant bioaugmentation enhanced the degradation of HCHs and DDTs to 81.18 and 85.4%, respectively. In order to develop a cleanup strategy for the polluted sites, a recent study conducted by Laquitaine et al. (2016) demonstrated the biodegradation of HCH in agricultural soils from Guadeloupe (French West Indies) and identified some of the degrading genes involved in the HCH degradation pathway.

The aerobic metabolism of HCHs has been reported for several species of bacteria (Nagata et al. 1999, Phillips et al. 2005, Raina et al. 2008, Cuozzo et al. 2009, Tabata et al. 2011, Camacho-Pérez et al. 2012). The biodegradation of HCH isomers consists of the progressive elimination of the chlorine and hydrogen atoms and the subsequent formation of double bonds; the chlorine atoms are possibly replaced by hydroxyls (Nagata et al. 2007). Due to their molecular structure, α-, β-, and γ-HCH show different and specific degradation pathways; therefore not all the degradation mechanisms are present in all the organisms studied to date (Raina et al. 2007, Heeb et al. 2015).

Pentachlorocyclohexene is often the product of the first step of HCH isomers degradation, and the degradation can continue through a variety of pathways, depending upon the HCH isomers and the microorganism involved, as determined by Singh and Kuhad (1999) and Cuozzo et al. (2009). Furthermore, more toxic intermediates such as chlorobenzenes and chlorophenols have been reported as the most persistent metabolites in aerobic bacterial degradation of HCH and in some cases, these were dead-end products, which presents a greater health risk. The ideal is to reach a complete aerobic degradation of HCH isomers. This fact has been reported for the Gram-negative bacterium *Sphingomonas* (Nagata et al. 2007, Heeb et al. 2014, 2015) and also for fungi like *Trametes hirsutus*, *Phanerochaete chrysosporium*, *Cyathus bulleri*, and *Phanerochaete sordida* (Mougin et al. 1999, Singh and Kuhad 1999, 2000). However, Raina et al. (2007) detected hydroxylated metabolites resistant to further bacterial degradation, in soils and water near pesticide production sites. It is, therefore, necessary to find versatile microorganisms able to degrade mixtures of HCH isomers without producing dangerous sub-products.

In particular, Benimeli et al. (2003) isolated *Streptomyces* sp. M7, which was able to remove organic and inorganic toxic compounds (Benimeli et al. 2008, Cuozzo et al. 2009, Polti et al. 2014). The mechanisms that govern the lindane degradation process in *Streptomyces* sp. M7 are currently under study. Recently, it was determined that *Streptomyces* sp. M7 is also able to remove different HCH isomers and the intermediate metabolites produced (Sineli et al. 2016).

Environmental factors such as temperature, pH, and salinity have a strong influence on microbial activity, as well as on the bioavailability of HCH isomers. Therefore, the optimization of these parameters is necessary in order to achieve substantial degradation of these pollutants. In this context, the authors describe the optimum conditions for α- and β-HCH aerobic removal by the actinobacterium *Streptomyces* sp. M7, as well as its ability to degrade a mixture of three HCH-isomers. In addition, the metabolic intermediates generated during α- and β-HCH degradation are also mentioned. This study represents a key step towards understanding the aerobic degradation of HCH isomers by actinobacteria, especially those belonging to the *Streptomyces* genus.

Capability of Actinobacteria to Use α- and β-HCH as Carbon Source

Actinobacteria have demonstrated great ability to remove and biodegrade different hexachlorocyclohexane isomers. Within actinobacteria, the genus *Streptomyces* has received considerable attention as an effective biotechnological approach to remove and degrade several chlorinated compounds (Benimeli et al. 2003, 2006, Cuozzo et al. 2009, Fuentes et al. 2013, 2016, Bourguignon et al. 2014). In addition to their metabolic diversity, *Streptomyces* strains may be well suited for soil inoculation as a consequence of their mycelial growth habit, relatively rapid growth rates, colonization of semi-selective substrates (Alvarez et al. 2017). On the other hand, as they are natural inhabitants of the soil, they are already adapted to this habitat (Shelton et al. 1996). Specifically, the strain *Streptomyces* sp. M7, isolated from a highly contaminated region has been extensively studied. Sineli et al. (2016) demonstrated that *Streptomyces* sp. M7 was able to grow in minimum medium supplemented with 1.66 mg l^{-1} of α- or β-HCH, showing similar growth profiles in both assays. The maximum biomass values for *Streptomyces* sp. M7 were obtained at 4 d of incubation in the presence of α- and β-isomers, with 0.32 and 0.33 mg ml^{-1}, respectively. Furthermore, no microbial growth inhibition was observed in the presence of the HCH isomers, since these biomass values represented twice the biomass of the biotic control (in the absence of pollutant). These results indicate that the xenobiotics α- and β-HCH are not toxic to this actinobacterium at the tested concentration (1.66 mg l^{-1}). Moreover, *Streptomyces* sp. M7 was able to grow in the presence of γ-HCH (100 μg l^{-1}) reaching a maximum value of biomass of 0.5 mg ml^{-1} (Benimeli et al. 2006). Also, in assays performed in minimal medium, *Streptomyces* sp. M7 was able to use chlordane as sole carbon source, achieving 0.36 mg ml^{-1} of biomass at 4 days of incubation (Cuozzo et al. 2012).

Regarding the removal of the HCH isomers by *Streptomyces* sp. M7, a 50% decrease in α-HCH concentration was observed after the 1st d of incubation. Then, at the 7th d of incubation, another significant decrease was observed, reaching 79% of the α-HCH removal. When the actinobacterium strain was incubated in the presence of the β-HCH isomer, a 78% decrease was detected at the 1st d of incubation, observing the same value at the end of incubation time. The concentration of both HCH isomers remained constant in the abiotic controls, thus demonstrating that the compounds were removed by the microbial activity (Sineli et al. 2016).

Optimal Conditions for the Removal of α-/β-HCH Isomers: Temperature, pH, and Isomer Concentration

Sineli et al. (2016) evaluated the removal of α-HCH at various temperatures, and demonstrated that the highest removal percentage was obtained at 30ºC (Table 1). As mentioned above, the elimination of chlorine atoms from the HCH molecule is an important step for its degradation. In this context, it is noteworthy that the release of chlorine ions presented the highest value (94%) after 7 d of incubation at 30ºC, indicating this was the optimum temperature for the removal and degradation of this isomer by *Streptomyces* sp. M7. Taking into account the different initial pH of the culture medium (5, 7, and 9), significant differences ($p < 0.05$) were observed for α-HCH removal between pH 5 and 7–9: 56%, 79%, and 83%, respectively, at 7 d (Table 1). The highest release of chloride ions (94%) and microbial biomass (0.32 mg l^{-1}) were both observed at pH 7; while the lowest values were observed at pH 5.

When increasing concentrations of the α-HCH isomer were assayed under optimal temperature and pH conditions (30°C and pH 7), the removal percentages were 79%, 85%, and 83% for the different concentrations (1.66, 8.33, and 16.6 mg l^{-1}) of α-HCH (Table 1) at 7 d of incubation, with chloride ion release percentages of 94%, 100%, and 63%, respectively (Sineli et al. 2016).

Similarly, β-HCH removal by *Streptomyces* sp. M7 was evaluated at different incubation temperatures and the highest removal percentage (78%), as well as the highest chloride ions release, was achieved at 30°C. Notably, the release of chloride ions at 30ºC was 100%, namely four times higher than the percentages reached at incubation temperatures of 25 and 35ºC. Conversely, no significant differences ($p < 0.05$) were observed for β-HCH isomer removal at pH 5 and 7 (79 and 78%, respectively), although it was significantly lower at

Table 1. Removal percentages of α- and β-HCH by *Streptomyces* sp. M7 at different culture conditions at seven days of incubation, and intermediate metabolites identified during the biodegradation process (Sineli et al. 2016).

Carbon source		Suggested compound[a]	Rt[b] (min)	Identity (%)	Removal (%) at different physicochemical conditions[c]		
α-HCH	Culture supernatant	α-HCH	30.2	99	*Temperature*	*pH*	*Concentration (mg l⁻¹)*
		1,2-dichlorobenzene	9.98	97	25°C: 22.8 ± 1.6	*5*: 55.6 ± 1.7	*1.66*: 78.9 ± 0.7
		1,3 or 1,4-dichlorobenzene	10.2	95	30°C: 78.9 ± 0.7	*7*: 78.9 ± 0.7	*8.33*: 85.6 ± 6.3
		Trichlorobenzene isomers	17.1	97	35°C: 36.8 ± 0.8	*9*: 82.7 ± 1.6	*16.6*: 83.1 ± 1.4
	Cell-free extract	α-HCH	30.2	99			
		Pentachlorocyclohexene	29.5	90			
		Dichlorobenzene isomers	10.2	81			
β-HCH	Culture supernatant	β-HCH	31.1	95	*Temperature*	*pH*	*Concentration (mg l⁻¹)*
		Dichlorobenzene isomers	10.2	81	25°C: 73.1 ± 3.6	*5*: 78.9 ± 5.2	*1.66*: 77.8 ± 0.6
	Cell-free extract	β-HCH	31.1	95	30°C: 77.8 ± 0.6	*7*: 77.8 ± 0.6	*8.33*: 71.2 ± 3.7
		Tetrachlorocyclohexene	26.6	59	35°C: 76.1 ± 2.5	*9*: 71.8 ± 4.2	*16.6*: 62.7 ± 2.7
		Dichlorobenzene isomers	10.2	81			

[a] Identification is based on mass spectrum compared with the WILEY7 NIST library. Nomenclature of compounds is according to IUPAC (International Union of Pure and Applied Chemistry). SIM: Selected Ion Monitoring.
[b] Retention time.
[c] Residual HCH isomers in the supernatants were determined by GC with micro electron capture detector (μECD).

pH 9 (72%) (Table 1). The release of chloride ions showed relative percentages of 23%, 100%, and 24% for the initial pH assayed of 5, 7, and 9, respectively. Therefore, based on this, Sineli et al. (2016) concluded that the best HCH-removal performance for *Streptomyces* sp. M7 was obtained at pH 7, similar to that the reported by Cuozzo et al. (2009) with *Streptomyces* sp. M7 and γ-HCH.

The degradation of β-HCH at the different concentrations (1.66, 8.33, and 16.6 mg l^{-1}) led to removal percentages of 78%, 71%, and 63% (Table 1), with chloride ions release percentages of 100%, 67%, and 32%, respectively.

The removal of both HCH isomers was optimum at pH 7. The pH 7 is generally considered optimal for metabolic activities of the *Streptomyces* genus (Cuozzo et al. 2009). In addition, Robinson et al. (2009) observed that an acidic environment was not favorable for aerobic bacterial dehalogenation.

It is important to highlight that the removal process not only involves the biodegradation, but also a combination of mechanisms such as adsorption, absorption, and desorption, as demonstrated by Abromaitis et al. (2016). Therefore, physicochemical conditions in the environment greatly influence the removal process of xenobiotic compounds (Kumar et al. 2016).

When the HCH isomer concentration was increased, *Streptomyces* sp. M7 was able to remove even the maximum β-HCH concentration from the culture medium, although the removal of β-HCH was 20% less than the removal obtained for α-HCH at 16.6 mg l^{-1}. This may indicate a possible inhibitory effect of the β-HCH on the removal capacity of the *Streptomyces* strain, considering that it is the most recalcitrant isomer according to Johri et al. (1998) and Phillips et al. (2005). Similarly, Rajashekara Murthy and Manonmani (2007) evaluated the effects of the initial concentrations of technical lindane on its degradation. These authors found that increasing the substrate concentration caused a decrease in the degradation rates for α- and β-HCH isomers of about 20% and 76%, respectively, by a defined microbial consortium consisting of ten bacterial isolates of the genera *Pseudomonas, Burkholderia, Flavobacterium,* and *Vibrio.*

Growth and Removal Capacity of Actinobacteria in a Mixture of HCH Isomers

Considering that multiple HCH isomers are normally present in contaminated sites, it is necessary to study degradation for a mixture of isomers. Thus, the growth of *Streptomyces* sp. M7 and its removal capacity were evaluated in a mixture of α-, β-, and γ-HCH. The maximum biomass value (0.55 mg ml^{-1}) was obtained at the first day of incubation, followed by a decrease along the time of 0.35 and 0.25 mg ml^{-1} at four and seven days of incubation, respectively (Sineli et al. 2016). Moreover, the removal percentages were 46%, 39%, and 45%, for α-, β-, and γ-HCH, respectively. This demonstrates the ability of *Streptomyces* sp. M7 to remove the three isomers together. However, the isomers removal resulted in a decrease of 33% and 39% for the individual removal of each isomer (79 and 78% for α- and β-, respectively). Furthermore, the β-HCH showed higher resistance to degradation than the other two isomers analyzed; its removal was approximately 5% lower at the 7th day of incubation.

Rajashekara Murthy and Manonmani (2007) also focused on the effect of a mixture of α-/β-HCH isomers and lindane on the degradation capacity, and they found that the degradation of lindane was not affected by the presence of the other isomers. However, those experiments were performed in the presence of a defined microbial consortium which could have produced a synergistic effect in the presence of the isomers mixture, possibly due to the presence of several degradation routes. Moreover, Manickam et al. (2006) carried out

HCH isomers biodegradation studies with the actinobacterium *Mycobacterium* sp. ITRC1. They described its ability to remove the isomers α-, β-, γ- and δ-HCH in liquid cultures and aged contaminated soil, obtaining around 90% of all the isomers in both systems.

Identification of Metabolites Produced During α- β- and γ-HCH Isomers Degradation Process by Actinobacteria

The identification of metabolic intermediaries from the HCH degradation process indicates the ability of a microorganism to biodegrade these compounds (Manickam et al. 2008, Raina et al. 2008). The microbial degradation of hexachlorocyclohexanes involves the removal of chlorine atoms from these molecules by the action of enzymes with dechlorinase activity (Cuozzo et al. 2009).

When Sineli et al. (2016) cultured *Streptomyces* sp. M7 with α-HCH, the GC-MS results from the culture supernatant obtained at 72 hr of incubation revealed the appearance of three peaks in the chromatogram (Table 1). These were identified as (i) 1,2-dichlorobenzene (1,2-DCB) with 97% identity (Rt 9.8 min); (ii) 1,4-dichlorobenzene (1,4-DCB) or 1,3-dichlorobenzene (1,3-DCB) with 97% identity (Rt 10.23 min); and (iii) trichlorobenzenes (1,2,3-trichlorobenzene, 1,2,4-trichlorobenzene, or 1,3,5-trichlorobenzene) with 97% identity (Rt 17.3 min).

In the case of *Streptomyces* sp. M7 cultured in the presence of β-HCH, dichlorobenzene isomers were identified in the culture supernatant with 82% identity. In both cultures, the isomer used as the carbon source (α- or β-HCH) was also identified, which indicates that these compounds were not completely degraded during the 72 hr of incubation (Sineli et al. 2016).

In the cell-free extracts of *Streptomyces* sp. M7 cultured with α-HCH, Sineli et al. (2016) identified the metabolic intermediates pentachlorocyclohexene (PCCH) with 90% identity (Rt 29.5 min) and dichlorobenzene isomers with 81% identity (Rt 12.2 min). The products identified from β-HCH degradation were tetrachlorocyclohexene (TCCH) with 59% identity (Rt 26.6 min) and dichlorobenzene isomers with 81% identity (Rt 12.2 min) (Table 1). In this study, the intermediate metabolites of α- and β-HCH were identified for the first time in culture supernatants and cell-free extracts of the genus *Streptomyces*.

De Paolis et al. (2013) have also been detected PCCH and TCCH as products of α- and β-HCH degradation by *Arthrobacter fluorescens* and *Arthrobacter giacomelloi*, respectively. These are the products resulting from the first and second chloride ion elimination reactions, as proposed in the degradation pathway for *Sphingobium japonicum* UT26 (Nagata et al. 2005).

The presence of intermediates in the supernatant would indicate that they are produced by the strain, as demonstrated by Cuozzo et al. (2009) in plates of *Streptomyces* sp. M7 through the presence of a clear halo around the colony indicating the capacity of this strain to degrade lindane. Furthermore, in this study the first intermediates of γ-HCH degradation were detected in cell-free extract of *Streptomyces* sp. M7, i.e., PCCH and 1,3,4,6-tetrachloro-1,4-cyclohexadiene (TCDN). These intermediates are product of the dehydrochlorinase enzyme LinA, which removes the chlorine atoms located in axial positions through bimolecular elimination mechanism (Lal et al. 2010). Shrivastava et al. (2015) characterized a novel LinA type 3 δ-hexachlorocyclohexane dehydrochlorinase from a HCH-contaminated soil sample using metagenomic approach. The gen *linA* involves the protein synthesis of the first enzyme of the γ-HCH degradation pathway, with dehydrochlorinase activity for all HCH isomers, except β-isomer (all the chlorine atoms in equatorial positions).

Saez et al. 2015 carried out a study of identification of γ-HCH degradation intermediates in minimal medium using a *Streptomyces* strains consortium (A5, A11, A2, M7). In this work, the pentachlorocyclohexene, 1,2- and 1,4-dichlorobenzene were identified in culture supernatant by GC-MS analysis. Moreover, the toxicity of these metabolic intermediates produced was analysed. Bioassays with *Lactuca sativa* demonstrated the effectiveness of the biodegradations by the *Streptomyces* consortium in liquid and soil slurry systems, since the lettuce seedlings showed an improvement in its biological parameters compared to abiotic controls, confirming a significant decrease in toxicity.

The α-HCH isomer exists in two enantiomeric forms, which are converted to its respective β-PCCH enantiomer: (+)-α-HCH is converted to β-(3S,4S,5R,6S)-1,3,4,5,6-PCCH, and (–)-α-HCH becomes β-(3R,4R,5S,6R)-1,3,4,5,6-PCCH. Interestingly, these two β-PCCH enantiomers are then metabolized to 1,2,4-trichlorobenzene (1,2,4-TCB) (Lal et al. 2010). Hence, the detection of these intermediate compounds suggests that the TCDN path could be followed, and two further rounds of dehydrochlorination appear to produce dead-end products in *Streptomyces* sp. M7 (Sineli et al. 2016). Few studies have reported on the toxic effects of HCH metabolites (Robles-González et al. 2012). It is known that 1,2,4-TCB is moderate to highly toxic to aquatic organisms, however, it is not classified as a human carcinogen. Conversely, lindane is classified as a carcinogen and neurotoxic; moreover, its persistence in soil is 5-fold shorter than that of lindane. In the case of 1,3-DCB, it is not classifiable as a carcinogenic to humans.

Notably, in the studies using *Streptomyces* strains for the degradation of HCH isomers (Cuozzo et al. 2009, Saez et al. 2015, Sineli et al. 2016) phenolic compounds were not detected, which could indicate that the process would have continued towards mineralization and the benzene rings could not be the final degradation products. On the opposite, Datta et al. (2000) identified six metabolic intermediates of γ-HCH produced by *Arthrobacter citreus* BI-100, including clorophenol and phenol. Manickam et al. (2006) detected and identified 2,5-dichlorophenol during the aerobic degradation of α-, β-, γ-, and δ-isomers of HCH by the actinobacterium *Microbacterium* sp. ITRC1.

Concluding Remarks

Lindane is a chlorinated cyclic saturated hydrocarbon used extensively for the control of agricultural pests and mosquitos. Due to its continuous use in the past decades throughout the world, lindane-contaminated sites are prominent, and thus there is an urgent need to develop cleanup strategies. Bioremediation is a technology that could be employed for the decontamination of soil/sites contaminated by pesticides. Microbial biotechnology possesses ample scope to work in this direction. The use of microorganisms to cleanup polluted environments is rapidly changing and expanding the field of environmental biotechnology. Although there is still much work to be done, our limited understanding of the potential contributions of biological approaches and their impact on ecosystems have been an obstacle to making these technologies safer and more reliable.

This chapter reports the ability of actinobacteria to remove and biodegrade the different hexachlorocyclohexane isomers. Within actinobacteria, the genus *Streptomyces* has shown a great ability to remove and degrade these chlorinated compounds. Specifically, the strain *Streptomyces* sp. M7, isolated from a highly contaminated region has the ability not only to use α- and β-HCH isomers as carbon sources but also to remove high concentrations of them from the culture medium. Moreover, this microorganism is able to remove a mixture composed of α-, β- and γ-HCH (lindane), which is important considering that multiple isomers are normally present in contaminated sites. Furthermore, *Streptomyces* sp. M7, as part of a defined consortium with other *Streptomyces* strains, has demonstrated the ability to degrade γ-HCH from soil and slurry systems. This is important in order to develop a real bioremediation process. Other actinobacteria genera which are able to degrade different HCH isomers in soil system are *Arthrobacter*, *Microbacterium*, and *Mycobacterium*, among others.

The data presented in this chapter have highlighted the ability of actinobacteria to remove α-, β-, and γ-HCH isomers from liquid and soil systems, under a diverse set of physicochemical conditions. Actinobacteria should be considered as a potential agent to bioremediate environments contaminated with organochlorine pesticides.

Acknowledgments

The authors acknowledge the financial support provided by Secretaría de Ciencia, Arte e Innovación Tecnológica of Universidad Nacional de Tucumán (SCAIT), Agencia Nacional de Promoción Científica y Tecnológica (ANPCyT), Consejo Nacional de Investigaciones Científicas y Técnicas (CONICET), and Universidad del Norte Santo Tomas de Aquino (UNSTA). We also thank Mr. G. Borchia for his technical assistance.

References Cited

Abhilash, P.C., S. Jamil, V. Singh, A. Singh, N. Singh and S.C. Srivastava. 2008. Occurrence and distribution of hexachlorocyclohexane isomers in vegetation samples from a contaminated area. Chemosphere 72(1): 79–86.

Abhilash, P.C. 2009. Monitoring of organochlorine pesticide (Lindane) in soil-plant system of a contaminated environment and its phytoremediation/bioremediation. Ph.D. Thesis, University of Lucknow, India.

Abhilash, P.C., S. Srivastava and N. Singh. 2011. Comparative bioremediation potential of four rhizospheric microbial species against lindane. Chemosphere 82(1): 56–63.

Abromaitis, V., V. Racys, P. van der Marel and R.J.W. Meulepas. 2016. Biodegradation of persistent organics can overcome adsorption-desorption hysteresis in biological activated carbon systems. Chemosphere 149: 183–189.

Alvarez, A., J.M. Saez, J.S. Davila Costa, V.L. Colin, M.S. Fuentes, S.A. Cuozzo et al. 2017. Actinobacteria: Current research and perspectives for bioremediation of pesticides and heavy metals. Chemosphere 166: 41–62.

ATSDR. 2011. Agency for Toxic Substances and Disease Registry. Hexachlorocyclohexane (HCH). Atlanta, USA.

Benimeli, C.S., M.J. Amoroso, A.P. Chaile and G.R. Castro. 2003. Screening of organochlorine pesticides removal by aquatic streptomycetes. Bioresour. Technol. 89: 133–138.

Benimeli, C.S., G.R. Castro, A.P. Chaile and M.J. Amoroso. 2006. Lindane removal induction by *Streptomyces* sp. M7. J. Basic Microbiol. 46(5): 348–357.

Benimeli, C.S., M.S. Fuentes, C.M. Abate and M.J. Amoroso. 2008. Bioremediation of lindane-contaminated soil by *Streptomyces* sp. M7 and its effects on *Zea mays* growth. Int. Biodeterior. Biodegradation 61: 233–239.

Beyer, A. and M. Matthies. 2001. Long-range transport potential of semivolatile organic chemicals in coupled air–water systems. Environ. Sci. Pollut. Res. 8(3): 173–179.

Bourguignon, N., M.S. Fuentes, C.S. Benimeli, S.A. Cuozzo and M.J. Amoroso. 2014. Aerobic removal of methoxychlor by a native *Streptomyces* strain: Identification of intermediate metabolites. Int. Biodeterior. Biodegradation 96: 80–86.

Camacho-Pérez, B., E. Ríos-Leal, N. Rinderknecht-Seijas and H.M. Poggi-Varaldo. 2012. Enzymes involved in the biodegradation of hexachlorocyclohexane: a mini review. J. Environ. Manage. 95: 306–318.

Chia, V.M., Y. Li, S.M. Quraishi, B.I. Graubard, J.D. Figueroa, J.P. Weber et al. 2010. Effect modification of endocrine disruptors and testicular germ cell tumour risk by hormone-metabolizing genes. Int. J. Androl. 33: 588–596.

Colt, J.S., N. Rothman, R.K. Severson, P. Hartge, J.R. Cerhan, N. Chatterjee et al. 2009. Organochlorine exposure, immune gene variation, and risk of non-Hodgkin lymphoma. Blood. 113: 1899–1905.

Concha-Graña, E., M.I. Turnes-Carou, S. Muniategui-Lorenzo, P. Lopez-Mahia, D. Prada-Rodriguez and E. Fernández-Fernández. 2006. Evaluation of HCH isomers and metabolites in soils, leachates, river water and sediments of a highly contaminated area. Chemosphere 64: 588–595.

Cuozzo, S.A., G. Rollan, C.M. Abate and M.J. Amoroso. 2009. Specific dechlorinase activity in lindane degradation by *Streptomyces* sp. M7. World J. Microbiol. Biotechnol. 25: 1539–1546.

Cuozzo, S.A., M.S. Fuentes, N. Bourguignon, C.S. Benimeli and M.J. Amoroso. 2012. Chlordane biodegradation under aerobic conditions by indigenous *Streptomyces* strains. Int. Biodeterior. Biodegradation 66(1): 19–24.

Datta, J., A.K. Maiti, D.P. Modak, P.K. Chakrabartty, P. Bhattacharyya and P.K. Ray. 2000. Metabolism of γ-hexachlorocyclohexane by *Arthrobacter citreus* strain BI-100: identification of metabolites. J. Gen. Appl. Microbiol. 46: 59–67.

De Paolis, M.R., D. Lippi, E. Guerriero, C.M. Polcaro and E. Donati. 2013. Biodegradation of α-, β-, and γ-Hexachlorocyclohexane by *Arthrobacter fluorescens* and *Arthrobacter giacomelloi*. Appl. Biochem. Biotechnol. 170: 514–524.

Fuentes, M.S., G.E. Briceño, J.M. Saez, C.S. Benimeli, M.C. Diez and M.J. Amoroso. 2013. Enhanced removal of a pesticides mixture by single cultures and consortia of free and immobilized *Streptomyces* strains. Biomed. Res. Int. ID 392573.

Fuentes, M.S., V.L. Colin, M.J. Amoroso and C.S. Benimeli. 2016. Selection of an actinobacteria mixed culture for chlordane remediation. Pesticide effects on microbial morphology and bioemulsifier production. J. Basic Microbiol. 56: 127–137.

Heeb, N.V., S.A. Wyss, B. Geueke, T. Fleischmann, H.P. Kohler and P. Lienemann. 2014. LinA2, a HCH-converting bacterial enzyme that dehydrohalogenates HBCDs. Chemosphere 107: 194–202.

Heeb, N.V., S.A. Wyss, B. Geueke, T. Fleischmann, H.P.E. Kohler, W.B. Schweizer et al. 2015. Stereochemistry of enzymatic transformations of (+)β- and (−)β-HBCD with LinA2—A HCH-degrading bacterial enzyme of *Sphingobium indicum* B90A. Chemosphere 122: 70–78.

Ingelido, A.M., A. Abballe, V. Marra, S. Valentini, A. Ferro, M.G. Porpora et al. 2009. Serum concentrations of beta-hexachlorocyclohexane in groups of the Italian general population: a human biomonitoring study. Annali dell'Istituto superiore di sanità 45: 401–408.

Johri, A.K., M. Dua, D. Tuteja, R. Saxena, D.M. Saxena and R. Lal. 1998. Degradation of alpha, beta, gamma and delta-hexachlorocyclohexanes by *Sphingomonas paucimobilis*. Biotechnol. Lett. 20: 885–887.

Kumar, D., A. Kumar and J. Sharma. 2016. Degradation study of lindane by novel strains *Kocuria* sp. DAB-1Y and *Staphylococcus* sp. DAB-1W. Bioresour. Bioprocess. 3: 53. DOI: 10.1186/s40643-016-0130-8.

Lal, R., G. Pandey, P. Sharma, K. Kumari, S. Malhotra, R. Pandey et al. 2010. Biochemistry of microbial degradation of hexachlorocyclohexane and prospects for and prospect of bioremediation. Microbiol. Mol. Biol. Rev. 74: 58–80.

Laquitaine, L., A. Durimel, L.F. de Alencastro, C. Jean-Marius, O. Gros and S. Gaspard. 2016. Biodegradability of HCH in agricultural soils from Guadeloupe (French West Indies): identification of the *lin* genes involved in the HCH degradation pathway. Environ. Sci. Pollut. Res. Int. 23(1): 120–127.

Liu, Z., H. Zhang, M. Tao, S. Yang, L. Wang, Y. Liu et al. 2010. Organochlorine pesticides in consumer fish and mollusks of Liaoning province, China: distribution and human exposure implications. Arch. Environ. Contam. Toxicol. 59: 444–453.

Mani, D. and C. Kumar. 2014. Biotechnological advances in bioremediation of heavy metals contaminated ecosystems: an overview with special reference to phytoremediation. Int. J. Environ. Sci. Technol. 11: 843–872.

Manickam, N., M. Mau and M. Schlömann. 2006. Characterization of the novel HCH-degrading strain, *Microbacterium* sp. ITRC1. Appl. Microbiol. Biotechnol. 69: 580–588.

Manickam, N., M.K. Reddy, H.S. Saini and R. Shanker. 2008. Isolation of hexachlorocyclohexane-degrading *Sphingomonas* sp. by dehalogenase assay and characterization of genes involved in γ-HCH degradation. J. Appl. Microbiol. 104: 952–960.

Manna, R.N., K. Zinovjev, I. Tunon and A. Dybala-Defratyka. 2015. Dehydrochlorination of hexachlorocyclohexanes catalyzed by the LinA dehydrohalogenase. A QM/MM Study. J. Phys. Chem. B 119: 15100–15109.

Mougin, C., C. Pericaud, C. Malosse, C. Laugero and M. Ashter. 1999. Biotransformation of the insecticide lindane by the white-rot basidiomycete *Phanerochaete chrysosporium*. Pept. Sci. 47: 51–59.

Nagata, Y., K. Miyauchi and M. Takagi. 1999. Complete analysis of genes and enzymes for γ-hexachlorocyclohexane degradation in *Sphingomonas paucimobilis* UT26. J. Ind. Microbiol. Biotechnol. 23: 380–390.

Nagata, Y., Z. Prokop, Y. Sato, P. Jerabek, Y. Ashwani Kumar, M. Ohtsubo et al. 2005. Degradation of β-hexachlorocyclohexane by haloalkane dehalogenase LinB from *Sphingomonas paucimobilis* UT26. Appl. Environ. Microbiol. 71: 2183–2185.

Nagata, Y., R. Endo, M. Ito, Y. Ohtsubo and M. Tsuda. 2007. Aerobic degradation of lindane (gamma-hexachlorocyclohexane) in bacteria and its biochemical and molecular basis. Appl. Microbiol. Biotechnol. 76: 741–752.

Nawab, A., A. Aleem and A. Malik. 2003. Determination of organochlorine pesticides in agricultural soil with special reference to γ-HCH degradation by *Pseudomonas* strains. Biores. Technol. 88: 41–46.

Okeke, B.C., T. Siddique, M.C. Arbestain and W.T. Frankenberger. 2002. Biodegradation of γ- hexachlorocyclohexane (Lindane) and γ-hexachlorocyclohexane in water and soil slurry by a *Pandoraea* species. J. Agric. Food Chem. 50: 2548–2555.

Österreicher-Cunha, P., T. Langenbach, J.P. Torres, A.L. Lima, T.M. de Campos, E.A. Vargas Jr. et al. 2003. HCH distribution and microbial parameters after liming of a heavily contaminated soil in Rio de Janeiro. Environ. Resour. 93: 316–327.

Pavlíková, N., L. Bláhová, P. Klán, S. Reddy Bathula, V. Sklenář, J.P. Giesy et al. 2012. Enantioselective effects of alpha-hexachlorocyclohexane (HCH) isomers on androgen receptor activity *in vitro*. Chemosphere 86: 65–69.

Phillips, T.M., A.G. Seech, H. Lee and J.T. Trevors. 2005. Biodegradation of hexachlorocyclohexane (HCH) by microorganisms. Biodegradation 16: 363–392.

Phillips, T.M., H. Lee, J.T. Trevors and A.G. Seech. 2006. Full-scale *in situ* bioremediation of hexachlorocyclohexane-contaminated soil. J. Chem. Technol. Biotech. 81: 289–298.

Polti, M.A., J.D. Aparicio, C.S. Benimeli and M.J. Amoroso. 2014. Simultaneous bioremediation of Cr(VI) and lindane in soil by actinobacteria. Int. Biodeterior. Biodegradation 88: 48–55.

Prakash, O., M. Suar, V. Raina, C. Dogra, R. Pal and R. Lal. 2004. Residues of hexachlorocyclohexane isomers in soil and water samples from Delhi and adjoining areas. Curr. Sci. 87: 73–77.

Quintero, J.C., M.T. Moreira, G. Feijoo and J.M. Lema. 2005. Anaerobic degradation of hexachlorocyclohexane isomers in liquid and soil slurry systems. Chemosphere 61: 528–536.

Quintero, J.C., T.A. Lú-Chau, M.T. Moreira, G. Feijoo and J.M. Lema. 2007. Bioremediation of HCH present in soil by the white-rot fungus *Bjerkandera adusta* in a slurry batch bioreactor. Int. Biodeterior. Biodegrad. 60: 319–326.

Raina, V., A. Hauser, H.R. Buser, D. Rentsch, P. Sharma, R. Lal et al. 2007. Hydroxylated metabolites of β- and δ-hexachlorocyclohexane: bacterial formation, stereochemical configuration, and occurrence in groundwater at a former production site. Appl. Environ. Microbiol. 41: 4292–4298.

Raina, V., D. Rentsch, T. Geiger, P. Sharma, H.R. Buser, C. Holliger et al. 2008. New metabolites in the degradation of α- and γ-hexachlorocyclohexane (HCH): pentachlorocyclohexenes are hydroxylated to cyclohexenols and cyclohexenediols by the haloalkane dehalogenase LinB from *Sphingobium indicum* B90A. J. Agric. Food Chem. 56: 6594–6603.

Rajashekara Murthy, H.M. and H.K. Manonmani. 2007. Aerobic degradation of technical hexachlorocyclohexane by a defined microbial consortium. J. Hazard. Mater. 149: 18–25.

Robinson, C., D.A. Barry, P.L. McCarty, J.I. Gerhard and I. Kouznetsova. 2009. pH control for enhanced reductive bioremediation of chlorinated solvent source zones. Sci. Total Environ. 407: 4560–4573.

Robles-González, I.V., E. Ríos-Leal, I. Sastre-Conde, F. Fava, N. Rinderknecht-Seijas and H.M. Poggi-Varaldo. 2012. Slurry bioreactors with simultaneous electron acceptors for bioremediation of an agricultural soil polluted with lindane. Process Biochem. 47: 1640–1648.

Saez, J.M., J.D. Aparicio, M.J. Amoroso and C.S. Benimeli. 2015. Effect of the acclimation of a *Streptomyces* consortium on lindane biodegradation by free and immobilized cells. Process Biochem. 50: 1923–1933.

Shelton, D.R., S. Khader, J.S. Karns and B.M. Pogell. 1996. Metabolism of twelve herbicides by *Streptomyces*. Biodegradation 7: 129–136.

Shrivastava, N., Z. Prokop and A. Kumar. 2015. Novel LinA type 3 δ-hexachlorocyclohexane dehydrochlorinase. Appl. Environ. Microbiol. 81(21): 7553–7559.

Simonich, S.L. and R.A. Hitéis. 1995. Organic pollutants accumulation in vegetation. Environ. Sci. Technol. 29(12): 2905–2914.

Sineli, P.E., G. Tortella, J.S. Dávila Costa, C.S. Benimeli and S.A. Cuozzo. 2016. Evidence of α-, β- and γ-HCH mixture aerobic degradation by the native actinobacteria *Streptomyces* sp. M7. World J. Microbiol. Biotechnol. 32(5): 81.

Singh, B.K. and R.C. Kuhad. 1999. Biodegradation of lindane (gamma-hexachlorocyclohexane) by the white-rot fungus *Trametes hirsutus*. Lett. Appl. Microbiol. 28: 238–241.

Singh, B.K. and R.C. Kuhad. 2000. Degradation of the insecticide lindane (γ-HCH) by white-rot fungi *Cyathus bulleri* and *Phanerochaete sordida*. Pest. Manag. Sci. 56: 142–146.

Sun, G., X. Zhang, Q. Hu, H. Zhang, D. Zhang and G. Li. 2015. Biodegradation of dichlorodiphenyltrichloroethanes (DDTs) and hexachlorocyclohexanes (HCHs) with plant and nutrients and their effects on the microbial ecological kinetics. Microbiol. Ecol. 69(2): 281–292.

Tabata, M., R. Endo, M. Ito, Y. Ohtsubo, A. Kumar, M. Tsuda et al. 2011. The *lin* genes for γ-hexachlorocyclohexane degradation in *Sphingomonas* sp. MM-1 proved to be dispersed across multiple plasmids. Biosci. Biotechnol. Biochem. 75: 466–472.

Vijgen, J. 2006. The legacy of lindane HCH isomer production. IHPA. http://www.ihpa.info/docs/library/reports/Lindane%20Main%20Report%20DEF20JAN06.pdf.

Vijgen, J., P.C. Abhilash, Y.F. Li, R. Lal, M. Forter, J. Torres et al. 2011. Hexachlorocyclohexane (HCH) as new Stockholm Convention POPs—a global perspective on the management of Lindane and its waste isomers. Environ. Sci. Pollut. Res. 18: 152–162.

Zhu, Y., H. Liu, Z. Xi, H. Cheng and X. Xu. 2005. Organochlorine pesticides (DDTs and HCHs) in soils from the outskirts of Beijing, China. Chemosphere 60: 770–778.

12

Mining and Mine Tailings

Characterization, Impacts, Ecology and Bioremediation Strategies

Ma. Laura Ortiz-Hernández,[1,*] *Patricia Mussali-Galante,*[1,a] *Enrique Sánchez-Salinas*[1,b] and *Efraín Tovar-Sánchez*[2]

Introduction

After agriculture, mining is traditionally regarded as the oldest and the most important activity of the world. Throughout history, mining has had important contributions to the people and has become one of the most important primary economic activities in several countries. It is oriented to the search, exploitation, and use of the various minerals taken directly from nature. It contributes to the economic development of the country and supplies inputs to different industries; besides, it provides direct employment to over 40 million people and indirect support for 200–250 million people (Cao 2007, Kossof et al. 2014). The mineral wealth of a country is part of its economic reserves, and when this wealth is greater, the country becomes richer. In this context, mining is an activity that converts the minerals of the earth's crust to other capital forms, which directly contributes to economic and social development (Davis and Tilton 2002).

Mining activities are easily recognizable. By nature, the mining industry, just like oil and gas industries, leaves behind a 'footprint', as well as an environmental, social, and economic impact (Al Rawashdeh et al. 2016). Each mining phase involves specific activities that have an effect on the environment (Tiwary 2001, Matsumoto et al. 2016). The general phases involved in the mining industry include exploration and development, exploitation, processing and closure, and rehabilitation (Ali 2009). After the huge quantity of earth and rock is moved, part of it is processed to recover valuable minerals, and the rest is discarded as waste (Schoenberger 2016).

[1] Centro de Investigación en Biotecnología, Universidad Autónoma del Estado de Morelos. Av. Universidad 1001, Col. Chamilpa. Cuernavaca Mor., México.
[a] Email: patricia.mussali@uaem.mx
[b] Email: sanchez@uaem.mx
[2] Centro de Investigación en Biodiversidad y Conservación, Universidad Autónoma del Estado de Morelos. Av. Universidad 1001, Col. Chamilpa. Cuernavaca Mor., México.
Email: efrain_tovar@uaem.mx
* Corresponding author: ortizhl@uaem.mx

Mining is an activity that generates a large amount of waste, producing about 65 billion tons annually; fifty-one billion tons is waste rock, and 14 billion tons, already stored globally, are particles with less than 120 μm, which are named "mine tailings" (Jones and Boger 2012, Kossof et al. 2014, Wang et al. 2014, Baghdasaryan 2016), which are deposited on land (Söderholm et al. 2015). Due to the current high prices of the metals and the growing demand, mine tailings production is also expanding (Jones and Boger 2012).

The main composition of mine tailings consists of minerals, heavy metals, coal or mineral fuels, fine rocks, loose sediments, ground waste, combustion dust, ashes, chemicals added during the recovery of metals, and fluids of diverse composition (Hudson-Edwards et al. 2011). It generally contains residual metal sulfides such as pyrite (FeS_2), pyrrhotite (Fe1-x S), galena (PbS), sphalerite (Zn), chalcopyrite ($CuFeS_2$) and arsenopyrite (FeAsS) which are a source of potentially toxic elements such as arsenic (As), cadmium (Cd), lead (Pb), copper (Cu), zinc (Zn) and iron (Fe), among others (Romero et al. 2007).

Lèbre et al. (2016) argue that it is necessary to establish a mine tailings management that considers them as a potential resource for the future, including a hierarchy of four main levels: reduce, reprocess and stockpile, down cycle, and dispose of. This management scheme could be a basic component of a global sustainability framework for the mining industry (Franks et al. 2011, Moran et al. 2014).

In this chapter, we address mining as an activity necessary to satisfy peoples' needs, but also consider the generated impacts, with special emphasis on both waste generation and environmental and health effects. It also addresses the remediation of contaminated sites with heavy metals, with special attention to the bioremediation and phytoremediation processes.

Impacts of Mining

Heavy Metals as Components of Mine Tailings

Mining activities release heavy metals into the environment, and nowadays, these metals are the major problem for the industrialized world. These pollutants are non-degradable, are persistent in the environment and cause serious effects on terrestrial and aquatic ecosystems (Boopathy 2000, Benzerara et al. 2011, Abbas et al. 2014, Chen et al. 2015, Ullah et al. 2015).

At a global level, heavy metal contamination is one of the major problems affecting human health (Gamalero et al. 2009, Fu and Wang 2011). Heavy metals are conventionally defined as elements with metallic properties such as conductivity, ductility, atomic number greater than 20, and whose density is greater than 5 g per cm^3. These elements constitute an important group since in high concentrations, they can be toxic to microorganisms, plants, animals, and humans, among others (Liao et al. 2003, Spain and Alm 2003).

Heavy metals can be found in the earth's crust as minerals, salts or other compounds and can be mobilized into the soil, air, and water by changes in redox state. In addition to this, they can be absorbed by plants and thus incorporated into the food chain (Rooney et al. 2006, Zhao et al. 2006, Young et al. 2013). Figure 1 shows the relationship of heavy metals with soil, air, water, microorganisms, plants, and animals. These interactions are complex because they depend on various factors such as physical and chemical properties of each component, as well as the prevailing environmental conditions.

Contamination with heavy metals is an important challenge for living organisms, since various metals such as Cu and Zn are essential micronutrients but toxic in high concentrations, while others such as Cd, Pb, and Mercury (Hg) are toxic even at low concentrations (Cervantes and Moreno 1999, Volke et al. 2004, Mancilla et al. 2012).

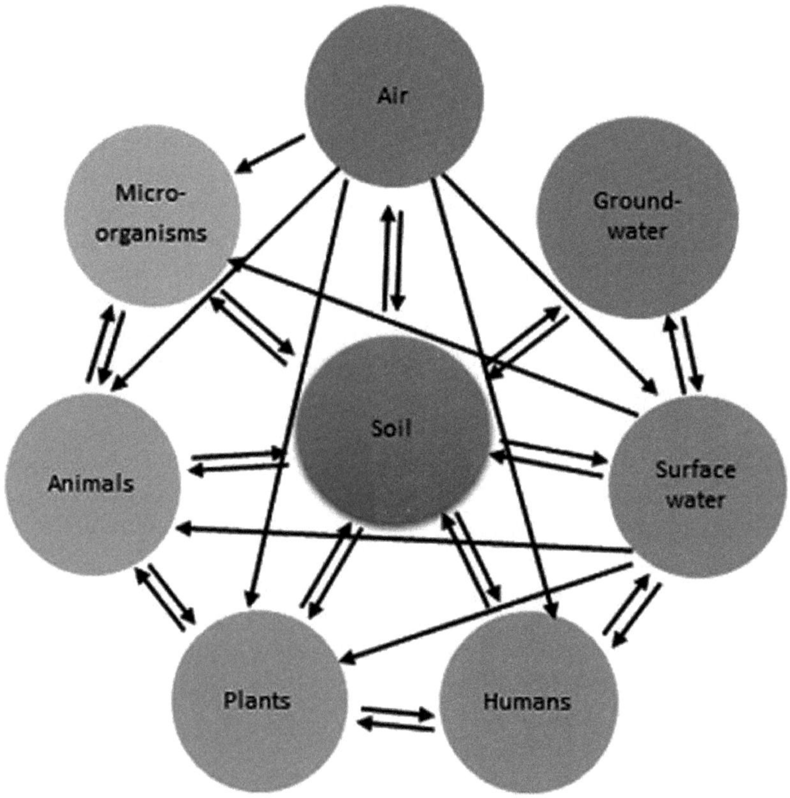

Figure 1. Heavy metals' relationship with soil, air, water, and biota (microorganisms, plants, and animals).

Heavy metals' toxicity is due to their ability to bind to organic molecules, so the toxic effects in biological systems depend on reactions with ligands that are essential for their assimilation. The major ligands of the heavy metals are sulfhydryl, amino, phosphate, carboxyl and hydroxyl groups. These ligand-metal bonds can damage cells, generating alterations at the protein level by inhibition or changes in their structure, displacement of essential elements of metabolism, generation of reactive oxygen species (ROS) or free radicals that cause oxidative stress, and therefore, alterations to DNA. DNA damage is discussed below (Hartwig 1998, Wilson and Thompson 2007, Navarro et al. 2008, Rojas 2009, Valavanidis et al. 2009, Mussali-Galante et al. 2013a). Each heavy metal can cause different effects on plants and microorganisms; some of them are shown in Table 1.

Because of heavy metal content in mine tailings and the environmental and human health damage, the strategies for the sustainable waste management are very important and should be taken into consideration from their generation until processing and/or final disposal.

Impacts on the Environment

Environmental impacts related to mining activities are generally determined both by the mining method employed in the extraction process and the mine's geographic location (Smith 2009). In addition to this, mining industry produces several billion tons of

Table 1. Heavy metals' effects on plants and microorganisms (modified from Ayangbenro and Babalola 2017).

Metal	Effects on plants	Effects on microorganisms
As	Damage on cell membrane, inhibition of growth, inhibits root extension and proliferation, interferes with critical metabolic processes, loss of fertility, yield and fruit production, oxidative stress, physiological disorders	Deactivation of enzymes
Cd	Chlorosis, decrease in plant nutrient content, growth inhibition, reduced seed germination	Nucleic acid alterations, denatures protein, inhibits cell division and transcription, inhibits carbon and nitrogen mineralization
Cr (III, VI)	Chlorosis, delayed, senescence, wilting, biochemical lesions, reduced biosynthesis germination, stunted growth, oxidative stress, root growth alteration, modifications in seedling growth and development, reduced biomass, induction of leaf chlorosis and necrosis, besides physiological and biochemical alterations (Jun et al. 2009, Shanker et al. 2009). In addition, Cr affects leaves by reducing its number, area and inducing structural abnormalities and affects grain yield of crop plants (Singh et al. 2013)	Elongation of lag phase, growth inhibition, inhibition of oxygen uptake
Cu	Chlorosis, oxidative stress, retarded growth	Disrupts cellular function, inhibits enzyme activities
Pb	Affects photosynthesis and growth, chlorosis, inhibits enzyme activities and seed germination, oxidative stress	Denatures nucleic acid and protein, inhibits enzymes activities and transcription
Hg	Affects antioxidative system and photosynthesis, enhances lipid peroxidation, induced genotoxic effects, inhibits plant growth, yield, nutrient uptake and homeostasis, oxidative stress	Decreases population size, denatures protein, disrupts cell membrane, inhibits enzyme function
Zn	Affects photosynthesis, inhibits growth rate, reduced chlorophyll content, germination rate and plant biomass	Decreases biomass, inhibits growth, death

mine tailings per year (Kossof et al. 2014) and is considered one of the main sources of environmental contamination (Romero et al. 2007). The main environmental impacts of mining industry are shown in Fig. 2.

Mining activity affects the natural resources available to communities, which could help them sustain their own future (Hilson 2002). The following requirements and consequences are detected during mining: use of polluting substances (e.g., cyanide or sulfuric acid), large energy inputs (e.g., 1,000,000 m^3 of natural gas d^{-1}), significant volumes of water for long time (350 l seg^{-1}), high traffic levels (e.g., one truck every ten minutes, 24 hr per day), and exposure of all components to the human population (Donadio 2009). Additionally, during the mining operation, there are different impacts on the environment, which are mentioned below.

During the exploration phase, the main impacts on the environment are the vegetation destruction, and removal and disabling of soils by covering them with the underground mineral material. In addition to this, during the operating phase, the surface is devastated, the morphology of the land is severely modified, lot of materials are stacked and exposed, and cultivated areas and other surface patrimonies are destroyed; this can alter watercourses and form large gaps for discarded material. Also, cutting of trees and vegetation burning located on the site are carried out. The use of heavy machinery is the most common way

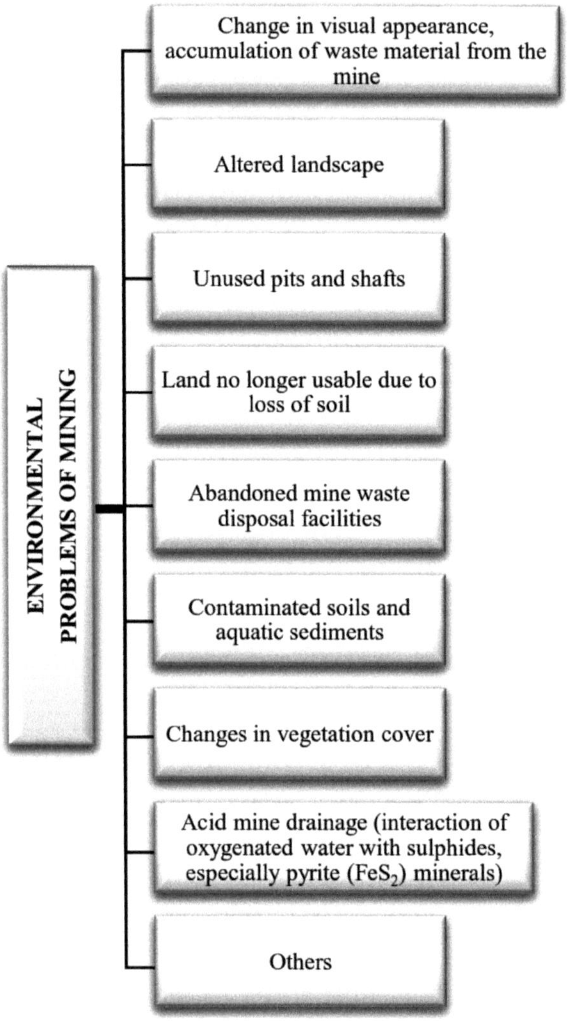

Figure 2. Environmental effects of mining sector (based on information from Mhlongo and Amponsah-Dacosta 2016).

to remove surpluses. In the case of open-pit mining, it frequently involves the removal of areas with native vegetation; therefore, this is one of the most destructive types of mining for the environment, especially in tropical forests. Other impacts on the environment include:

a) Air pollution: Cyanide vapors or gases are emitted, as well as Hg and sulfur dioxide, which are contained in waste gases and incomplete combustion processes.
b) Contamination of surface water: To separate the metal from the rock and thus obtain the commercially valuable ore, the water used during the process is combined with chemical reagents (cyanide, arsenic, foaming reagents, and depressants, among others). Because of this process, wastes of crushed rock, water, and residual chemical reagents are generated, which could reach rivers, lakes, and lagoons located in the areas surrounding the mining operations, thus generating water contamination. Acid drainage is the most important impact on water quality.

c) Contamination of groundwater: Water contaminated with solid waste, as well as rainwater, can reach the groundwater. Additionally, when groundwater is used during the mining process, the groundwater levels decrease.
d) Impact on soil: The main effect is the land use change and land clearing activities. Other important impacts are the temporary and permanent deposit of materials and wastes resulting from the process that causes disabling of soils and diminution of the capacity of water infiltration.
e) Impact on wildlife: Mining activities involve the partial or total elimination of the vegetation from operating or surrounding areas. Consequently, changes in the wildlife habitat associated with this vegetation occur and protected species could be at risk. Wildlife is disturbed and/or driven away by noise, as well as by air and water pollution. The clearing will affect mammals, reptiles, and birds; heavy metals from wastes generate mutagenic effects.
f) Impact on climate change: Deforestation decreases carbon sequestration capacity, contributing to greenhouse gas emissions and, therefore, to global warming (Sánchez-Salinas and Ortiz-Hernández 2014, Adiansyah et al. 2015).

The main environmental problem associated with the mine tailings is the acid drainage generation and its dispersion through surface runoff (Jiang et al. 2011). Acid drainage is generated by the oxidation of metal sulphides and therefore, solutions with low pH values are generated (Lin et al. 1997, Johnson et al. 2000, Johnson and Hallberg 2005). When these solutions are transported, it can become a severe environmental problem by contaminating soils, sediments, surface waters and groundwater (Bain et al. 2000, Armienta et al. 2001, Jung 2001).

Oxidation in mine tailings may be carried out when reactive metal sulphides, as well as the appropriate climatic conditions (air and water or humid atmosphere), are present. During the mine operation, metal sulphides oxidation in the mine tailings is generally very limited and is carried out slowly, only after the deposit accumulation is completed and its porosity allows the diffusion of the atmospheric oxygen. Before the tailings oxidation occurs, a visible alteration cannot be observed and they are usually grey in color. When the oxidation of metallic sulphides occurs, the tailings have a brown, yellow or red coloration due to pH values ranging from 2.4 to 3.7 (Martín and Gutierrez 2010).

Impacts on Human Health

The human health hazard risk derived from mine tailings is mainly due to their high content of metals and metalloids (Tovar-Sánchez et al. 2012). Some metals are considered as macronutrients for living organisms, like sodium (Na), potassium (K), calcium (Ca), and magnesium (Mg), which play important physiological roles. However, some metals such as Fe, Cu, Zn, cobalt (Co), molybdenum (Mb), and manganese (Mn), even though considered as trace elements, they can be toxic in high concentrations. Other metals like Hg, Pb, As, and Cd are not necessary for biological functions and are very toxic at small concentrations (Fraga 2005).

Consequently, metal toxicity has been widely analyzed, and it has been recognized that the relationship between exposure and disease is a multistage process, which includes external exposure, internal dose, early biological effects, altered structure and function and finally, clinical changes or disease (Vanden-Heuvel and Davis 1999, Bernard 2008, Mussali-Galante et al. 2013a).

Among the common effects resulting from metal exposure, carcinogenicity, neurotoxicity, and immunotoxicity may be listed (Florea and Busselberg 2006).

In general, metals have the capacity to use the normal physiological pathways to enter the cell and alter different cell constituents. Metals cause cellular damage by two general mechanisms. First, the metal binds to distinct macromolecules, driving to a conformational change or the metal may replace trace metals from their active sites or binding sites, altering cell homeostasis. These effects result from the great chemical similarity between the metal in question with trace divalent cations like Zn, Ca, Fe, Mg (Leonard et al. 2004). Second, exogenous metals act as catalytic sites in redox reactions, which form ROS. Reactive oxygen species are intermediates formed during metabolic oxidation processes. These include the superoxide anion radical (O_2.-), the hydroxyl radical (–OH) and the hydrogen peroxide (H_2O_2) (Valko et al. 2005). High ROS concentrations may cause oxidative damage to lipids, proteins, and DNA. Both mechanisms may alter cellular signaling, resulting in altered cell cycle (Harris and Shi 2003, Valko et al. 2006). Hence, many metals have the capacity to alter DNA structure, directly or indirectly, causing DNA single and double strand breaks (Valko et al. 2006, Mussali-Galante et al. 2007, Frenzilli et al. 2009). In addition to this, they are capable of forming chromosomal aberrations, sister chromatid exchanges, micronucleus, as well as alterations in DNA repair mechanisms, among others (Fenech et al. 1999, Valavanidis et al. 2009). All these alterations result in genetically unstable diseases such as cancer and neurodegenerative disorders. For example, certain compounds of hexavalent chromium are human carcinogens, causing tumor in the respiratory system (Quievryn et al. 2002, Jomova and Valko 2011). Adverse health effects seen in individuals exposed to Cr(VI) include gastrointestinal symptoms, hypotension, and hepatic and renal failure (Barceloux 1999). Cadmium is a potent human carcinogen causing prostate, lung, kidney, and pancreas cancer. The effect of environmental exposure to cadmium due to mining activities on cancer incidence in the environmentally contaminated North-East Belgium (the neighborhood of zinc smelters) has been extensively investigated (Sartor et al. 1992). The evidence shows an association between risk of cancer and cadmium exposure. Likewise, environmental arsenic exposure is definitively associated with human cancer (lung, skin, urinary, and liver). Skin lesions, peripheral neuropathy, and anemia are hallmarks of chronic arsenic exposure. Arsenic is also a potential risk factor for atherosclerosis. In addition to this, cardiovascular disorders following oral exposure to arsenic are well documented (Navas-Acien et al. 2005). There is some evidence from epidemiological trials that inhaled inorganic arsenic can also affect the cardiovascular system (Das et al. 2010). Furthermore, the estimation of relative risks for coronary disease, for stroke, and for peripheral arterial disease has been conducted.

Regarding neurotoxicity, Pb is the best example. The inability of the organism to manage and eliminate Pb effectively causes this metal to accumulate in the body. Though the half-life of Pb in blood is only 35 d, in the brain it is approximately 2 yr, and in bones it can last for decades. Lead is considered as a neurotoxin; its effects in humans are well established and the exposure to Pb can cause IQ deficit to peripheral neuropathy (Waalkes et al. 2000). Specifically, Pb exposure is much more dangerous during the nervous system development. This is one of the reasons why damage caused by Pb poisoning in children is more severe than in adults, causing severe impairment in their motor and cognitive abilities (Tong et al. 2000, Canfield et al. 2003). Infantile Pb exposure has been correlated with irreversible decrease in intelligence. Other alterations that have been reported are reading disorders, dyscalculia, deficits in short-term memory, and visual agnosia (Garza et al. 2006). Similarly, Hg is considered as an immunotoxin. The immunotoxicity of Hg is difficult to explain; it can have both immunosuppression and immunopotentiation effects and Hg is one of the few suspected environmental agents where an association between exposure and autoimmune disease has been suggested (Waalkes et al. 2000). Table 2 summarizes other impacts of heavy metals on health.

Overall, the above mentioned metals are the most common and toxic metals found in mine tailings and abandoned mine sites. Therefore, it becomes necessary to develop bioremediation strategies for mine sites in order to mitigate the detrimental effects of different metals that arise from active or inactive mining sites and pose a risk to human beings living near these areas.

Table 2. Heavy metals' effects on health (modified from Ayangbenro and Babalola 2017).

Metal	Effects on health
As	Brain damage, cardiovascular and respiratory disorder, conjunctivitis, dermatitis, skin cancer
Cd	Bone disease, coughing, emphysema, headache, hypertension, itai-itai, kidney diseases, lung and prostate cancer, lymphocytosis, microcytic hypochromic anemia, testicular atrophy, vomiting
Cr	Bronchopneumonia, chronic bronchitis, diarrhea, emphysema, headache, skin irritation, itching of respiratory tract, liver diseases, lung cancer, nausea, renal failure, reproductive toxicity, vomiting
Cu	Abdominal pain, anemia, diarrhea, headache, liver and kidney damage, metabolic disorders, nausea, vomiting
Pb	Anorexia, chronic nephropathy, damage to neurons, high blood pressure, hyperactivity, insomnia, learning deficits, reduced fertility, renal system damage, risk factor for Alzheimer's disease, shortened attention span
Hg	Ataxia, attention deficit, blindness, deafness, decreased rate of fertility, dementia, dizziness, dysphasia, gastrointestinal irritation, gingivitis, kidney problems, loss of memory, pulmonary edema, reduced immunity, sclerosis
Zn	Ataxia, depression, gastrointestinal irritation, hematuria, icterus, impotence, kidney and liver failure, lethargy, macular degeneration, metal fume fever, prostate cancer, seizures, vomiting

Interactions and Biological Process between Mine Tailings Components and Biota

Interactions and Biological Processes with Microorganisms

Some minerals, abundant in nature, offer huge reactive surfaces and can sorb or release metal and metalloid pollutants, depending on the chemical, physical and microbiological conditions prevailing in the soil (Benzerara et al. 2011). Further, microbial activity is affected by physicochemical and environmental parameters. The study of the microorganism's role in the fundamental geological process is named "geomicrobiology" (Gadd 2010). Microorganisms are widespread even in extreme environments (acid, hot, saline, anoxic, cold, dry, and those with high concentrations of metals). In addition to this, microorganisms participate in biogeochemical transformations on micro and macrogeographical scales, occurring in aquatic and terrestrial ecosystems, with consequences for soil fertility and human health (Gadd 2004, 2007).

The importance of understanding how microbial metabolism can affect the mobility and transformation of metals under different environmental conditions allows the prediction of their stability. Because of their small size and diverse metabolic capabilities, microorganisms (bacteria, archaea, yeast, fungi, some algae, and protozoa) are able to interact intimately with metal ions of soil, sediments or water (Gadd 2010, François et al. 2012). Some metals are required for metabolism and are taken into the cell through various mechanisms; then they are incorporated into the physiological pathways and biosynthetic structures (Douglas and Beveridge 1998). On the other hand, Pérez-de-Mora et al. (2006) point out that the soil microorganism's population and their activity are indicators of soil quality.

In nature, the ability of organisms to transform contaminants to either simpler or more complex chemical forms is very diverse; thus, microbial metabolic processes can increase the toxicity of the contaminants (Boopathy 2000). In addition to this, bacterial diversity is quite low in contaminated environments and may reflect in part the toxicity of heavy metals (Bruneel et al. 2006).

Microorganisms can catalyze a wide number of reactions, including redox reactions, which can contribute to metal speciation and modify their mobility in the environment (Benzerara et al. 2011). Research on extracellular polymer substances shows that bacterial biofilms are essential in leaching processes due to an increase in redox potential (Sand and Gehrke 2006). This knowledge has allowed the development of bio-mining, where acidophilic microorganisms are used for metal recovery from sulfide minerals in the bioleaching and bio-oxidation processes (Das et al. 2011, Brune and Bayer 2012). On the other hand, metal and mineral transformations by microorganisms can result in spoilage and destruction of building materials, metals corrosion, among others, all with important health, environment, social, and economic effects (Gadd 2010, Wells and Melchers 2014).

Microbiological strategies for resistance to toxicity

Microbial populations inhabiting environments contaminated with metals may have developed specific strategies for resistance to toxicity (Hegler et al. 2008, Phoenix and Konhauser 2008). In general, there are three methods of resistance (Das et al. 2016):

a) Efflux of metals outside the cell by using transporters: It normally involves members of the heavy-metal protein superfamily with efflux pumps.
b) Transformation of the metals into non-toxic/less toxic forms (biotransformation): Bacteria have the ability to transform metals by reduction, metal precipitation mineralization, among others.
c) Biosorption: It is the binding of heavy metals onto the cellular surface; in bacteria, metallothioneins are involved as well as other functional groups from the cell wall. A metal can be absorbed into the cell, and then metals are precipitated as salts by enzymes.

Interactions and Biological Processes of Animals and Plants with the Mine Tailings Components

The environment is continuously loaded with toxic substances, mostly from anthropogenic origin, which causes significant perturbation. Consequently, increasing attention has been dedicated to better understanding the long-term ecological effects of chronically exposed populations, communities and ecosystems (Mussali-Galante et al. 2014). This perspective is based on the information that effects of environmental chemical pollution can be observed at all levels of biological organization.

However, we need to take into account that when analyzing ecotoxicological effects of chemical pollution, threats to populations and communities rising from chronic exposures to mixtures of chemical agents at lower doses (realistic exposures) are the point of interest (Depledge 1994). At higher levels of biological organization, population level has been the most studied in terms of ecological consequences of chronic metal exposure, arising from a bad disposal of mine tailings.

At the population level, genetic structure and diversity may be affected by exposure to metal contamination. Increased genetic variation may result from new mutations induced directly by the genotoxic agent(s), such as heavy metals. Decreased genetic variation may arise due to bottlenecks or selection that will also alter allele and genotype frequencies in these populations. Both responses are the result of population adaptation to the polluted environment (Bickham et al. 2000, Gardeström et al. 2008, Durrant et al. 2011).

In this context, studies conducted by Mussali-Galante et al. (2013b) analyzed the effect of metal bioaccumulation levels on genetic parameters (population and individual genetic diversity and structure) in a small mammal species (*Peromyscus melanophrys*) that inhabits mine tailings and has been considered as a sentinel organism. It was found that the highest values of genetic differentiation (F_{ST} and R_{ST}) and the lowest number of migrants per generation (*Nm*) were registered among the exposed populations. Genetic distance analyses showed that the most polluted population was the most genetically distant among the five populations examined. Moreover, a negative and a significant relationship was detected between genetic diversity (*He*, *IR*) and each metal concentration and for the bioaccumulation index in *P. melanophrys*. This is an example that highlights that metal stress is a major factor affecting the distribution and genetic diversity levels of small mammal populations living inside mine tailings. Therefore, the use of genetic population changes at micro-geographical scales (permanent biomarkers) is a suitable approach to explore population health effects arising because of mine tailing exposure. Other studies that have reported that animal population exposed to metal stress have decreased genetic variability in field conditions are: Ma et al. (2000) (*Balanus glandula*); Kim et al. (2003) (*Littorina brevicula*); Eeva et al. (2006) (*Ficedula hypoleuca*); Gardeström et al. (2008) (*Attheyella crassa*) and Ungherese et al. (2010) (*Talitrus saltador*).

On the contrary, there are also studies that report that metal contamination arising from mining activities have increased the genetic diversity levels of the exposed populations, for example: Yauk et al. (2000) (*Larus argentatus*); Peles et al. (2003) (*Lumbricus rubellus*); Eeva et al. (2006) (*Parus major*); Jordaens et al. (2006) (*Cepaea nemoralis*) and Giska et al. (2015) (*Staphylinus erythropterus*).

To date, scientific evidence suggests that the causes of decreased genetic diversity in exposed populations are mainly because of genotypic selection which affects genetic populations pool. Also, changes in population size may result in genetic bottlenecks and possibly genetic drift of the population, along with alteration in demographic patterns and reduced migration rates which can also lead to reduced genetic diversity (Van Straalen and Timmermans 2002). Since genetic variability is the basis of adaptation by natural selection, it is usually accepted that the loss of genetic diversity makes it more difficult for a population to adapt to future environmental changes. Any reduced variation can lead to an increased extinction rate (Bickham et al. 2000, Tremblay et al. 2008). This last scenario is known as "genetic erosion" and is one of the most important ecological effects seen at the population level. On the other hand, causes of increased genetic diversity levels may be the result of increased mutation rates. Several authors have evidenced that mutations accumulate more rapidly in more polluted environments (Peles et al. 2003, Gardeström et al. 2008). Therefore, in populations exposed to genotoxins such as metals, it is expected that the mutational load will increase and the population fitness will decrease, a fact that probably results in population extinction (Lynch et al. 1995).

Other approaches used to analyze ecological consequences of metal exposure due to an improper deposition of mine tailings are studies regarding the effects of chronic metal exposure on plants. Plants are an excellent study system since they represent the first step of the trophic chain (primary consumers) and they play a key role in transferring metals from the soil to their aerial biomass and to other trophic levels (biomagnification). Alterations in exposed plants may have ecological consequences in two general directions; first, because of the great variety of phytotoxic effects of metals, and second, because of metal bioaccumulation in plants that contributes to biomagnification through the trophic chain.

Phytotoxic effects of metals

It is well known that metal exposure exerts phytotoxic effects, resulting in cellular, physiological and morphological alterations (Nagajyoti et al. 2010). Xu and Shi (2000) mention that non-trace metals in plants, like As, Cd and Pb, may affect the absorption and transport of essential elements like Fe, Mg and Zn, a fact that may have consequences at the physiological and morphological levels, resulting in reduced biomass, leaf chlorosis and inhibited root growth, often leading to plant death at excessive exposures (Lin and Aarts 2012).

Plants inhabiting abandoned mine sites or near tailing sites are capable of absorbing metals from soils which may be accumulated in roots or may be translocated to aerial parts. One of the most common metals found in mine sites is chromium. Therefore, its phytotoxic effects (Table 1) have been widely studied. Also, studies analyzing metal contamination report that plants growing in vanadium (V) contaminated soils were smaller compared to the plants growing in areas further away from V mines (Panichev et al. 2006). Similarly, declines in root lengths and area, as well as less developed roots in *Phaseolus vulgaris*, were reported by Olness et al. (2005) and by Saco et al. (2013) in *Cuphea* sp. under V contamination.

Compared to other heavy metals, Cu is not readily bioaccumulated, and so it is rarely dangerous for human health, but it is highly toxic to plants (Fernandes and Henriques 1991). Enhanced mining activities have contributed to the increasing occurrence of Cu in ecosystems. Mining activities generate a lot of waste rocks and tailings, which are deposited at the surface, causing harm to plants, resulting in growth retardation, leaf chlorosis and excess production of ROS which causes oxidative damage to macromolecules, including DNA (Nagajyoti et al. 2010). Similar results have been reported in plants grown in soil containing high levels of Mn (Bachman and Miller 1995), Pb (Sharma and Dubey 2005) and Cd (Mohanpuria et al. 2007, Guo et al. 2008).

Because of the above mentioned studies, ecological consequences such as plant-herbivore interactions may be affected (positively or negatively for plant survival) in polluted environments. Scientific evidence provided by Zvereva and Kozlov (2001), who compared compensatory responses to herbivory in natural populations of two willow species growing at different distances from two Cu-Ni smelters, revealed that plant responses to herbivory may be modified by metal pollution. On the contrary, some authors suggest that there are beneficial effects for plants growing in multimetal contaminated soils. Recently, it has been suggested that plants absorb high concentrations of metals from soils as a self-defense mechanism against pathogens and herbivores. This metal defense hypothesis is very attractive because it gives an understanding of the adaptation mechanisms of metal hyperaccumulator plant species (Poschenrieder et al. 2006). This idea was initially formulated based on an observation that fewer insects feed on Ni hyperaccumulators (Martens and Boyd 1994).

Metal biomagnification through the trophic chain

Metals are expected to cause toxicity not only at the base of trophic chains but also in top consumers if the assimilated metals bioaccumulate through the trophic chain. For example, organic Hg is most probable to biomagnify because organisms efficiently assimilate methyl-Hg and eliminate it very slowly in proportion to biomass (Croteau et al. 2005). Therefore, studies on aquatic ecosystems regarding Hg biomagnification are numerous. In addition to it, other metals like selenium have been reported to biomagnify under special circumstances. However, metals like Cd, Pb, and Ag could biomagnify but the evidence is not conclusive (Stewart et al. 2004). For example, some crops in southern Poland have been polluted with elevated concentrations of Cd, Pb, and Zn as a result of emissions from two Pb-Zn smelters active in the area (Dudka and Adriano 1997), a fact that may pose an ecological risk due to the biomagnification of metals trough the trophic chain.

A study conducted by Ping et al. (2009) analyzed the bioaccumulation and transfer of Pb, Zn, Cu, and Cd along a soil-plant-insect-chicken trophic chain at contaminated sites by a Pb-Zn mine. Cd and Pb concentrations gradually declined with increasing trophic level. However, Zn and Cu concentrations increased from plants to insects. The highest metal concentrations were recorded in the chicken muscle. This study reveals that plants growing on metal polluted soils from miming activities bioaccumulate metals which biomagnify through the food chain. Another interesting study investigated the effects of hyperaccumulation on metal concentrations across trophic levels in a soil dominated by the nickel hyperaccumulator plant *Alyssum pintodasilvae*. Grasshoppers, spiders and other invertebrates sampled from sites where *A. pintodasilvae* was common had significantly higher Ni concentrations, compared to sites where this hyperaccumulator was absent. This study highlights that Ni passes to herbivore and carnivore trophic levels, where the presence of the hyperaccumulator *A. pintodasilvae* plays a key role in the metal transfer across the trophic chain (Peterson et al. 2003).

Finally, chronic exposures to multimetal-contaminated soils have ecological consequences at all levels of biological organization. At higher trophic levels, populations may suffer shifts in their genetic diversity pools, sex proportion alterations, age structure alterations, low reproductive success and inbreeding, low fitness and ultimately, the population declines. At the community level, there may be shifts in diversity and species richness, changes in dominant species, along with changes in species composition leading to biodiversity loss. At the ecosystem level, alterations in energy and nutrient cycles and especially trophic chain modifications may take place (Mussali-Galante et al. 2013b). Hence, the information presented here is useful for predicting and preventing potential risks to ecosystem health originating from an improper deposition of mine tailings and for planning bioremediation strategies.

Strategies for Mine Tailing Remediation

Unlike organic pollutants, heavy metals cannot be biologically, physically or chemically decomposed; therefore, their removal from contaminated sites is limited to waste confinement or solubility, mobility and/or toxicity alteration (Volke et al. 2005) through changes in its valence state, which favors its immobilization (chelation) and/or mobilization (dissolution). Depending on their nature, remediation methods for sites contaminated with heavy metals are classified as:

1. Physicochemical: Those that use the physical and chemical properties of the pollutants to transform, separate, or immobilize them (Dermont et al. 2008).

2. Biological: Those that take advantage of the metabolic potential of living organisms (bacteria, fungi, and plants) to clean contaminated sites (Gadd 2010).

The most commonly used physicochemical treatments include metals oxidation/reduction (transformation), soil washing and solidification/extraction (metals immobilization). These treatments are relatively rapid but costly, cause adverse effects on soil and secondary pollution, incomplete metal removal, as well as high reagent and energy requirements. Other disadvantages are that they require excavation, followed by a treatment or final disposal process, which generates an economic and environmental cost (Dermont et al. 2008, Wuana and Okieimen 2011).

The alternative is the biological treatment (bioremediation), which is an economically viable and environmentally acceptable technology, has a high public acceptance and preserve natural soil properties. In general, it is possible to identify three major strategies of bioremediation of sites contaminated by heavy metals, which are (1) by using microorganisms; (2) by using plants and (3) by using a combined treatment with microorganisms and plants.

Bioremediation by Using Microorganisms

The microorganisms used in heavy metal bioremediation processes are bacteria and fungi (Covarrubias and Peña 2017). The microbial biochemical mechanism is based on an oxidation state change of the metal for its detoxification.

The microorganisms play an important role in the uptake of toxic metals from the environment through mechanisms of biotransformation, bioaccumulation, biosorption (bioadsorption) and immobilization (Dixit 2015). These mechanisms may be present in biogeochemical cycles, in which microbial strains capable of withstanding the metals harmful effects, are involved (Marques 2016). Some of these mechanisms are described below.

Biosorption (passive uptake)

Heavy metals can be passively adsorbed to the cellular components of bacteria, mainly due to electrostatic attractions. The negative charges of the carboxyl, phosphoryl and amine groups present in the cell walls, membranes, and extracellular material of bacteria and fungi, attract the positive charge of metals. Once adhered to the cell surface, some heavy metals are internalized into the cell, where they can be bound or precipitated within vacuoles to minimize their toxicity (Gadd 2004). Biosorption can be carried out by living cells, cells fragments, tissues and dead biomass. It is important for the removal and/or recovery of pollutants, which utilizes materials of biological origin, including bacteria, fungi, algae, and plants. This is a rapid passive metal sequestration process (such as adsorption, surface complexation, ion exchange, or surface precipitation) by the biomass/adsorbents, due to the higher affinity of absorbent-adsorbate metals species (François et al. 2012, Abbas et al. 2014, Fomina and Gadd 2014). The sorption process (desorption-adsorption, absorption) depends on the organism involved, as well as the environmental conditions (Gadd 2004).

Bioaccumulation (active uptake)

Heavy metals must enter the cytoplasm and their pass depends on a variety of physical, chemical, and biological mechanisms. This process is dependent on the microorganism's tolerance, at high concentrations, of heavy metals, which are capable of transforming them into less toxic forms (Ayangbenro and Babalola 2017). Once the heavy metals are incorporated into the cytoplasm, they are sequestered by the presence of proteins rich in sulfhydryl groups called metallothioneins, or they can also be compartmentalized within a vacuole. Some examples of this process are very interesting, for example, uranium accumulation by *Pseudomonas aeruginosa*, which was detected entirely in the cytoplasm, as well as in *Saccharomyces cerevisiae* (Lloyd and Lovley 2001).

Bioprecipitation

Some heavy metals can be immobilized efficiently due to a microbial reduction process, decreasing their bioavailability. This phenomenon is an important metabolic process that controls the heavy metals' transport and destination (Volke et al. 2005).

Bioleaching (heterotrophic and autotrophic)

It consists of the medium acidification by the protons' mobilization through the plasmatic membrane, causing a release of metals through several routes, for example, the competition between protons and metals in a metal-anion complex generates the release of metallic cations. In addition to this, the heterotrophic metabolism can cause leaching due to the production of organic acids and siderophores (Gadd 2004).

Biomineralization

Microorganisms can form minerals for capturing pollutants within relatively stable solid phases. Numerous studies have shown that microorganisms produce extracellular polymers that participate in the capture of metals. Prokaryotes with a metabolic diversity are able to catalyze the formation of chemically diverse minerals, including oxides, oxi-hydroxides, carbonates, phosphates, sulphates, and sulphides, which results in the trapping of various metal and metalloids. As a result, their impact on the metal global mobility and biogeochemical cycling in the environment is likely to be stronger in comparison with eukaryotes (Gadd 2010, Benzerara et al. 2011).

All these microbiological processes are used to apply different bioremediation technologies, which are described in Table 3.

Phytoremediation

Phytoremediation is a process that involves the use of plants to remove, transfer, stabilize, decompose and/or degrade soil, sediment and water contaminants, such as solvents, pesticides, polyaromatic hydrocarbons, heavy metals, explosives, radioactive elements, and fertilizers. This technology is inexpensive and is a promising solution for cleaning up sites contaminated by a variety of metals, although it also has a number of limitations (Pilon-Smits 2005, Mench et al. 2009).

Phytoremediation requires the use of plants that desirably produce sufficient biomass in the aerial part, with a high accumulation capacity of heavy metals, which would allow harvesting of the parts where the metals accumulate, in addition to well-developed roots, and a high growth rate (Barceló and Poschenrieder 2002).

Plants have different strategies to contend with the effects of heavy metals. These organisms must be able to properly control the intracellular concentration of each metal

Table 3. Bioremediation technologies derived from the knowledge of interaction processes between heavy metals and microorganisms.

Technology	Short description	References
Bioaugmentation	Technology that involves the addition of indigenous or exogenous microbial cultures into the contaminated site. It has been used to accelerate bioremediation.	Kumar et al. 2011, Mani and Kumar 2014, Dixit et al. 2015
Composting	Technique that involves mixing contaminated soil with manure, food wastes or agricultural wastes. The wastes are degraded by supporting a microbial population development and elevated temperature, characteristic of composting.	Amir et al. 2005, Kumar et al. 2011, Mani and Kumar 2014
Landfarming	The contaminated site is excavated up to 30 cm and the indigenous microorganisms are stimulated to facilitate the aerobic activity. The costs of monitoring and maintenance are reduced, as well as the treatment liabilities.	Lynch and Moffat 2005, Kumar et al. 2011, Mani and Kumar 2014
Biopiles	The combination of landfarming and composting results in the biopiles technique. It allows the surface treatment, controls leaching, and volatilization; stimulates aerobic and anaerobic microbial degradation.	Lynch and Moffat 2005, Kumar et al. 2011, Mani and Kumar 2014
Bioventing	Is the most common *in situ* treatment. The microorganisms of a contaminated soil receive nutrients, which stimulate the biodegradation and minimizes the volatilization of pollutants to the atmosphere.	Lynch and Moffat 2005, Kumar et al. 2011, Mani and Kumar 2014
Biostimulation	Involves the addition of supplements to a contaminated site with the objective of stimulating the indigenous microbial population growth, capable of degrading the contaminants from soil and aquatic environment.	Mani and Kumar 2014, Dixit et al. 2015
Bioreactors	Bioreactors are used for *ex situ* treatment of contaminated soil, sediments, sludge or water through an engineered containment system. A bioreactor is a containment vessel where the biodegradation-controlled conditions of the contaminants are established in an aqueous suspension.	Lloyd and Lovley 2001, Santos and Judd 2010, Kumar et al. 2011, Mani and Kumar 2014
Intrinsic bioremediation	This biotechnological process takes advantage of indigenous microbial populations and provides nutrients and oxygen to increase its degradative metabolic activity.	Kumar et al. 2011, Mani and Kumar 2014
Biosorption	In general, the mechanism is based on a number of metal-binding processes which include: chemisorption, complexation, hydroxide condensation onto the biosurface, ion exchange, microprecipitation, surface and pore adsorption-complexation, and surface adsorption.	Kumar et al. 2011, Abbas et al. 2014, Dixit et al. 2015, Yang et al. 2016

because its excess is toxic, whereby they have two mechanisms. The first one is the strategic accumulation of metals in cellular compartments, such as cell walls or vacuoles. In addition, they can be conjugated with some organic acids or proteins. The second one is the heavy metal exclusion, in which metals are quickly transported to the shoots and leaves and are subsequently volatilized or discarded by senescence of the leaf (Llugany et al. 2007, Ali et al. 2013).

Among the main constraints of phytoremediation are the metal toxicity to plants and microorganisms, the risk of possible entry of metals into the trophic chain and the long time required. Depending on the recovery strategy used, these processes will result in the immobilization or removal of metals from the soil. Taking into account that the used plant influences the remediation, the use of native species is easier for its establishment (Pilon-Smits 2005, Mench et al. 2009). Phytoremediation includes different mechanisms, which are shown in Table 4.

Table 4. Phytoremediation technologies derived from the knowledge of interaction processes between heavy metals and plants.

Technology	Short description	Reference
Rhizodegradation	The degradation of pollutants, mainly organic in the soil, is carried out through the microbial biomass associated to the rhizosphere.	Ghosh and Singh 2005, Tangahu et al. 2011, Ali et al. 2013, Kavitha et al. 2013, Sharma et al. 2015
Phytovolatilization	Absorption and transpiration of a metal through the vegetal tissues; release of the metal to the atmosphere. It can be in its original or a modified form.	Lone et al. 2008, Jadia and Fulekar 2009, Karami and Shamsuddin 2010, Tangahu et al. 2011, Ali et al. 2013, Kavitha et al. 2013, Dixit et al. 2015, Sharma et al. 2015
Phytoextraction (also known as phytoaccumulation, phytoabsorption or phytosequestration)	The plant roots uptake/absorb and translocate metals from soil or water to the aboveground biomass.	Ghosh and Singh 2005, Lone et al. 2008, Jadia and Fulekar 2009, Karami and Shamsuddin 2010, Tangahu et al. 2011, Ali et al. 2013, Kavitha et al. 2013, Dixit et al. 2015, Olowu et al. 2015, Sharma et al. 2015
Phytostabilization (also known as phytoimmobilization)	Plants immobilize metals from the soil through absorption and accumulation in plant tissues, sorption by roots, precipitation, complexation or metal valence reduction within the rhizosphere, preventing their migration into the soil, as well as their dispersion by the wind and/or water erosion. This technique is used to reduce the mobility and bioavailability of pollutants in the environment.	Lone et al. 2008, Jadia and Fulekar 2009, Karami and Shamsuddin 2010, Tangahu et al. 2011, Wuana and Okieimen 2011, Ali et al. 2013, Kavitha et al. 2013, Dixit et al. 2015, Olowu et al. 2015, Sharma et al. 2015
Rhizofiltration	Adsorption or precipitation onto plant roots or absorption and sequestration into the roots of metals that are in solution of the rhizosphere.	Lone et al. 2008, Jadia and Fulekar 2009, Karami and Shamsuddin 2010, Tangahu et al. 2011, Ali et al. 2013, Kavitha et al. 2013, Dixit et al. 2015, Sharma et al. 2015
Phytofiltration	Removal of pollutants from contaminated surface water or wastewater by plants.	Gardea-Torresdey et al. 2004, Wuana and Okieimen 2011, Ali et al. 2013, Dixit et al. 2015, Olowu et al. 2015

Social Aspects Regarding Mining

Besides the environmental impacts discussed, mining activities have been associated with social and cultural conflicts, as well as local economic instability (Prno and Slocombe 2012, Al Rawashdeh et al. 2016). The aggressive activities of mining industry have generated conflict between poor and indigenous communities; moreover, mining corporations are typically foreign (Gordon and Webber 2008). Some social impacts consist of indigenous and peasant movements, which, in some cases, have succeeded in causing a political crisis and achieving changes in environmental regulations, with a great national impact. Resistance strategies include legal lawsuits, activist-scientific collaborations, and local referendums to reject mining projects. However, in most cases, opposition to mining has been violently repressed (Andreucci and Kallis 2017, Conde 2017). On the other hand, Szablowski (2007) argues that the conflicts regulation between transnational mining investment and local communities has been recently examined through transnational, national and local legal processes.

Due to structural adjustments, lax environmental regulations, rising mineral prices, strong equity markets and low domestic interest rates, the commodity frontier in mining has been expanded especially to the South (Conde 2017). For example, the Canadian company participation in the exploration market of bigger companies in Latin America and the Caribbean has grown steadily since the beginning of the 90s (Gordon and Webber 2008). There are more than 1,000 mining companies listed on the Canadian stock exchanges, more than any other country, and represent the most important source of investment in the mining of Africa (Campbell 2008). In the United States of America and Canada, mining managers have expressed concern about stricter environmental regulation, which, combined with delays in issuing licenses for mining, favors that the mining companies start operations in developing countries (Söderholm et al. 2015).

In other countries, similar situations are prevailing, and some examples are mentioned below:

1. In the Republic of Armenia, derived from a weak legislative system, private interests and inefficient management, heavy metals are at high concentration in mine tailings; therefore, they spread to nearby territories and cause soil, water, and air pollution; then, they enter the food chain, and adversely affect the human health (Baghdasaryan 2016).
2. In Peru, since 1990, 89% of the ever-claimed area for mining has been given as concessions to transnational mining companies. Until 2008, the country did not have a Ministry of Environment; thus, responsibilities such as granting mining concessions, promoting mining, regulating of mining environmental and social impacts, and approving the companies, all corresponded to the Ministry of Energy and Mines, creating a clear conflict of interests (Bebbington and Bury 2009).
3. In Ghana, the continuous expansion of economic globalization through the adoption of structural adjustment policies in the 1980s led to the influx of multinational mining companies. This increase in the mining economy was facilitated by the search for both international financial institutions and governments to promote foreign investment, resulting in a "race to the bottom", a phenomenon in which governments relax environmental laws to attract multinational companies in order to extract resources from their economies (Essah and Andrews 2016).

In general, existing environmental legislation largely does not apply to abandoned mines and has not developed mechanisms to address them, except through emergency response of the government that uses public funds to remedy the problem.

Söderholm et al. (2015) suggest that there is no simple and direct balance between the environment and competitiveness, which could be achieved without compromising the industry competitiveness through the implementation of different public policies and environmental results that are more favorable. Therefore, developed and developing countries' governments are striving to improve their national environmental policy and there is a trend towards increasingly stringent environmental requirements that must be applied to mining activities (Cao 2007). On the other hand, Edraki et al. (2014) point that in order to optimize the environmental, social and economic results of mine tailings management, it is important to develop technologies to achieve their complete characterization, planning, processing, disposal, reprocessing, recycling and reuse.

Nevertheless, the evidence of the positive effects of the mining sector is weak, especially in rural areas. Indeed, mining has led to increased social conflicts (Bebbington and Bury 2009). In this regard, Bastida (2008) mentions that the mining policy's prevailing patterns, mining legislation, as well as the contractual arrangements are undergoing an intense scrutiny faced with society's changing expectations, and mining's contribution to sustainable development is increasingly being questioned.

In the context of sustainable development, civil society and local communities around the world, where mining industries had been established, have emerged as relevant actors. These communities demand greater benefits and participation in decision-making. As a result, and to avoid social conflicts, the mining sector needs to gain a "social license to operate" (Prno and Slocombe 2012, Prno 2013). Owen and Kemp (2013) argue that the mining industry holds up the idea of a "social license to operate" as a plausible and viable construct. The mining industry's social dimension is acknowledged as important; however, it is the least understood aspect of the sustainable development concept (Solomon et al. 2008). Despite these concerns, a globalized world society is economically, socially, and culturally dependent on metals (Prior et al. 2012).

Legislation Regarding Mining and Mine Tailings

Ali et al. (2006) suggest that because of differences in the strictness of environmental protection among countries, industries that cause high levels of pollution are established in nations where regulations are laxer. Developing countries often have less stringent environmental standards than developed countries. The World Bank played a very influential role in reforming the legal, fiscal and institutional legislation of the mining sector; the strategies formulated by the World Bank for Africa and for Latin America were clear and similar in their objectives (Bastida 2008). However, regarding the mining industry and its impacts, Bridge (2004) pointed out that the environmental regulations were not enough to satisfy the demands of civil society. To achieve balanced development in developing countries, effective regulation of extractive industries is very important (Sing 2015).

Globally, more than 30 countries have included development requirements of the community in their mining laws. Since the mid-1980s, the approach to mining regulation included actions to mitigate negative impacts on local communities, as well as ensuring real social and economic benefits (Dupuy 2014, Al Rawashdeh et al. 2016). The poor environmental reputation of the mining industry has forced it to face regulatory and community/social pressures, which has caused greening in their production chains (Kusi-Sarpong et al. 2016). However, not all mining producer countries have adopted these laws.

The social demand's inclusion in legislative requirements is very recent and rare. Agreements developed by communities (ADC) are becoming increasingly popular within the mining sector in the entire world. According to O'Faircheallaigh (2013), Owen and Kemp (2013) and Nwapi (2017), the ADC, legislatively mandated, can also help to promote equity and transparency on its creation and implementation, and can be more successful. However, developing countries need to promote foreign investment, but they cannot properly enforce environmental legislation, which has resulted in innumerable opportunities for multinational mining companies (Hilson and Haselip 2004). In addition to this, the developing countries' government perceptions have been that the economic benefit of mining is greater than social cost and environmental impacts. This position has resulted in various conflicts between the communities and the government (Sing 2015). In Latin America, the mining industry has undergone significant expansion due to the microelectronics market growth and the crisis in others markets. In this continent, Mexico and Chile are the main producers of silver and copper, respectively, and Brazil is the third largest producer of iron. The national governments have created conditions for these companies to have mining paradises, through the modification of mining legislation (Núñez 2015).

Concluding Remarks

In this chapter, mining has been approached from an integral point of view, including environmental, toxicological, ecological, health, economic, political, social and legal aspects. As it has been shown, mining is a complex problem and must be analyzed from all points of view to reduce the impacts mentioned.

One of the major problems of mining is the generation of waste or mine tailings, which are frequently deposited near the mines without the required care. Its dangerousness lies in its content, especially of heavy metals, which cause various impacts to biota, including human beings. In developing countries, this waste is one of the major challenges to be addressed, so it is necessary to consider the sustainable management of these wastes as a resource to take advantage of its components.

Once the metals are incorporated into the environment, there are different technologies that can be applied, and among the most accepted are those using biological tools, in which microorganisms and plants can be used. However, it is necessary to have specialists capable of conducting diagnostic studies and integrating the remediation aspects suitable for each particular case.

References Cited

Abbas, S.H., I.M. Ismail, T.M. Mostafa and A.H. Sulaymon. 2014. Biosorption of heavy metals: A review. J. Chem. Sci. Tech. 3(4): 74–102.

Adiansyah, J.S., M. Rosano, S. Vink and G. Keir. 2015. A framework for a sustainable approach to mine tailings management: disposal strategies. J. Clean. Prod. 108: 1050–1062.

Al Rawashdeh, R., G. Campbell and A. Titi. 2016. The socio-economic impacts of mining on local communities: The case of Jordan. Extr. Ind. Soc. 3(2): 494–507.

Ali, A., H. Fujiono and K.H. Jin. 2006. Competitiveness Factor-Potter Analysis in the Copper Industry, Unpublished report, Colorado School of Mines, USA.

Ali, A. 2009. Implications of trade liberalisation to Malaysia's mining industry. Bull. Geol. Soc. Malaysia 55(1): 1–6.

Ali, H., E. Khan and M.A. Sajad. 2013. Phytoremediation of heavy metals—concepts and applications. Chemosphere 91(7): 869–81.

Amir, S., M. Hafidi, G. Merlina and J.C. Revel. 2005. Sequential extraction of heavy metals during composting of sewage sludge. Chemosphere 59(6): 801–810.

Andreucci, D. and G. Kallis. 2017. Governmentality, development and the violence of natural resource extraction in Peru. Ecol. Econ. 134: 95–103.

Armienta, M.A., G. Villaseñor, R. Rodriguez, L.K. Ongley and H. Mango. 2001. The role of arsenic-bearing rocks in groundwater pollution at Zimapan Valley, Mexico. Environ. Geol. 40: 571–581.

Ayangbenro, A.S. and O.O. Babalola. 2017. A new strategy for heavy metal polluted environments: A review of microbial biosorbents. Int. J. Environ. Res. Public Health. 14(1): 94.

Bachman, G. and W. Miller. 1995. Iron chelate inducible iron/manganese toxicity in zonal geranium. J. Plant. Nutri. 18: 1917–1929.

Baghdasaryan, T. 2016. Assessment of the environmental impact of tailings in the Republic of Armenia. Master's Thesis. Universidad de Coruña.

Bain, J.G., D.W. Blowes, W.D. Robertson and E.O. Frind. 2000. Modelling of sulfide oxidation with reactive transport at a mine drainage site. J. Contam. Hydrol. 41: 23–47.

Barceló, J. and C. Poschenrieder. 2002. Fast root growth responses, root exudates, and internal detoxifiation as clues to the mechanisms of aluminium toxicity and resistance: a review. Environ. Exp. Bot. 48: 75–92.

Barceloux, D.G. 1999. Chromium. J. Toxicol. Clin. Toxicol. 37: 173–194.

Bastida, A.E. 2008. Mining law in the context of development. An overview. pp. 101–136. *In*: Andrews-Speed, P. [ed.]. International Competition for Resources: The Role of Law, the State and of Markets. Dundee University Press, Dundee, UK.

Bebbington, A.J. and J.T. Bury. 2009. Institutional challenges for mining and sustainability in Peru. PNAS 106(41): 17296–17301.

Benzerara, K., J. Miot, G. Morin, G. Ona-Nguema, F. Skouri-Panet and C. Ferard. 2011. Significance, mechanisms and environmental implications of microbial biomineralization. C. R. Geosci. 343(2): 160–167.

Bernard, A. 2008. Biomarkers of metal toxicity in population studies: research potential and interpretation issues. J. Toxicol. Environ. Health A. 71: 1259–1265.

Bickham, J., S. Sandhu, P. Hebert, L. Chikhi and R. Athwal. 2000. Effects of chemical contaminants on genetic diversity in natural populations: implications for biomonitoring and ecotoxicology. Mutat. Res. 463: 33–51.

Boopathy, R. 2000. Factors limiting bioremediation technologies. Bioresour. Technol. 74(1): 63–67.

Bridge, G. 2004. Contested terrain: mining and the environment. Annu. Rev. Environ. Resour. 29: 205–259.

Brune, K.D. and T. Bayer. 2012. Engineering microbial consortia to enhance biomining and bioremediation. Front. Microbiol. 3: 203.

Bruneel, O., R. Duran, C. Casiot, F. Elbaz-Poulichet and J.C. Personne. 2006. Diversity of microorganisms in Fe-As-Rich acid mine drainage waters of Carnoules, France. Appl. Environ. Microbiol. 72: 551–556.

Campbell, B. 2008. Regulation & legitimacy in the mining industry in Africa: Where does Canada stand? Rev. Afr. Polit. Econ. 35(117): 367–385.

Canfield, R.L., C.R. Henderson, D.A. Cory-Slechta, C. Cox, T.A. Jusko and B.P. Lanphear. 2003. Intellectual impairment in children with blood lead concentrations below 10 microg per deciliter. N. Engl. J. Med. 348: 1517–26.

Cao, X. 2007. Regulating mine land reclamation in developing countries: The case of China. Land use policy. 24: 472–483.

Cervantes, C. and R. Moreno. 1999. Contaminación ambiental por metales pesados: impacto en los seres vivos. México, AGT Publisher.

Chen, M., P. Xu, G. Zeng, C. Yang, D. Huang and J. Zhang. 2015. Bioremediation of soils contaminated with polycyclic aromatic hydrocarbons, petroleum, pesticides, chlorophenols and heavy metals by composting: applications, microbes and future research needs. Biotechnol. Adv. 33(6): 745–755.

Conde, M. 2017. Resistance to mining. A review. Ecol. Econ. 132: 80–90.

Covarrubias, S.A. and J.C. Peña. 2017. Contaminación ambiental por metales pesados en México: problemática y estrategias de fitorremediación. Rev. Int. Contam. Ambie. 33: 7–21.

Croteau, M., S. Luoma and A. Stewart. 2005. Trophic transfer of metals along freshwater food webs: Evidence of cadmium biomagnification in nature. Limnol. Oceanogr. 50: 1511–1519.

Das, S., H.R. Dash and J. Chakraborty. 2016. Genetic basis and importance of metal resistant genes in bacteria for bioremediation of contaminated environments with toxic metal pollutants. Appl. Microbiol. Biotech. 100(7): 2967–2984.

Das, A.K., R. Sahu, T.K. Dua, S. Bag, M. Gangopadhyay, M.K. Sinha et al. 2010. Arsenic-induced myocardial injury: protective role of *Corchorus olitorius* leaves. Food Chem. Toxicol. 48: 1210–1217.

Das, A.P., L.B. Sukla, N. Pradhan and S. Nayak. 2011. Manganese biomining: a review. Bioresour. Technol. 102(16): 7381–7387.

Davis, G.A. and J.E. Tilton. 2002. Should developing countries renounce mining? A perspective on the debate. Report prepared for the International Council on Mining and Metals (ICMM). London: ICMM.

Depledge, M. 1994. Genotypic toxicity: implications for individuals and populations. Environ. Health Perspect. 102: 101–104.

Dermont, G., G. Bergeron, G. Mercier and M. Richer-Lafleche. 2008. Soil washing for metal removal: A review of physical/chemical technologies and field applications. J. Hazard. Mater. 152(1): 1–31.

Dixit, R., D. Malaviya, K. Pandiyan, U.B. Singh, A. Sahu, R. Shukla et al. 2015. Bioremediation of heavy metals from soil and aquatic environment: an overview of principles and criteria of fundamental processes. Sustainability 7(2): 2189–2212.

Donadio, E. 2009. Ecólogos y mega-minería, reflexiones sobre porque y como involucrarse en el conflicto minero-ambiental. Ecol. Austral. 19(3): 247–254.

Douglas, S. and T.J. Beveridge. 1998. Mineral formation by bacteria in natural microbial communities. FEMS Microbiol. Ecol. 26(2): 79–88.

Dudka, S. and D. Adriano. 1997. Environmental impacts of metal ore mining and processing: A review. J. Environ. Qual. 26: 590–602.

Dupuy, K.E. 2014. Community development requirements in mining laws. Extr. Ind. Soc. 1(2): 200–215.

Durrant, C., J. Stevens, C. Hogstrand and N. Bury. 2011. The effect of metal pollution on the population genetic structure of brown trout (*Salmo trutta* L.) residing in the River Hayle, Cornwall, UK. Environ. Pollut. 159: 3595–3603.

Edraki, M., T. Baumgartl, E. Manlapig, D. Bradshawb, D. Franks and C. Moran. 2014. Designing mine tailings for better environmental, social and economic outcomes: a review of alternative approaches. J. Clean. Prod. 84: 411–420.

Eeva, T., E. Belskii and B. Kuranov. 2006. Environmental pollution affects genetic diversity in wild bird populations. Mutat. Res. 608: 8–15.

Essah, M. and N. Andrews. 2016. Linking sustainable mining practices and corporate social responsibility? Insights from Ghana. Resour. Policy. 50: 75–85.

Fenech, M., N. Holland, W. Chang, E. Zeiger and S. Bonassi. 1999. The HUman Micro-Nucleus Project—an international collaborative study on the use of the micronucleus technique for measuring DNA damage in humans. Mutat. Res. 428: 271–283.
Fernandes, J. and F. Henriques. 1991. Biochemical, physiological, and structural effects of excess copper in plants. Bot. Rev. 57: 246–273.
Florea, A.M. and D. Büsselberg. 2006. Occurrence, use and potential toxic effects of metals and metal compounds. BioMetals. 19(4): 419–427.
Fomina, M. and G.M. Gadd. 2014. Biosorption: current perspectives on concept, definition and application. Bioresour. Technol. 160: 3–14.
Fraga, C.G. 2005. Relevance, essentiality and toxicity of trace elements in human health. Mol. Aspects Med. 26: 235–244.
François, F., C. Lombard, J.M. Guigner, P. Soreau, F. Brian-Jaisson, G. Martino et al. 2012. Isolation and characterization of environmental bacteria capable of extracellular biosorption of mercury. Appl. Environ. Microbiol. 78(4): 1097–1106.
Franks, D.M., D.V. Boger, C.M. Côte and D.R. Mulligan. 2011. Sustainable development principles for the disposal of mining and mineral processing wastes. Resour. Policy. 36(2): 114–122.
Frenzilli, G., M. Nigroa and B. Lyons. 2009. The comet assay for the evaluation of genotoxic impact in aquatic environments. Celebrating the 20th Anniversary of the invention of the comet assay. Mutat. Res. Rev. 681: 80–92.
Fu, F. and Q. Wang. 2011. Removal of heavy metal ions from wastewaters: a review. J. Environ. Manage. 92(3): 407–418.
Gadd, G.M. 2004. Microbial influence on metal mobility and application for bioremediation. Geoderma. 122(2): 109–119.
Gadd, G.M. 2007. Geomycology: biogeochemical transformations of rocks, minerals, metals and radionuclides by fungi, bioweathering and bioremediation. Mycol. Res. 111(1): 3–49.
Gadd, G.M. 2010. Metals, minerals and microbes: geomicrobiology and bioremediation. Microbiology 156(3): 609–643.
Gamalero, E., G. Lingua, G. Berta and B.R. Glick. 2009. Beneficial role of plant growth promoting bacteria and arbuscular mycorrhizal fungi on plant responses to heavy metal stress. C. J. Microbiol. 55(5): 501–514.
Gardea-Torresdey, J.L., G. De La Rosa and J.R. Peralta-Videa. 2004. Use of phytofiltration technologies in the removal of heavy metals: a review. Pure Appl. Chem. 76(4): 801–813.
Gardeström, J., U. Dahl, O. Kotsalainen, A. Maxson, T. Elfwing, M. Grahn et al. 2008. Evidence of population genetic effects of long-term exposure to contaminated sediments: a multi-endpoint study with copepods. Aquat. Toxicol. 86: 426–436.
Garza, A., R. Vega and E. Soto. 2006. Cellular mechanisms of lead neurotoxicity. Med. Sci. Monit. 12: RA57–65.
Ghosh, M. and S. Singh. 2005. A review on phytoremediation of heavy metals and utilization of it's by products. As. J. Energy Env. 6(4): 214–231.
Giska, I., B. Wieslaw, C.A.M. Van Gestel, N.M. Van Straalen and R. Laskowski. 2015. Genome-wide genetic diversity of rove beetle populations along a metal pollution gradient. Ecotoxicol. Environ. Saf. 119: 98–105.
Gordon, T. and J.R. Webber. 2008. Imperialism and resistance: Canadian mining companies in Latin America. Third World Q. 29(1): 63–87.
Guo, J., X. Dai, W. Xu and M. Ma. 2008. Over expressing GSHI and AsPCSI simultaneously increases the tolerance and accumulation of cadmium and arsenic in *Arabidopsis thaliana*. Chemosphere 72: 1020–1026.
Harris, G.K. and X. Shi. 2003. Signaling by carcinogenic metals and metal-induced reactive oxygen species. Mutat. Res./Fundamental Mol. Mech. Mutagenesis 533: 183–200.
Hartwig, A. 1998. Carcinogenicity of metal compounds: possible role of DNA-repair inhibition. Toxicol. Lett. 102-103: 235–239.
Hegler, F., N.R. Posth, J. Jiang and A. Kappler. 2008. Physiology of phototrophic iron (II)-oxidizing bacteria: implications for modern and ancient environments. FEMS Microbiol. Ecol. 66: 250–260.
Hilson, G. 2002. An overview of land use conflicts in mining communities. Land Use Policy. 19(1): 65–73.
Hilson, G. and J. Haselip. 2004. The environmental and socioeconomic performance of multinational mining companies in the developing world economy. Miner. Energy Raw Mater. Rep. 19(3): 25–47.
Hudson-Edwards, K.A., H.E. Jamleson and B.G. Lottermoser. 2011. Mine wastes: past, present, future. Elements. 7(6): 375–380.
Jadia, C.D. and M.H. Fulekar. 2009. Phytoremediation of heavy metals: recent techniques. Afr. J. Biotechnol. 8(6): 921–928.
Jiang, Z.F., S.Z. Huang, J.Z. Zhao and J.J. Fu. 2011. Physiological response of Cu and Cu mine tailing remediation of *Paulownia fortunei* (Seem) Hemsl. Ecotoxicol. 21(3): 759–767.
Johnson, D.B. and K.B. Hallberg. 2005. Acid mine drainage remediation options: a review. Sci. Total Environ. 338(1): 3–14.

Johnson, R.H., D.W. Blowes, W.D. Robertson and J.L. Jambor. 2000. The hydrogeochemistry of the Nickel Rim mine tailings impoundment, Sudbury, Ontario. J. Contam. Hydrol. 41: 49–80.

Jomova, K. and M. Valko. 2011. Advances in metal-induced oxidative stress and human disease. Toxicology 283: 65–87.

Jones, H. and D.V. Boger. 2012. Sustainability and waste management in the resource industries. Ind. Eng. Chem. Res. 51(30): 10057–10065.

Jordaens, K., H. De Wolf, N. Van Houtte, B. Vandecasteele and T. Backeljau. 2006. Genetic variation in two land snails *Cepaea nemoralis* and *Succinea putris* (Gastropoda, Pulmonata), from sites differing in heavy metal content. Genetica 128: 227–239.

Jun, R., T. Ling and Z. Guanghua. 2009. Effects of chromium on seed germination, root elongation and coleoptile growth in six pulses. Int. J. Environ. Sci. Tech. 6: 571–578.

Jung, M.C. 2001. Heavy metal contamination of soils and waters in and around the Imcheon Au-Ag mine, Korea. Appl. Geochem. 16: 1369–1375.

Karami, A. and Z.H. Shamsuddin. 2010. Phytoremediation of heavy metals with several efficiency enhancer methods. Afr. J. Biotechnol. 9(25): 3689–3698.

Kavitha, B., P. Jothimani, S. Ponmani and R. Sangeetha. 2013. Phytoremediation of heavy metals—A review. Int. J. Res. Stud. Biosci. 1(2): 17–23.

Kim, S., M. Rodriguez, J. Suh and J. Song. 2003. Emergent effects of heavy metal pollution at a population level: *Littorina brevicula* a study case. Mar. Pollut. Bull. 46: 74–80.

Kossoff, D., W.E. Dubbin, M. Alfredsson, S.J. Edwards, M.G. Macklin and K.A. Hudson-Edwards. 2014. Mine tailings dams: characteristics, failure, environmental impacts, and remediation. Appl. Geochem. 51: 229–245.

Kumar, A., B.S. Bisht, V.D. Joshi and T. Dhewa. 2011. Review on bioremediation of polluted environment: A management tool. Int. J. Environ. Sci. 1(6): 1086–1100.

Kusi-Sarpong, S., J. Sarkis and X. Wang. 2016. Assessing green supply chain practices in the Ghanaian mining industry: A framework and evaluation. Int. J. Prod. Econ. 181: 325–341.

Lèbre, E., G.D. Corder and A. Golev. 2016. Sustainable practices in the management of mining waste: A focus on the mineral resource. Miner. Eng. 107: 34–42.

Leonard, S.S., J.J. Bower and X. Shi. 2004. Metal-induced toxicity, carcinogenesis, mechanisms and cellular responses. Mol. Cell Biochem. 255: 3–10.

Liao, J., X. Lin, Z. Cao, Y. Shi and M. Wong. 2003. Interactions between arbuscular mycorrhizae and heavy metals under sand culture experiment. Chemosphere 50(6): 847–853.

Lin. 1997. Mobilization and retention of heavy metals in mill-tailings from Garpenberg sulfide mines, Sweden. Sci. Total Environ. 198: 13–31.

Lin, Y. and M. Aarts. 2012. The molecular mechanism of zinc and cadmium stress response in plants. Cell. Mol. Life Sci. 69: 3187–3206.

Lloyd, J.R. and D.R. Lovley, 2001. Microbial detoxification of metals and radionuclides. Curr. Opin. Biotechnol. 12: 248–253.

Llugany, M., R. Tolrà, C. Poschnrieder and J. Barceló. 2007. Hiperacumulación de metales ¿una ventaja para la planta y para el hombre? Ecosistemas 16(2): 4–9.

Lone, M.I., Z. He, P.J. Stoffella and X. Yang. 2008. Phytoremediation of heavy metal polluted soils and water: Progresses and perspectives. J. Zhejiang Univ. Sci. B. 9(3): 210–220.

Lynch, J.M. and A.J. Moffat. 2005. Bioremediation—prospects for the future application of innovative applied biological research. Ann. Appl. Biol. 146(2): 217–221.

Lynch, M., J. Conery and R. Bürger. 1995. Mutation accumulation and the extinction of small population. Am. Nat. 146(4): 489–518.

Ma, X., D. Cowles and R. Carter. 2000. Effect of pollution on genetic diversity in the Bay Mussel *Mytilus galloprovincialis* and the acorn bernacle *Balanus gladula*. Mar. Environ. Res. 50: 559–563.

Mancilla-Villa, Ó.R., H.M. Ortega-Escobar, C. Ramírez-Ayala, E. Uscanga-Mortera, R. Ramos-Bello and A.L. Reyes-Ortigoza. 2012. Metales pesados totales y arsénico en el agua para riego de Puebla y Veracruz, México. Rev. Int. Contam. Ambie. 28(1): 39–48.

Mani, D. and C. Kumar. 2014. Biotechnological advances in bioremediation of heavy metals contaminated ecosystems: an overview with special reference to phytoremediation. Int. J. Environ. Sci. Technol. 11: 843–872.

Marques, C.R. 2016. Bio-rescue of marine environments: On the track of microbially-based metal/metalloid remediation. Sci. Total Environ. 565: 165–180.

Martens, S. and R. Boyd. 1994. The ecological significance of nickel hyperaccumulation-a plant-chemical defense. Oecologia. 98: 379–384.

Martín, R.F. and R.M. Gutierrez. 2010. Estudio comparativo de la peligrosidad de jales en dos zonas mineras localizadas en el sur y centro de México. Bol. Soc. Geol. Mex. 62(1): 43–53.

Matsumoto, S., H. Shimada and T. Sasaoka. 2016. The key factor of acid mine drainage (AMD) in the history of the contribution of mining industry to the prosperity of the United States and South Africa: A review. Nat. Resour. 7(07): 445.
Mench, M., J. Schwitzguebel, P. Schroeder, V. Bert, S. Gawronski and S. Gupta. 2009. Assessment of success full experiments and limitations of phytotechnologies: contaminant uptake, detoxification and sequestration, and consequences for food safety. Env. Sci. and Poll. Res. 16: 876–900.
Mhlongo, S.E. and F. Amponsah-Dacosta. 2016. A review of problems and solutions of abandoned mines in South Africa. Int. J. Min. Reclam. Environ. 30(4): 279–294.
Mohanpuria, P., N. Rana and S. Yadav. 2007. Cadmium induced oxidative stress influence on glutathione metabolic genes of *Camella sinensis* (L.). O Kuntze. Environ. Toxicol. 22: 368–374.
Moran, C.J., S. Lodhia, N.C. Kunz and D. Huisingh. 2014. Sustainability in mining, minerals and energy: new processes, pathways and human interactions for a cautiously optimistic future. J. Clean. Prod. 84: 1–15.
Mussali-Galante, P., E. Tovar-Sánchez and T. Fortoul. 2007. Cell cycle, P53 and metals. pp. 9–13. *In*: Fortoul, I.T. [ed.]. Metals and Toxicological Implications in Health. Research Signpost. Kerala, India.
Mussali-Galante, P., E. Tovar-Sánchez, M. Valverde and E. Rojas. 2013a. Biomarkers of exposure for assessing environmental metal pollution: from molecules to ecosystems. Rev. Int. Contam. Ambie. 29: 117–140.
Mussali-Galante, P., E. Tovar-Sánchez, M. Valverde, L. Valencia-Cuevas and E. Rojas. 2013b. Evidence of population genetic effects in *Peromyscus melanophrys* chronically exposed to mine tailings in Morelos, Mexico. Environ. Sci. Pollut. Res. 20: 7666–7679.
Mussali-Galante, P., E. Tovar-Sánchez, M. Valverde and E. Rojas. 2014. Genetic structure and diversity of animal populations exposed to metal pollution. Rev. Environ. Contam. Toxicol. 227: 79–106.
Nagajyoti, P., K. Lee and T. Sreekanth. 2010. Heavy metals, occurrence and toxicity for plants: a review. Environ. Chem. Lett. 8: 199–216.
Navarro, M.C., C. Pérez-Sirvent, M.J. Martínez-Sánchez, J. Vidal, P.J. Tovar and J. Bech. 2008. Abandoned mine sites as a source of contamination by heavy metals: a case study in a semi-arid zone. J. Geochem. Explor. 96(2): 183–193.
Navas-Acien, A., A. Richey Sharrett, E.K. Silbergeld, B.S. Schwartz, K.E. Nachman, T.A. Burke and E. Guallar. 2005. Arsenic exposure and cardiovascular disease: a systematic review of the epidemiologic evidence. Am. J. Epidemiol. 162: 1037–1049.
Núñez, R.V. 2015. Mining in Mexico under the accumulation by dispossession. NERA 28: 132–148.
Nwapi, C. 2017. Legal and institutional frameworks for community development agreements in the mining sector in Africa. Extr. Ind. Soc. 4(1): 202–215.
O'Faircheallaigh, C. 2013. Community development agreements in the mining industry: an emerging global phenomenon. Community Dev. 44(2): 222–238.
Olness, A., R. Gesch, F. Forcella, D. Archer and J. Rinke. 2005. Importance of vanadium and nutrient ionic ratios on the development of hydroponically grown Cuphea. Ind. Crops Prod. 21: 165–171.
Olowu, R.A., G.O. Adewuyi, O.J. Onipede, O.A. Lawal and O.M. Sunday. 2015. Concentration of heavy metals in root, stem and leaves of acalypha indica and panicum maximum jacq from three major dumpsites in Ibadan metropolis, South West Nigeria. Am. J. Chem. 5(1): 40–48.
Owen, J.R. and D. Kemp. 2013. Social licence and mining: A critical perspective. Resour. Policy. 38(1): 29–35.
Panichev, N., K. Mandiwana, D. Moema, R. Molatlhegi and P. Ngobeni. 2006. Distribution of vanadium (V) species between soil and plants in the vicinity of vanadium mine. J. Hazard. Mater. A137: 649–653.
Peles, J., W. Towler and S. Guttman. 2003. Population genetic structure of earthworms (*Lumbricus rubelluz*) in soils contaminated by heavy metals. Ecotoxicol. 12: 379–386.
Pérez-de-Mora, A., P. Burgos, E. Madejón, F. Cabrera, P. Jaeckel and M. Schloter. 2006. Microbial community structure and function in a soil contaminated by heavy metals: effects of plant growth and different amendments. Soil Biol. Biochem. 38: 327–341.
Peterson, L., V. Trivett, A. Baker, C. Aguiar and J. Pollard. 2003. Spread of metals through an invertebrate food chain as influenced by a plant that hyperaccumulates nickel. Chemoecology 13: 103–108.
Phoenix, V.R. and K.O. Konhauser. 2008. Benefits of bacterial biomineralization. Geobiology 6: 303–308.
Pilon-Smits, E. 2005. Phytoremediation. Annu. Rev. Plant Biol. 56: 15–39.
Ping, Z., Z. Huiling and S. Wensheng. 2009. Biotransfer of heavy metals along a soil-plant-insect-chicken food chain: Field study. J. Environ. Sci. 21: 849–853.
Poschenrieder, Ch., R. Tolra and J. Barcelo. 2006. Can metals defend plants against biotic stress? Trends Plant Sci. 11(6): 288–295.
Prior, T., D. Giurco, G. Mudd, L. Mason and J. Behrisch. 2012. Resource depletion, peak minerals and the implications for sustainable resource management. Global Environ. Change. 22(3): 577–587.
Prno, J. and D.S. Slocombe. 2012. Exploring the origins of 'social license to operate' in the mining sector: Perspectives from governance and sustainability theories. Resour. Policy. 37(3): 346–357.
Prno, J. 2013. An analysis of factors leading to the establishment of a social licence to operate in the mining industry. Resour. Policy. 38(4): 577–590.

Quievryn, G., J. Messer and A. Zhitkovich. 2002. Carcinogenic chromium (VI) induces cross-linking of vitamin C to DNA *in vitro* and in human lung A549 cells. Biochemistry 41: 3156–3167.
Rojas, E. 2009. Special issue on the 20th anniversary of the comet assay. Mutat. Res. 681(1): 1–2.
Romero, F.M., M.A. Armienta and G. González-Hernández. 2007. Solid-phase control on the mobility of potentially toxic elements in an abandoned lead/zinc mine tailings impoundment, Taxco, Mexico. Appl. Geochem. 22(1): 109–127.
Rooney, C., F. Zhao and S. McGrath. 2006. Soil factors controlling the expression of copper toxicity to plants in a wide range of European soils. Environ. Toxicol. Chem. 25: 726–732.
Saco, D., S. Martín and P. San Jose. 2013. Vanadium distribution in roots and leaves of *Phaseolus vulgaris:* morphological and ultrastructural effects. Biol. Plant. 57: 128–132.
Sánchez-Salinas, E. and M.L. Ortiz-Hernández. 2014. Escenarios ambientales y sociales de la minería a cielo abierto. Inventio. 20: 27–34.
Sand, W. and T. Gehrke. 2006. Extracellular polymeric substances mediate bioleaching/biocorrosion via interfacial processes involving iron (III) ions and acidophilic bacteria. Res. Microbiol. 157: 49–56.
Santos, A. and S. Judd. 2010. The fate of metals in wastewater treated by the activated sludge process and membrane bioreactors: a brief review. J. Environ. Monitor. 12(1): 110–118.
Sartor, F.A., D.J. Rondia, F.D. Claeys, J.A. Staessen, R.R. Lauwerys, A.M. Bernard et al. 1992. Impact of environmental cadmium pollution on cadmium exposure and body burden. Arch. Environ. Health. 47: 347–353.
Schoenberger, E. 2016. Environmentally sustainable mining: The case of tailings storage facilities. Resour. Policy. 49: 119–128.
Shanker, A., M. Djanaguiraman and B. Venkateswarlu. 2009. Chromium interactions in plants: current status and future strategies. Metallomics 1: 375–383.
Sharma, P. and R. Dubey. 2005. Lead toxicity in plants. Braz. J. Plant Physiol. 17: 35–52.
Sharma, S., B. Singh and V.K. Manchanda. 2015. Phytoremediation: role of terrestrial plants and aquatic macrophytes in the remediation of radionuclides and heavy metal contaminated soil and water. Environ. Sci. Pollut. Res. Int. 22(2): 946–62.
Sing, J. 2015. Regulating mining resource investments towards sustainable development: The case of Papua New Guinea. Extr. Ind. Soc. 2(1): 124–131.
Singh, H., P. Mahajan, K. Shalinder, D. Batish and R. Kohli. 2013. Chromium toxicity and tolerance in plants. Environ. Chem. Lett. 11: 229–254.
Smith, S. 2009. Mining and the environment. Amicus Books NSW, Australia.
Söderholm, K., P. Söderholm, H. Helenius, M. Pettersson, R. Viklund V. Masloboev et al. 2015. Environmental regulation and competitiveness in the mining industry: Permitting processes with special focus on Finland, Sweden and Russia. Resour. Policy. 43: 130–142.
Solomon, F., E. Katz and R. Lovel. 2008. Social dimensions of mining: Research, policy and practice challenges for the minerals industry in Australia. Resour. Policy. 33(3): 142–149.
Spain, A. and E. Alm. 2003. Implications of microbial heavy metals tolerance in the environment. Rev. Undergraduate Res. 2: 1–6.
Stewart, A., S. Luoma, C. Schlekat, M. Doblin and K. Hieb. 2004. Food web pathway determines how selenium affects aquatic ecosystems. Environ. Sci. Technol. 38: 4519–4526.
Szablowski, D. 2007. Transnational law and local struggles, mining, communities, and the World Bank, Oxford & Portland: Hart Publishing. 337.
Tangahu, B.V., S.R. Sheikh Abdullah, H. Basri, M. Idris, N. Anuar and M. Mukhlisin. 2011. A Review on heavy metals (As, Pb, and Hg) uptake by plants through phytoremediation. Int. J. Chem Eng. 2011: 31.
Tiwary, R.K. 2001. Environmental impact of coal mining on water regime and its management. Water Air Soil Pollut. 132(1-2): 185–199.
Tong, S., Y.E. Von Schirnding and T. Prapamontol. 2000. Environmental lead exposure: a public health problem of global dimensions. Bull. World Health Organ. 78: 1068–77.
Tovar-Sánchez, E., L.T. Cervantes, C. Martínez, E. Rojas, M. Valverde, M.L. Ortiz-Hernández et al. 2012. Comparison of two wild rodent species as sentinels of environmental contamination by mine tailings. Environ. Sci. Pollut. Res. 19: 1677–1686.
Tremblay, A., D. Lesbarreres, T. Merritt and C. Wilson. 2008. Genetic structure and phenotypic plasticity of yellow perch (*Perca flavescens*) populations influences by habitat, predation, and contamination gradients. Integr. Environ. Assess. Manag. 4: 264–266.
Ullah, A., S. Heng, M.F.H. Munis, S. Fahad and X. Yang. 2015. Phytoremediation of heavy metals assisted by plant growth promoting (PGP) bacteria: a review. Environ. Exp. Bot. 117: 28–40.
Ungherese, G., A. Mengoni, S. Somigli, D. Baroni, S. Focardi and A. Ugolini. 2010. Relationship between heavy metals pollution and genetic diversity in Mediterranean population of the sandhopper *Talitrus saltator* (Montagu) (Crustaceae, Amphipoda). Environ. Pollut. 158: 1638–1643.
Valavanidis, A., T. Vlachogianni and C. Fiotakis. 2009. 8-hydroxy-2-deoxyguanosine (8-OHdG): a critical biomarker of oxidative stress and carcinogenesis. J. Environ. Sci. Health Part C. 27: 120–139.

Valko, M., H. Morris and M. Cronin. 2005. Metals, toxicity and oxidative stress. Curr. Med. Chem. 12: 1161–1208.
Valko, M., C. Rhodes, J. Moncol, M. Izakovic and M. Mazura. 2006. Free radicals, metals and antioxidants in oxidative stress-induced cancer. Chem. Biol. Interact. 160: 1–40.
Van Straalen, N. and M. Timmermans. 2002. Genetic variation in toxicant-stressed populations: an evaluation of the "genetic erosion" hypothesis. Hum. Ecol. Risk Assess. 8: 983–1002.
Vanden-Heuvel, J. and J. Davis. 1999. Molecular approaches to identify exposure and risk to specific environmental pollutants. Biomarkers 4: 93–105.
Volke, S.T., T.J. Velasco and P.D. De la Rosa. 2005. Suelos contaminados por metales y metaloides: muestreo y alternativas para su remediación. Secretaría del Medio Ambiente y Recursos Naturales. Instituto Nacional de Ecología, México. 144.
Volke, S.T.L., T.J. Velasco, P.D. De la Rosa and O.G. Solórzano. 2004. Evaluación de tecnologías de remediación para suelos contaminados con metales. Etapa I. Secretaría de Medio Ambiente y Recursos Naturales, México. 44.
Waalkes, M., D.A. Fox, J. Christopher, S.R. Patierno and M.J. McCabe. 2000. Metals and disorders of cell accumulation: modulation of apoptosis and cell proliferation. Toxicol. Sci. 56: 255–261.
Wang, C., D. Harbottle, Q. Liu and Z. Xu. 2014. Current state of fine mineral tailings treatments: a critical review on theory and practice. Miner. Eng. 58: 113–131.
Wells, T. and R.E. Melchers. 2014. An observation-based model for corrosion of concrete sewers under aggressive conditions. Cem. Concr. Res. 61: 1–10.
Wilson, D.M. and L.H. Thompson. 2007. Molecular mechanisms of sister-chromatid exchange. Mut. Res.-Fund. Mol. M. 616(1): 11–23.
Wuana, R.A. and F.E. Okieimen. 2011. Heavy metals in contaminated soils: a review of sources, chemistry, risks and best available strategies for remediation. ISRN Ecology. 2011, Article ID 402647.
Xu, Q. and G. Shi. 2000. The toxic effects of single Cd and interaction of Cd with Zn on some physiological index of [*Oenanthe javanica* (Blume) DC]. J. Nanjing Norm. Univ. (Nat. Sci.) 23: 97–100.
Yang, J., X. Pan, C. Zhao, S. Mou, V. Achal, F. Al-Misned et al. 2016. Bioimmobilization of heavy metals in acidic copper mine tailings soil. Geomicrobiol. J. 33(3-4): 261–266.
Yauk, C., G. Fox, B. McCarry and J. Quinn. 2000. Induced minisatellite germline mutations in herring gulls (*Larus argentatus*) living near steel mills. Mutat. Res. 452: 211–218.
Young, S.D. 2013. Chemistry of heavy metals and metalloids in soils. pp. 51–95. *In*: Alloway, B.J. [ed.]. Heavy Metals in Soils: Trace Metals and Metalloids in Soils and Their Bioavailability. Springer, Netherlands.
Zhao, F., C. Rooney, H. Zhang and P. McGrath. 2006. Comparison of soil solution speciation and diffusive in thin-films measurement as an indicator of copper bioavailability to plants. Environ. Toxicol. Chem. 25(3): 733–742.
Zvereva, E. and M. Kozlov. 2001. Effects of pollution-induced habitat disturbance on the response of willows to simulated herbivory. J. Ecol. 89(1): 21–30.

13

Phytomanagement of Metal-Rich and Contaminated Soils

Implicated Factors and Strategies for its Improvement

Cristina Becerra-Castro,[†] *Vanessa Álvarez-López,*[a] *Tania Pardo,*[b] *Beatriz Rodríguez-Garrido,*[c] *Andrea Cerdeira-Pérez,*[d] *Ángeles Prieto-Fernández*[e] and *Petra S. Kidd*[*]

Introduction

Soil formation is an extremely slow process which results in a complex medium performing many vital functions. Soil generates numerous ecosystem services which are classified as (i) supporting (soil formation, nutrient cycling, habitat), (ii) regulating (regulation of elemental cycles, C sequestration, water purification and storage, adsorption and transformation of pollutants), (iii) provisioning (raw materials and biomass, physical stability and support of plants), and (iv) cultural (heritage sites, archaeological archive) (Lorenz 2013). Contamination leads to a decline in soil quality and biodiversity, which in turn reduces the soil's capacity to perform ecological functions and provide these essential ecosystem services.

Soil Contaminants: Sources and Environmental Associated Problems

Soil contamination refers to the presence (in elevated concentrations) of chemical substance(s) (normally of anthropogenic origin) which alter soil quality and functions, and can negatively affect water quality, biodiversity, food security or human health. Soil contaminants are generally divided into inorganic and organic substances, and the most

Instituto de Investigaciones Agrobiológicas de Galicia (IIAG), Consejo Superior de Investigaciones Científicas (CSIC), Santiago de Compostela 15705, Spain.
[a] Email: vanessa@iiag.csic.es
[b] Email: tpardo@iiag.csic.es
[c] Email: beatriz@iiag.csic.es
[d] Email: acerdeira@iiag.csic.es
[e] Email: apf@iiag.csic.es
[*] Corresponding author: pkidd@iiag.csic.es
[†] Deceased

frequent types are trace elements (TE), mineral oil and polycyclic aromatic hydrocarbons (Panagos et al. 2013). The large volumes of industrial, municipal and agricultural wastes and widespread use of chemicals during past decades have left numerous contaminated sites worldwide (Panagos et al. 2013).

Trace elements, which include metals and metalloids, are present in the lithosphere in concentrations of < 0.1%. Most TE are classified as heavy metals, with a density of > 5 g cm^{-3} (Kabata-Pendias 2010). Although some TE are required by organisms at low concentrations (essential metals such as Cu, Fe, Mn, Zn, and Ni) when they are present in concentrations exceeding those normally found in nature they may produce genotoxic effects and negatively affect metabolic processes of organisms (Adriano 2001). Furthermore, they can alter the structure and composition of populations, communities and ecosystems. Unlike organic pollutants, metals do not biodegrade and their residence time in the soil may prolong for thousands of years (Adriano 2001).

Natural TE concentrations or background levels are strongly dependent on the soil parent material and acting weathering processes, while the main anthropogenic sources of metal contamination include application of metal-based pesticides and metal-enriched sewage sludges in agriculture, combustion of fossil fuels, mining activities, metallurgical and electronic industries, military training, etc. (Adriano 2001). Mining activities are amongst the most polluting activities and are the main source of TE contamination. Metal(loid) ore processing frequently leads to a multi-elemental contamination of the environment (Dudka and Adriano 1997) and generates large amounts of tailings and rock deposits which can cover extensive areas (Salomons 1995).

Organic contaminants include polycyclic aromatic hydrocarbons (PAHs), persistent organic pollutants (POPs), volatile organic contaminants (VOCs) and petroleum hydrocarbons (Swartjes 2011). The majority of PAHs (e.g., naphthalene, phenanthrene, benzo(a)anthracene, benzo(a)pyrene, etc.) derives from anthropogenic activities such as the processing or burning of fossil fuels and wood treatment, although there are also some natural sources of PAHs such as forest fires and volcanic eruptions (Swartjes 2011). POPs are toxic compounds resistant to degradation and many of them are banned obsolete pesticides which persist in the environment. According to the Stockholm Convention some of the main POPs are isomers of 1,2,3,4,5,6-hexachlorocyclohexane (including the γ-isomer known as lindane), aldrin, chlordane, DDT, polychlorinated biphenyls (PCBs), as well as some PAHs. Volatile organic contaminants are typically generated from metal degreasing, gasoline, and wood preserving processes, e.g., trichloroethylene, tetrachloroethylene, trichloroethane and BTEX (benzene, toluene, ethylbenzene, and xylenes). Petroleum hydrocarbons (often called total petroleum hydrocarbons or TPH) are actually complex mixtures of contaminants, mainly hydrocarbons and additives (e.g., benzene, toluene, xylenes, naphthalene, and fluorine) (Swartjes 2011). Most of these organic compounds cause acute toxicity to living organisms and the exposure to them, even at low concentrations, results in accumulation of toxic concentrations in tissues (Ridolfi et al. 2014).

Phytomanagement of Metal-Rich and Contaminated Soils

Remediation techniques can be classified into two groups which are either aimed at contaminant containment (immobilizing contaminants in the substrate) or actual decontamination (eliminating or removing contaminants from the soil). These treatments can be based on physical (thermal treatments, adsorption, containment cells, etc.), chemical (oxidative or reductive degradation, chemical binding), or biological processes (such as bioremediation or phytoremediation) (van Liedekerke et al. 2014).

Gentle remediation options (GRO) have been developed as eco-friendly alternatives to traditional, civil-engineering methods of soil remediation (Kidd et al. 2015). Gentle remediation options include *in situ* stabilization (inactivation) and plant-based (generally termed as *phytoremediation*) options, and are addressed to decreasing the labile (bioavailable) and/or the total content of contaminants. These techniques are mainly based on the use of plants and soil microorganisms, also aided by agronomic management, which effectively reduce pollutant linkages while preserving the soil resource and restoring ecological functions (Vangronsveld et al. 2009). The use of contaminated land for the production of valuable biomass (such as the production of timber, bioenergy crops, biofortified products, etc.) falls within the concept of phytomanagement (Robinson et al. 2009) and is considered essential for the commercial success of these phytotechnologies (Conesa et al. 2012). Different options for the phytomanagement of contaminated soils are described below and presented in Figs. 1 and 2.

- Phytostabilization uses tolerant plant species to establish a vegetation cover and progressively stabilize and/or reduce the availability of soil pollutants (Mench et al. 2006, Ruttens et al. 2006a, 2006b, Vangronsveld et al. 2009, Dary et al. 2010). The incorporation of amendments into the soil or use of microbial inoculation (aided phytostabilization) (Mench et al. 2010) can further decrease the bioavailability and phytotoxicity of pollutants, while improving plant establishment. Phytostabilization does not lead to the actual removal of contaminants but reduces pollutant

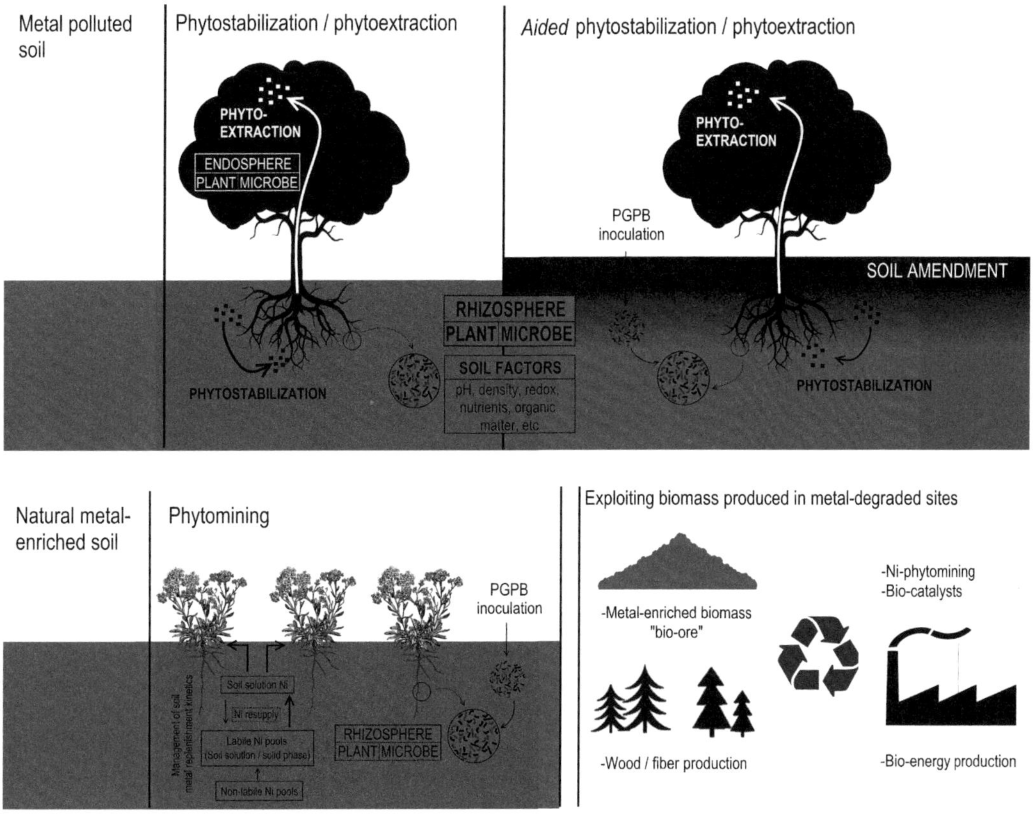

Figure 1. Schematic representation of phytomanagement options targeting trace elements-contaminated soils and simplified overview of plant-microbe-soil interactions involved in the remediation process.

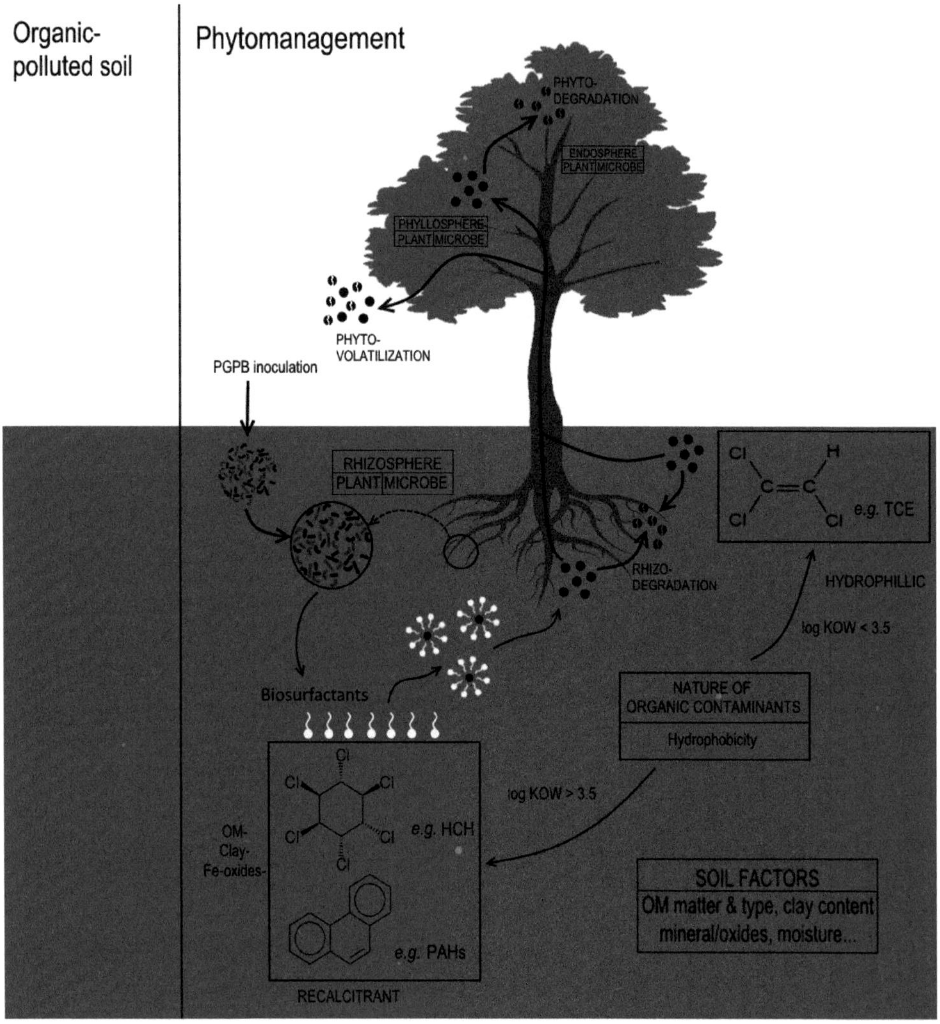

Figure 2. Schematic representation of phytomanagement options targeting soils contaminated with organic compounds and simplified overview of plant-microbe-soil interactions involved in the remediation process.

bioavailability and transfer to other environmental compartments. The mechanical action of the plant roots reduces soil erosion and transport of soil particles, while evapotranspiration minimizes leaching and therefore contaminant dissemination. In addition, the adsorption, precipitation, and accumulation of the contaminants in the rhizosphere (in collaboration with microorganisms associated with plant roots) entail their immobilization (Mench et al. 2010).

- Phytoextraction is based on the use of plants that take up contaminants (mainly TE) from the soil and accumulate them in their harvestable aboveground biomass (Vangronsveld et al. 2009). Phytoextraction can be aided by soil amendments and soil microoroganisms (aided phytoextraction). When marketable TE (such as Ni, Au, etc.) are recovered from the plant biomass (bio-ores) it is known as phytomining (Chaney et al. 2007).
- Phytovolatilization exploits the ability of plants to transform pollutants into volatile compounds or to absorb and transport volatile compounds from the soil to the

aboveground biomass where they can then be released to the atmosphere (Wenzel 2009). When the contaminant is transformed and released directly from the soil surrounding plant roots (rhizosphere), it is usually termed as rhizovolatilization (Zhang and Frankenberger 2000).
- Phytodegradation or phytotransformation uses plants (and their associated microorganisms) to degrade organic contaminants to non-toxic metabolites (Weyens et al. 2009b). When the degradation takes place in the rhizosphere of plants (due to microbial activity or release of enzymes from plants), terms such as phytostimulation or rhizodegradation are more correct (Becerra-Castro et al. 2013c).
- Rhizofiltration is based on the use of aquatic plants to absorb and/or adsorb contaminants present in water, sediments or aqueous wastes in their roots. The use of aquatic macrophytes as biofilters in natural and constructed wetlands and wastewater treatment facilities has gained interest due to their well-known bioaccumulation properties (Salem et al. 2014).

Factors Influencing Phytomanagement Success

The remediation of contaminated soils by phytotechnologies is considered an environmentally-friendly, esthetically pleasing and economically viable alternative to harsher civil engineering-based methods, which can be applied *in situ* and on a large scale. However, phytoremediation techniques do of course present a series of limitations and still require optimization before they can become fully implemented on a wide-scale. In addition to the inherent problems associated with any agronomical practice (such as the dependence on climate and season, outbreaks of pests or disease, etc.), a major problem associated with these techniques is the length of time required for the clean-up process (of particular concern in phytoextraction). Several authors have suggested that to be realistically viable the clean-up time should not exceed 10 years (Robinson et al. 2009, Vangronsveld et al. 2009). The establishment and growth of plants on contaminated sites are other major obstacles (Tordoff et al. 2000, Mendez and Maier 2008). In addition to the phytotoxic concentrations of pollutants, contaminated soils usually present edaphic conditions which can severely limit plant growth (nutrient deficiency, poor soil structure, low organic matter, etc.). The careful selection of tolerant plant species is vital for the long-term success of phytomanagement strategies (Batty 2005, Clemente et al. 2012, Parraga-Aguado et al. 2014). The use of deep-rooting plants, mycorrhizal plants or more novel biotechnological methods can enhance plant growth and GRO efficiency (Kidd et al. 2009). Additional aspects that should be considered include the degree of soil contamination, the bioavailability and accessibility of the contaminants, and the capacity of the plants and their associated microorganisms to adsorb, accumulate and/or degrade the contaminants (Vangronsveld et al. 2009).

Selection of Adequate Plant Species

The screening of plant species for selecting appropriate candidates for different scenarios has received much attention (Conesa et al. 2009, Becerra-Castro et al. 2012, Martínez-López et al. 2014, Monterroso et al. 2014). Within the same plant species, different ecotypes, cultivars, varieties or clones can vary greatly in their response to the presence of contaminants (Kidd et al. 2015). Whenever possible, it is preferable to use native plants species which, as well as avoiding potential negative impacts of alien species, are adapted to the edaphoclimatic conditions of the affected area (pH, salinity, soil structure, temperature, water content, etc.) (Clemente et al. 2012, Pardo et al. 2014d, Parraga-Aguado et al. 2014).

Metalliferous soils, both natural and from anthropogenic origin, host metal-tolerant plant species of useful application in phytoremediation, and many studies have explored the biodiversity of these sites (Batty 2005, Conesa et al. 2009, Otones et al. 2011a, 2011b, Parraga-Aguado et al. 2014). For example, in mining areas, although diversity is often low, pollution stress facilitates the selection of tolerant populations (Conesa et al. 2009). Otones et al. (2011a) found a high tolerance to As in the plant community of an abandoned mining area, and highlighted the potential of *Salix atrocinerea*, *Scirpus holoschoenus*, *Agrostis castellana* and *Genista scorpius* for their use in phytostabilization. Del Río et al. (2002) and Gardea-Torresdey et al. (2004) emphasized the ability of *Convolvulus arvensis* to accumulate As, Cu, Cd, Cr, and Pb in populations growing on contaminated soils.

The metal-excluding phenotype is of particular interest for phytostabilization processes. In Europe, metallicolous populations of metal-excluding grasses, such as *Agrostis* spp., *Festuca* spp. or *Poa* spp., have shown a good ability to colonize Pb-, Zn- and Cu-contaminated soils (Mench et al. 2010). Short rotation coppice (SRC) systems with fast-growing high-yielding woody crops (such as *Salix* and *Populus*) showed to be well suited for either phytostabilization (e.g., Cu and Pb) or phytoextraction (Cd and Zn) after selection of TE-excluding or accumulating clones (Pierzynski et al. 2002, Pulford and Watson 2003, Lewandowski et al. 2006, Witters et al. 2009) due to their deep root systems, tolerance to TE and nutrient deficiency, and the ability to re-sprout from the stumps after harvests (Dimitriou et al. 2012, Kidd et al. 2015). Moreover, the harvested biomass can be used for bioenergy, timber production or biofortified products, etc. (Witters et al. 2009, Conesa et al. 2012, Chalot et al. 2012).

Native halophyte plant species usually present high tolerance to TE and low TE accumulation in their tissues, and have been reported as promising candidates for phytostabilization purposes (Parraga-Aguado et al. 2014, Pardo et al. 2016a, Pardo et al. 2017a). Their adaptation to TE stress is considered to be related to specific physiological mechanisms for tolerance to salt and drought (Van Oosten and Maggio 2015). The genus *Atriplex*, and in particular halophyte shrubs like *Atriplex halimus*, have been identified as suitable species for the phytomanagement of mine soils due to their fast growth and low water requirement (Walker et al. 2014), low TE translocation and ability to develop a sustainable vegetative cover (Clemente et al. 2012, Pardo et al. 2014c, Pardo et al. 2016a, Pardo et al. 2017a).

Hyperaccumulators are of special interest for phytoextraction; these plants accumulate extreme amounts of trace metals (between 100 and 1000 times higher than non-accumulating plants) in their aboveground biomass when growing in metal-enriched habitats (van der Ent et al. 2013). However, their efficiency can be limited due to their frequent slow growth, low biomass and/or shallow root systems (with a few exceptions such as *Berkheya coddii*). As a result, a high number of cropping cycles is required for clean-up (Robinson et al. 2009).

Plant cropping patterns can also improve plant growth and performance and, depending on the phytotechnology, can be designed so as to enhance or mitigate metal availability, uptake and accumulation (Kidd et al. 2015). Polycultures offer several benefits over monocultures, in terms of plant productivity and nutrition, soil protection and quality, enhancing biodiversity, or even in pest control (Facknath and Lalljee 2000, Lasat 2000, Mench et al. 2010). Beneficial effects are due to changes in the root environment (Carrillo-Garcia et al. 1999), protection from solar radiation (Zúñiga et al. 2005), soil temperature buffering, enhanced water holding capacity (Drezner 2006), and/or nutrient availability (e.g., transfer of fixed N_2 or mobilization of P by acidification) (Bonanomi et al. 2011). As a consequence of these root interactions, intercropped species may alter metal bioavailability in their rhizospheres (Tang et al. 2012).

Soils contaminated with organic compounds can also be a source of tolerant populations of plant species (Kaimi et al. 2007, Abhilash et al. 2008). The bioavailability of organic contaminants strongly influences their phytotoxicity and absorption (Chaudhry et al. 2005, Wenzel 2009). However, several studies have shown that some plants are capable of taking up significant amounts of hydrophobic contaminants, and accumulating them in their tissues despite their high hydrophobicity (Chhikara et al. 2010, White 2010, Namiki et al. 2015), probably through the exudation of organic acids that increase their availability or the synthesis of transport proteins (White et al. 2003, Namiki et al. 2015). In addition, plants may also release enzymes that degrade organic contaminants (Wenzel 2009) or alternatively, they may stimulate the activity of degrading rhizosphere microorganisms (phytostimulation). Alvarez et al. (unpublished results) analyzed the degradation of 1,2,3,4,5,6-hexachlorocyclohexane isomers (HCHs) by the strains *Sphingobium* sp. D4 and *Streptomyces* sp. A11 *in vitro*, in the presence and absence of corn exudates. The exudates accelerated the dissipation of HCHs by the two strains tested. The effect was particularly marked for D4, which is able to mineralize HCH isomers. Plants often produce and exude natural chemicals such as phenolic compounds, terpenes, etc., with a structure close to that of xenobiotic compounds which have been shown to stimulate the growth of degrading microorganisms or induce the synthesis of enzymes involved in the degradation process in the rhizosphere (Donnelly et al. 1994).

Over the last few decades numerous studies have focused on improving the genetic properties of plants for increasing their efficiency in phytotechnologies. Traditional techniques based on the selection of parental lines have offered good results for example for the optimization of Co and Ni phytoextraction in species of the genus *Alyssum* (Chaney et al. 2007, 2014). However, the potential improvements obtained with this traditional approach are limited by the finite natural genetic diversity within the study species. Genetic engineering techniques involving the transfer of genes responsible for the characteristics of interest have also been applied with some degree of success. For example, Gisbert et al. (2003) obtained a high accumulation of Pb and Cd and a significant increase in the length of *Nicotiana glauca* roots through the overexpression of the TaPCS1 gene (responsible for the formation of phytochelatins). Hsieh et al. (2009), by expression of the *mer*P gene from *Bacillus megaterium* in *Arabidopsis thaliana*, succeeded in improving the tolerance and accumulation of Hg in this species. Nevertheless, the application of transgenic plants is limited because of potential ecological problems (e.g., loss of diversity, hybridization with natural species, etc.) and important social and legal objections.

Contaminant Bioavailability

Bioavailability can be defined as the fraction of the contaminant that can be taken-up or transformed by living organisms and react with their metabolism (Semple et al. 2003). It is largely dependent upon the nature of the compound, its physical and chemical properties (solubility, specific ionic radius, charge, hydrophobicity, diffusion and mass transport, etc.), but also by the dominating edaphic properties (content and type of organic matter and clays, pH, redox potential, cation exchange capacity, etc.) (Kidd et al. 2009, Wenzel 2009). Contaminant bioavailability can be substantially modified in the rhizosphere as a result of edaphic processes, root exudates or microbial activity, which in turn influences phytoremediation efficiency (Kidd et al. 2009). Increasing contaminant bioavailability would improve processes such as phytoextraction, phytodegradation or rhizoremediation, whereas its reduction could substantially improve phytostabilization (Wenzel 2009).

The legislation available in most industrialized countries, regulating local soil contamination, and guidelines for assessing potentially contaminated soils, are generally based on total contaminant concentrations. In recent years, more sophisticated risk-based approaches to deal with the local effects of soil pollution have been developed, which include the concept of pollutant linkage (contaminant-receptor-pathway). Decision makers and regulatory organizations have accepted that bioavailability of soil contaminants is a key variable to be taken into consideration in risk assessment, regulation policies and soil remediation (Naidu et al. 2015). These risk orientated policies focus on the abandonment of policies aimed at restoring soils to their original 'clean' state. Some national trigger values classifying soils as contaminated or requiring remediation now have bioavailability explicitly (e.g., in the UK, Belgium, Switzerland) or implicitly (trigger values set according to the main influencing soil physicochemical properties, e.g., soil pH, granulometry, organic matter content) embedded within them. Several phytomanagement options are aimed at removing the bioavailable contaminant fraction ("bioavailable stripping"), a target which significantly reduces the length of time required for rehabilitation.

Influence of trace metal bioavailability on phytomanagement

Metal(loid) availability to plants is governed by a pseudo-equilibrium between the elements in the soil solution and in the solid phase (Kidd et al. 2009) and is therefore determined not only by the concentration in the soil solution and the assimilation mechanisms of the different organisms, but also by the capacity of the solid phase to resupply the element (Peijnenburg and Jager 2003, Pardo et al. 2016a). In the soil solution, TE are present as free uncomplexed ions, ion pairs, and ions complexed with organic anions, organic macromolecules, and inorganic colloids. The most important metal(loid) pools in the solid phase include the exchange complex, metal(loid)s complexed by organic matter, sorbed onto or occluded within oxides and clay minerals, co-precipitated with secondary pedogenic minerals (e.g., Al, Fe, Mn oxides, carbonates, phosphates, and sulfides) or as part of the crystal lattices of primary minerals (Adriano 2001). The distribution of TE onto these fractions and their modification depend on the interactions between the TE and the components of soil solution and solid matrix, highlighting the reactions of ionic exchange, adsorption, precipitation and complexation (Peijnenburg and Jager 2003). These processes are influenced by the edaphic conditions (such as pH, cation-exchange capacity, soil redox potential, amount and type of organic matter, Fe, Mn and Al oxides, clay minerals, size and superficial area of soil particles, temperature, etc.) (Kabata-Pendias 2010), but also by the biological activity of the living organisms in soil, especially in the rhizosphere (Wenzel 2009).

Plants mobilize soil nutrients and/or trace metals by actively changing the soil physicochemical properties in the rhizosphere. Between 10 and 20% of the photosynthetic carbon that roots receive are released as rhizodeposits in the rhizosphere (Singer et al. 2003). These exudates are believed to be implicated in the mobilization of soil elements through the weathering of minerals (Hinsinger et al. 2009). Moreover, root-induced reactions such as chelation, precipitation or changes in pH and redox conditions also influence soil element availability, as well as the activity of plant-associated rhizosphere microorganisms (Uren and Reisenauer 1988, Tao et al. 2004). Metabolic processes of microorganisms can lead to changes in contaminant availability and these processes can be exploited in phytoremediation techniques (Kidd et al. 2009, Wenzel 2009). Metal mobilization by bacteria can be achieved by acidolysis, chelation or complexation, chemical transformation and/or redox reactions, while metal immobilization can occur by sorption to cell components or exopolymers, transport and intracellular sequestration, release of metal binding compounds or precipitation as insoluble organic or inorganic molecules (Gadd 2004).

Bacterial strains isolated from the rhizosphere of Ni-hyperaccumulating subspecies of *Alyssum serpyllifolium* have been characterized for Ni-mobilizing capacity (Becerra-Castro et al. 2013a). Strain *Arthrobacter* spp. LA44 (indoleacetic acid (IAA)-producer) was found to be efficient at mobilizing Ni *in vitro* from ground ultramafic rock, apparently solubilizing Ni associated with Mn oxides through oxalate exudation. *Arthrobacter* spp. SBA82 (siderophore-producer, PO_4-solubiliser and IAA-producer) also led to release of Ni and Mn, albeit to a much lower extent. In this case, the concurrent mobilization of Fe and Si indicates preferential weathering of Fe oxides and serpentine minerals, possibly related to siderophore production. In further inoculation experiments strain LA44 significantly enhanced phytoextraction of Ni by the hyperaccumulator *A. serpyllifolium* growing in ultramafic soils (Becerra-Castro et al. 2013a, Cabello-Conejo et al. 2014). The capacity of rhizobacterial strains associated with hyperaccumulating species to mobilize metals, such as Ni, from less labile fractions can help the replenishment of labile metal pools leading to enhanced metal uptake by these plants and an overall improvement in the efficiency of the phytoextraction (or phytomining) process.

Influence of organic pollutant bioavailability on phytomanagement

The behavior of organic compounds in the soil is principally governed by their hydrophobicity (lipophilicity) (Reid et al. 2000, Semple et al. 2003). Hydrophobicity is habitually indicated by the compound solubility and the log octanol/water partition coefficient, log K_{OW}. High values of log K_{OW} (log $K_{OW} > 3.5$) indicate a higher hydrophobicity, a higher resistance to degradation or transformation, a higher persistence, thus tending to bioaccumulate in the food chain. Apart from the physicochemical properties of the contaminant, soil properties, such as organic matter content, clay minerals and oxides, pH, moisture and cation exchange capacity are also directly implicated in the degree to which these compounds are retained (sorbed) within the soil. It is generally accepted that the main critical factors determining these soil-compound interactions are the amount and nature of the soil organic matter, although clay minerals and Fe oxides also play an important role. With time, availability of organic contaminants in soils progressively decreases ("aging" effect), as they are sequestered in the organic matter or sorbed to surfaces within nano- and micropores in soils containing little organic matter (Semple et al. 2003).

Bioavailability of organic pollutants in soils seems to be an important and restrictive factor for effective phytoremediation (Schnoor et al. 1995, Chaudhry et al. 2005). This is particularly the case for contaminants with a low water solubility and high hydrophobicity which impedes microbial degradation or their uptake and absorption within plants. Bacteria can be useful for promoting the bioavailability of organic contaminants, and thus facilitating their degradation and/or uptake (Pandey et al. 2009, Wenzel 2009). The main mechanisms operating in this process include the production of biosurfactants, the formation of biofilms or chemotaxis.

Biosurfactants can increase the solubility of organic contaminants in water through the formation of micelles, or alternatively, they can modify the hydrophobicity of the bacterial cells membranes and facilitate the attachment of hydrophobic compounds (Hickey et al. 2007). For instance, the addition of rhamnolipids increased the degradation of phenanthrene by *Sphingomonas* sp. GF2B (Pei et al. 2010). Hickey et al. (2007) found that biosurfactants significantly increased desorption of PAHs from soil and enhanced their biodegradation by *Pseudomonas alcaligenes* PA-10. The addition of biosurfactants also enhanced the phytoremediation of PAH-contaminated soils inoculated with a PAH-degrading bacterial strain (Liu et al. 2010, Zhang et al. 2010). Becerra-Castro et al. (2011) isolated rhizobacterial and endophytic bacterial strains from *Cytisus striatus* (Hill) Rothm., a plant which grows spontaneously in HCH-contaminated sites and proposed as a candidate for the rhizoremediation of these compounds (Kidd et al. 2008). Cerdeira-Pérez et al. (unpublished results) selected from this isolate collection biosurfactant-producing strains and evaluated their capacity for mobilizing HCH isomers in soils. Strains were grown in two culture media formulated so as to stimulate biosurfactant production according to Saimmai et al. (2012). The biosurfactant production was measured using an emulsification test and the cell-free culture filtrates were used to extract HCH isomers from two soils with contrasting organic matter content. Strain *Rhodococcus erythropolis* ET54b enhanced the bioavailability of HCH isomers, including the more lipophilic and recalcitrant β-HCH isomer (Fig. 3). HCH mobilization in the more organic soil (soil A), which presumably has a higher HCH retention capacity, was only achieved using cell-free culture filtrates from the medium which also induced higher biosurfactant production (containing glucose and sodium nitrate). The strain ET54b and another biosurfactant producing strain (*Streptomyces* sp. M7) were also used in combination with the HCHs degrader *Sphingobium* D4 (isolated from HCH-contaminated soil) for the removal of HCH isomers in slurries prepared with soils of contrasting organic matter content. Both ET54b and M7 accelerated the degradation of HCHs. Moreover, in the slurry with high organic matter content, the stimulation was more pronounced for the β-isomer. *Rhodococcus erythropolis* ET54b strain was also shown to increase tolerance and growth of *C. striatus* when growing in HCH-contaminated substrates. Becerra-Castro et al. (2013c) inoculated *C. striatus* with a combination of the ET54b and D4 strains and observed decreased HCH phytotoxicity and improved plant growth in HCH-spiked perlite:sand mixtures. HCH-exposed plants inoculated with the combination presented an increase of up to 160% in root and shoot biomass and a decrease in the activities of enzymes involved in antioxidative defense. When used as a combined inoculant of *C. striatus* growing in two soils, HCH dissipation was significantly improved in soils with inoculated plants compared to non-inoculated controls (Becerra-Castro et al. 2013b). Abhilash and Singh (2010) also found that inoculating the HCH-tolerant plant *Withania somnifera* with the lindane-degrading rhizobacteria *Staphyylococcus cohnii* subsp. *urealyticus* led to an enhanced dissipation of this organochlorine and also improved plant growth.

Fertilization Regimes and Soil Amendments

Phytomanagement requires the use of appropriate agronomic and crop management practices, and may be assisted through the application of soil amendments (generally termed as *aided phytoremediation*). Contaminated soils are frequently nutrient deficient, with a low organic matter content, poor structure and sometimes very acidic pH and/or high salinity, and therefore, the selection of an adequate combination of amendments to facilitate plant establishment is a critical issue for the success of phytotechnologies (Bernal et al. 2007, Clemente et al. 2015, Pardo et al. 2016a). The use of organic and inorganic amendments may optimize plant growth and performance by improving soil physicochemical properties, fertility and microbial activity and diversity (Bolan et al. 2011, Pardo et al. 2014b, 2014c). In addition, amendments directly or indirectly influence the availability and mobility contaminants through the modification of soil physicochemical and biological conditions (pH, redox conditions, concentration of chelating and complexing

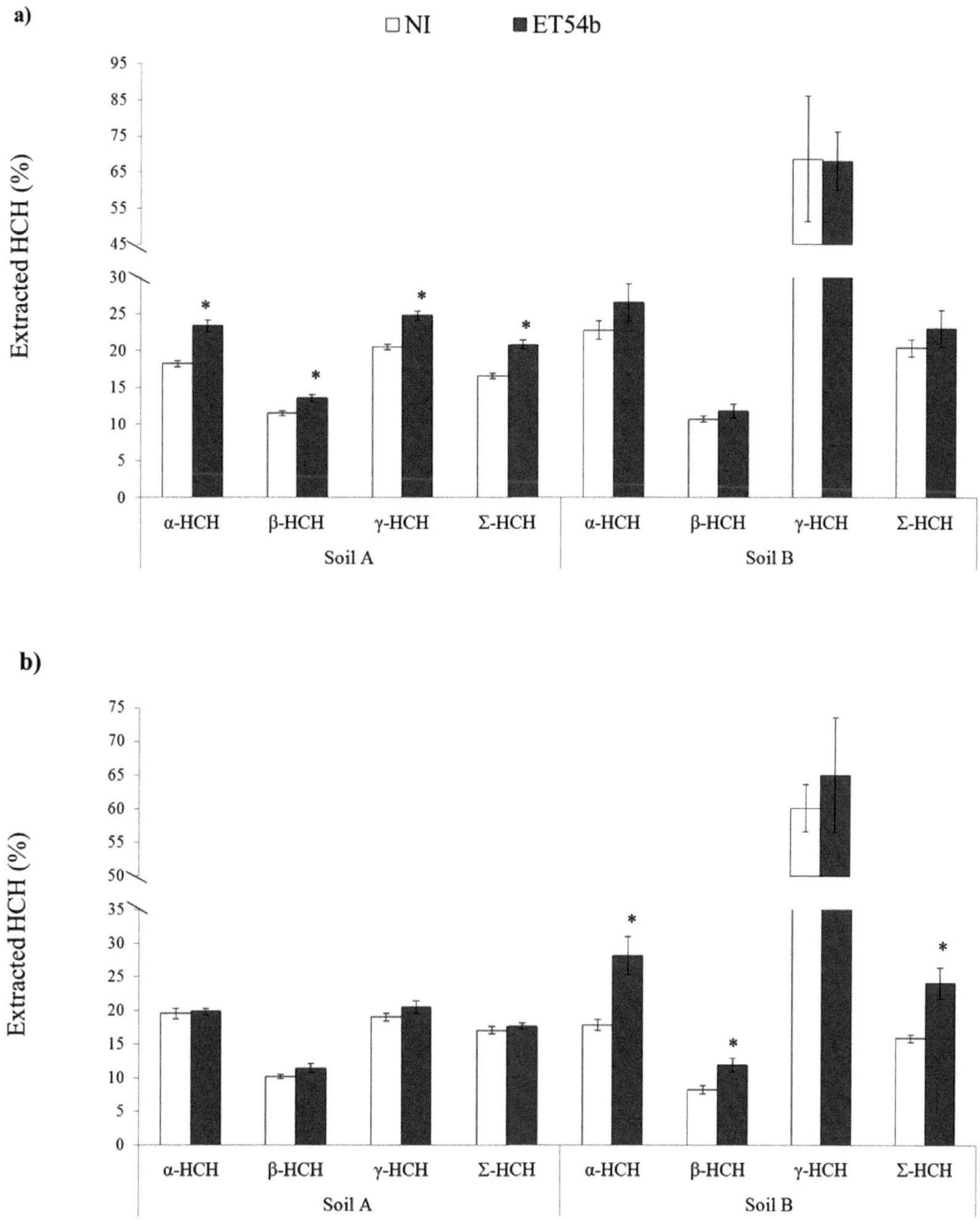

Figure 3. Percentage of α-, β-, γ- and total-HCH extracted from Soil A and B using cell-free culture filtrates of *Rhodococcus* sp. ET54b strain cultivated in two media including (a) glucose and $NaNO_3$, or (b) glycerol and sodium glutamate.

agents, cation exchange capacity, and biological activity) (Pérez-De-Mora et al. 2006, Kidd et al. 2015, Pardo et al. 2016a).

Common inorganic amendments used into phytostabilization options are phosphorus or phosphate compounds and derivatives (Boisson et al. 1999), inorganic fertilizers (Pardo et al. 2014d), clay minerals (Mench et al. 1998, Castaldi et al. 2005), byproducts and industrial wastes rich in Fe, Mn and Al oxyhydroxides (Hartley et al. 2004, Fresno et al. 2016), and alkalinizing agents such as calcium carbonate or calcium hydroxide (Pardo et al. 2011, 2014b). These amendments can successfully reduce metal(loid) solubility by promoting the formation of insoluble precipitates or by increasing soil binding capacity, thus reducing phytotoxicity and facilitating the revegetation of contaminated sites during aided phytostabilization (Fig. 1). The application of red muds (an alkaline and Fe/Al oxide-rich by-product of the aluminium industry) led to a significant decrease in soil Zn, Cd and Ni extractability in a 15-month pot study (Friesl-Hanl et al. 2006). Pardo et al. (2016b) reported that the addition of iron sulfate and iron nano-oxides significantly increased the formation of iron plaque on roots of *Phragmites australis*, which enhanced the As removal capacity of this plant and improved the rhizofiltration efficiency.

The addition of different types of organic byproducts or wastes as a source of organic matter is also a common practice in phytoremediation to improve soil physicochemical and biological properties, which simultaneously offers an environmentally-friendly alternative for the recycling of such wastes (Clemente et al. 2015, Pardo et al. 2017b). In addition to reducing TE bioavailability, the incorporation of effective amendments can restore soil quality by correcting pH, adding organic matter, increasing water holding capacity, re-establishing microbial communities, and alleviating compaction. Several studies have demonstrated that the application of wastes or byproducts of industrial and agri-food processes, such as those from the production of olive oil (de la Fuente et al. 2011, Pardo et al. 2017b), sawdust and rice husk (Singh and Prasad 2014), winery wastes (Mosse et al. 2012), sugar beet residues (Madejón et al. 2006), spent mushroom substrates (García-Delgado et al. 2013), or rapeseed residue (Lee et al. 2013), are able to reduce the toxicity of contaminated soils by reducing TE availability. Lindsay et al. (2011) found in a three-year experiment that spent brewing grain led to sulfate reduction and effective TE removal, meanwhile municipal biosolids and conifer-derived peat were ineffective. Lee et al. (2013) reported that the application of rapeseed residue and eggshell waste to a soil adjacent to a mine site in Korea successfully buffered soil acidity and immobilized the metals in soil. However, the use of fresh organic waste usually implies some risks as they can cause problems due to high nitrogen and phosphorus content, high electrical conductivity, the presence of metals and certain xenobiotics (like antibiotics or pesticides), or phytotoxic compounds like phenolic acids (Clemente et al. 2015, Pardo et al. 2017b). Organic wastes can also be applied after a composting process which generates neutral or alkaline substrates with high richness in organic matter and available essential nutrients (Alburquerque et al. 2006, Hachicha et al. 2009). Composted organic residues present a high proportion of humified organic matter which decreases metal mobility through metal binding to exchange sites, adsorption, and the formation of stable organo-metal complexes (Soler-Rovira et al. 2010). Many studies demonstrated the suitability of organic waste composts for enhancing plant establishment and improving plant growth and nutritional status, which has been related to the liming effect, the reduction of TE mobility and the long term supply of nutrients (Nogales and Benítez 2006, de la Fuente et al. 2011, Alvarenga et al. 2014, Curaqueo et al. 2014, Pardo et al. 2014a, Pardo et al. 2014b, Pardo et al. 2014c, Kohler et al. 2015, Touceda-González et al. 2017). Nevertheless, undesired mobilization of Cu, As or Tl in the soil (Burgos et al. 2010, Pardo et al. 2011, Clemente et al. 2012, Moreno-Jiménez et al. 2013, Beesley et al. 2014, Pardo et al. 2014a, Pardo et al. 2016a, Pardo et al. 2017a) and the leaching of nutrients (Pardo et al. 2014a) has also been reported.

Recently, much attention has been paid to the production and use of biochar from different organic wastes. Biochar is a C-rich organic material produced during slow exothermic decomposition of biomass at temperatures $\leq$ 700°C under low oxygen conditions (Lehman and Joseph 2009). The application of biochar usually enhances soil cation exchange capacity and increases soil pH (due to its alkaline nature and high surface area), binding the positively charged metal ions (due to its richness of negatively charged functional groups) and slowly releasing nutrients (due to their recalcitrant nature) (Dong et al. 2011, Olmo et al. 2014, Alburquerque et al. 2015, Hmid et al. 2015). However, its high sorption capacity can also lead to the immobilization of essential nutrients. Nevertheless, the effect of biochar addition depends on the characteristics of the soil and of the raw material of the biochar (Beesley et al. 2014).

Several studies have also focused on the combination of organic and inorganic amendments. Amendments rich in Fe, Mn and Al oxyhydroxides combined with compost, fertilizers, beringite, cyclonic ashes, or lime have successfully enhanced plant growth in TE-contaminated soils (Bes and Mench 2008, Vangronsveld et al. 2009, Fresno et al. 2016). The combination of iron grit with lime and compost was more effective in reducing Cu concentrations in soil pore water than individual amendments (Bes and Mench 2008). Application of cyclonic ashes (formerly known as beringite) which are rich in clay minerals induced a strong and long term (over 12 years) decrease in plant-available Pb, Zn, and Cd concentrations, and restored vegetation, in the Zn-smelter affected area (Vangronsveld et al. 2000a, Vangronsveld et al. 2000b). Zornoza et al. (2012) assessed the long-term effect of the combined addition of three levels of pig manure or sewage sludge with marble waste to a mine soil, and identified the treatment with pig slurry as the best approach for the reactivation of biogeochemical cycles and increased the plant establishment. Pardo et al. (2016a) reported that the application of a red mud derivate or calcium carbonate in combination with compost or digestate to two mine tailings successfully decrease TE solubility (95–99%) and change TE speciation as consequence of the increase of soil pH and soluble and total organic C. In addition, the amendments combination with compost improved plant growth and nutritional status of *Atriplex halimus* and reduced TE accumulation in shoots.

Finally, synthetic or natural organic ligands such as diethylenetriaminepentaacetic acid (DTPA), glycol-etherdiaminetetraacetic acid (EDGA), ethylenediaminetetraacetic acid (EDTA), ethylenediamine-N,N′-disuccinic acid (EDDS), nitrilotriacetic acid (NTA), low molecular weight organic acids (LMWOAs) and humic substances (HSs) (Shahid et al. 2012) or acidifying amendments like sulfur-containing materials (de la Fuente et al. 2008), have been used extensively for improving the efficiency of phytoextraction techniques (called *chelate-induced or aided phytoextraction*). These materials may desorb the metal(loid)s from soil matrix, increase their mobility and facilitate their uptake and accumulation in the harvestable plant tissues (Gómez-Sagasti et al. 2012). However, metal-chelate complexes are frequently toxic to plants and soil microorganisms, present low degradability, and pose a considerable environmental risk due to contaminants leaching.

Application of Plant Associated Microorganisms

Phytoremediation processes are governed by the interactions between three key players: soil, plants and microorganisms (Figs. 1 and 2). The last few years have seen a growing interest in the influence of microorganisms on plant growth and contaminant bioavailability and degradation, and the role of plant-associated microorganisms in improving phytoremediation success (Sessitsch et al. 2013, Afzal et al. 2014, Lenoir et al. 2016, Thijs et al. 2016, Benizri and Kidd 2017, Deng and Cao 2017, Feng et al. 2017, Kidd et al. 2017). Rhizosphere and endophytic organisms that have received much attention because of their beneficial effects on plant growth health and resistance to stress are the plant growth-promoting bacteria (PGPB), mycorrhizal and endophytic fungi (Mendes et al. 2013, Coninx et al. 2017). Plants can promote the abundance of beneficial microorganisms in the rhizosphere (Vessey 2003, Thijs et al. 2016). Such plant-microbial symbioses are mutualistic: the plant provides exudates and creates habitats for microorganisms and, in return, bacteria and fungi promote plant growth through various direct or indirect methods (Raaijmakers et al. 2002, Mendes et al. 2013). Beneficial plant-associated microorganisms have been isolated from a wide range of plant species, ranging from tree species to herbaceous crop plants (Glick 2010). Microorganisms can increase the availability of essential plant nutrients, such as nitrogen (N_2-fixing organisms), phosphorus (by solubilization or mineralization through the production of organic acids and/or phosphatases) or iron (by releasing Fe(III)-specific chelating agents or siderophores). Arbuscular mycorrhizal fungi form mutualistic associations with the roots of 80–90% of vascular plant species and may constitute up to 50% of the total soil microbial biomass (Rajtor and Piotrowska-Seget 2016). Mycorrhizal inoculation can also greatly aid plant establishment and growth in contaminated soils under the combined stresses of nutrient deficiency and chemical toxicity (Foulon et al. 2016b). Plant growth-promoting bacteria can also directly influence plant growth and physiology through the production of phytohormones (e.g., IAA or by reducing stress ethylene levels in plants through the production of the enzyme 1-aminocyclopropane-1-carboxylate deaminase). Finally, some bacteria, often referred to as biocontrol agents, can inhibit or reduce plant diseases indirectly by competing for nutrients and space (niche exclusion), producing antimicrobial compounds or through the induction of plant defence mechanisms (Compant et al. 2005, Lemanceau et al. 2007).

Natural TE-enriched areas or TE and organic-contaminated sites are not only a source of tolerant plant species but also of microorganisms. Numerous metal-tolerant and PGP microbial strains have been isolated from TE-polluted mine tailings and their use as inoculants in phytoextraction is documented (Batty 2005, Becerra-Castro et al. 2012, Pereira et al. 2015). Serpentine soils are a source of potentially beneficial microorganisms for Ni phytomining (Cabello-Conejo et al. 2014, Álvarez-López et al. 2016a, Durand et al. 2016). Hyperaccumulating plant species have also been shown to harbour a specific microbiome (Benizri and Kidd 2017) and several studies have suggested that rhizosphere microbiota play an important role in the accumulation of TE by these plants (Lebeau et al. 2008, Kidd et al. 2009, Glick 2010, Becerra-Castro et al. 2013a, Sessitsch et al. 2013, Glick 2014, Muehe et al. 2015).

The presence of organic contaminants leads to the selective-enrichment of resistant or degrading microorganisms and contaminated soils are generally noted for their biodegradation potential (Urbance et al. 2003). The majority of isolated bacteria from oak and ash trees, growing on a trichloroethylene (TCE)-contaminated site, showed tolerance to TCE and toluene, and TCE degradation capacity was observed in some rhizosphere strains (Weyens et al. 2009a). Siciliano et al. (2001) showed that plants growing in soils contaminated with petroleum hydrocarbons and nitroaromatics naturally recruited root endophytes with the necessary contaminant-degrading genes.

Exploiting these plant-microbial partnerships in phytomanagement is generally based on the capacity of the bacteria to (1) improve plant growth and survival in stress conditions of contaminated soils (plant growth promotion) and/or (2) degrade contaminants or alter contaminant bioavailability (Ryan et al. 2008, Weyens et al. 2009b, 2009c) (Figs. 1 and 2). For example, microbial-induced mobilization of TE and plant metal(loid) accumulation has been related to: the release of chelating agents (such as organic acids, phenolic compounds and siderophores), acidification or redox changes in the rhizosphere (Gadd 2010). In contrast, TE immobilization occurs through sorption to cell components or exopolymers, transport and intracellular sequestration, release of metal(loid) binding compounds or precipitation as insoluble organic or inorganic molecules.

At a bench level, examples of bacterial-induced plant growth promotion and metal accumulation in a phytoextraction context can be found in a wide array of plant species, including crop plants, hyperaccumulators and woody tree species (Kidd et al. 2017 and references therein). Moreover, this type of inoculant has been tested in various types of contaminated soils or natural metal-enriched soils (Kidd et al. 2017 and references therein). Most of these studies have found bacterial inoculants to be more successful in promoting plant growth and biomass production (hence increasing the metal yield and metal removal from the soil), rather than increasing the metal concentration in shoots (Sessitsch et al. 2013). The phytostabilization process can also be more effective after inoculating plants with PGPB, due to an enhanced plant metal tolerance, growth and survival (Petrisor et al. 2004, Grandlic et al. 2008, Dary et al. 2010).

A few studies have evaluated the effects of combining soil amendments together with microbial inoculants. The use of inoculants can potentially reduce the requirement for amendments and associated economic costs. Wang et al. (2013) found some indications that combining arbuscular mycorrhizal fungal inoculation with organic amendments enhanced the success of revegetation. The combination of urban organic waste compost and a native mycorrhizal fungus was effective at promoting the establishment of the legume shrub *Anthyllis cytisoides* under field conditions (Kohler et al. 2015). Álvarez-López et al. (2017) found that the combined effects of soil bacterial inoculants (strains *Massilia niastensis* P87, *Pseudomonas costantinii* P29, *Rhodococcus erythropolis* P30, *Tsukamurella strandjordii* P75), together with compost amendments, on biomass production of tobacco were more beneficial than when either of these methods was used on their own. Bacterial inoculants effectively improved plant growth, modified soil metal availability, and increased plant metal accumulation and yield.

Microbe-assisted phytoremediation is particularly effective for organic pollutants. Numerous studies demonstrate significant enhanced dissipation and/or mineralization of organic pollutants at the root-soil interface or rhizosphere (Azaizeh et al. 2011, Lenoir et al. 2016). This plant-induced effect is generally attributed to stimulation in microbial density, diversity and/or activity in the rhizosphere, or the selective enrichment of degrading microorganisms (Anderson et al. 1993, Chaudhry et al. 2005, Kidd et al. 2008, Azaizeh et al. 2011). Due to the release of plant rhizodeposits, bacterial metabolic biodiversity in the rhizosphere can be greater than in non-vegetated soil and can additionally stimulate contaminant degradation through co-metabolism. Plant uptake and mineralization of xenobiotics, such as PCBs, is limited due to the recalcitrance and low bioavailability of these molecules, thus making rhizoremediation the most suitable strategy for the remediation of soils contaminated with organic hydrophobic compounds. The extraordinary natural metabolic diversity presented by many microorganisms is a characteristic that can be exploited in phytoremediation methods. Microorganisms promote plant growth under the prevailing stress conditions, while root exudates promote the responsible pathway for microbial oxidative metabolism of xenobiotics, thereby improving the overall degradation performance (Fester et al. 2014, Vergani et al. 2017). Plant growth limitations can be overcome by inoculating plants with strains of PGPB that can increase their tolerance to contaminants and improve growth (de-Bashan et al. 2012 and references herein). Microbial-assisted phytoremediation of such recalcitrant organic compounds is influenced by several factors, including the chemical structure and properties of the pollutant, the compatibility of microbial symbionts with their host plants, soil water content and nutrient status, and survival and proliferation of the inoculated microbial strains.

Field Case Studies of Phytomanagement

Long-term field experiments monitoring the efficiency of phytomanagement, and particularly of TE contaminated sites, are limited. However, studies under field conditions are necessary if these processes are to reach full-scale deployment.

Phytostabilization has been proposed as a suitable technique to decrease the environmental risks associated with metal(loid)-enriched mine tailings (Clemente et al. 2012, Conesa et al. 2012, Pardo et al. 2014c, 2014d). This type of phytomanagement has been shown to reduce TE mobility by altering speciation, as well as to improve soil physicochemical properties and fertility (Clemente et al. 2012, Zornoza et al. 2012, Pardo et al. 2014d, 2016a, 2017a). Moreover, it can lead to an increase in microbial diversity and restore functionality of TE-contaminated sites in the long term (Kumpiene et al. 2009, Pardo et al. 2014c, Xue et al. 2015).

Several field-based trials in TE-contaminated soils have shown the benefits of organic-based amendments for recovery of soil biological fertility. Zanuzzi et al. (2009) amended acidic mine tailings with pig manure or sewage sludge in combination with blanket application of marble wastes. Two years after soil amendment, soil pH was increased (from pH 2.7 to 7.4) alongside a build-up in total organic carbon, permitting the establishment of a vegetation cover. After five years, microbial biomass and soil enzymatic activities were higher in the amended soils than in the untreated mine tailings (Zanuzzi et al. 2009, Zornoza et al. 2012). Touceda-González et al. (2017) amended highly acidic Cu mine tailings with composted municipal sewage wastes and established both, an SRC system with *Populus nigra* and *Salix viminalis*, and a grassy cover (*Agrostis capillaris* cv. Highland). Compost amendment improved soil properties such as pH, cation exchange capacity (CEC) and fertility, and decreased soil Cu availability, permitting the development of healthy vegetation cover of both woody plants and grassy species. Aided phytostabilization stimulated microbial activity and led to the establishment of vital biogeochemical cycles (reflected through soil enzymatic activities). Both compost-amendment and plant root activity induced important shifts in the bacterial community structure over time. The beneficial effects were maintained at least three years after treatment. Pardo et al. (2014d) successfully used olive-mill waste compost as a soil amendment to promote the growth of a native legume (*Bituminaria bituminosa*) in a mine-affected soil from a semi-arid area (Southeast Spain) contaminated with trace elements (As, Cd, Cu, Mn, Pb and Zn). Compost addition increased the total soil organic C content and maintained elevated available P concentrations throughout two years. In addition, the treatment applied also increased the spontaneous growth of native species and percentage ground cover. Clemente et al. (2012) and Pardo et al. (2014c) found that the addition of olive mill waste compost and pig slurry to a mine soil increased soil pH and reduced TE availability, improved the growth of *Atriplex halimus* allowing the development of a sustainable vegetation cover, stimulated soil microorganisms (increasing microbial biomass, activity and functional diversity, and reducing stress) two and a half years after their application. Increases in the water soluble organic C, total N, and available-K in the soil were crucial for such improvements. Both amendments, but especially compost, reduced the risk of contamination of surface water and groundwater (decrease of toxicity for *Thamnocephalus platyurus* and *Vibrio fischeri*), while also decreasing the direct toxicity for plants (*Lactuca sativa, Lepidium sativum,* and *Zea mays*) in the long-term. Clemente et al. (2007) reported initial low values of microbial biomass after the addition of solid olive mill waste to a soil indirectly affected by mine activity, but they significantly increased with time, while values of microbial stress declined. This positive evolution of microbial activity was associated with the gradual degradation of toxic compounds supplied by the amendment (such as polyphenols) and the gradual re-oxidation of Mn (II) to Mn (IV) in the soil.

Successful phytomining trials have also been reported in Europe, for example, Bani et al. (2015) showed that native populations of *Alyssum murale* in Albania could phytoextract an economically-viable crop of Ni after incorporating inorganic fertilization. These authors achieved a biomass production of 9 t ha^{-1} with a Ni concentration of 11.5 g kg^{-1} which resulted in a total amount of phytoextracted Ni of 105 kg Ni ha^{-1} per year. The estimated costs of the phytomining cycle were around \$1000 ha^{-1} yr^{-1} and, based on the price of Ni at the time of the study the net return was calculated as \$1055 ha^{-1} yr^{-1}. These calculations were based on prices of gross Ni metal, however, other Ni products (such as Ni salts like $Ni(NH_4)_2(SO_4)_2.6H_2O$) with a high purity and higher value and which can be easily obtained from the biomass of *A. murale*, would substantially increase the profit of the crop (Barbaroux et al. 2012). Other Ni-hyperaccumulating genera such as *Leptoplax* or *Bornmuellera* have also been suggested as potential candidates for Ni recovery (Li et al. 2003, Zhang et al. 2014). A field experiment evaluating the use of *Noccaea caerulescens* (Viviez population from South France) to phytoextract Cd and Zn from an agricultural soil amended with TE rich-sewage sludge addition found two successive crops simultaneously extracted about 9% of the total Cd and 7% of the total Zn (Schwartz et al. 2003). Similarly, the same hyperaccumulator extracted up to 5% of total Cd and 11% of total Zn from an industrial soil affected by a former coking plant, and the available Zn fraction (NH_4NO_3-extractable) was reduced by up to 70%. The As-hyperaccumulator *Pteris vittata* decreased total As concentration in contaminated paddy soils by up to 11.4% after nine months of growth, while As pore water concentrations decreased up to 77% (Ye et al. 2011). Recently, van der Ent et al. (2013) suggested the potential for phytomining of rare elements, such as Au or rare earth elements such as Ce or La.

High yielding crops are considered viable alternatives to hyperaccumulating plant species for phytoextraction purposes, particularly in approaches targeting the bioavailable metal fraction in the soil. Phytoextraction with somaclonal variants of tobacco and sunflower mutant lines with enhanced metal uptake and tolerance were shown to be a sustainable alternative to conventional destructive decontamination methods (Herzig et al. 2014). After five years of a sunflower-tobacco rotation scheme, soil bioavailable ($NaNO_3$-extractable) Zn concentrations were significantly reduced by up to 70% in top soils of an agricultural soil polluted with deposits of a former hot-dip Zn factory at Bettwiesen in eastern Switzerland (Herzig et al. 2014). Sunflower-tobacco rotations and tobacco–sunflower-vetiver rotation were also successfully implemented in Cu-contaminated soils at a former wood preservation site (Kolbas 2012). Janssen et al. (2015) carried out a field experiment in a former agricultural area diffusely contaminated with Pb, Cd, and Zn due to smelter activities to select the best-performing clones of poplar or *Salix* for phytoextraction purposes. After three years of experiment, two clones were highlighted for high Cd/Zn extraction potential, uniformity of growth and low mortality rate: a *Salix viminalis* clone and a *Salix alba* x *alba* clone. These studies also revealed that harvesting the trees with leaves could double the amount of Cd phytoextracted in comparison with the yield obtained harvesting only twigs (Vangronsveld et al. 2009). At the same site, increased soil microbial biomass and respiration, functional gene richness and diversity, and soil enzymatic activities were observed in plots planted with the willow clone "Tora" (Xue et al. 2015). In field trials evaluating the Cd/Zn extraction potential of *Salix smithiana* BOKU 03 CZ-001 from mine spoils, Álvarez-López et al. found similar increases in soil enzymatic activities under *Salix* cover compared to unplanted mine-soil (unpublished results). Kubátová et al. (2016) evaluated short rotation coppice plantations using two *Salix* clones and two poplar clones for phytoextraction efficiency on moderately Cd-, Pb- and Zn-contaminated agricultural soil. Sewage sludge applications were found to be a viable source of nutrients for *Salix* clones, however, during the first year the competition with weeds negatively affected the plantation. The best-developing plantations showed suitability for the phytoextraction of Cd and partly for Zn. Maxted et al. (2007) also performed field experiments with *Salix* clones in sewage sludge-contaminated soils. The best clones were found to remove 15–20% of the available soil Cd concentration within four years. In agreement, Courchesne et al. (2017) found willows which could reduce soil Cd and Zn in moderately contaminated soils within a decadal timeframe. These authors emphasized the importance of monitoring the soil TE concentrations during phytoextraction as opposed to plant metal yields. Changes in labile soil TE are quite inconsistent across different studies. In some cases, labile TE concentrations decreased during phytoextraction, while in other cases concentrations remained constant or even increased (Hartley et al. 2011, Dimitriou et al. 2012, Greger and Landberg 2015). The decomposition of willow leaves can also have an impact on soil TE mobility by increasing dissolved organic carbon and reducing soil solution pH (Hartley et al. 2011). On the whole, the available field studies indicate that phytoextraction based on SRC systems with high-accumulating clones can be effective in soils with low to moderate TE contamination, although the time required will vary according to TE type, initial TE concentration, decontamination target, soil properties, and plant behavior (Courchesne et al. 2017).

Despite arguments in favor of multi-species or multi-cultivar plant stands (Batty and Dolan 2013, Kidd et al. 2015), there is a lack of field trials assessing the potential benefits of such intercropping within phytoextraction systems. In a field trial, Álvarez-López et al. (unpublished results) monitored the effects of intercropping *Salix smithiana* BOKU 03 CZ-001 with *Alnus glutinosa* in mine-spoils on soil physico and biochemical properties, plant growth, nutritive status, and Cd/Zn accumulation. Alder is an important pioneer species and is recognized as a soil improving plant, through its symbiosis with N_2 fixing microorganisms (e.g., *Frankia*), ectomycorrhizae and arbuscular mycorrhizal fungi (Kidd et al. 2015). Although after three years no beneficial effects of intercropping were observed on tree height or nutrition, both stem concentrations and the bioconcentration factor of Cd and Zn were higher when *S. smithiana* was intercropped with *A. glutinosa* (unpublished results). At the same site, the hyperaccumulator *Noccaea caerulescens* was intercropped with the legume *Lotus conriculatus*. *Lotus* spp. are recognized for their tolerance to metal contamination and have been recommended for cultivation in degraded or marginal soils (Escaray et al. 2012). Intercropping with the legume had a beneficial effect on shoot nutrient concentrations and on shoot Cd and Zn concentrations of the hyperaccumulator (unpublished results; Fig. 4). Cd and Zn yields were also higher (albeit not significantly) when the hyperaccumulator was cultivated together with the legume (Fig. 4). The phytoextraction process led to significant reductions in the soil labile Zn (but not Cd) fraction over the 18-month experiment (Fig. 4). Soil enzymatic activities were also higher in soils managed by phytoextraction than in unplanted soils (unpublished results).

Soil contamination significantly impacts both bacterial and microbial community structure, with potential negative impacts on key ecological functions such as soil organic matter decomposition and nutrient mineralization (Bell et al. 2014, Azarbad et al. 2015, Pagé et al. 2015, Zappelini et al. 2015). The recovery of these key soil processes and soil microbial community functionality is a crucial step in the phytomanagement process,

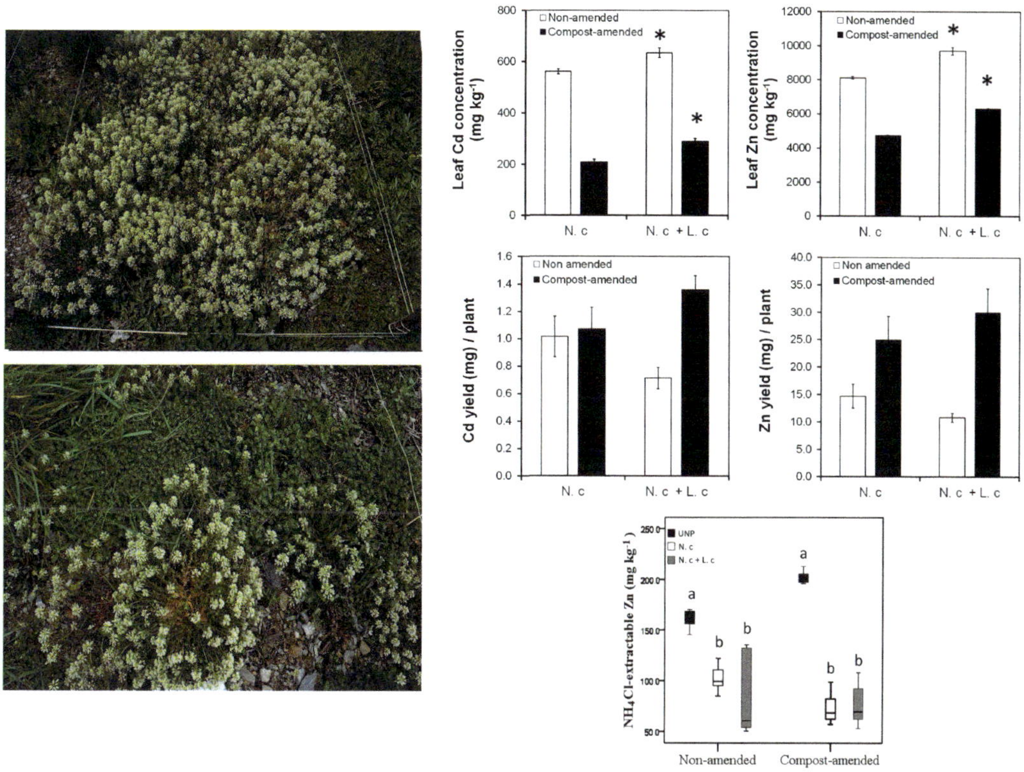

Figure 4. Results of a field experiment carried out in mine tailings to evaluate the effects of plant cropping pattern and organic amendments on Cd and Zn phytoextraction by *Noccaea caerulescens*. The top photo shows *N. caerulescens* (N. c) in monoculture and the bottom photo shows the hyperaccumulator intercropped with *Lotus corniculatus* (L. c). The graphs present shoot Cd and Zn concentrations/yields and soil NH_4Cl-extractable Zn concentrations.

governing its long-term effectiveness and sustainability. A five-ha field demonstration trial was established in 2007 at Pierrelaye (France) with 14 poplar cultivars grown under either SRC management (planting density of 1000 trees ha^{-1}) or VSRC (10000 trees ha^{-1}) (Foulon et al. 2016a). The site is an example of polymetallic contamination (principally Zn, Cd, Pb, and Cu) as a result of wastewater irrigation for more than 100 years (Pottier et al. 2015). Seven years after plantation, significant reductions in $CaCl_2$-extractable TE concentrations (particularly Cd and Zn) under poplar-planted soils were observed compared to unplanted soils or initial values before implementing phytomanagement. Alongside this reduction in metal bioavailability and phytotoxicity was an improvement in soil fertility (in terms of K, Mg, N and P) which was not observed in control soils where no phytomanagement regime was used (Foulon et al. 2016a). Moreover, these changes in soil physicochemical properties were accompanied by shifts in microbial community composition. Higher bacterial diversity and richness indices were observed in poplar soils compared to unplanted soils. Poplars were associated with an increased proportion of symbiotic/endophytic fungi in the rhizosphere (Foulon et al. 2016a). Plant-induced stimulations in bacterial diversity were also observed in soil under a *Miscanthus* plantation at the Pierrelaye site (Bourgeois et al. 2015) a crop which is increasingly being adopted for bioenergy purposes in Europe and North America (Zegada-Lizarazu and Monti 2011).

Concluding Remarks

Several phytotechnologies have been developed for the treatment of soils contaminated with organic compounds and/or trace elements, or for a sustainable exploitation of metalliferous soils inadequate for conventional agroforestry. These plant-based techniques can allow for a sustainable and profitable management of such sites, within the concept of phytomanagement. A careful consideration of implicated factors affecting the success of these technologies and the adequate optimization of their implementation is of vital importance before they can be fully developed. When choosing the most appropriate technique, special attention should be given to the nature and bioavailability of the contaminants, additional edaphic or biotic limitations, as well as, defining the objectives of the treatment in terms of the final land-use and aimed endpoint contaminant concentration(s). Native species naturally growing at the targeted contaminated sites or endemic to metalliferous soils frequently show optimal survival and performance and avoid the risks associated with the introduction of alien plants. The application of soil amendments, fertilizers, and optimal cropping patterns can ameliorate adverse soil properties and permit successful long-term plant establishment and growth. The role of plant-associated microorganisms, such as those inhabiting the rhizosphere, has been longer recognized for the remediation of organic contaminants, particularly in the case of hydrophobic compounds. More recently new strategies based on the application of plant-associated microorganisms with growth promoting properties (e.g., mycorrhiza and plant growth promoting bacteria) or the ability to modify contaminant bioavailability have been incorporated into phytotechnologies targeting both TE-enriched soils and soils contaminated with organics. A growing number of phytoremediation-orientated field based trials can be found in the literature, but field assessments of the potential benefits of novel strategies able to improve phytomanagement success, such as the use of plant-associated microbial inoculants, is still required.

Acknowledgements

This review is in memory of our colleague and friend, Cristina Becerra Castro, who was taken from us too soon and will be dearly missed. We had the fortune to share several years together and will always remember her enthusiasm and generosity. We acknowledge the financial support of the Spanish Ministerio de Economía y Competitividad (CTM2015-66439-R) and FEDER, as well as the Interreg Sudoe Programme [PhytoSUDOE Project (SOE1/P5/E0189)].

References Cited

Abhilash, P., S. Jamil, V. Singh, A. Singh, N. Singh and S. Srivastava. 2008. Occurrence and distribution of hexachlorocyclohexane isomers in vegetation samples from a contaminated area. Chemosphere 72: 79–86.

Abhilash, P.C. and N. Singh. 2010. *Withania somnifera* dunal-mediated dissipation of lindane from simulated soil: implications for rhizoremediation of contaminated soil. J. Soils Sed. 10: 272–282.

Adriano, D.C. 2001. Trace elements in terrestrial environments: Biogeochemistry, bioavailability, and risks of metals. Springer, New York.

Afzal, M., Q.M. Khan and A. Sessitsch. 2014. Endophytic bacteria: Prospects and applications for the phytoremediation of organic pollutants. Chemosphere 117: 232–242.

Alburquerque, J.A., J. Gonzálvez, D. García and J. Cegarra. 2006. Composting of a solid olive-mill by-product ("alperujo") and the potential of the resulting compost for cultivating pepper under commercial conditions. Waste Manage. 26: 620–626.

Alburquerque, J.A., M. Cabello, R. Avelino, V. Barrón, M.C. del Campillo and J. Torrent. 2015. Plant growth responses to biochar amendment of Mediterranean soils deficient in iron and phosphorus. J. Plant Nutr. Soil Sci. 178: 567–575.

Alvarenga, P., A. de Varennes and A.C. Cunha-Queda. 2014. The effect of compost treatments and a plant cover with *Agrostis tenuis* on the immobilization/mobilization of trace elements in a mine-contaminated soil. Int. J. Phytoremediation 16: 138–154.

Álvarez-López, V., Á. Prieto-Fernández, C. Becerra-Castro, C. Monterroso and P.S. Kidd. 2016a. Rhizobacterial communities associated with the flora of three serpentine outcrops of the Iberian Peninsula. Plant Soil 403: 233–252.

Álvarez-López, V., Á. Prieto-Fernández, S. Roiloa, B. Rodríguez-Garrido, R. Herzig, M. Puschenreiter et al. 2017. Evaluating phytoextraction efficiency of two high-biomass crops after soil amendment and inoculation with rhizobacterial strains. Environ. Sci. Pollut. Res. Int. 24: 7591–7606.

Anderson, T.A., E.A. Guthrie and B.T. Walton. 1993. Bioremediation in the rhizosphere. Environ. Sci. Technol. 27: 2630–2636.

Azaizeh, H., P.M.L. Castro and P. Kidd. 2011. Biodegradation of organic xenobiotic pollutants in the rhizosphere. pp. 191–215. *In*: Schröder, P. and C.D. Collins [eds.]. Organic Xenobiotics and Plants: From Mode of Action to Ecophysiology. Springer, Netherlands, Dordrecht.

Azarbad, H., M. Niklińska, R. Laskowski, N.M. van Straalen, C.A.M. van Gestel, J. Zhou et al. 2015. Microbial community composition and functions are resilient to metal pollution along two forest soil gradients. FEMS Microbiol. Ecol. 91: 1–11.

Bani, A., G. Echevarria, S. Sulçe and J.L. Morel. 2015. Improving the agronomy of *Alyssum murale* for extensive phytomining: a five-year field study. Int. J. Phytoremediation 17: 117–127.

Batty, L.C. 2005. The potential importance of mine sites for biodiversity. Mine Water Environ. 24: 101–103.

Batty, L.C. and C. Dolan. 2013. The potential use of phytoremediation for sites with mixed organic and inorganic contamination. Crit. Rev. Environ. Sci. Technol. 43: 217–259.

Becerra-Castro, C., P.S. Kidd, Á. Prieto-Fernández, N. Weyens, M.J. Acea and J. Vangronsveld. 2011. Endophytic and rhizoplane bacteria associated with *Cytisus striatus* growing on hexachlorocyclohexane-contaminated soil: Isolation and characterisation. Plant Soil 340: 413–433.

Becerra-Castro, C., C. Monterroso, Á. Prieto-Fernández, L. Rodríguez-Lamas, M. Loureiro-Viñas, M.J. Acea et al. 2012. Pseudometallophytes colonising Pb/Zn mine tailings: a description of the plant-microorganism-rhizosphere soil system and isolation of metal-tolerant bacteria. J. Hazard. Mater. 217-218: 350–359.

Becerra-Castro, C., P. Kidd, M. Kuffner, A. Prieto-Fernández, S. Hann, C. Monterroso et al. 2013a. Bacterially induced weathering of ultramafic rock and its implications for phytoextraction. Appl. Environ. Microbiol. 79: 5094–5103.

Becerra-Castro, C., P.S. Kidd, B. Rodríguez-Garrido, C. Monterroso, P. Santos-Ucha and Á. Prieto-Fernández. 2013b. Phytoremediation of hexachlorocyclohexane (HCH)-contaminated soils using *Cytisus striatus* and bacterial inoculants in soils with distinct organic matter content. Environ. Pollut. 178: 202–210.

Becerra-Castro, C., Á. Prieto-Fernández, P.S. Kidd, N. Weyens, B. Rodríguez-Garrido, M. Touceda-González et al. 2013c. Improving performance of *Cytisus striatus* on substrates contaminated with hexachlorocyclohexane (HCH) isomers using bacterial inoculants: Developing a phytoremediation strategy. Plant Soil 362: 247–260.

Beesley, L., O.S. Inneh, G.J. Norton, E. Moreno-Jimenez, T. Pardo, R. Clemente et al. 2014. Assessing the influence of compost and biochar amendments on the mobility and toxicity of metals and arsenic in a naturally contaminated mine soil. Environ. Pollut. 186: 195–202.

Bell, T.H., S. El-Din Hassan, A. Lauron-Moreau, F. Al-Otaibi, M. Hijri, E. Yergeau et al. 2014. Linkage between bacterial and fungal rhizosphere communities in hydrocarbon-contaminated soils is related to plant phylogeny. ISME J. 8: 331–343.

Benizri, E. and P.S. Kidd. 2017. The role of the rhizosphere and microbes associated with hyperaccumulator plants in metal accumulation. *In*: Echevarria, G., A. van der Ent, J.L. Morel and A.J.M. Baker [eds.]. Agromining: Farming for Metals. Extracting Unconventional Resources using Plants. Springer Nature (in press).

Bernal, M.P., R. Clemente and D.J. Walker. 2007. The role of organic amendments in the bioremediation of heavy metal-polluted soils. pp. 1–57. *In*: Gore, R.B. [ed.]. Environmental Research at the Leading Edge. Nova, New York.

Bes, C. and M. Mench. 2008. Remediation of copper-contaminated topsoils from a wood treatment facility using *in situ* stabilisation. Environ. Pollut. 156: 1128–1138.

Boisson, J., A. Ruttens, M. Mench and J. Vangronsveld. 1999. Evaluation of hydroxyapatite as a metal immobilizing soil additive for the remediation of polluted soils. Part 1. Influence of hydroxyapatite on metal exchangeability in soil, plant growth and plant metal accumulation. Environ. Pollut. 104: 225–233.

Bolan, N.S., D.C. Adriano, A. Kunhikrishnan, T. James, R. McDowell and N. Senesi. 2011. Dissolved organic matter: biogeochemistry, dynamics, and environmental significance in soils. vol 110.

Bonanomi, G., G. Incerti and S. Mazzoleni. 2011. Assessing occurrence, specificity, and mechanisms of plant facilitation in terrestrial ecosystems. Plant Ecol. 212: 1777.

Bourgeois, E., S. Dequiedt, M. Lelièvre, F. van Oort, I. Lamy, P.-A. Maron et al. 2015. Positive effect of the *Miscanthus* bioenergy crop on microbial diversity in wastewater-contaminated soil. Environ. Chem. Letters 13: 495–501.

Burgos, P., P. Madejón, F. Cabrera and E. Madejón. 2010. By-products as amendment to improve biochemical properties of trace element contaminated soils: effects in time. Int. Biodeterior. Biodegrad. 64: 481–488.

Cabello-Conejo, M.I., C. Becerra-Castro, Á. Prieto-Fernández, C. Monterroso, A. Saavedra-Ferro, M. Mench et al. 2014. Rhizobacterial inoculants can improve nickel phytoextraction by the hyperaccumulator *Alyssum pintodasilvae*. Plant Soil 379: 35–50.

Carrillo-Garcia, Á., J.L.L. De La Luz, Y. Bashan and G.J. Bethlenfalvay. 1999. Nurse plants, mycorrhizae, and plant establishment in a disturbed area of the Sonoran Desert. Restor. Ecol. 7: 321–335.

Castaldi, P., L. Santona and P. Melis. 2005. Heavy metal immobilization by chemical amendments in a polluted soil and influence on white lupin growth. Chemosphere 60: 365–371.

Clemente, R., C. de la Fuente, R. Moral and M.P. Bernal. 2007. Changes in microbial biomass parameters of a heavy metal-contaminated calcareous soil during a field remediation experiment. J. Environ. Qual. 36: 1137–1144.

Clemente, R., D.J. Walker, T. Pardo, D. Martínez-Fernández and M.P. Bernal. 2012. The use of a halophytic plant species and organic amendments for the remediation of a trace elements-contaminated soil under semi-arid conditions. J. Hazard. Mater. 223–224: 63–71.

Clemente, R., T. Pardo, P. Madejón, E. Madejón and M.P. Bernal. 2015. Food byproducts as amendments in trace elements contaminated soils. Food Res. Int. 73: 176–189.

Compant, S., B. Duffy, J. Nowak, C. Clément and E.A. Barka. 2005. Use of plant growth-promoting bacteria for biocontrol of plant diseases: principles, mechanisms of action, and future prospects. Appl. Environ. Microbiol. 71: 4951–4959.

Conesa, H.M., A.B. Moradi, B.H. Robinson, G. Kühne, E. Lehmann and R. Schulin. 2009. Response of native grasses and *Cicer arietinum* to soil polluted with mining wastes: Implications for the management of land adjacent to mine sites. Environ. Exp. Bot. 65: 198–204.

Conesa, H.M., M.W.H. Evangelou, B.H. Robinson and R. Schulin. 2012. A critical view of current state of phytotechnologies to remediate soils: Still a promising tool? Scientific World Journal. 2012: 10.

Coninx, L., V. Martinova and F. Rineau. 2017. Mycorrhiza-assisted phytoremediation. pp. 127–188. *In*: Cuypers, A. and J. Vangronsveld [eds.]. Phytoremediation. Advances in Botanical Research. Academic Press, USA.

Courchesne, F., M.-C. Turmel, B. Cloutier-Hurteau, S. Constantineau, L. Munro and M. Labrecque. 2017. Phytoextraction of soil trace elements by willow during a phytoremediation trial in Southern Québec, Canada. Int. J. Phytoremediation 19: 545–554.

Curaqueo, G., M. Schoebitz, F. Borie, F. Caravaca and A. Roldán. 2014. Inoculation with arbuscular mycorrhizal fungi and addition of composted olive-mill waste enhance plant establishment and soil properties in the regeneration of a heavy metal-polluted environment. Environ. Sci. Pollut. Res. 21: 7403–7412.

Chalot, M., D. Blaudez, Y. Rogaume, A.-S. Provent and C. Pascual. 2012. Fate of trace elements during the combustion of phytoremediation wood. Environ. Sci. Technol. 46: 13361–13369.

Chaney, R.L., J.S. Angle, C.L. Broadhurst, C.A. Peters, R.V. Tappero and D.L. Sparks. 2007. Improved understanding of hyperaccumulation yields commercial phytoextraction and phytomining technologies. J. Environ. Qual. 36: 1429–1443.

Chaney, R., R. Reeves, I. Baklanov, T. Centofanti, L. Broadhurst, A. Baker et al. 2014. Phytoremediation and phytomining: Using plants to remediate contaminated or mineralized environments. pp. 365–391. *In*: Rajakaruna, N., Boyd, R.S. and Harris, T.B. [eds.]. Plant Ecology and Evolution in Harsh Environments. Nova Science Publishers, New York.

Chaudhry, Q., M. Blom-Zandstra, S.K. Gupta and E. Joner. 2005. Utilising the synergy between plants and rhizosphere microorganisms to enhance breakdown of organic pollutants in the environment. Environ. Sci. Pollut. Res. 12: 34–48.

Chhikara, S., B. Paulose, J.C. White and O.P. Dhankher. 2010. Understanding the physiological and molecular mechanism of persistent organic pollutant uptake and detoxification in cucurbit species (Zucchini and Squash). Environ. Sci. Technol. 44: 7295–7301.

Dary, M., M.A. Chamber-Pérez, A.J. Palomares and E. Pajuelo. 2010. "*In situ*" phytostabilisation of heavy metal polluted soils using *Lupinus luteus* inoculated with metal resistant plant-growth promoting rhizobacteria. J. Hazard. Mater. 177: 323–330.

de-Bashan, L.E., J.-P. Hernandez and Y. Bashan. 2012. The potential contribution of plant growth-promoting bacteria to reduce environmental degradation—A comprehensive evaluation. Appl. Soil Ecol. 61: 171–189.

de la Fuente, C., R. Clemente and M.P. Bernal. 2008. Changes in metal speciation and pH in olive processing waste and sulphur-treated contaminated soil. Ecotoxicol. Environ. Saf. 70: 207–215.

de la Fuente, C., R. Clemente, I. Martínez-Alcalá, G. Tortosa and M.P. Bernal. 2011. Impact of fresh and composted solid olive husk and their water-soluble fractions on soil heavy metal fractionation; microbial biomass and plant uptake. J. Hazard. Mater. 186: 1283–1289.

Del Río, M., R. Font, C. Almela, D. Vélez, R. Montoro and A. De Haro Bailón. 2002. Heavy metals and arsenic uptake by wild vegetation in the Guadiamar river area after the toxic spill of the Aznalcóllar mine. J. Biotechnol. 98: 125–137.

Deng, Z. and L. Cao. 2017. Fungal endophytes and their interactions with plants in phytoremediation: A review. Chemosphere 168: 1100–1106.

Dimitriou, I., B. Mola-Yudego, P. Aronsson and J. Eriksson. 2012. Changes in organic carbon and trace elements in the soil of willow short-rotation coppice plantations. BioEnergy Res. 5: 563–572.

Dong, X., L.Q. Ma and Y. Li. 2011. Characteristics and mechanisms of hexavalent chromium removal by biochar from sugar beet tailing. J. Hazard. Mater. 190: 909–915.

Donnelly, P.K., R.S. Hegde and J.S. Fletcher. 1994. Growth of PCB-degrading bacteria on compounds from photosynthetic plants. Chemosphere 28: 981–988.

Drezner, T.D. 2006. Plant facilitation in extreme environments: The non-random distribution of saguaro cacti (*Carnegiea gigantea*) under their nurse associates and the relationship to nurse architecture. J. Arid Environ. 65: 46–61.

Dudka, S. and D.C. Adriano. 1997. Environmental impacts of metal ore mining and processing: A review. J. Environ. Qual. 26: 590–602.

Durand, A., S. Piutti, M. Rue, J.L. Morel, G. Echevarria and E. Benizri. 2016. Improving nickel phytoextraction by co-cropping hyperaccumulator plants inoculated by plant growth promoting rhizobacteria. Plant Soil 399: 179–192.

Escaray, F.J., A.B. Menendez, A. Gárriz, F.L. Pieckenstain, M.J. Estrella, L.N. Castagno et al. 2012. Ecological and agronomic importance of the plant genus *Lotus*. Its application in grassland sustainability and the amelioration of constrained and contaminated soils. Plant Sci. 182: 121–133.

Facknath, S. and B. Lalljee. 2000. Allelopathic strategies for eco-friendly crop protection. pp. 33–46. *In*: Narwal, S.S., R.E. Hoagland, R.H. Dilday and M.J. Reigosa [eds.]. Allelopathy in Ecological Agriculture and Forestry. Springer Netherlands, Dordrecht.

Feng, N.-X., J. Yu, H.-M. Zhao, Y.-T. Cheng, C.-H. Mo, Q.-Y. Cai et al. 2017. Efficient phytoremediation of organic contaminants in soils using plant–endophyte partnerships. Sci. Total Environ. 583: 352–368.

Fester, T., J. Giebler, L.Y. Wick, D. Schlosser and M. Kästner. 2014. Plant–microbe interactions as drivers of ecosystem functions relevant for the biodegradation of organic contaminants. Curr. Opin. Biotechnol. 27: 168–175.

Foulon, J., C. Zappelini, A. Durand, B. Valot, D. Blaudez and M. Chalot. 2016a. Impact of poplar-based phytomanagement on soil properties and microbial communities in a metal-contaminated site. FEMS Microbiol. Ecol. 92: fiw163.

Foulon, J., C. Zappelini, A. Durand, B. Valot, O. Girardclos, D. Blaudez et al. 2016b. Environmental metabarcoding reveals contrasting microbial communities at two poplar phytomanagement sites. Sci. Total Environ. 571: 1230–1240.

Fresno, T., E. Moreno-Jiménez and J.M. Peñalosa. 2016. Assessing the combination of iron sulfate and organic materials as amendment for an arsenic and copper contaminated soil. A chemical and ecotoxicological approach. Chemosphere 165: 539–546.

Friesl-Hanl, W., J. Friedl, K. Platzer, O. Horak and M.H. Gerzabek. 2006. Remediation of contaminated agricultural soils near a former Pb/Zn smelter in Austria: batch, pot and field experiments. Environ. Pollut. 144: 40–50.

Gadd, G.M. 2004. Microbial influence on metal mobility and application for bioremediation. Geoderma 122: 109–119.

Gadd, G.M. 2010. Metals, minerals and microbes: geomicrobiology and bioremediation. Microbiology 156: 609–643.

García-Delgado, C., N. Jiménez-Ayuso, I. Frutos, A. Gárate and E. Eymar. 2013. Cadmium and lead bioavailability and their effects on polycyclic aromatic hydrocarbons biodegradation by spent mushroom substrate. Environ. Sci. Pollut. Res. 20: 8690–8699.

Gardea-Torresdey, J.L., J.R. Peralta-Videa, M. Montes, G. de la Rosa and B. Corral-Diaz. 2004. Bioaccumulation of cadmium, chromium and copper by *Convolvulus arvensis* L.: impact on plant growth and uptake of nutritional elements. Bioresour. Technol. 92: 229–235.

Gisbert, C., R. Ros, A. De Haro, D.J. Walker, M.P. Bernal, R. Serrano et al. 2003. A plant genetically modified that accumulates Pb is especially promising for phytoremediation. Biochem. Biophys. Res. Commun. 303: 440–445.

Glick, B.R. 2010. Using soil bacteria to facilitate phytoremediation. Biotechnol. Adv. 28: 367–374.

Glick, B.R. 2014. Bacteria with ACC deaminase can promote plant growth and help to feed the world. Microbiol. Res. 169: 30–39.

Gómez-Sagasti, M.T., I. Alkorta, J.M. Becerril, L. Epelde, M. Anza and C. Garbisu. 2012. Microbial monitoring of the recovery of soil quality during heavy metal phytoremediation. Water Air Soil Pollut. 223: 3249–3262.

Grandlic, C.J., M.O. Mendez, J. Chorover, B. Machado and R.M. Maier. 2008. Plant growth-promoting bacteria for phytostabilization of mine tailings. Environ. Sci. Technol. 42: 2079–2084.

Greger, M. and T. Landberg. 2015. Novel field data on phytoextraction: Pre-cultivation with Salix reduces cadmium in wheat grains. Int. J. Phytoremediation 17: 917–924.

Hachicha, S., J. Cegarra, F. Sellami, R. Hachicha, N. Drira, K. Medhioub et al. 2009. Elimination of polyphenols toxicity from olive mill wastewater sludge by its co-composting with sesame bark. J. Hazard. Mater. 161: 1131–1139.

Hartley, W., R. Edwards and N.W. Lepp. 2004. Arsenic and heavy metal mobility in iron oxide-amended contaminated soils as evaluated by short- and long-term leaching tests. Environ. Pollut. 131: 495–504.

Hartley, W., P. Riby, N.M. Dickinson, B. Shutes, S. Sparke and M. Scholz. 2011. Planting woody crops on dredged contaminated sediment provides both positive and negative effects in terms of remediation. Environ. Pollut. 159: 3416–3424.

Herzig, R., E. Nehnevajova, C. Pfistner, J.P. Schwitzguebel, A. Ricci and C. Keller. 2014. Feasibility of labile zn phytoextraction using enhanced tobacco and sunflower: Results of five- and one-year field-scale experiments in Switzerland. Int. J. Phytoremediation 16: 735–754.

Hickey, A.M., L. Gordon, A.D.W. Dobson, C.T. Kelly and E.M. Doyle. 2007. Effect of surfactants on fluoranthene degradation by *Pseudomonas alcaligenes* PA-10. Appl. Microbiol. Biotechnol. 74: 851–856.

Hinsinger, P., A.G. Bengough, D. Vetterlein and I. Young. 2009. Rhizosphere: biophysics, biogeochemistry and ecological relevance. Plant Soil 321: 117–152.

Hmid, A., Z. Al Chami, W. Sillen, A. De Vocht and J. Vangronsveld. 2015. Olive mill waste biochar: a promising soil amendment for metal immobilization in contaminated soils. Environ. Sci. Pollut. Res. 22: 1444–1456.

Hsieh, J.-L., C.-Y. Chen, M.-H. Chiu, M.-f. Chein, J.-S. Chang, G. Endo et al. 2009. Expressing a bacterial mercuric ion binding protein in plant for phytoremediation of heavy metals. J. Hazard. Mater. 161: 920–925.

Janssen, J., N. Weyens, S. Croes, B. Beckers, L. Meiresonne, P. Van Peteghem et al. 2015. Phytoremediation of metal contaminated soil using willow: Exploiting plant-associated bacteria to improve biomass production and metal uptake. Int. J. Phytoremediation 17: 1123–1136.

Kabata-Pendias, A. 2010. Trace Elements in Soils and Plants. 4th edn. CRC Press LLC, Boca Raton, Florida.

Kaimi, E., T. Mukaidani and M. Tamaki. 2007. Screening of twelve plant species for phytoremediation of petroleum hydrocarbon-contaminated soil. Plant Prod. Sci. 10: 211–218.

Kidd, P.S., A. Prieto-Fernández, C. Monterroso and M.J. Acea. 2008. Rhizosphere microbial community and hexachlorocyclohexane degradative potential in contrasting plant species. Plant Soil 302: 233–247.

Kidd, P., J. Barceló, M.P. Bernal, F. Navari-Izzo, C. Poschenrieder, S. Shilev et al. 2009. Trace element behaviour at the root–soil interface: Implications in phytoremediation. Environ. Exp. Bot. 67: 243–259.

Kidd, P., M. Mench, V. Álvarez-López, V. Bert, I. Dimitriou, W. Friesl-Hanl et al. 2015. Agronomic practices for improving gentle remediation of trace element-contaminated soils. Int. J. Phytoremediation 17: 1005–1037.

Kidd, P.S., V. Álvarez-López, C. Becerra-Castro, M. Cabello-Conejo and Á. Prieto-Fernández. 2017. Potential role of plant-associated bacteria in plant metal uptake and implications in phytotechnologies. pp. 87–126. *In*: Cuypers, A. and J. Vangronsveld [eds.]. Phytoremediation. Advances in Botanical Research. Academic Press, USA.

Kohler, J., F. Caravaca, R. Azcón, G. Díaz and A. Roldán. 2015. The combination of compost addition and arbuscular mycorrhizal inoculation produced positive and synergistic effects on the phytomanagement of a semiarid mine tailing. Sci. Total Environ. 514: 42–48.

Kolbas, A. 2012. Phenotypic traits and development of plants exposed to trace elements; use for phytoremediation and biomonitoring. PhD thesis, University of Bordeaux, France.

Kubátová, P., M. Hejcman, J. Száková, S. Vondráčková and P. Tlustoš. 2016. Effects of sewage sludge application on biomass production and concentrations of Cd, Pb and Zn in shoots of *Salix* and *Populus* clones: Improvement of phytoremediation efficiency in contaminated soils. BioEnergy Res. 9: 809–819.

Kumpiene, J., G. Guerri, L. Landi, G. Pietramellara, P. Nannipieri and G. Renella. 2009. Microbial biomass, respiration and enzyme activities after *in situ* aided phytostabilization of a Pb- and Cu-contaminated soil. Ecotoxicol. Environ. Saf. 72: 115–119.

Lasat, M. 2000. Phytoextraction of metals from contaminated soil: a review of plant/soil/metal interaction and assessment of pertinent agronomic issues. J. Hazard. Sub. Res. 2: 1–25.

Lebeau, T., A. Braud and K. Jezequel. 2008. Performance of bioaugmentation-assisted phytoextraction applied to metal contaminated soils: a review. Environ. Pollut. 153: 497–522.

Lee, S.S., J.E. Lim, S.A.M.A. El-Azeem, B. Choi, S.-E. Oh, D.H. Moon et al. 2013. Heavy metal immobilization in soil near abandoned mines using eggshell waste and rapeseed residue. Environ. Sci. Pollut. Res. 20: 1719–1726.

Lehman, J. and S. Joseph. 2009. Biochar for environmental management: an introduction. pp. 1–12. *In*: Lehman, J. and S. Joseph [eds.]. Biochar for Environmental Management: Science and Technology. Earthscan, London, UK.

Lemanceau, P., M. Maurhofer and G. Défago. 2007. Contribution of studies on suppressive soils to the identification of bacterial biocontrol agents and to the knowledge of their modes of action. pp. 231–267. *In*: Gnanamanickam, S.S. [ed.]. Plant-Associated Bacteria. Springer, Netherlands.

Lenoir, I., A. Lounes-Hadj Sahraoui and J. Fontaine. 2016. Arbuscular mycorrhizal fungal-assisted phytoremediation of soil contaminated with persistent organic pollutants: a review. Eur. J. Soil Sci. 67: 624–640.

Lewandowski, I., U. Schmidt, M. Londo and A. Faaij. 2006. The economic value of the phytoremediation function—Assessed by the example of cadmium remediation by willow (*Salix* ssp.). Agr. Syst. 89: 68–89.
Li, Y.-M., R. Chaney, E. Brewer, R. Roseberg, J.S. Angle, A. Baker et al. 2003. Development of a technology for commercial phytoextraction of nickel: economic and technical considerations. Plant Soil 249: 107–115.
Lindsay, M.B.J., K.D. Wakeman, O.F. Rowe, B.M. Grail, C.J. Ptacek, D.W. Blowes et al. 2011. Microbiology and geochemistry of mine tailings amended with organic carbon for passive treatment of pore water. Geomicrobiol. J. 28: 229–241.
Liu, W.W., R. Yin, X.G. Lin, J. Zhang, X.M. Chen, X.Z. Li et al. 2010. Interaction of biosurfactant-microorganism to enhance phytoremediation of aged polycyclic aromatic hydrocarbons (PAHs) contaminated soils with alfalfa (*Medicago sativa* L.). Huanjing Kexue/Environmental Science 31: 1079–1084.
Lorenz, K. 2013. Ecosystem carbon sequestration. pp. 39–63. *In*: Lal, R., K. Lorenz, R.F. Hüttl, B.U. Schneider and J. von Braun [eds.]. Ecosystem Services and Carbon Sequestration in the Biosphere. Springer, Dordrecht Heidelberg, New York, London.
Madejón, E., A.P. de Mora, E. Felipe, P. Burgos and F. Cabrera. 2006. Soil amendments reduce trace element solubility in a contaminated soil and allow regrowth of natural vegetation. Environ. Pollut. 139: 40–52.
Martínez-López, S., M.J. Martínez-Sánchez, C. Pérez-Sirvent, J. Bech, M.C. Gómez Martínez and A.J. García-Fernandez. 2014. Screening of wild plants for use in the phytoremediation of mining-influenced soils containing arsenic in semiarid environments. J. Soils Sed. 14: 794–809.
Maxted, A.P., C.R. Black, H.M. West, N.M. Crout, S.P. McGrath and S.D. Young. 2007. Phytoextraction of cadmium and zinc from arable soils amended with sewage sludge using *Thlaspi caerulescens*: development of a predictive model. Environ. Pollut. 150: 363–372.
Mench, M., J. Vangronsveld, N.W. Lepp and R. Edwards. 1998. Physicochemical aspects and efficiency of trace element immobilization by soil amendments. pp. 151–182. *In*: Vangronsveld, J. and S.D. Cunningham [eds.]. Metal-contaminated Soils: *In Situ* Inactivation and Phytorestoration. Springer Verlag, Berlin, Heidelberg.
Mench, M., G. Renella, A. Gelsomino, L. Landi and P. Nannipieri. 2006. Biochemical parameters and bacterial species richness in soils contaminated by sludge-borne metals and remediated with inorganic soil amendments. Environ. Pollut. 144: 24–31.
Mench, M., N. Lepp, V. Bert, J.-P. Schwitzguébel, S. Gawronski, P. Schröder et al. 2010. Successes and limitations of phytotechnologies at field scale: outcomes, assessment and outlook from COST Action 859. J. Soils Sed. 10: 1039–1070.
Mendes, R., P. Garbeva and J.M. Raaijmakers. 2013. The rhizosphere microbiome: significance of plant beneficial, plant pathogenic, and human pathogenic microorganisms. FEMS Microbiol. Rev. 37: 634–663.
Mendez, M.O. and R.M. Maier. 2008. Phytostabilization of mine tailings in arid and semiarid environments—an emerging remediation technology. Environ. Health Perspect. 116: 278.
Monterroso, C., F. Rodríguez, R. Chaves, J. Diez, C. Becerra-Castro, P.S. Kidd et al. 2014. Heavy metal distribution in mine-soils and plants growing in a Pb/Zn-mining area in NW Spain. Appl. Geochem. 44: 3–11.
Moreno-Jiménez, E., R. Clemente, A. Mestrot and A.A. Meharg. 2013. Arsenic and selenium mobilisation from organic matter treated mine spoil with and without inorganic fertilisation. Environ. Pollut. 173: 238–244.
Mosse, K.P.M., A.F. Patti, R.J. Smernik, E.W. Christen and T.R. Cavagnaro. 2012. Physicochemical and microbiological effects of long- and short-term winery wastewater application to soils. J. Hazard. Mater. 201–202: 219–228.
Muehe, E.M., P. Weigold, I.J. Adaktylou, B. Planer-Friedrich, U. Kraemer, A. Kappler et al. 2015. Rhizosphere microbial community composition affects cadmium and zinc uptake of the metal-hyperaccumulating plant *Arabidopsis halleri*. Appl. Environ. Microbiol. 81: 2173–2181.
Naidu, R., R. Channey, S. McConnell, N. Johnston, K.T. Semple, S. McGrath et al. 2015. Towards bioavailability-based soil criteria: past, present and future perspectives. Environ. Sci. Pollut. Res. 22: 8779–8785.
Namiki, S., T. Otani, N. Seike and S. Satoh. 2015. Differential uptake and translocation of β-HCH and dieldrin by several plant species from hydroponic medium. Environ. Toxicol. Chem. 34: 536–544.
Nogales, R. and E. Benítez. 2006. Absorption of zinc and lead by *Dittrichia viscosa* grown in a contaminated soil amended with olive-derived wastes. Bull. Environ. Contam. Toxicol. 76: 538–544.
Olmo, M., J.A. Alburquerque, V. Barrón, M.C. del Campillo, A. Gallardo, M. Fuentes et al. 2014. Wheat growth and yield responses to biochar addition under Mediterranean climate conditions. Biol. Fertility Soils 50: 1177–1187.
Otones, V., E. Álvarez-Ayuso, A. García-Sánchez, I. Santa Regina and A. Murciego. 2011a. Arsenic distribution in soils and plants of an arsenic impacted former mining area. Environ. Pollut. 159: 2637–2647.
Otones, V., E. Álvarez-Ayuso, A. García-Sánchez, I. Santa Regina and A. Murciego. 2011b. Mobility and phytoavailability of arsenic in an abandoned mining area. Geoderma 166: 153–161.
Pagé, A.P., É. Yergeau and C.W. Greer. 2015. *Salix purpurea* stimulates the expression of specific bacterial xenobiotic degradation genes in a soil contaminated with hydrocarbons. PLOS ONE 10: e0132062.
Panagos, P., M. Van Liedekerke, Y. Yigini and L. Montanarella. 2013. Contaminated sites in Europe: Review of the current situation based on data collected through a European network. J. Environ. Public Health.

Pandey, J., A. Chauhan and R.K. Jain. 2009. Integrative approaches for assessing the ecological sustainability of *in situ* bioremediation. FEMS Microbiol. Rev. 33: 324–375.

Pardo, T., R. Clemente and M.P. Bernal. 2011. Effects of compost, pig slurry and lime on trace element solubility and toxicity in two soils differently affected by mining activities. Chemosphere 84: 642–650.

Pardo, T., M.P. Bernal and R. Clemente. 2014a. Efficiency of soil organic and inorganic amendments on the remediation of a contaminated mine soil: I. Effects on trace elements and nutrients solubility and leaching risk. Chemosphere 107: 121–128.

Pardo, T., R. Clemente, P. Alvarenga and M.P. Bernal. 2014b. Efficiency of soil organic and inorganic amendments on the remediation of a contaminated mine soil: II. Biological and ecotoxicological evaluation. Chemosphere 107: 101–108.

Pardo, T., R. Clemente, L. Epelde, C. Garbisu and M.P. Bernal. 2014c. Evaluation of the phytostabilisation efficiency in a trace elements contaminated soil using soil health indicators. J. Hazard. Mater. 268: 68–76.

Pardo, T., D. Martínez-Fernández, R. Clemente, D.J. Walker and M.P. Bernal. 2014d. The use of olive-mill waste compost to promote the plant vegetation cover in a trace-element-contaminated soil. Environ. Sci. Pollut. Res. 21: 1029–1038.

Pardo, T., C. Bes, M.P. Bernal and R. Clemente. 2016a. Alleviation of environmental risks associated with severely contaminated mine tailings using amendments: Modeling of trace element speciation, solubility, and plant accumulation. Environ. Toxicol. Chem. 35: 2874–2884.

Pardo, T., D. Martínez-Fernández, C. de la Fuente, R. Clemente, M. Komárek and M.P. Bernal. 2016b. Maghemite nanoparticles and ferrous sulfate for the stimulation of iron plaque formation and arsenic immobilization in *Phragmites australis*. Environ. Pollut. 219: 296–304.

Pardo, T., M.P. Bernal and R. Clemente. 2017a. Phytostabilisation of severely contaminated mine tailings using halophytes and field addition of organic and inorganic amendments. Chemosphere (in press).

Pardo, T., P. Bernal and R. Clemente. 2017b. The use of olive mill waste to promote phytoremediation. pp. 183–204. *In*: Galanakis, C.M. [ed.]. Olive Mill Waste. Academic Press, USA.

Parraga-Aguado, I., J.I. Querejeta, M.N. González-Alcaraz, F.J. Jiménez-Cárceles and H.M. Conesa. 2014. Usefulness of pioneer vegetation for the phytomanagement of metal(loid)s enriched tailings: Grasses vs. shrubs vs. trees. J. Environ. Manage. 133: 51–58.

Pei, X.-H., X.-H. Zhan, S.-M. Wang, Y.-S. Lin and L.-X. Zhou. 2010. Effects of a biosurfactant and a synthetic surfactant on phenanthrene degradation by a *Sphingomonas* strain. Pedosphere 20: 771–779.

Peijnenburg, W. and T. Jager. 2003. Monitoring approaches to assess bioaccessibility and bioavailability of metals: matrix issues. Ecotoxicol. Environ. Saf. 56: 63–77.

Pereira, S.I.A., L. Barbosa and P.M.L. Castro. 2015. Rhizobacteria isolated from a metal-polluted area enhance plant growth in zinc and cadmium-contaminated soil. Int. J. Environ. Sci. Technol. 12: 2127–2142.

Pérez-De-Mora, A., P. Burgos, E. Madejón, F. Cabrera, P. Jaeckel and M. Schloter. 2006. Microbial community structure and function in a soil contaminated by heavy metals: Effects of plant growth and different amendments. Soil Biol. Biochem. 38: 327–341.

Petrisor, I.G., S. Dobrota, K. Komnitsas, I. Lazar, J.M. Kuperberg and M. Serban. 2004. Artificial inoculation-perspectives in tailings phytostabilization. Int. J. Phytorem. 6: 1–15.

Pierzynski, G.M., J.L. Schnoor, A. Youngman, L. Licht and L.E. Erickson. 2002. Poplar trees for phytostabilization of abandoned Zinc-Lead smelter. Pract. Periodical Hazard., Toxic, Radioact. Waste Manage. 6: 177–183.

Pottier, M., V.S. García de la Torre, C. Victor, L.C. David, M. Chalot and S. Thomine. 2015. Genotypic variations in the dynamics of metal concentrations in poplar leaves: A field study with a perspective on phytoremediation. Environ. Pollut. 199: 73–82.

Pulford, I.D. and C. Watson. 2003. Phytoremediation of heavy metal-contaminated land by trees: a review. Environ. Int. 29: 529–540.

Raaijmakers, J.M., M. Vlami and J.T. de Souza. 2002. Antibiotic production by bacterial biocontrol agents. Antonie Van Leeuwenhoek 81: 537–547.

Rajtor, M. and Z. Piotrowska-Seget. 2016. Prospects for arbuscular mycorrhizal fungi (AMF) to assist in phytoremediation of soil hydrocarbon contaminants. Chemosphere 162: 105–116.

Reid, B.J., K.C. Jones and K.T. Semple. 2000. Bioavailability of persistent organic pollutants in soils and sediments—a perspective on mechanisms, consequences and assessment. Environ. Pollut. 108: 103–112.

Ridolfi, A.S., G.B. Álvarez and M.E. Rodríguez Girault. 2014. Organochlorinated contaminants in general population of Argentina and other Latin American countries. pp. 17–40. *In*: Alvarez, A. and M.A. Polti [eds.]. Bioremediation in Latin America: Current Research and Perspectives. Springer International Publishing, Cham.

Robinson, B.H., G. Bañuelos, H.M. Conesa, M.W.H. Evangelou and R. Schulin. 2009. The phytomanagement of trace elements in soil. Crit. Rev. Plant Sci. 28: 240–266.

Ruttens, A., J.V. Colpaert, M. Mench, J. Boisson, R. Carleer and J. Vangronsveld. 2006a. Phytostabilization of a metal contaminated sandy soil. II: Influence of compost and/or inorganic metal immobilizing soil amendments on metal leaching. Environ. Pollut. 144: 533–539.

Ruttens, A., M. Mench, J.V. Colpaert, J. Boisson, R. Carleer and J. Vangronsveld. 2006b. Phytostabilization of a metal contaminated sandy soil. I: Influence of compost and/or inorganic metal immobilizing soil amendments on phytotoxicity and plant availability of metals. Environ. Pollut. 144: 524–532.

Ryan, R.P., K. Germaine, A. Franks, D.J. Ryan and D.N. Dowling. 2008. Bacterial endophytes: recent developments and applications. FEMS Microbiol. Lett. 278: 1–9.

Saimmai, A., O. Rukadee, V. Sobhon and S. Maneerat. 2012. Biosurfactant production by *Bacillus subtilis* TD4 and *Pseudomonas aeruginosa* SU7 grown on crude glycerol obtained from biodiesel production plant as sole carbon source. J. Sci. Ind. Res. 71: 396–406.

Salem, Z.B., X. Laffray, A. Ashoour, H. Ayadi and L. Aleya. 2014. Metal accumulation and distribution in the organs of Reeds and Cattails in a constructed treatment wetland (Etueffont, France). Ecol. Eng. 64: 1–17.

Salomons, W. 1995. Environmental impact of metals derived from mining activities: Processes, predictions, prevention. J. Geochem. Explor. 52: 5–23.

Schnoor, J.L., L.A. Light, S.C. McCutcheon, N.L. Wolfe and L.H. Carreia. 1995. Phytoremediation of organic and nutrient contaminants. Environ. Sci. Technol. 29: 318A–323A.

Schwartz, C., G. Echevarria and J.L. Morel. 2003. Phytoextraction of cadmium with *Thlaspi caerulescens*. Plant Soil 249: 27–35.

Semple, K.T., A.W.J. Morriss and G.I. Paton. 2003. Bioavailability of hydrophobic organic contaminants in soils: fundamental concepts and techniques for analysis. Eur. J. Soil Sci. 54: 809–818.

Sessitsch, A., M. Kuffner, P. Kidd, J. Vangronsveld, W.W. Wenzel, K. Fallmann et al. 2013. The role of plant-associated bacteria in the mobilization and phytoextraction of trace elements in contaminated soils. Soil Biol. Biochem. 60: 182–194.

Shahid, M., E. Pinelli and C. Dumat. 2012. Review of Pb availability and toxicity to plants in relation with metal speciation; role of synthetic and natural organic ligands. J. Hazard. Mater. 219–220: 1–12.

Siciliano, S.D., N. Fortin, A. Mihoc, G. Wisse, S. Labelle, D. Beaumier et al. 2001. Selection of specific endophytic bacterial genotypes by plants in response to soil contamination. Appl. Environ. Microbiol. 67: 2469–2475.

Singer, A.C., D.E. Crowley and I.P. Thompson. 2003. Secondary plant metabolites in phytoremediation and biotransformation. Trends Biotechnol. 21: 123–130.

Singh, A. and S.M. Prasad. 2014. Effect of agro-industrial waste amendment on Cd uptake in Amaranthus caudatus grown under contaminated soil: an oxidative biomarker response. Ecotoxicol. Environ. Saf. 100: 105–113.

Soler-Rovira, P., E. Madejón, P. Madejón and C. Plaza. 2010. *In situ* remediation of metal-contaminated soils with organic amendments: Role of humic acids in copper bioavailability. Chemosphere 79: 844–849.

Swartjes, F.A. 2011. Introduction to contaminated site management. pp. 3–89. *In*: Swartjes, F.A. [ed.]. Dealing with Contaminated Sites: From Theory towards Practical Application. Springer Netherlands, Dordrecht.

Tang, Y.-T., T.-H.-B. Deng, Q.-H. Wu, S.-Z. Wang, R.-L. Qiu, Z.-B. Wei et al. 2012. Designing cropping systems for metal-contaminated sites: A review. Pedosphere 22: 470–488.

Tao, S., W. Liu, Y. Chen, F. Xu, R. Dawson, B. Li et al. 2004. Evaluation of factors influencing root-induced changes of copper fractionation in rhizosphere of a calcareous soil. Environ. Pollut. 129: 5–12.

Thijs, S., W. Sillen, F. Rineau, N. Weyens and J. Vangronsveld. 2016. Towards an enhanced understanding of plant-microbiome interactions to improve phytoremediation: Engineering the metaorganism. Front. Microbiol. 7: 341.

Tordoff, G.M., A.J.M. Baker and A.J. Willis. 2000. Current approaches to the revegetation and reclamation of metalliferous mine wastes. Chemosphere 41: 219–228.

Touceda-González, M., V. Álvarez-López, Á. Prieto-Fernández, B. Rodríguez-Garrido, C. Trasar-Cepeda, M. Mench et al. 2017. Aided phytostabilisation reduces metal toxicity, improves soil fertility and enhances microbial activity in Cu-rich mine tailings. J. Environ. Manage. 186: 301–313.

Urbance, J.W., J. Cole, P. Saxman and J.M. Tiedje. 2003. BSD: the biodegradative strain database. Nucleic Acids Res. 31: 152–155.

Uren, N. and H. Reisenauer. 1988. The role of root exudates in nutrient acquisition. pp. 79–114. *In*: Tinker, P. and A. Lauchli [eds.]. Advances in Plant Nutrition, vol 3. Praeger, New York.

van der Ent, A., A.M. Baker, R. Reeves, A.J. Pollard and H. Schat. 2013. Hyperaccumulators of metal and metalloid trace elements: Facts and fiction. Plant Soil 362: 319–334.

van Liedekerke, M., G. Prokop, S. Rabl-Berger, M. Kibblewhite and G. Louwagie. 2014. Progress in the management of contaminated sites in Europe. Institute for Environment and Sustainability; European Commission.

Van Oosten, M.J. and A. Maggio. 2015. Functional biology of halophytes in the phytoremediation of heavy metal contaminated soils. Environ. Exp. Bot. 111: 135–146.

Vangronsveld, J., A. Ruttens, J. Colpaert and D. Van der Lelie. 2000a. *In situ* fixation and phytostabilization of metals in polluted soils. Hangzhou, China.

Vangronsveld, J., A. Ruttens, M. Mench, J. Boisson, N. Lepp, R. Edwards et al. 2000b. *In situ* inactivation and phytoremediation of metal/metalloid contaminated soils: field experiments. pp. 859–884. *In*: Wise, D., D.J.

Trantolo, E.J. Cichon, H.I. Inyang and U. Stottmeister [eds.]. Bioremediation of Contaminated Soils. Marcel Dekker, New York.
Vangronsveld, J., R. Herzig, N. Weyens, J. Boulet, K. Adriaensen, A. Ruttens et al. 2009. Phytoremediation of contaminated soils and groundwater: lessons from the field. Environ. Sci. Pollut. Res. Int. 16: 765–794.
Vergani, L., F. Mapelli, E. Zanardini, E. Terzaghi, A. Di Guardo, C. Morosini et al. 2017. Phyto-rhizoremediation of polychlorinated biphenyl contaminated soils: An outlook on plant-microbe beneficial interactions. Sci. Total Environ. 575: 1395–1406.
Vessey, J.K. 2003. Plant growth promoting rhizobacteria as biofertilizers. Plant Soil 255: 571–586.
Walker, D.J., S. Lutts, M. Sánchez-García and E. Correal. 2014. *Atriplex halimus* L.: Its biology and uses. J. Arid Environ. 100–101: 111–121.
Wang, F.Y., Z.Y. Shi, X.F. Xu, X.G. Wang and Y.J. Li. 2013. Contribution of AM inoculation and cattle manure to lead and cadmium phytoremediation by tobacco plants. Environ. Sci. Process Impacts 15: 794–801.
Wenzel, W.W. 2009. Rhizosphere processes and management in plant-assisted bioremediation (phytoremediation) of soils. Plant Soil 321: 385–408.
Weyens, N., S. Taghavi, T. Barac, D. van der Lelie, J. Boulet, T. Artois et al. 2009a. Bacteria associated with oak and ash on a TCE-contaminated site: characterization of isolates with potential to avoid evapotranspiration of TCE. Environ. Sci. Pollut. Res. 16: 830–843.
Weyens, N., D. van der Lelie, S. Taghavi, L. Newman and J. Vangronsveld. 2009b. Exploiting plant–microbe partnerships to improve biomass production and remediation. Trends Biotechnol. 27: 591–598.
Weyens, N., D. van der Lelie, S. Taghavi and J. Vangronsveld. 2009c. Phytoremediation: plant–endophyte partnerships take the challenge. Curr. Opin. Biotechnol. 20: 248–254.
White, J.C., M.I. Mattina, W.-Y. Lee, B.D. Eitzer and W. Iannucci-Berger. 2003. Role of organic acids in enhancing the desorption and uptake of weathered p,p'-DDE by *Cucurbita pepo*. Environ. Pollut. 124: 71–80.
White, J.C. 2010. Inheritance of p,p'-DDE Phytoextraction ability in hybridized *Cucurbita pepo* cultivars. Environ. Sci. Technol. 44: 5165–5169.
Witters, N., S. Van Slycken, A. Ruttens, K. Adriaensen, E. Meers, L. Meiresonne et al. 2009. Short-Rotation coppice of willow for phytoremediation of a metal-contaminated agricultural area: A sustainability assessment. BioEnergy Res. 2: 144–152.
Xue, K., J.D. van Nostrand, J. Vangronsveld, N. Witters, J.O. Janssen, J. Kumpiene et al. 2015. Management with willow short rotation coppice increase the functional gene diversity and functional activity of a heavy metal polluted soil. Chemosphere 138: 469–477.
Ye, W.-L., M.A. Khan, S.P. McGrath and F.-J. Zhao. 2011. Phytoremediation of arsenic contaminated paddy soils with *Pteris vittata* markedly reduces arsenic uptake by rice. Environ. Pollut. 159: 3739–3743.
Zanuzzi, A., J.M. Arocena, J.M. van Mourik and A. Faz Cano. 2009. Amendments with organic and industrial wastes stimulate soil formation in mine tailings as revealed by micromorphology. Geoderma 154: 69–75.
Zappelini, C., B. Karimi, J. Foulon, L. Lacercat-Didier, F. Maillard, B. Valot et al. 2015. Diversity and complexity of microbial communities from a chlor-alkali tailings dump. Soil Biol. Biochem. 90: 101–110.
Zegada-Lizarazu, W. and A. Monti. 2011. Energy crops in rotation. A review. Biomass Bioenergy 35: 12–25.
Zhang, J., R. Yin, X. Lin, W. Liu, R. Chen and X. Li. 2010. Interactive effect of biosurfactant and microorganism to enhance phytoremediation for removal of aged polycyclic aromatic hydrocarbons from contaminated soils. J. Health Sci. 56: 257–266.
Zhang, X., V. Houzelot, A. Bani, J.L. Morel, G. Echevarria and M.-O. Simonnot. 2014. Selection and combustion of Ni-hyperaccumulators for the phytomining process. Int. J. Phytoremediation 16: 1058–1072.
Zhang, Y. and W.T. Frankenberger. 2000. Formation of dimethylselenonium compounds in soil. Environ. Sci. Technol. 34: 776–783.
Zornoza, R., A. Faz, D.M. Carmona, S. Martínez-Martínez and J.A. Acosta. 2012. Plant cover and soil biochemical properties in a mine tailing pond five years after application of marble wastes and organic amendments. Pedosphere 22: 22–32.

14

Bioremediation of Heavy Metals by Immobilized Microbial Cells and Metabolites

María F. Castro,[1,a] *José O. Bonilla,*[1,2] *Claudio D. Delfini*[1,b] and *Liliana B. Villegas*[1,2,*]

Introduction

Heavy metals exist in nature as dilute components of the geochemical cycles. In general, soils have heavy metals as a result of geological and edaphic processes. In the last few decades, concentration of these metals increased due to certain human activities. Soil heavy metal pollution derives from an increase in the heavy metal concentration regarding the background level (Nakić et al. 2007, Villegas et al. 2013). All ecosystems, mainly soil, are important habitats for millions of microorganisms including a large variety of fungi, algae, protozoa and different types of bacteria. Microorganisms are an essential part of the living soil and of paramount importance for soil health. In soil, microorganisms are key players in the cycle of nitrogen, sulfur and phosphorus, and in the organic waste decomposition. Also, microbial communities from these ecosystems can interact and utilize heavy metals. Often these metals are present in very high concentrations, and there also exist mixed contamination by the presence of numerous heavy metals at the same time. Unlike organic contaminants, heavy metals cannot be degraded by physicochemical and biological processes, so when they are released into the environment, they remain there indefinitely. For decades, these persistent pollutants have accumulated in the environment and are still being released. This situation is the reason to apply a suitable solution to effectively detoxify and reduce their exposure to the environment. Additionally, metals can leach into groundwater aquifers and contaminate drinking water, and are unreachable by many methods actually available to decontaminate surface soils. Many developing countries are faced with the challenge of reducing human exposure to heavy metals, mainly due to their limited economic capacities to use advanced technologies for their removal (Chowdhury et al. 2016). For this reason, heavy metals provide several unique challenges for remediation.

[1] Instituto de Química San Luis (INQUISAL) CONICET, Chacabuco 917, 5700, San Luis, Argentina.
[a] Email: fercastro_mfc@hotmail.com
[b] Email: cdelfini47@hotmail.com
[2] Facultad de Química, Bioquímica y Farmacia, Universidad Nacional de San Luis, Argentina, Chacabuco 917, 5700, San Luis, Argentina.
Email: jobonilla@unsl.edu.ar
* Corresponding author: lbvilleg@gmail.com

Heavy metals profoundly affect biological systems, either positively because they are essential or negatively because they are toxic, when they are present in excessive amounts. Living organisms incorporate and use these metals; for example, a significant number of proteins require metals for their correct catalytic activity and/or to maintain their structure (Waldron et al. 2009). Clearly, heavy metals are directly or indirectly involved in all aspects of cell growth, metabolism, and cellular differentiation.

The importance of heavy metals to human health as well as their impact on the environment has stimulated research on the action of these on biological systems with the aim of understanding their response to the presence of heavy metals and the tolerance mechanisms in many prokaryotic and eukaryotic microorganisms. Moreover, numerous reports have shown the microorganism's abilities to tolerate and remove heavy metals from pure culture or microbial consortium (Villegas et al. 2008, 2009, Piñón-Castillo et al. 2010, Villegas et al. 2013, Bonilla et al. 2016).

Several techniques are used in order to minimize the impact that high concentrations of heavy metals cause to the ecosystems, the most used being the physicochemical methods. These conventional methods applied to the removal of heavy metals are limited to the waste confinement or to the modification of its solubility and toxicity through changes in oxidation state. Heavy metals removal using these methods requires high levels of energy and chemical reagents and the cleanup process is generally incomplete. Moreover, they also generate a toxic waste or scrap that may become a secondary environmental pollutant. In this context, microbial bioremediation is an attractive eco-friendly technology due to a moderate-cost and greater efficiency than traditional techniques (Wagner-Dobler 2003).

Bioremediation is a well-described technique which involves the utilization of organisms or part of them in order to transform some toxic pollutants in inert or less-toxic compounds (Dua et al. 2002, Okoh and Trejo-Hernandez 2006, Luqueño et al. 2011). Bioremediation is the most effective management tool to recover polluted environment (Kurmar et al. 2011). Currently, a variety of strategies utilize microbial biomass for the treatment of contaminated sites with different toxins. Although the use of microorganisms to clean up heavy metals contaminated environments or effluents is a proposal that overcomes the use of physico-chemical methods, the application of free microorganisms' presents many limitations.

Immobilized cells have been extensively used as an important tool in biotechnological processes, especially for the production of useful chemicals because the immobilization extends the useful life and increases cellular stability and survival (Saifuddin and Raziah 2007). However, little is known about the application of immobilized cells or microbial metabolites in the treatment of wastewaters or bioremediation of contaminated sites with heavy metals. Therefore, the use of immobilized systems as a promising tool to clean up heavy metals-contaminated sites is considered in this chapter.

Immobilization Technology: Advantages and Disadvantages

Immobilization is a general term describing both cells and particles confinement by a wide variety of natural or artificial means. It can be applied to basically all types of biocatalysts systems including enzymes, cellular organelles, specific metabolites, even cell free crude extract. Immobilization involves the physical confinement in a defined region, under conditions in which cells or particles maintain their functionality. The immobilization of enzymes and microbial cells are the most used, and have been applied for the production of useful compounds such as antibiotics, amino acids, organic acids, biodiesel and bioethanol production or fermentation processes (Hasan et al. 2009, Ivanova et al. 2011, Shyamkumar et al. 2014).

On the other hand, the use of immobilized enzymes and cells has been investigated as an alternative technology for environmental applications. Several studies show the application of immobilized yeasts, bacteria, fungi, algae-microalgae and bacterial consortium in the treatment of organic and inorganic contaminants (Cabuk et al. 2006, 2007, Angelim et al. 2013, Dellagnezze et al. 2016, Sharma and Malaviya 2016). Moreover, the microbial cell immobilization technology offers many advantages over their free cell counterparts as showed by Tsekova et al. (2010), who compared the elimination of several heavy metals from industrial wastewater with free and immobilized cells of *Aspergillus niger*. Similar results were obtained with immobilized cells of *Serratia marcecens* on active carbon for hexavalent chromium reduction (Campos et al. 2013). This benefit is due in particular to the high density of biomass and metabolic activity, stability and stronger resistance to harsh environments such as salinity, metal toxicity, and pH of immobilized cells because they are protected in the support material (Hall-Stoodley et al. 2004, Tripathi et al. 2010). Moreover, this technology could be cost effective since it can be used several times without significant loss of activity and presents easier solid–liquid separation and minimal clogging in continuous systems and it is easier to handle generally (Poopal and Laxman 2008, Liu et al. 2012).

Obviously, this technique has some limitations, for example, the support must be conducive to cell viability as well as possess proper permeability to allow sufficient diffusion and transport of oxygen, essential nutrients, and metabolic waste and secreted products across the polymer network. These limitations are easy to overcome and therefore, immobilization is an eco-friendly promising technique for bioremediation processes.

Cell Immobilization Methods

Cell immobilization emerged as an alternative to enzyme immobilization because it eliminates long and costly procedures involving the separation and purification of enzymes. Immobilization of cells has also been proposed to increase the biosorption capacity and bioactivity of the biomass. It allows bioprocesses with higher cell densities and also easy harvesting of biomass from its liquid environment. Working with immobilized microbial cells is advantageous mainly when the desired enzymes are intracellular and purified enzymes become unstable after immobilization, when the microorganism does not contain interfering enzymes or when substrates and interest products do not have a high molecular mass and can diffuse through the cell membrane (Stolarzewicz et al. 2011, Martins et al. 2013).

Immobilization techniques of microbial cells can be divided into two groups: "Passive or reversible", using the natural tendency of microorganism to grow and attach to different types of adsorbent materials surfaces, also called carrier, or "active or irreversible", using flocculants agents, chemical attachment and gel encapsulation (Moreno-Garrido 2008, de-Bashan and Bashan 2010).

Below, each technique of immobilization is considered in detail.

Passive or Reversible Immobilization

As the name suggests, cells can be easily removed from the support by changing the conditions that influence the interaction (pH, ionic strength, temperature or polarity of the solvent). The reversible immobilization includes the formation of biofilm or adsorption to synthetic or natural surfaces by electrostatic bonds.

Biofilms

The biofilm formation is a natural phenomenon of cell immobilization (microalgae, archaea, bacteria, fungi, and/or protozoa). With the objective to provide a terminology that is usable, without any confusion in the various areas, recommendations of International Union of Pure Applied Chemistry (IUPAC) are used to define the biofilms as "aggregates of microorganisms in which cells are frequently embedded in a self-produced matrix of extracellular polymeric substances (EPS) that are adherent to each other and/or to a surface". It is also clarified in this publication that a biofilm is a fixed system that can be adapted internally to environmental conditions by its inhabitants. The self-produced matrix of EPS is a conglomeration composed of extracellular biopolymers in various structural forms (Vert et al. 2012). Microbial cells in biofilms can be considered to be habitat formers, owing to their generation of a matrix that forms the physical foundation of the biofilm. The matrix is composed of EPS that provides architecture and stability to the biofilm (Flemming et al. 2016).

Biofilms are complex systems that have high cell density, ranging from 1×10^8 to 1×10^{11} cells per g weight, and typically comprise many species (Morgan-Sastume et al. 2008). The term "aggregate" in biofilms definition accounts for the fact that most cells in multilayered biofilms experience cell-to-cell contact. In surface-attached biofilms, only one layer is in direct contact with the substratum. These properties are summarized in Fig. 1, and include localized gradients that provide habitat diversity, resource capture by sorption, enzyme retention that provides digestive capabilities, social interactions and the ability, through tolerance and/or resistance, to survive exposure to antibiotics. The formation of the matrix is a dynamic process and depends on nutrient availability, the synthesis and secretion of extracellular material, shear stress, and social competition with other organisms (Flemming et al. 2016). The cells that are part of the biofilms are more active than their free-living counterparts and exhibit differences in gene expression. Thus, intercellular interactions together with the properties of the matrix make biofilm lifestyle clearly distinct and this is not predictable from that of free-living microbial cells (Konopka 2009).

The formation of biofilms causes terrible problems in the field of agriculture or livestock because they are associated with numerous diseases in plants and animals and even in medicine since they contaminate medical devices and implants, seriously affecting

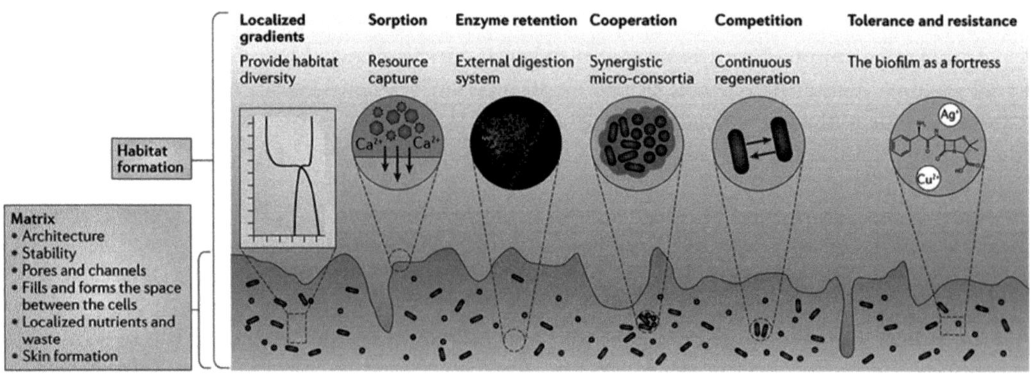

Figure 1. Emergent properties of biofilms and habitat formation (Flemming et al. 2016, reproduced by permission of Nature Publishing Group License N° 4063201365016).

the human health. Clinically, biofilms are responsible for many persistent and chronic infections due to their inherent resistance to antimicrobial agents and the selection for phenotypic variants (Wei and Ma 2013). However, there are numerous biotechnological applications that use biofilms as a tool, such as potable water filtration, solid waste degradation, and biocatalysis in biotechnological processes, such as the production of bulk, fine chemicals and biofuels.

Without the intention of going into detail, the biofilms formation can be summarized in two main steps. The first involves the adsorption of a cell to a surface. This step is reversible and the microorganisms can be removed from the surface by gentle washing. In the second step, after the initial adsorption, the cells can bind to the surface by the production of extracellular polysaccharides. Biofilm develops as a result of both adsorption and adhesion of new planktonic or free cells, combined with the continuous growth of those already adhered.

Adsorption

Not all microbial strains have the ability to form a consistent biofilm. For example, a work with several strains of *Listeria monocytogenes*, an important foodborne pathogen, found that all these cells had the ability to adhere to a surface for a short period but the majority of strains did not form biofilms (Kalmokoff et al. 2001). In these cases, the cellular immobilization is performed by means of a physical process, similar to the adsorption of colloidal particles. The immobilization by natural adsorption of microorganisms onto porous and inert support is the elementary and probably the simplest method of reversible immobilization and involves adhesion of microorganism to the surface of the carrier via several weak non-covalent and reversible interactions such as hydrogen bond, Van Der Waal's interactions and hydrophobic interaction (Górecka et al. 2011, Bayat et al. 2015).

In this process, different structures such as fimbria, capsules, and cells wall components are responsible for natural attachment of the microorganism to a surface. An alternative is the electrostatics binding, similar to physical adsorption, but which requires the wash off of the surface carrier with a buffer solution to obtain a hydrophilic surface that can attract the negatively charged cells. In this way, the microorganisms' leaking are lower. Like biofilms, the immobilization of microorganisms into adsorbents appropriately stimulates microbial metabolism, protects cells from unfavorable agents and preserves their physiological activity. Unlike that, it is not necessary to maintain the viability of the cells, since processes with dead cells are feasible.

Some advantages of adsorption immobilization are: easy preparation because naturally, most of the microbial cells have the tendency to adhere to the surface of support matrix, is very simple and cost-effective, reloading of the support is possible, generally the cell productivity is not affected and a wide range of both organic and inorganic materials can be used as a support. Among most the common supports, we can find carboxymethylcellulose, starch, collagen, modified sepharose, ion exchange resins, active charcoal, silica gel, clay, aluminum oxide, titanium, diatomaceous earth, hydroxyapatite, ceramic, zeolite or treated porous glass (Górecka et al. 2011).

However, this method has many disadvantages and very few applications. Adsorption of the cells is used for exploratory work over short periods of time. The weakness of the physical interaction causes the cells to be desorbed from the carrier material; therefore, the immobilization efficiency decreases and the reproducibility is also low. In this technique, the binding forces are weak, reversible and susceptible to physical parameters or to changes in environmental conditions such as pH, temperature or ionic strength (Nisha et al. 2012, Nabweteme et al. 2016).

The selection of support is one of the crucial decisions to take in the course of the preparation of the immobilization processes and it depends on the use that is required for the immobilized cells. In general, it can be said that the material must meet the following requirements: it should be available in sufficient quantity and at a low price; it must be mechanically, chemically and thermally stable under processing and storage conditions; it must contain a sufficient number of functional groups to bind the cells; and it must be capable of recycling or safe disposal. In the case of viable cells, the matrix must have a volume large enough to house new cells and the material should not reduce cell activity or initiate cell lysis.

Active or Irreversible Immobilization

Reversible immobilized technology cannot fulfill the aim to maintain the long-term process. On the contrary, irreversible immobilization involves strong chemical bonds and particularly serves to maintain the reasonable stability of cells and enzymes over a long period (Krishnamoorthi et al. 2015).

Covalent binding or cross-linking

These methods are based on covalent bond formation between inorganic support and cells reactive groups (e.g., $-NH_2$ or $-COOH$) (covalent coupling) or in the presence of a binding (cross-linking) agent. In the covalent coupling method, it is necessary to activate the support to achieve a more efficient immobilization. On the other hand, the cross-linking is different from other techniques in the sense that it does not require a support for the immobilization (Krishnamoorthi et al. 2015).

One advantage of this method is that coupling and cross-linking leads to an increase in the stability of microorganism but the bioactivity of microorganism decreases rapidly after the post-operation process. These methods are simple but very difficult to control. These techniques are rarely applied to cell immobilization because the agent used for the formation of covalent bonds is usually cytotoxic (Ramakrishna et al. 1999, Das and Adholeya 2015).

Entrapment

This method consists of cells entrapped in porous polymer carriers such as polysaccharide gels like alginates, agar, chitosan and polygalacturonic acid or other polymeric matrixes like gelatin, collagen, and polyvinyl alcohol. This technique creates a protective barrier around the immobilized microbes, ensuring their prolonged viability during not only processing but also storage in polymers. Due to the porous structure of polymers, the pollutants and various metabolites can easily diffuse through the matrix (Górecka and Jastrzębska 2011, Das and Adholeya 2015). However, one important problem of cell entrapment within a porous matrix is the ability of cells located on the outer surface of the beads to multiply and be released from the inclusion bead. This leads to a system consisting of immobilized and free cells. To avoid this problem, double layer beads have been developed where hydrogel beads possess an internal core that contains the cells and an external layer, which prevents the cells from the core to escape (Kourkoutas et al. 2004).

Encapsulation

It is based on entrapment of cells in a semi-permeable capsule, which is a selective membrane that allows those small molecules (e.g., oxygen and nutrients) and reaction products to freely diffuse, retaining large molecules and cells inside the capsule. This allows maintaining the viability of the cells. This method can be achieved by enveloping the biological components within various forms of spherical semi-permeable membranes with a selectively controlled permeability. A wide variety of synthetics and natural polymers such as polyacrylamide, polyvinyl alcohol, collagen, agar, agarose, and chitosan are used to encapsulate microorganisms. The encapsulation of cells is divided into two types based on the size of the polymeric bead produced: macroencapsulation and microencapsulation. Macroencapsulation is the entrapment of cells in polymeric structures of a larger size, ranging from few millimeters to centimeters. This method is generally applied to the encapsulation of animal cells or tissues and may occur with organic materials or inert inorganic polymeric materials (John et al. 2011). On the other hand, microencapsulation allows the immobilization of cells on a very small scale, where capsules ranging from less than one micrometer to hundreds of microns are used. The ratio of pore size of membrane and size of the core material is a significant factor in this phenomenon. This limited size availability inside of microcapsule is one of the main advantages of microencapsulation. Due to the protection of the cell from the extreme conditions, it prevents cells leakage, increasing the process efficiency (Park and Chang 2000, Bayat et al. 2015).

Recently, some authors have described a system of microbial encapsulation using electrospinning (ES) method, which entails incorporating bacteria within microtubes of polymer on a nano scale. ES is a fiber production method which uses electric force to draw charged threads of polymer solutions or polymer melts up to fiber diameters in the order of some hundred nanometers. ES is probably the most researched top-down method to form nanofibres from a remarkable range of organic and inorganic materials. This method allows for the creation of a natural environment for the immobilized cells, mainly due to the viscosity of the water-soluble core polymer (Luo et al. 2012). Additionally, adjustment of the properties of the shell polymers allows for tuning the mechanical properties, cell attachment and cell division (Letnik et al. 2017). An original process was reported by Gensheimer et al. (2011) where physically cross-linked poly (vinyl alcohol)-hydrogel microparticles were utilized for encapsulation of bacteria. This process is characterized by simplicity, versatility and cost-effectiveness. Electrospun nanofibers and webs display a variety of unique properties such as an extremely high surface-to-volume ratio which can provide a large specific surface area for highly efficient immobilization and a nanoscale pore size.

Flocculation (cell aggregation)

Cell flocculation has been defined by many authors as an aggregation of cells that form a larger unit or "floc" which sediments rapidly. The word floc derives from the Latin word *floccus*, which means a tuft of wool. The cells with the ability to form flocs are called flocculants, while the cells that are not able to form flocs are usually known as powdery (Soares 2011). Flocculation can be considered as an immobilization technique given that the large size of the aggregates makes them potentially to be used in reactors. This method takes advantage of the natural ability of some types of cells to form aggregates such as some fungi members. It is affected by many factors including cell wall composition, pH, dissolved oxygen and medium composition. For example, yeast flocculation is a property of major importance for the brewing industry and it affects fermentation productivity and beer quality in addition to yeast removal and recovery. However, artificial flocculating agents or cross-linkers also can be used to enhance aggregation in cell cultures that do not naturally flocculate (Kourkoutas et al. 2004, Covizzi et al. 2007).

Flocculation is used in biotechnology applications in conjunction with microfiltration to improve the efficiency of biological feeds. The addition of synthetic flocculants to the bioreactor can increase the average particle size making microfiltration more efficient. When flocculants are not added, cakes form and accumulate causing low cell viability. Positively charged flocculants work better than negatively charged ones since cellular surfaces often have negative charge (Han et al. 2003). Kourkoutas et al. (2004) summarized the basic methods of cell immobilization clearly and they can be seen in Fig. 2.

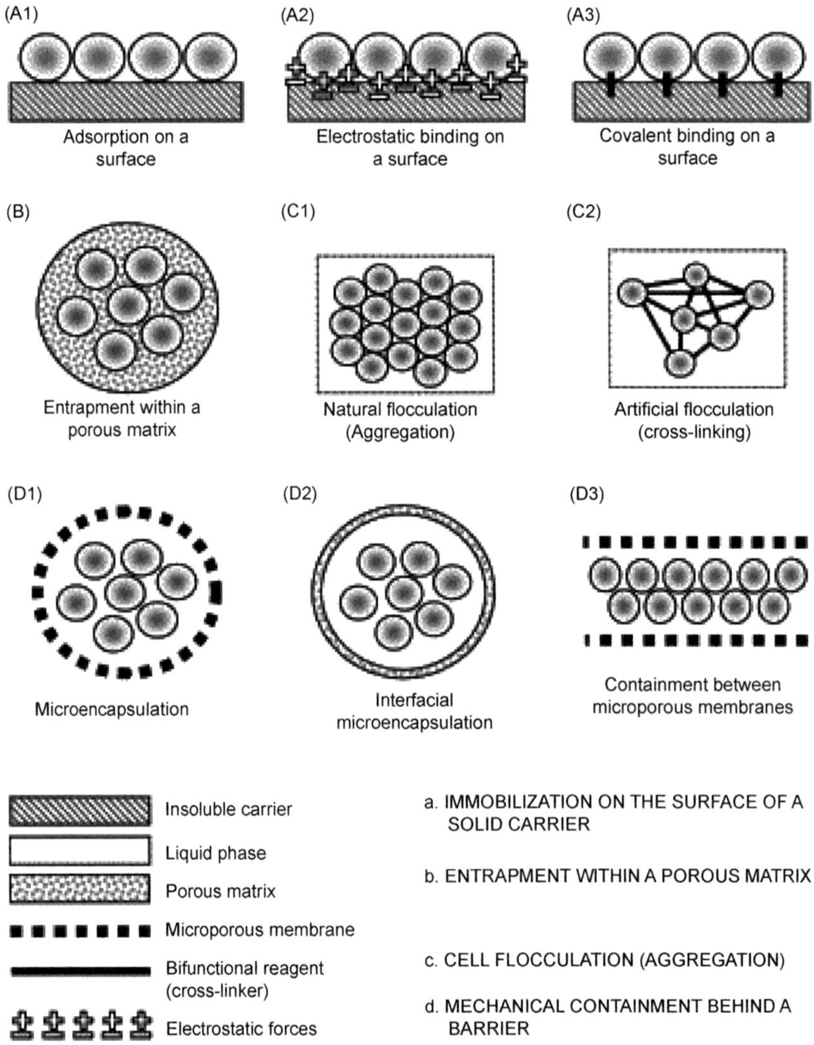

Figure 2. Basic methods of cell immobilization (Kourkoutas et al. 2004, reproduced by permission of Elsevier, License N° 4064151179113).

Carriers Used in Immobilization: Requirements and Classification

As was mentioned above, cells immobilization behind a barrier can be achieved by the use of microporous membrane filters or by entrapment of cells in a microcapsule or by the immobilization of cells on a surface. The choice of carrier is a fundamental factor in the immobilization process preparation. The material used as carrier must meet certain requisites:

1. A support must provide a wide surface, with many functional groups for cells to adhere to. Also, it must have high biomass retention capacity, an adequate porosity and should preferably be inexpensive.
2. The carrier must be insoluble, non-biodegradable, non-toxic, non-pollutant and easy to regenerate.
3. The carrier should retain good mechanical, chemical, thermal and biological stability and not be easily degraded by enzymes, solvents, pressure changes or shearing forces.

The support porosity should be uniform and controllable, allowing free exchange of substrates, products, cofactors and gasses (Kourkouta 2004, Górecka and Jastrzębska 2011). The shape of the carrier can be classified into two types, i.e., irregular and regular shapes such as beads, fibers, hollow spheres, thin films, discs, and membranes. Selection of the geometric properties for immobilization is largely dependent on the peculiarity of certain applications.

Supports can be classified as inorganic or organic, according to their chemical composition. Both types of materials are used for immobilization of viable microbial cells.

Inorganic Carrier

Inorganic carriers have a high chemical, physical, biological resistance and have good thermo stability performance. Various inorganic support materials are used such as zeolite, clay, anthracite, porous glass, activated charcoal, and ceramics A significant disadvantage of these carriers is the presence of small number of functional groups, which prevents sufficient bonding of the biocatalyst, their low adsorption capacities and relatively weak interactions with metallic ions and separation difficulties and the regeneration of some of them from water (Kourkoutas et al. 2004, Verma et al. 2006 cited by Martins et al. 2013, Bayat et al. 2015).

Organics Carriers

In general, the organic supports are hydrophilic, biodegradable, biocompatible and inexpensive. The organic polymeric carriers are more abundant than inorganic carriers and can be subdivided into natural and synthetic polymers. Various synthetic polymers (acrylamide, polyacrylamide, silica gel, polyurethane and polyvinyl resins) and natural polymers (alginate, carrageenan, agar, collagen, and chitosan) have been used as matrices for immobilization, although these last ones are more attractive because of their biocompatibility (Bayat et al. 2015). It is important to mention that chitin is an abundant waste in the fishing industry and the second most abundant polymer in nature. The most commonly natural polymer used is alginate because it is easy to handle, non-toxic to humans or to the environment, can be used to entrap microorganisms, is legally safe for human uses, is available in large quantities, and is inexpensive. However, these natural polymers are less stable in waste water than synthetic polymers. The major disadvantage of these materials is the low resistance to biodegradation, sensitivity to organic solvents and stability to narrow pH range (Moreno-Garrido 2008, Stolarzewicz et al. 2011, Martins et al. 2013).

Effects of Immobilization on Microbial Cells

Some types of alterations produced during the immobilization processes are mostly associated with physiological and metabolic activities, cell growth and metabolic pathway changes. Krishnan et al. (2001) have reported a change in the metabolic pathway of *Lactobacillus plantarum*, immobilized in a chitosan matrix, from homofermentative to the heterofermentative. This change was accompanied by variations in the microorganism morphology. In wastewater treatment environments, changes on the microbial growth, metabolic activity, nucleic acid content and diversity of mixed cultures have been registered. Free and entrapped cells systems using two different supports (alginate and polivynyl alcohol, PVA) were studied in parallel in a batch system to analyze the immobilization influence. In this study, significant changes in morphology, expression of stress-related genes (rpoS and hmpA), and nucleic acid content (DNA and RNA) were reported. A morphological change from bacilli to coccoid was observed in immobilized cells on alginate, while the PVA-entrapped cells had a slim morphology when it was compared to non-entrapped cells. The expression of genes rpoS and hmpA was higher in the alginate entrapped cells, the increase of DNA and RNA content was detected in alginate and PVA immobilized cells, respectively (Pramanik et al. 2011). A similar study was performed on immobilized mixed cultures using three different supports (alginate, carrageenan, and PVA) and three cell-to-matrix ratios. The results indicated that the immobilization process, the type of support used and the cell-to-matrix ratio had significant effects on the growth and metabolic activity of the mixed culture cells (Pramanik and Khan 2008).

Evaluating the viability and understanding the physiological, morphological and growth changes that occur in immobilized cells is fundamental for their application in the biotechnology process. Cell viability preservation has become the main goal for immobilization; therefore, monitoring physiological and metabolic changes is crucial and decisive to ensure successful of biological treatments. Main aspects related to the use of immobilized cell systems to remove heavy metals are considered below.

Immobilized Cell Systems in Heavy Metals' Removal

Heavy metals' removal in aqueous solutions by immobilized cells has been studied with a wide selection of metals and microorganisms. The proposed system combines the biosorption properties of the matrix where the microbial cells are immobilized and the ability of these cells to remove the metal. The success of this tool in bioremediation processes is based on the suitability of the matrix to immobilize cells, as well as its stability and the ability of the microbial cells to remove contaminants.

As was previously mentioned, the biofilm is an aggregate of microorganisms embedded in an organic polymer derived from the microbial origin and attached to solid surfaces. In addition to this, the biofilm polymer matrix plays a crucial role in the biosorption of heavy metals. Numerous anionic groups allow the immobilization of metal cations in the biofilm. In Fig. 3, mechanisms influencing the mobility and biotransformation of heavy metals in presence of biofilms are shown. These characteristics and properties of the biofilms can effectively be utilized to develop strategies for bioremediation processes of

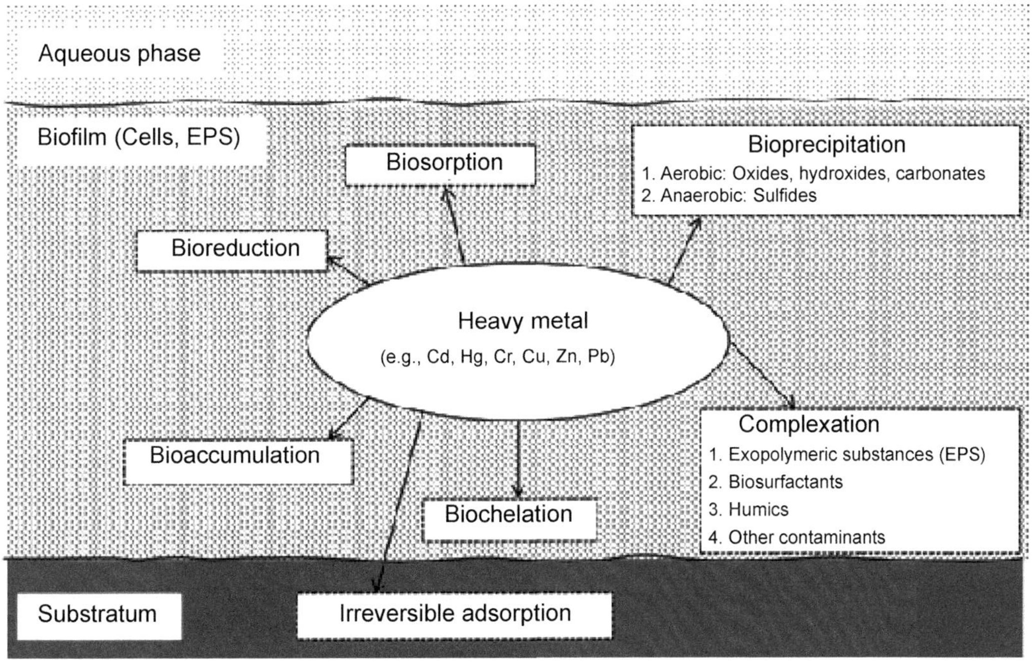

Figure 3. Mechanisms influencing the mobility and biotransformation of heavy metals in the presence of biofilms (Edwards and Kjellerup 2013, reproduced by permission of Springer, License N° 4070211295545).

inorganic compounds. There are numerous reports of the application of biofilms for heavy metals' removal. In a study performed by White and Gadd (2000) with sulphate-reducing bacteria (SRB) biofilms exposed to a medium containing 20–200 mM Cu, SRB biofilms were capable of accumulating it as copper sulfide. Costley and Wallis (2001) reported the ability of microbial biofilms to remove Cu, Zn, and Cd in successive sorption-desorption cycles in a rotating biological contactor (RBC). Passive oxidation of arsenic and iron by biofilms was successful in gold-quartz mining sites (Guezennec et al. 2012). Selenium has been reduced and subsequently concentrated in biofilms grown in tubes containing nutrients (Williams et al. 2013). A Pb-resistant bacterial strain, of *Bacillus* genus, has been isolated from highly polluted marine sediments at the Sarno River (Italy). These bacteria form a brown and compact biofilm when it is grown in Pb(II) presence, and it was demonstrated that this biofilm was able to sequester 14% of Pb(II) from the inoculated medium (Pepi et al. 2016). These instances exemplify the diverse potential use of biofilm-mediated remediation.

Biosorption processes using many biomaterials including fungal biomass, marine algae, bacteria, yeast and waste sludge, have been extensively investigated during the last decade for heavy metals' and dyes' removal because of their low cost, high efficiency, reduction in the amount of chemical and biological sludge, regeneration of the bio-sorbent and possibility for metal recovery.

The efficiency of heavy metals' removal by immobilized biomass in different systems using alginate as a matrix has also been studied. Alginate is a skeletal component of marine algae and it is commercially available as a sodium salt of alginic acid. Sinha et al. (2012) showed that *Bacillus cereus* immobilized in this material was able to absorb 104 mg mercury per g dry cells through a biosorption phenomenon in a continuous column. Similar results were obtained in batch systems with *Candida utilis* and *Candida tropicalis* immobilized in the same carrier. This work showed biosorption of high amounts of zinc ions from aqueous solution by both immobilized species. However, *C. utilis* showed better biosorption capacity for zinc ions as compared to *C. tropicalis* (Ahmad et al. 2015).

Another carrier used for microbe immobilization is chitosan, the only pseudo-natural cationic biopolymer derived from chitin, which is an excellent adsorbent due to amine ($-NH_2$) and hydroxyl (OH^-) groups' presence. This polymer has shown greater stability, remediation efficiency and renewability (de-Bashan and Bashan 2010). Chitosan also presents metal uptake ability and it can interact with various metallic species through ion exchange and/or chelation mechanism as was summarized by Miretzky and Cirelli (2009). The combination of the chitosan with microorganisms have shown excellent results, for example, Harsi et al. (2010) reported the adsorption of Pb(II) ions by chitosan immobilized *Aspergillus niger*, isolated from shrimp shells waste. These authors suggest that this adsorbent can be a good candidate for adsorption of not only Pb(II) ions but also other heavy metal ions in the wastewater stream.

Surprisingly, Saifuddin and Dinara (2012) produced a new adsorbent combining the sorption benefits of three components: chemically modified chitosan, immobilized microbial cells, and a magnetic nanoparticle into a single adsorbent. They called it as *Saccharomyces cerevisiae*-cross linked chitosan-magnetic nanoparticle (SC-CTS-ECH-MNP). The combination of magnetic nanoparticles into chitosan and immobilization of *S. cerevisiae* cells onto the matrix was carried out using microwave irradiation for both the grafting of the microbial cell onto the chitosan and the rapid synthesis of magnetic nanoparticles. The result of batch adsorption experiment with SC-CTS-ECH-MNP showed that adsorption of hexavalent uranium [U(VI)] from aqueous solution was fast. In addition to this, more than 90% of U(VI) was removed within the first 20 min. In a recent work, Tong (2017) evaluated the efficiency of alginate-chitosan microcapsule on the viability of bacteria and the capacity of *Bacillus subtilis*/alginate-chitosan microcapsule to remove the uranium ion from aqueous solution using batch adsorption experiments. The results showed that the alginate-chitosan microcapsule significantly improved the viability of *B. subtilis* and that the *B. subtilis*/alginate-chitosan microcapsule strongly removed uranium. Therefore, the author affirms that the *B. subtilis*/alginate-chitosan microcapsule has great potential as a low-cost uranium ion bio-sorbent.

Living cells of a Cr(VI)-reducing fungus (*Rhizopus* sp.) were immobilized and applied in continual reduction of Cr(VI). Different concentrations of a synthetic matrix (polyvinyl alcohol, PVA) and the natural material (sodium alginate, SA) were combined to immobilize spores for Cr(VI) reduction experiments. Spores immobilized with 3% PVA and 3% SA produced the most stable and efficient bio-beads. The bio-beads could be reused to reduce Cr(VI) for more than 30 cycles during an 82-d operation period. Afterwards, bio-beads were washed and transferred to fresh medium. Additionally, these results showed that the use of immobilized live fungus for the removal of Cr(VI) was stable for a long-term treatment, it was easy to reuse and showed less biomass leakage (Liu et al. 2012).

Several studies have been reported about the use of electrospun nanofibers/membranes as carrier materials applied to heavy metals' removal (Haider et al. 2009, Tian et al. 2011, Zhou et al. 2011, Horzum et al. 2013, Chauhan et al. 2014). However, studies regarding the use of microorganisms integrated to electrospun fibers/membranes for heavy metal removal have recently begun to appear in the literature (Lenik et al., 2017). These authors studied the enhanced capability of *Micrococcus luteus* to bind copper under three configurations: free cells on an agar surface and cells encapsulated in alginate or in electrospun polymer composites. Electrospun and alginate-encapsulated bacteria showed the highest capability to bind copper without finding significant differences between these techniques in relation to copper removal. However, the fibers have some advantages like insolubility of polymer in water, providing the opportunity to use it in water-based applications such as a sewage filter. In addition to this, such polymer composites could potentially be used in applications such as treatment of wastewater with a high concentration of copper or other heavy metals.

In the last decade, the use of industrial waste as a carrier in the immobilization process has been of interest because the high cost of culture growth medium is a big problem faced in the scaling-up of biological processes involved in wastewater treatment. Therefore, waste and agricultural by-products are good and cheap candidates to be used as supports in the removal of heavy metals from wastewater. For example, Ahmad et al. (2015) used sugarcane bagasse as a nutrient and support material for the Cr (VI) reducing biofilm and found that this waste can be used as a cost-effective alternative for growth and cell immobilization.

Many polluted areas often contain diverse pollutants in a simultaneously form, thereby, it is very important to consider remediation procedures utilizing microbial consortia formed by microorganisms with an enhanced ability to remove different pollutants. In nature, most processes are performed by a plethora of microorganisms, many of which have not been even cultured yet. Indeed, microbial consortia are implicated in practices fundamental to humans such as food digestion. Microbial mediated environmental protection and remediation processes involve multiple microbial consortia which can perform functions that are difficult or even impossible for individual strains or species. A microbial consortium is defined as the natural association of two or more microbial populations of different species, which act together as a community in a complex system where everyone benefits from the activities of others (Ochoa Carreño and Montoya 2010). If this microbial consortium with excellent characteristics applied to a bioremediation process is immobilized, the efficiency increases. Von Canstein et al. (2002) worked with packed-bed bioreactors for treatment of mercury-contaminated wastewater. Mixed culture biofilms were able to retain a higher amount of mercury and higher diversity in the presence of rapidly changing mercury concentrations compared with monoculture biofilms. These results showed that more diverse biofilms were more efficient for metals bioremediation.

Another study employed a mixed population of SRB treated contaminated waters with mildly acidic metal (Cu, Zn, Ni, Fe, Al, and Mg), arsenic and sulfate in a bench-scale up flow anaerobic packed bed reactor filled with silica sand (Jong and Parry 2003). The activity of SRB increased the water pH from 4.5 to 7.0 and enhanced the removal of sulfate and metals in comparison to controls not inoculated with SRB (Jong and Parry 2003). Similar results were obtained in the treatment of synthetic acid-mine drainage (AMD) with another mixture of SRB. A column reactor was inoculated with sulfate reducing bacteria and fed with a solution containing sulfate and heavy metals [As(V), Cd, Cr(VI), Cu and Zn]. At steady state, the sulfate reduction was 50%, while the metals were completely eliminated. The reduction of AMD toxicity was evaluated using the nematode *Caenorhabditis elegans* as a test organism. A toxicity lethality test was performed before and after treatment, showing that only 5% of the animals were alive after 48 hr in the presence of contaminants, while the percentage increased to 73% when the nematodes were exposed to the solution eluted from the column (Cruz Viggi et al. 2010).

Increased microbial diversity can result in the establishment of a network of ecological and metabolic niches allowing biofilm associated microorganisms to survive in the midst of rapidly changing environmental stress, which results in greater performance whether that takes place in a bioreactor or *in situ* in the environment (Edwards and Kjellerup 2013).

Microbial Enzymes and Metabolites Immobilized in the Removal of Heavy Metals

Enzymes immobilization

Although the recent publications refer to immobilization of microbial cells to environmental applications, enzyme immobilization has been carried out since the 19th century and has been perfected and diversified from that time to present times (Brena et al. 2013). Immobilization of enzymes is one of the important facets of biotechnology and it is a technique that refers to the physical confinement of enzymes within matrices or supports that help to maintain their catalytic activity and stability, allowing them to be used continuously and repeatedly (Katchalski-Katzir 1993, Aggarwal and Sahni 2012). Immobilized enzymes have many advantages in relation to free enzymes such as cost reduction of processes because they can be reused many times, are easily separated from reaction mixtures, and have the possibility of using higher enzyme activity per unit volume in the reactor (Munjal and Sawhney 2002). Enzyme immobilization techniques are the same as those mentioned above for cell immobilization. Therefore, natural and synthetic support materials have been used for cell and enzyme immobilization. Some authors speculate that chromate reductases found in chromium resistant microorganisms are a good candidate to develop an environmental friendly process for the remediation of contaminated sites with this heavy metal. The reduction of Cr(VI) to trivalent form followed by precipitation is actually the most common remediation method.

Chromate reductases catalyse the reduction of Cr(VI) to Cr(III). The reduction can either be aerobic, anaerobic or sometimes both. Membrane-bound reductases are involved in anaerobic reduction. These proteins can be part of electron transport systems and use chromate as the terminal electron acceptor. On the other hand, aerobic Cr(VI) reduction is generally associated with soluble proteins localized in the cytoplasm and requires NAD(P)H as the electron donor (Shen and Wang 1993). Many of the microorganisms that catalyze redox reactions use the metals or metalloids as terminal electron acceptors in anaerobic respiration (Thatoi et al. 2014).

Elangovan et al. (2010) evaluated chromium reduction by immobilized Chromium reductase associated with the cell-free extracts of *Arthrobacter rhombi*-RE. Among the various immobilization matrices screened, calcium alginate beads were the most effective one. Also, this work showed that immobilized enzyme in alginate beads can be reused in many operation cycles, though there was 37% reduction in efficiency by the end of the fourth cycle. Oxydoreductases enzymes found in bacteria that catalyze Cr(VI) reduction could be flavin reductases, NADH/NADPH-dependent reductases, nitroreductases, quinone reductases, iron reductase and dehydrogenases (Thatoi et al. 2014). All these enzymes could be important in the development of bioremediation processes.

Though many partially purified and crude enzymes are reported to have significant Cr(VI) reduction potential, their application in Cr(VI)-contaminated wastewater treatment is limited due to the expensive and tedious enzymes' separation. Many reports are available on the use of whole cell reactors for the treatment of Cr(VI)-contaminated wastewater. However, data on the application of enzyme-immobilized reactors for the same is scanty, thereby, this topic needs to be studied and explored in the near future.

Immobilized surfactants for heavy metals' removal

A good adsorbent for heavy metals must possess certain characteristics such as bind the metal strongly in environmental conditions, should have an affinity for specific metals and release the bound metals when conditions change, with the objective of recovering the adsorbent to be used in a new cycle of remediation.

Some adsorbents widely used for heavy metals' remediation are clays, especially montmorillonite, which possesses chemical and mechanical stability, high specific surface area, high cation exchange and swelling ability. Clays possess a net negative charge resulting from isomorphic substitution of cations in the crystal lattice. However, they have an inherent negative surface charge with weak affinity for divalent metal ions. Some studies were focused on the modification of clays with organic cations to alter the surface to improve the capability of heavy metals' adsorption. A very common modification is to immobilize cationic surfactants such as the quaternary ammonium surfactant hexadecyl-trimethyl-ammonium bromide (HDTMA-Br), followed by chromate adsorption at the integral framework of the clay, to create a surfactant-immobilized interlayer species bonded to montmorillonite clay (SIIS-clay), useful for adsorption of inorganic and organic species (Krishna et al. 2004, Mahadevaiah et al. 2011).

This kind of modified clays have been used to remove Pb(II) ions from aqueous solution (Krishna et al. 2004) and permanganate in the pH range (pH = 1.0–4.0) by adsorption (Mahadevaiah et al. 2011). Krishna et al. (2004) recovered the SIIS-clay by using 0.02 M ethylendiamine tetra acetic acid (EDTA) solution and reused the adsorbent without problems in a new performance of lead ion adsorption.

Most of the surfactants actually used are synthetic. However, surfactants can be produced by eukaryotic or prokaryotic microorganisms. The use of bio-surfactants could be more appropriate than synthetic surfactants in bioremediation processes given that bio-surfactants are biodegradables and present low toxicity (Christofi and Ivshina 2002). In recent years, interest in the use of bio-emulsifiers as washing agents that can enhance desorption of soil-bound metals has been increasing. Most of the bio-surfactants are anionic or neutral. However, a few of them are cationic-containing majorly amine groups and they can probably be immobilized into clays such as HDTMA-Br (Bustamante et al. 2012).

Gnanamania et al. (2010) have hypothesized that Cr(VI) remediation carried out by marine *Bacillus* sp. MTCC 5514 occurs through a two-step process: the extracellular enzymatic reduction of the hexavalent species to Cr(III), and after that, the entrapment of the Cr(III) by a biosurfactant produced by the same strain. However, a work on bioemulsifier production by an actinobacterium was able to mediate direct Cr(VI) recovery without prior reduction (Colin et al. 2013).

The immobilization of biosurfactants appears promising for the development of soil remediation technologies for heavy metal based upon direct use of these microbial emulsifiers.

Siderophore activity: Bacterial siderophores are produced when soluble iron in environment is limited and they act as solubilizing agents for insoluble Fe(III) from minerals or organic compounds. Siderophores are able to bind other metals other than iron and allow their solubility. Heavy metals' presence in the environment can enhance siderophore biosynthesis by bacteria (Rajkumar et al. 2010). On the other hand, the presence of high concentrations of soluble iron can inhibit siderophores biosynthesis and bacteria can suffer toxicity of some other heavy metals (Braud et al. 2007).

In bioremediation processes applied to heavy metals contaminated sites, it is common to produce a bioaugmentation process to solubilize heavy metals and then proceed to a phytoextraction from the environment. For the bioaugmentation process, exogenous microorganisms capable to synthesize siderophores are very important. Microbial survival of inoculated microorganisms should be guaranteed and this can be possible if microorganisms are immobilized into carriers (alginate, clays, etc.) to protect them against natural competition with endogenous microorganisms and predation (Gentry et al. 2004, Braud et al. 2007). On the other hand, it is necessary to provide conditions to favor siderophores biosynthesis.

Braud et al. (2006) immobilized *Pseudomonas aeruginosa* and *P. fluorescens* into Ca-alginate beads to create a deficiency in minerals and Fe inside beads and increased siderophore activity under these conditions. They also demonstrated that cell-immobilization of *P. aeruginosa* along with skim milk (C and N substrate) can give a competitive advantage to the inoculated bacterium against endogenous microorganisms. They found that μ_{max} of immobilized cells cultivated with skim milk was higher than that of free cells and siderophore activity remained constant under different conditions such as presence or absence of soluble iron with or without other toxic metals in the culture media (Braud et al. 2007). This cultivation method can help to maintain the close limits of environmental conditions inside beads even when external conditions of environment change and can help to keep siderophore activity in narrow limits with a culture medium containing (or not) Fe and other metals. These benefits could be relevant to the stabilization of bioremediation performances in contaminated soils with changing conditions, the main reason for the failure of these biological processes.

Concluding Remarks

Numerous biotechnological activities have benefited from the use of technologies involving the immobilization of cells or enzymes, especially in the food industry and alcoholic beverages. This fits well with the concept of "White Biotechnology", which advises the optimization of biotechnological processes with the reduction of waste and energy consumption and contributes to a more sustainable future. But in the field of environmental biotechnology or "Grey Biotechnology", much remains to be done. A better understanding of these processes and the physiology of immobilized microorganisms will contribute to the implementation of this technology in other processes such as effluent treatment or bioremediation.

Bioremediation is a potential strategy for the decontamination of the natural environment, modified mainly by anthropogenic activities. Its application *in situ* is not very widespread, except in some cases of decontamination of environments contaminated with petroleum hydrocarbons where it has been used with good results.

The use of immobilized microbial cells, as well as their metabolites capable of modifying toxic substances in natural environments, offers a promising alternative and is more environmental friendly than the strategies used so far.

It is hoped that the compilation of the current literature about bioremediation of heavy metals using immobilized microbial cells and metabolites will encourage the scientific community to approach this issue for its prompt application. This strategy must be strongly promoted as it constitutes a harmless and compatible environmental sanitation tool for a healthier planet.

Acknowledgments

The support of CONICET and Agencia Nacional de Promoción Científica y Tecnología Argentina (PICT 2013 N° 3170) are greatly appreciated.

References Cited

Aggarwal, S. and S. Sahni. 2012. The commercial exploitation of immobilized enzymes. Int. Proc. Chem. Biol. Environ. Eng. 41: 18–22.

Ahmad, W.H.W., J.B. Chyan, Z.A. Zakaria and W.A. Ahmad. 2015. Sugarcane bagasse as nutrient and support material for Cr (VI)-reducing biofilm. Int. Biodeter. Biodegr. 102: 3–10.

Angelim, A.L., S.P. Costa, B.C.S. Farias, L.F. Aquino and V.M.M. Melo. 2013. An innovative bioremediation strategy using a bacterial consortium entrapped in chitosan beads. J. Environ. Manag. 127: 10–17.

Bonilla, J.O., E.A. Callegari, C.D. Delfini, M.C. Estevez and L.B. Villegas. 2016. Simultaneous chromate and sulfate removal by *Streptomyces* sp. MC1. Changes in intracellular protein profile induced by Cr(VI). J. Basic. Microbiol. 56: 1212–1221.

Braud, A., K. Jézéquel, M.A. Léger and T. Lebeau. 2006. Siderophore production by using free and immobilized cells of two *Pseudomonads* cultivated in a medium enriched with Fe and/or toxic metals (Cr, Hg, Pb). Biotechnol. Bioeng. 94: 1080–1088.

Braud, A., K. Jézéquel and T. Lebeau. 2007. Impact of substrates and cell immobilization on siderophore activity by *Pseudomonads* in a Fe and/or Cr, Hg, Pb containing-medium. J. Hazard. Mater. 144: 229–239.

Brena, B., P. González-Pombo and F. Batista-Viera. 2013. Immobilization of Enzymes: A Literature Survey. Chapter 2. En J. Guisan, Immobilization of Enzymes and Cells: Third Edition, Methods in Molecular Biology, 1051. New York: Springer Science+Business Media.

Bustamante, M., N. Durán and M.C. Diez. 2012. Biosurfactants are useful tools for the bioremediation of contaminated soil: a review. J.S.S.P.N. 12(4): 667–687.

Cabuk, A., T. Akar, S. Tunali and Ö. Tabak. 2006. Biosorption characteristics of *Bacillus* sp. ATS-2 immobilized in silica gel for removal of Pb(II). J. Hazard. Mater. 136: 317–323.

Cabuk, A., T. Akar, S. Tunali and S. Gedikli. 2007. Biosorption of Pb(II) by industrial strain of *Saccharomyces cerevisiae* immobilized on the biomatrix of cone biomass of Pinusnigra: equilibrium and mechanism analysis. Chem. Eng. J. 131: 293–300.

Campos, V., R. Moraga, I. Fernández, F. Yáñez, A. Valenzuela and M.A. Mondaca. 2013. Reduction of hexavalent cromium by *Serratia marcecens* immobilized on active carbon and their potential use in bioremediation. Gayana (Concepción) 7: 61–63. https://dx.doi.org/10.4067/S0717-65382013000100008.

Chauhan, D., J. Durivedi and N. Sankaramakrishnan. 2014. Novel chitosan/PVA/zerovalent iron biopolymeric nanofibers with enhanced arsenic removal applications. Environ. Sci. Pollut. Res. 21: 9430–9442.

Chowdhury, S., M.A. Mazumder, O. Al-Attas and T. Husain. 2016. Heavy metals in drinking water: Occurrences, implications, and future needs in developing countries. Sci. Total Environ. 569-570: 476–88.

Christofi, N. and I.B. Ivshina. 2002. Microbial surfactants and their use in field studies of soil remediation. J. Appl. Microbiol. 93: 915–929.

Colin, V.L., M.F. Castro, M.J. Amoroso and L.B. Villegas. 2013. Production of bioemulsifiers by *Amycolatopsis tucumanensis* DSM 45259 and their potential application in remediation technologies for soils contaminated with hexavalent chromium. J. Hazard Mater. 261: 577–583.

Costley, S.C. and F.M. Wallis. 2001. Bioremediation of heavy metals in a synthetic wastewater using a rotating biological contactor. Water Res. 35: 3715–3723.

Covizzi, L.G., E.G. Giese, E. Gomes, R.F.H. Dekker and R. Da Silva. 2007. Immobilization of microbial cells and their biotechnological applications. Semina: Ciencias Exatas e Tecnológicas, Londrina. 28: 143–160.

Cruz Viggi, C., F. Pagnanelli, A. Cibati, D. Uccelletti, C. Palleschi and L. Toro. 2010. Biotreatment and bioassessment of heavy metal removal by sulphate reducing bacteria in fixed bed reactors. Water Res. 44: 151–158.

Das, M. and A. Adholeya. 2015. Potential uses of immobilized bacteria, fungi, algae, and their aggregates for treatment of organic and inorganic pollutants in wastewater. ACS Symposium Series, Chapter 15, 1206–1319.

de-Bashan, L.E and Y. Bashan. 2010. Immobilized microalgae for removing pollutants: Review of practical aspects. Bioresour. Technol. 101: 1611–1627.

Dellagnezze, B.M., S. Vasconcellos, A. Angelim, V.M.M. Melo, S. Santisi, S. Cappello et al. 2016. Bioaugmentation strategy employing a microbial consortium immobilized in chitosan beads for oil degradation in mesocosm scale. Mar. Pollut. Bull. 107: 107–117.

Dua, M., A. Sing, N. Sethunathan and A. Johri. 2002. Biotechnology and bioremediation: successes and limitations. Mini-review. Appl. Microbiol. Biotechnol. 59: 143–152.

Edwards, S.J. and V. KjellerupBirthe. 2013. Applications of biofilms in bioremediation and biotransformation of persistent organic pollutants, pharmaceuticals/personal care products, and heavy metals. Appl. Microbiol. Biotechnol. 97: 9909–9921.

Elangovan, R., L. Philip and K. Chandraraj. 2010. Hexavalent chromium reduction by free and immobilized cell-free extract of *Arthrobacter rhombi*-RE. Appl. Biochem. Biotechnol. 160: 81–97.

Flemming, H.C., J. Wingender, U. Szewzyk, P. Steinberg, S.A. Rice and S. Kjelleberg. 2016. Biofilms: an emergent form of bacterial life. Nat. Rev. Microbiol. 14: 563–575.

Gensheimer, M., A. Brandis-Heep, S. Agarwal, R.K. Thauer and A. Greiner. 2011. Polymer/bacteria composite nanofiber non-wovens by electrospinning of living bacteria protected by hydrogel microparticles. Macromol. Biosci. 11: 333–337.
Gentry, T.J., C. Rensing and I.L. Pepper. 2004. New approaches for bioaugmentation as a remediation technology. Crit. Rev. Environ. Sci. Technol. 34: 447–494.
Gnanamani, A., V. Kavitha, N. Radhakrishnan, G. Suseela Rajakumar, G. Sekaran and A.B. Mandal. 2010. Microbial products (biosurfactant and extracellular chromatereductase) of marine microorganism are the potential agents reduce the oxidative stress induced by toxic heavy metals. Colloids Surf. B. 79: 334–339.
Górecka, E. and M. Jastrzębska. 2011. Immobilization techniques and biopolymer carriers. Biotechnol. Food Sci. 75: 65–86.
Guezennec, A.-G., C. Michel, C. Joulian, M.C. Dictor and F. Battaglia-Brunet. 2012. Treatment of arsenic contaminated mining water using biofilms. Interfaces Against Pollution, Nancy, France. <Hal-00691189>
Haider, S. and S.Y. Park. 2009. Preparation of the electrospun chitosan nanofibres and their applications to the adsorption of Cu(II) and Pb(II) ions from an aqueous solution. J. Membrane Sci. 328: 90–96.
Hall-Stoodley, L., J.W. Costerton and P. Stoodley. 2004. Bacterial biofilms: From the natural environment to infectious diseases. Nat. Rev. Microbiol. 2: 95–108.
Han, B., S. Akeprathumchai, S.R. Wickramasinghe and X. Qian. 2003. Flocculation of biological cells: Experiment vs. theory. AIChE Journal 49: 1687–1701.
Hasan, F., S. Khan, A.A. Shah and A. Hameed. 2009. Production of antibacterial compounds by free and immobilized *Bacillus pumilus* SAF1. Pak. J. Bot. 41: 1499–1510.
Hasri, M. and N.H.A. Roto. 2010. Immobilization of *Aspergillus niger* biomass on chitosan and its application as an adsorbent for Pb(II) metal ion. pp. 85–90. *In*: Jumina, Siswanta, Kartini and Sudiono [eds.]. The 2nd International Conference on Chemical Sciences Proceeding (Yogyakarta): Chemistry Departement Universitas Gadjah Mada.
Horzum, N., M.M. Demir, M. Nairatc and T. Shahwan. 2013. Chitosan fibre-supported zero-valent iron nanoparticles as a novel sorbent for sequestration of inorganic arsenic. RSC Adv. 3: 7828–7837.
Ivanova, V., P. Petrova and J. Hristov. 2011. Application in the ethanol fermentation of immobilized yeast cells in matrix of alginate/magnetic nanoparticles, on chitosan-magnetite microparticles and cellulose-coated magnetic nanoparticles. Int. Rev. Chem. Eng. 3: 289–299.
John, R.P., R.D. Tyagi, S.K. Brar, R.Y. Surampalli and D. Prévost. 2011. Bioencapsulation of microbial cells for targeted agricultural delivery. Crit. Rev. in Biotechn. 31: 211–226.
Jong, T. and D.L. Parry. 2003. Removal of sulfate and heavy metals by sulfate reducing bacteria in short-term bench scale upflow anaerobic packed bed reactor runs. Water Res. 37: 3379–3389.
Kalmokoff, M.L., J.W. Austin, X.D. Wan, G. Sanders, S. Banerjee and J.M. Farber. 2001. Adsorption, attachment and biofilm formation among isolates of *Listeria monocytogenes* using model conditions. J. Appl. Microbiol. 91: 725–734.
Katchalski-Katzir, E. 1993. Immobilized enzymes: learning from past successes and failures. Trends Biotechnol. 11: 471–478.
Konopka, A. 2009. What is microbial community ecology? ISME J. 3: 1223–1230.
Kourkoutas, Y., A. Bekatorou, I.M. Banat, R. Marchant and A.A. Koutinas. 2004. Immobilization technologies and support materials suitable in alcohol beverages production: a review. Food Microbiol. 21: 377–397.
Krishna, B.S., N. Mahadevaiah, D.S.R. Murty and B.S. Jai Prakash. 2004. Surfactant immobilized interlayer species bonded to montmorillonite as recyclable adsorbent for lead ions. J. Colloid Interface Sci. 271: 270–276.
Krishnamoorthi, S., A. Banerjee and A. Roychoudhury. 2015. Immobilized enzyme technology: Potentiality and prospects. J. Enzymol. Metab. 1: 1–11.
Krishnan, S., L.R. Gowda, M.C. Misra and N.G. Karanth. 2001. Physiological and morphological changes in immobilized *Lactobacillus plantarum* NCIM 2084 cells during repeated batch fermentation for production of lactic acid. Food Biotechnol. 15: 193–200.
Kumar, A., B.S. Bisht, V.D. Joshi and T. Dhewa. 2011. Review on bioremediation of polluted environment: A management tool. Int. J. Environ. Sci. 1: 1079–1093.
Letnik, I., R. Avrahami, R. Port, A. Greiner, E. Zussman, J. Rokem et al. 2017. Biosorption of copper from aqueous environments by *Micrococcus luteus* in cell suspension and when encapsulated. Int. Biodet. Biodeg. 116: 64–72.
Liu, H., L. Guo, S. Liao and G. Wang. 2012. Reutilization of immobilized fungus *Rhizopus* sp. LG04 to reduce toxic chromate. J. Appl. Microbiol. 112: 651–659.
Luo, C.J., S.D. Stoyanov, E. Stride, E. Pelan and M. Edirisinghe. 2012. Electrospinning versus fibre production methods: from specifics to technological convergence. Chem. Soc. Rev. 41: 4708–4735.
Luqueño, F., V.C. Encinas, R. Marsch, C. Martínez-Suárez, E. Vázquez-Núñez and L. Dendooven. 2011. Microbial communities to mitigate contamination of PAHs in soil possibilities and challenges: a review. Environ. Sci. Pollut. Res. 18: 12–30.

Mahadevaiah, N., B. Vijayakumar, K. Hemalatha and B.S. Jai Prakash. 2011. Uptake of permanganate from aqueous environment by surfactant modified montmorillonite batch and fixed bed studies. Bull. Mater. Sci. 34: 1675–1681.

Martins, S.C.S., C.M. Martins, L.M.C. Guedes Fiúza and S.T. Santaella. 2013. Immobilization of microbial cells: A promising tool for treatment of toxic pollutants in industrial wastewater. Afr. J. Biotechnol. 12: 4412–4418.

Miretzky, P.A. and F. Cirelli. 2009. Hg(II) removal from water by chitosan and chitosan derivatives: A review. J. Hazard. Mater. 167: 10–23.

Moreno-Garrido, I. 2008. Microalgae immobilization: current techniques and uses. Bioresour. Technol. 99: 3949–3964.

Morgan-Sastume, F., P. Larsen, J.L. Nielsen and P.H. Nielsen. 2008. Characterization of the loosely attached fraction of activated sludge bacteria. Water Res. 42: 843–854.

Munjal, N. and S.K. Sawhney. 2002. Stability and properties of mushroom tyrosinase entrapped in alginate, polyacrylamide and gelatin gels. Enzyme. Microb. Technol. 30: 613–619.

Nabweteme, R., H.-S. Kwon, S. Park, L. Chang-Ha and A. Ik-Sung. 2016. Immobilized culture of *Sulfurovumlithotrophicum* 42BKTT in polyurethane foam cubes. J. Ind. Eng. Chem. 148: 200–208.

Nakić, Z., K. Posavec and A. Bacani. 2007. A visual basic spreadsheet macro for geochemical background analysis. Ground Water. 45: 642–647.

Nisha, S., S. Arun Karthick and N. Gobi. 2012. A review on methods application and properties of immobilized enzyme. Chem. Sci. Rev. 1: 148–155.

Ochoa Carreño, D.C. and R.A. Montoya. 2010. Microbial consortia: a biological metaphor applied to enterprise association in agricultural production chains. Rev. Fac. Cienc. Econ. 18: 55–74.

Okoh, A.I. and M.R. Trejo-Hernandez. 2006. Remediation of petroleum hydrocarbon polluted systems: exploiting the bioremediation strategies. Afr. J. Biotechnol. 5: 25.

Park, J.K. and H.N. Chang. 2000. Microencapsulation of microbial cells. Biotechnol. Adv. 18: 303–319.

Pepi, M., M. Borra, S. Tamburrino, M. Saggiomo, A. Viola and E. Biffali. 2016. A *Bacillus* sp. isolated from sediments of the Sarno River mouth, Gulf of Naples (Italy) produces a biofilm biosorbing Pb(II). Sci. Total Environ. 562: 588–595.

Piñón-Castillo, H.A., E.M. Brito, M. Goñi-Urriza, R. Guyoneaud, R. Duran, G.V. Nevarez-Moorillon et al. 2010. Hexavalent chromium reduction by bacterial consortia and pure strains from an alkaline industrial effluent. J. Appl. Microbiol. 109: 2173–2182.

Poopal, A.C. and R.S. Laxman. 2008. Hexavalent chromate reduction by immobilized *Streptomyces griseus*. Biotechnol. Lett. 30: 1005–1010.

Pramanik, S. and E. Khan. 2008. Effects of cell entrapment on growth rate and metabolic activity of mixed cultures in biological wastewater treatment. Enzyme Microb. Technol. 43: 245–251.

Pramanik, S., R. Khanna, K. Katti, J. McEvoy and E. Khan. 2011. Effects of entrapment on nucleic acid content, cell morphology, cell surface property, and stress of pure cultures commonly found in biological wastewater treatment. Appl. Microbiol. Biotechnol. 92: 407–418.

Rajkumar, M., A. Noriharu, M.N.V. Prasad and H. Freitas. 2010. Potential of siderophore-producing bacteria for improving heavy metal phytoextraction. Trends Biotechnol. 28: 142–149.

Ramakrishna, S.V. and R.S. Prakasha. 1999. Microbial fermentations with immobilized cells. Curr. Sci. 77: 87–100.

Saifuddin, N. and A.Z. Raziah. 2007. Removal of heavy metals from industrial effluent using *Saccharomyces cerevisiae* (Baker's Yeast) immobilized in chitosan/lignosulphonate. Matrix. J. Appl. Sci. Res. 3: 2091–2099.

Saifuddin, N. and S. Dinara. 2012. Immobilization of *Saccharomyces cerevisiae* onto cross-linked chitosan coated with magnetic nanoparticles for adsorption of Uranium (VI) ions. Adv. Nat. Appl. Sci. 6: 249–267.

Sharma, S. and P. Malaviya. 2016. Bioremediation of tannery wastewater by chromium resistant novel fungal consortium. Ecol. Eng. 91: 419–425.

Shen, H. and Y.T. Wang. 1993. Characterization of enzymatic reduction of hexavalent chromium by *Escherichia coli* ATCC 33456. Appl. Environ. Microbiol. 59: 3771–3777.

Shyamkumar, R., I.M. Moorthy, K. Ponmurugan and R. Baskar. 2014. Production of L-glutamic Acid with *Corynebacterium glutamicum* (NCIM 2168) and *Pseudomonas reptilivora* (NCIM 2598): A study on immobilization and reusability. Avicenna J. Med. Biotechnol. 6: 163–168.

Sinha, A., K.K. Pant and S.K. Khare. 2012. Studies on mercury bioremediation by alginate immobilized mercury tolerant *Bacillus cereus* cells. Int. Biodeterior. Biodegradation 71: 1–8.

Soares, E.V. 2011. Flocculation in *Saccharomyces cerevisiae*: a review. J. Appl. Microbiol. 110: 1–18.

Stolarzewicz, I., E. Bialecka-Florjańczyk, E. Majewska and J. Krzyczkowska. 2011. Immobilization of yeast on polymeric supports. Chem. Biochem. Eng. 25: 135–144.

Thatoi, H., S. Das, J. Mishra, B.P. Rath and N. Das. 2014. Bacterial chromate reductase, a potential enzyme for bioremediation of hexavalent chromium: A review. J. Environ. Manage. 146: 383–399.

Tian, Y., M. Wu, R. Liu, Y. Li, D. Wang, J. Tan et al. 2011. Electrospun membrane of cellulose acetate for heavy metal ion adsorption in water treatment. Carbohyd. Polym. 83: 743–748.

Tong, K. 2017. Preparation and biosorption evaluation of *Bacillus subtilis*/alginate–chitosan microcapsule. Nanotechnol. Sci. Appl. 10: 35–43.

Tripathi, A., H. Sami, S.R. Jain, M. Viloria-Cols, N. Zhuravleva, G. Nilsson et al. 2010. Improved bio-catalytic conversion by novel immobilization process using cryogel beads to increase solvent production. Enzyme Microb. Technol. 47(1–2): 44–51.

Tsekova, K., D. Todorova and S. Ganeva. 2010. Removal of heavy metals from industrial wastewater by free and immobilized cells of *Aspergillus niger*. Int. Biodeter. Biodegradation 64: 447–451.

Vert, M., Y. Doi, K.H. Hellwich, M. Hess, P. Hodge, P. Kubisa et al. 2012. Terminology for biorelated polymers and applications (IUPAC Recommendations 2012). Pure Appl. Chem. 84: 377–410.

Villegas, L.B., P.M. Fernández, M.J. Amoroso and L.I. de Figueroa. 2008. Chromate removal by yeasts isolated from sediments of a tanning factory and a mine site in Argentina. Biometals 21: 591–600.

Villegas, L.B., M.J. Amoroso and L.I. de Figueroa. 2009. Responses of *Candida fukuyamaensis* RCL-3 and *Rhodotorula mucilaginosa* RCL-11 to copper stress. J. Basic Microbiol. 49: 395–403.

Villegas, L.B., C. Pereira, V.L. Colin and C.M. Abate. 2013. The effect of sulphate and phosphate ions on Cr(VI) reduction by *Streptomyces* sp. MC1, including studies of growth and pleomorphism. Int. Biodeter. Biodegr. 82: 149–156.

Von Canstein, H., S. Kelly, Y. Li and I. Wagner-Döbler. 2002. Species diversity improves the efficiency of mercury-reducing biofilms under changing environmental conditions. Appl. Environ. Microbiol. 68: 2829–2837.

Wagner-Döbler, I. 2003. Pilot plant for bioremediation of mercury-containing industrial wastewater. Appl. Microbiol. Biotechnol. 62: 124–133.

Waldron, K.J., J.C. Rutherford, D. Ford and N.J. Robinson. 2009. Metalloproteins and metal sensing. Nature 460: 823–830.

Wei, Q. and L.Z. Ma. 2013. Biofilm matrix and its regulation in *Pseudomonas aeruginosa*. Int. J. Mol. Sci. 14: 20983–21005.

White, C. and G.M. Gadd. 2000. Copper accumulation by sulfate-reducing bacterial biofilms. FEMS Microbiol. Left. 183: 313–318.

Williams, K.H., M.J. Wilkins, A.L. N'Guessan, B. Arey, E. Dodova, A. Dohnalkova et al. 2013. Field evidence of selenium bioreduction in a uranium-contaminated aquifer. Environ. Microbiol. 5: 444–452.

Zhou, W., J. He, S. Cui and W. Gao. 2011. Preparation of electrospun silk fibroin/cellulose acetate blend nanofibers and their applications to heavy metal ions adsorption. Fiber Polym. 12: 431–437.

15

Hexavalent Chromium Removal Related to Scale-up Studies

Pablo M. Fernández,[1,2,*] *Silvana C. Viñarta,*[1,3,a] *Anahí R. Bernal*[1,b] and *Lucía I. Castellanos de Figueroa*[1,4,c]

Introduction

Contamination of soil, air, and water with heavy metals and toxic chemicals is one of the major problems facing the world today. Environmental pollution by industrial effluents, especially those containing heavy metals, was previously reported (Klimaviciute et al. 2010). Discharge of heavy metals became a worldwide severe socio-environmental problem due to the risks they can generate for ecosystems and human health. Many heavy metals (lead, arsenic, chromium, zinc, copper, cadmium, cobalt, antimony mercury, nickel, etc.) in their elemental forms or in various chemical combinations are considered toxic (Gavrilescu 2004, Wang and Chen 2006), but some of these are useful in low concentration. They are non-degradable toxic pollutants (Pavel et al. 2012, Modoi et al. 2014), thus persistent in nature that accumulate in the food chain, which, with time, reach detrimental levels in living systems, resulting in serious health diseases such as irritation and/or cancer in lungs and digestive tract, low growth rates in plants and death of animals (Cheung and Gu 2007, Orozco et al. 2008).

Chromium is a geochemical element of anthropogenic origin and is widely distributed in rocks, minerals soils, and fresh water. Environmental risk due to Cr(VI) contamination is caused by different industrial applications and commercial processes, which eventually lead to environment pollution (soil, surface water or atmosphere), in both developing and

[1] Planta Piloto de Procesos Industriales Microbiológicos PROIMI-CONICET. Av. Belgrano y Pje. Caseros, San Miguel de Tucumán, T4001MVB, Tucumán, Argentina.

[a] Email: scvinarta@hotmail.com

[b] Email: anahirbernal@gmail.com

[c] Email: proimiunt@gmail.com

[2] Calidad de Productos II, Gestión de Empresas Agroindustriales, Universidad de San Pablo – T, Av. Solano Vera y Camino a Villa Nougués, San Pablo, Tucumán, Argentina.

[3] Microbiología General, Facultad de Ciencias Agrarias, Universidad Nacional de Catamarca, Av. Belgrano 300, 4700, Catamarca, Argentina.

[4] Microbiología Superior, Facultad de Bioquímica, Química y Farmacia, Universidad Nacional de Tucumán, Ayacucho 450, 4000, Tucumán, Argentina.

* Corresponding author: pfernandez@proimi.org.ar

developed countries. Effluents from electroplating, wood preservation, leather, mining industries, and others, are responsible for the release of chromium into the air, soil and water (Tekerlepoulou et al. 2013). Trivalent [Cr(III)] and hexavalent [Cr(VI)] forms are the dominant chromium oxidation states in the environment. Cr(VI) compounds (mainly CrO_4^{2-} at neutral pH or alkaline conditions) are toxic to living organisms, causing allergies, irritations and respiratory diseases, being also mutagenic and carcinogenic due to their oxidizing nature (Costa 2003). This heavy metal has been designated a priority pollutant in many countries and by the United States Environmental Protection Agency-USEPA (Ksheminska et al. 2003, Juvera-Espinosa et al. 2006). Its toxicity to biological systems is due to the fact that Cr(VI) complexes can easily cross cellular membranes and undergo immediate reduction, leading to a variety of oxidative intermediates harmful to organelles, proteins and nucleic acids (Morales-Barrera and Cristiani-Urbina 2006). Cr(VI) may also exert its toxicity as a redox active metal that can participate in Fenton reactions, being able to generate Reactive Oxygen Species (ROS) which causes cell oxidative stress (Jamnik and Raspor 2003). On the other hand, Cr(III) is more stable and it constitutes a trace element conventionally considered an essential micro-nutrient for living organisms related to cellular membrane stability, synthesis and stability of nucleic acids and proteins (Poljsak et al. 2010). From the toxicological point of view, Cr(VI) is soluble in water, being 100 times more toxic and 1,000 times more mutagenic than Cr(III) (Chojnacka 2010). However, at high concentrations, Cr(III) can complex with organic compounds interfering with metallo-enzyme systems (Krishna and Philip 2005, Poljsak et al. 2010), may also cause health problems, for example, lung cancer (Costa 2003), birth deficiency and the decrease of reproductive health (Marsh and McInerny 2001). Di Bona et al. (2011) recently reported that Cr(III) can no longer be considered as a dietary supplement because rats subjected to a diet with low content of trivalent chromium, suffered no adverse consequences when they were compared with rats subjected to a diet with a sufficient dose of Cr(III).

Keeping in mind the toxic effects of chromium and tolerance limits for disposal of waste (0.5 to 270 mg l^{-1}), the concerned industries must minimize the total chromium level in wastewaters. The removal of Cr(VI) from wastewater requires serious and immediate attention. The most widely used methods are the conventional physicochemical processes, for example, the electrochemical process, ion exchange, adsorption on activated carbon, excavation and solidification/stabilization, etc. (Witek-Krowiak 2013). These technologies are suitable to control contamination but they present cost-effectiveness limitations, generating unsafe by-products or inefficiency, high energy consumption or incomplete removal of the pollutant (Bahi et al. 2012). Numerous studies are conducted for the development of cheaper and more effective technologies, this being necessary to develop methods that are more economic, safe, and environmental friendly to remove Cr(VI) ions from industrial wastewaters. Biological methods solve these drawbacks since they are easy to operate, do not produce secondary pollution, and they are less expensive and highly efficient even at low heavy metal concentrations (Ahluwalia 2012, Chojnaka 2010). In this context, isolation and characterization of Cr-resistant microorganisms (e.g., bacteria, fungi, and yeasts) led to hypothesize that biological removal methods would be a sustainable alternative technology of lower impact on the environment. Different microorganisms have been isolated and identified as having the capacity to remove Cr(VI) contamination by different biological methods (biosorption, bioaccumulation, bioreduction). Since they use native biomass, they are considered an eco-friendly and economic option. The microbes that retain such properties have been isolated from a diverse range of environments, both those contaminated and uncontaminated with Cr(VI).

Microbes may protect themselves from toxic substances in the environment by transforming toxic compounds through oxidation, reduction or methylation into more volatile, less toxic or readily precipitating forms (Ahmad et al. 2010). Also, they can be scaled up to pilot or large scales in various contacting systems bioreactors to ensure optimal operating conditions (hydrodynamic, mass transfer and growing conditions). The aim of this chapter is to present the state of the art related to Cr(VI) removal processes applied in different types of bioreactors, pilot scale studies and their potential to be applied at industrial scale to diminish the contents of Cr(VI) in their effluents till acceptable levels.

Microbial Mechanism for Cr(VI) Resistance

Microbial remediation is defined as the process by which microorganisms are stimulated to rapidly degrade the hazardous contaminants to environmentally safe levels in soil, subsurface materials, water, sludge and residues (Asha and Sandeep 2013). The microorganisms may be indigenous to the polluted area or they may be isolated from elsewhere and brought to the contaminated site (Kumar et al. 2011). For that reason, development of efficient biological process (accompanied by studies on molecular biology and ecology) offers numerous opportunities in the biological treatment of environmental heavy metal pollution (Kumar et al. 2011).

Knowledge about the interaction between microorganisms and heavy metals has an increasing interest since microorganisms have developed various strategies for their survival in heavy metal-polluted sites. Different detoxifying mechanisms developed by these microorganisms include the metal uptake which can either occur actively (bioaccumulation) or passively (biosorption), and/or the biotransformation (reduction or oxidation) changing the oxidative state of the metals. Among them, the microbial biotransformation (i.e., bioreduction) of the highly toxic, water soluble, and mobile Cr(VI) to the less toxic, insoluble, and immobile Cr(III) is an interesting option. Many researchers reported that biosorption can be a promising and efficient method for the removal of heavy metals due to its reusability, low operating cost and the absence of associated contamination (Kumar et al. 2008, Uluozlu et al. 2008, Wu et al. 2008, Konczyk et al. 2010). The Cr(VI) microbial detoxifying mechanisms are considered in more detail below.

Cr(VI) Biosorption

Biosorption is a physico-chemical process (sorption, ion exchange, surface complexation) between metal species (sorbate) and biological material (biosorbent) (Ahluwalia and Goyal 2007). The microbial cell wall provides structural integrity to the cell and offers many functional groups (such as carboxylate, hydroxyl, amino and phosphate) that can bind heavy metal ions (Scott and Karanjkar 1992). The importance of using microorganisms as biosorbents is due to the high surface/volume ratio and the low cost of producing biomass. Particularly, biosorption of heavy metals by fungi has received much attention, since fungi can tolerate and adapt themselves to various environmental conditions. This is a passive process and independent from cell activity due to the fact that metals are attached to the cell surface just by a physical mechanism. This process occurs in both living and non-living organisms. After that, a sorption of metals across cell onto internal parts of the cell body takes place. That active sorption demands energy consumption and occurs only in viable body. The presence of metal-binding proteins is also associated with the second process (Doble and Kumar 2005). The active sorption of metals can be very selective and also irreversible (Hawari and Mulligan 2006).

In fungi, carboxyl, phosphate, amine, amide and alkane groups are involved in chromium binding (Ahluwalia and Goyal 2010). Chitin and chitosan present on the cell wall of *Rhizopus arrizus* are also involved in Cr absorption (Ismael et al. 2004). A study carried out by Iram et al. (2013) revealed that the *Aspergillus fumigatus* fungal isolated from the contaminated site has good biosorption capacity towards selected heavy metals. Mathiyazhagan and Natarajan (2011) studied the bioremediation on effluents from magnasite and bauxite mines using *Thiobacillus* spp. and *Pseudomonas* spp. The results of biosorption process showed that *T. ferrooxidans* reduced and/or absorbed some heavy metals from mines (Cd, Ca, Zn, Cr, Mn, and Pb) while *P. aeruginosa* absorbed most of the metals that *T. ferrooxidans* could not absorb. Both species effectively absorbed Cd, Ca, Zn followed by Pb. Bahafid et al. (2013) found that Cr(VI) removal by *Pichia anomala* involves adsorption onto functional groups (e.g., amide I, amide II, amide III, polysaccharides, sulfonate and carboxyl) of cell surfaces, followed by accumulation inside the cell and reduction of Cr(VI) to Cr(III). The same authors also reported on three yeasts (*Cyberlindnera fabianii, Wickerhamomyces anomalus,* and *Candida tropicalis*) able to remove Cr(VI) from contaminated sites via adsorption mechanism (Bahafid et al. 2013).

Cr(VI) Bioaccumulation

Different researchers summarized the results achieved with bacteria, fungi, algae, and other plant derived biomass for the uptake of heavy metals for the aqueous solution (Gavrilescu 2004, Cheung and Gu 2007, Wang and Chen 2009, Chojnacka 2010). Metal uptake by microorganisms depends on initial metal concentration and contact time and is a metabolism-dependent mechanism. However, the major limitation is the inhibition of the cell growth when the metal concentration is high (Donmez and Aksu 1999). The understanding of the mechanism by which some microorganisms accumulate Cr(VI) is determinant to the development of processes for concentration, removal, and recovery from contaminated solutions. Ksheminska et al. (2008) suggested that Cr(VI) compounds (analogous to anions SO_4^{2-} and PO_4^{2-}) enter cells through non-selective and oxidative state-sensitive anion channel via facilitated diffusion. Cellular membranes are often impermeable to Cr(III), possibly because they form complexes that have low solubility. This mechanism is already unknown, and it is unclear if the known metal transport systems are responsible for the intracellular accumulation of Cr(III). It is also unclear as to whether there is a specific system to transport this cation in yeasts. Ksheminska et al. (2008) identified several yeasts as one of the best organisms to study bioremediation due to their capability of growing in matrices and bind or accumulate metal ions into cells and transform them via chelation to less toxic forms. According to this, Amoroso et al. (2001) determined Cr(VI) bioaccumulation by two *Streptomyces* strains, which were able to accumulate Cr(VI) 5–10 mg g^{-1} of cell in minimal medium.

Cr(VI) Biotransformation

The biotransformation of Cr(VI) to Cr(III) is considered as a detoxification mechanism because Cr(III) is considered less toxic than Cr(VI). Chromium biotransformation from hexavalent to trivalent form was slightly higher in nutrient/biological synthetic media than that of industrial effluent as there might be some constituents in effluent hampering transformation process or necessary enzyme production. Indeed, many yeast strains are known to biotransform Cr(VI) to Cr(III); examples include *Saccharomyces cerevisiae,*

Rhodotorula pilimanae, Yarrowia lipolytica and *Hansenula polymorpha* (Ksheminska et al. 2006), *Pichia guilliermondii* (Ksheminska et al. 2008), *Rhodotorula mucilaginosa* (Chatterjee et al. 2012), and *Cyberlindnera jadinii* and *Wickerhamomyces anomalus* (Fernández et al. 2013). Acevedo-Aguilar et al. (2006) studied the reduction of Cr(VI) by *Aspergillus* sp. Ed8 isolated from contaminated industrial area. This strain has the capability to reduce Cr(VI) present in the growth medium without accumulating it in the biomass whereas Cr(III), which was not initially present, appeared in aqueous solution.

In Cr(VI) resistant yeast of *Candida maltosa*, NAD-dependent chromate-reducing activity was discovered to take place mainly in the soluble protein fraction, with the membrane fraction being less active (Ramírez-Ramírez et al. 2004). Cr(VI) detoxification occurs via extracellular reducing substances that are secreted by the yeast cells (Ksheminska et al. 2006) such as sulphate and riboflavin (Fedorovych et al. 2009).

Jeyasingh and Philip (2005) had revealed that bacterial strains that were isolated and enriched from the contaminated site show high Cr(VI) reduction performance. They evaluated Cr(VI) reduction both in aerobic and anaerobic conditions. Though the aerobic system performed better than the anaerobic one, they carried further study in the anaerobic condition due to its economic viability. When initial Cr(VI) concentration was higher, the reduction was not complete even after 108 hr; however, specific Cr(VI) reduction was greater at higher concentration. It was found that a microbial concentration of 15 ± 1.0 mg per g of soil (wet weight), and 50 mg of molasses as carbon source per g of soil were required for the maximum Cr(VI) reduction. The system operated at these conditions could reduce entire Cr(VI) (5.6 mg per g of soil) in 20 d. This study showed that bioremediation is a viable, environmental friendly technology. Moreover, the removal of Cr(VI) with cell-free extracts of *Bacillus cereus* was 92.7%, close to 96.85% reported with whole cells, suggesting that chromium reductase is primarily intracellular (Zhao et al. 2012). Similarly, Martorell et al. (2012) described Cr(VI) reduction by cell-free extracts by *Pichia jadinii* and *Pichia anomala* isolated from textile-dye factory effluents. However, the principal reason that yeasts are resistant to chromium is related more to their limited ion uptake than to the biological reduction of Cr(VI) to Cr(III) (Ksheminska et al. 2005).

Types of Bioreactors

The research related to bioremediation technologies to remove Cr(VI) should be oriented towards practical applications of larger scale technologies. For this, it is necessary to work at higher volumes in different systems (reactors) and to establish the optimal operating conditions such as hydrodynamics, mass transfer and growth conditions. These parameters could vary according to the type of microorganism and the heavy metals used. According to some published articles, bacteria species have been successfully tested in different types of bioreactors for Cr(VI) removal but in the case of yeasts, the information is scarce.

The typical categories of systems used in this scope are:

Stirred Tank Bioreactors (STRs)

The design of this reactor is equipped with a stirrer that maintains the biomass in suspension. STR can work in batch or continuous modes. The solid phase can be separated from the liquid phase by sedimentation or by filtration in a separate process unit. STRs are considered less complicated than other systems but the operational costs are more elevated because of the energy used for the agitation (Roman et al. 1992, Hlihor et al. 2014).

Fixed-Bed Reactors (FXRs)

It is a system typically utilized for Cr(VI) biosorption. The biosorbent is placed in a fixed bed on a column through which passes the solution contaminated with the metal. The adsorptive is fed from the bottom of the fixed-bed. Packed bed columns (PBCs) are the most used ones to perform biosorption studies. The processes can be carried out in a counter current flow and in continuous or in batch settings. The underlying advantages are the use of large particles for biosorbents immobilization and the simplicity of construction and operation (Rosca et al. 2015).

Fluidized-Bed Bioreactors (FBRs)

FBRs are based on the development of a microbial biofilm on solid particles that support microbial growth. The bed is maintained in suspension by water flowing into the reactor. In the fluidized state, the particles are in continuous movement and constantly migrating in the entire volume of the column. In this type of reactor, a biosorbent is placed inside a column and a solution with metals is supplied in the counter-flow system. This allows the retention of biomass within the reactor and therefore operation at short hydraulic retention time. The process with a movable bed is carried out continuously; the biosorbent is fed from the top, whereas the adsorptive is fed counter currently from the bottom of the column (Rosca et al. 2015).

Air Lift Reactors (ALRs)

This bioreactor is generally used to overcome the problems associated with the mold morphology and the rheology of the broth. They are often employed in bioprocesses where gas-liquid contact is important. The air bubbles forced through the sparger are responsible for the induced turbulent liquid mixing and the accompanying mass transfer. If needed, additional external liquid circulation is added to obtain the required mixing pattern. The main advantages of ALRs are: no moving parts, low power consumption, high heat and mass transfer, maintenance of solid form in suspension, homogeneous shear, rapid mixing, increased oxygen solubility, high homogenization efficiency and low shear stress to cells, easy sterilization, and low contamination risk (Gavrilescu and Tudose 1998, Cozma and Gavrilescu 2010, 2011). ALRs are considered as practical and sustainable alternatives for stirred tank reactors, particularly for numerous bioprocesses such as treatment of contaminated fluid fluxes (wastewater) (Choi et al. 2007, Cozma and Gavrilescu 2010). Numerous studies on treatment technologies based on biosorption, biofilm-mediated bioremediation, suspended microorganism processes highlighted the significant advantages of ALRs to remove different pollutants from contaminated fluid (wastewater and gaseous) streams (Nikakhtari and Hill 2008, Vergara-Fernández et al. 2008).

The bioreduction of Cr(VI) to Cr(III) was made possible by pure and enriched mixed cultures of microorganisms under aerobic and anaerobic conditions. Fernández et al. (2016) reported that the intermittent reseeding of the biological system is necessary for the Cr(VI) continuous removal on a long-term basis. In recent publications, bioreactors of continuous-flow and fixed-film have been employed for Cr(VI) bioreduction. Shen and Wang (1995) demonstrated Cr(VI) reduction by *Escherichia coli* in continuous-flow suspended growth bioreactor system. In a first stage, cells grown completely in a mixed reactor were pumped into the second-stage plug-flow reactor to reduce Cr(VI). Chirwa and Wang (2001) were the first to demonstrate the potential of a fixed-film bioreactor for reduction of Cr(VI). They reported biological Cr(VI) reduction in a continuous-flow laboratory-scale biofilm reactor without the intermittent addition of fresh biomass to the system. In these reactors, electron donors were added to the wastewater depending upon the necessity. At an influent Cr(VI) concentration of 5 mg l^{-1}, the Cr(VI) reduction efficiency was nearly 100%. Dermou et al. (2005) reported the reduction of 30 mg l^{-1} Cr(VI) in an aerobic trickling filter inoculated with a consortium containing *Acinetobacter* sp. using sodium acetate as carbon source. Chang and Kim (2007) developed two-stage packed bed reactor to remedy electroplating wastewater under anaerobic conditions.

Additionally, performances of various bioreactors under different operating conditions with respect to Cr(VI) reduction were evaluated. *Arthrobacter rhombi*, a Cr(VI) reducing strain enriched and isolated from chromium contaminated soil, was used in a continuous reactor using three different systems with: (i) aerobic suspended growth, (ii) aerobic attached growth, and (iii) anoxic attached growth. All the bioreactors were designed for Cr(VI) biotransformation. Aerobic suspended and anoxic attached growth systems performed worse compared to aerobic attached growth system. The aerobic suspended growth system was able to achieve 95% Cr(VI) reduction (initial concentration: 20 mg l^{-1}). Cr(VI) reduction efficiency achieved using an aerobic attached growth system was 98%. However, under the same operating conditions, Cr(VI) reduction efficiency of 95–98% was achieved in the anoxic attached growth system (Elangovan and Philip 2009).

A comparative study on adsorption of Cr(VI) ions in different types of reactors, operated under identical conditions, indicated that the maximum removal efficiency was obtained for the stirred tank reactor, followed by the fluidized reactor and packed bed reactor when a chemically modified and polysulfone-immobilized biomass of the fungus *Rhizopus nigricans* was used (Bai and Abraham 2005).

Pilot Scale Studies

Promising results obtained at laboratory scale does not necessarily equate to be reached in large scale operating conditions. Generally, a transfer of the result from lab to industry is a relatively slow process. For that reason, a very limited number of industrial processes in the heavy metal bioremediation area have been implemented. Thus, research for future technology transfer is focused on the evaluation of the feasibility of the process at higher scale and the robustness of the system (Ahluwalia 2014).

Several studies of Cr(VI) removal were performed at pilot-scale bioreactor. In this perspective, treatment of Cr(VI)-containing waters using sulfate-reducing bacteria (SRB) in hydrogen-fed bioreactors at pilot scale has been reported by Battablia-Brunet et al. (2006). A 200 dm^3 pilot bioreactor was designed using fixed bed column filled with pozzolana and inoculated with *Desulfomicrobium norvegicum*. A removal percentage of Cr(VI) of 100% during the first 18 d of the experimental assay was achieved at Cr(VI) initial concentration of 15 mg l^{-1}. Further, Barros et al. (2007) observed the average removal percentage of chromium from the reactor inoculated with municipal wastewater sludge. An average removal percentage of 99.9% was obtained, varying from 100 to 99.3% at pilot scale during the first 30 d at 10 mg l^{-1} initial concentration. Similarly, the results obtained by Quintelas et al. (2009) using biofilm of *Arthrobacter viscosus* supported on granular activated carbon (GAC) at the pilot-scale reactor were very promising for environmental applications. Data obtained showed an average of Cr(VI) removal of 99.9 and 72%, during the first 30 d, for the initial Cr(VI) concentration of 10 mg l^{-1} and 100 mg l^{-1}, respectively. Uptake values of

11.35 mg g^{-1} and 14.55 mg g^{-1} were obtained, respectively, for the initial concentration of 10 and 100 mg l^{-1}. The volume of chromium solution treated was of 8,140 l for the assay with the initial concentration of 10 mg l^{-1} and 3,732 l for the more concentrated solution. Rehman et al. (2009) reported the removal of chromium through biological mechanisms in dual stage process. The first stage was of attached growth using plastic media for biofilm formation and the second one was of suspended growth process, operating under aerobic conditions. The high percentage conversion of Cr(VI) to Cr(III) reduction in a volume of 200 l by a bacterial system (*Acinetobacter haemolyticus*), immobilized onto carrier material, demonstrated a good decontamination system. The enzymatic reduction of Cr(VI) to Cr(III) by chromium resistant bacteria, *A. haemolyticus*, immobilized onto carrier material inside a bioreactor (200 l) constitutes the ChromeBac™ system. For this volume, liquid pineapple wastewater utilized as a nutrient is an excellent example of the substitution of a cheap and available industrial waste to replace an expensive growth medium and could be a significant factor in the commercial use of a process such as this (Ahmad et al. 2010). Bioremediation process may involve the study of many biochemical and physical parameters, including media formulation and culture parameters (Fernández et al. 2016).

In other publications, pilot scale studies were carried out by Jeyasingh et al. (2012) to evaluate the suitability of bioremediation of Cr(VI) contaminated aquifers using bio-barrier, employing chromium reducing bacteria. Results showed that a 10 cm thick biobarrier with a concentration of 0.44 mg of initial biomass per g of soil was able to completely contain a Cr(VI) plume of 50 mg l^{-1} concentration. A mathematical model was proposed for simulating the bioremediation process and predicts the overall trends observed in the experiments, which is limited by the assumption of homogeneous conditions. Biobarrier system was able to contain Cr(VI) plumes even for the case of high Cr(VI) concentrations when appropriate conditions such as inoculum concentration, the thickness of barrier, injection wells number, and enough flow velocity were maintained (Jeyasingh et al. 2012).

Based on the results of the laboratory investigation, a 24,000 l fixed-film pilot bioreactor was designed (Williams et al. 2014). Similar to the laboratory column study, steady-state conditions could be achieved, resulting in maintaining the reduction of Cr(VI) (> 99%). Research results presented in this paper provide the basis for a low-cost and low-maintenance strategy for the biological treatment of Cr(VI)-contaminated water. Stoichiometrically balancing terminal electron acceptors in the feed water with a selected electron donor, directed reactor balance for complete Cr(VI) reduction. The presence of a suitably adapted microbial community (including *Enterobacter cloacae*, *Flavobacterium* sp. and *Ralstonia* sp.) effective for chromium reduction is essential for this system. This represents the first up-scaled, effective demonstration of a biological Cr(VI) bioremediation system in South Africa.

Another strategy is the slurry-phase bioremediation. This *ex situ* system is a controlled treatment of soil in a bioreactor. This processing involves the separation of stones and rubbles from the contaminated soil mixed with water. After that, the next procedure is the disposition of the soil and further treatment of the resulting fluids (Kulshreshtha et al. 2014).

Continuous-flow and fixed-film bioreactor offer the most reliable mode of application to be used at industrial scale due to its ease of handling and simple operation. The technique using fixed-film bioreactors for Cr(VI) reduction was first reported by Chirwa and Wang (1997) where a laboratory-scale biofilm reactor at continuous-flow was effective to reduce Cr(VI) without the need to constantly supply fresh biomass.

Concluding Remarks

In this study, we summarized the methods and devices for heavy metals' removal from contaminated effluents. Fungal mechanisms of interaction with chromium in the context of environmental biotechnology are biosorption, bioaccumulation, and biotransformation; of these, the least understood at the biochemical level is the biotransformation.

Additional studies at bioreactor scale are needed to improve the biotreatment of industrial effluents or bioremediation of contaminated sites. It could be concluded that laboratory/pilot scale trials do show their potential for commercialization since they possess good chromium-removal capacity.

There are few studies related to chromium detoxification mediated by fungi, so it is expected that the application of proteomics, genomics, and metabolomics contribute to the understanding of the changes associated with responses to the toxicity of ion and/or of importance for the detoxification thereof.

Acknowledgments

This work was supported by Agencia Nacional de Promoción Científica y Tecnológica and Fondo para la Investigación Científica y Tecnológica FONCYT (PICT 1154/2013; PICT 3639/2015); Consejo Nacional de Investigaciones Científicas y Técnicas (PIO CONICET YPF 133201301000022CO); Secretaría de Ciencia, Arte e Innovación Tecnológica de la Universidad Nacional de Tucumán SCAIT UNT (PIUNT D509).

References Cited

Acevedo-Aguilar, F.J., A.E. Espino-Saldan, I.L. León-Rodŕguez and M.E. Rivera-Cano. 2006. Hexavalent chromium removal *in vitro* and from industrial wastes, using chromate-resistant strains of filamentous fungi indigenous to contaminated wastes. Can. J. Microbiol. 52: 809–815.

Ahluwalia, S.S. and D. Goyal. 2007. Microbial and plant derived biomass for removal of heavy metals from wastewater. Bioresour. Technol. 98: 2243–2257.

Ahluwalia, S.S. and D. Goyal. 2010. Removal of Cr(VI) from aqueous solution by fungal biomass. Eng. Life Sci. 10: 480–485.

Ahluwalia, S.S. 2012. Waste biomaterials for removal of heavy metals: An overview. pp. 62–67. *In*: Rath, C. [ed.]. Dynamic Biochemistry, Process Biotechnology and Molecular Biology. UK, Ikenobe, Japan.

Ahluwalia, S.S. 2014. Microbial removal of hexavalent chromium and scale up potential. Int. J. Curr. Microbiol. Appl. Sci. 3: 383–398.

Ahmad, W.A., Z.A. Zakaria, A.R. Khasim, M.A. Alias and S.M.H. Shaik Ismail. 2010. Pilot-scale removal of chromium from industrial wastewater using the ChromeBac™ system. Bioresour. Technol. 101: 4371–4378.

Amoroso, M.J., G.R. Castro, A. Durán, O. Peraud, G. Oliver and R.T. Hill. 2001. Chromium accumulation by two *Streptomyces* sp. isolated from riverine sediments. J. Ind. Microbiol. Biotechnol. 26: 210–15.

Asha, L.P. and R.S. Sandeep. 2013. Review on bioremediation—potential tool for removing environmental pollution. Int. J. Basic Appl. Chem. Sci. ISSN: 2277–2073.

Bahafid, W., N.T. Joutey, H. Sayel and N.E.L. Ghachtouli. 2013. Mechanism of hexavalent chromium detoxification using *Cyberlindnera fabianii* yeast isolated from contaminated site in Fez (Morocco). J. Mater. Environ. Sci. 4: 840–847.

Bahi, J.S., O. Radziah, A.W. Samsuri, H. Aminudin and S. Fardin. 2012. Bioleaching of heavy metals from mine tailing by *Aspergillus fumigatus*. Bioremediation J. 16: 57–65.

Bai, S.R. and T.E. Abraham. 2005. Continuous adsorption and recovery of Cr(VI) in different types of reactors. Biotechnol. Prog. 21: 1692–1699.

Barros, A.J.M., S. Prasad, V.D. Leite and A.G. Souza. 2007. Biosorption of heavy metals in upflow sludge columns. Bioresour. Technol. 98: 1418–1425.

Battaglia-Brunet, F., S. Touzé, C. Michel and I. Ignatiadis. 2006. Treatment of chromate polluted groundwater in a 200 dm^3 pilot bioreactor fed with hydrogen. J. Chem. Technol. Biotechnol. 81: 1506–1513.

Chang, I.S. and B.H. Kim. 2007. Effect of sulfate reduction activity on biological treatment of hexavalent chromium [Cr(VI)] contaminated electroplating wastewater under sulfate-rich condition. Chemosphere 68: 218–226.

Chatterjee, S., N.C. Chatterjee and S. Dutta. 2012. Bioreduction of chromium (VI) to chromium (III) by a novel yeast strain *Rhodotorula mucilaginosa* (MTCC 9315). Afr. J. Biotechnol. 11: 14920–14929.

Cheung, K.H. and J.D. Gu. 2007. Mechanisms of hexavalent chromium detoxification by microorganisms and bioremediation application potential: A review. Int. Biodeterior. Biodegradation 59: 8–15.

Chirwa, E.M. and Y.T. Wang. 1997. Hexavalent chromium reduction by *Bacillus* sp. in a packed-bed bioreactor. Environ. Sci. Technol. 31: 1446–1451.

Chirwa, E.M. and Y.T. Wang. 2001. Simultaneous chromium (VI) reduction and phenol degradation in a fixed-film co-culture bioreactor: reactor performance. Water Res. 35: 1921–1932.

Choi, H., E. Stathatos and D.D. Dionysiou. 2007. Photocatalytic TiO_2 films and membranes for the development of efficient wastewater treatment and reuse systems. Desalination 202: 199–206.

Chojnacka, K. 2010. Biosorption and bioaccumulation—the prospects for practical applications. Environ. Int. 36: 299–307.

Costa, M. 2003. Potential hazards of hexavalent chromate in our drinking water. Toxicol. Appl. Pharmacol. 118: 15.

Cozma, P. and M. Gavrilescu. 2010. Airlift reactors: Hydrodynamics, mass transfer and applications in environmental remediation. Environ. Eng. Manag. J. 9: 681–702.

Cozma, P. and M. Gavrilescu. 2011. Airlift reactors: application in wastewater treatment. Environ. Eng. Manag. J. 11: 1505–1515.

Dermou, E., A. Velissariou, D. Xenos and D.V. Vayenas. 2005. Biological chromium(VI) reduction using a trickling filter. J. Hazard. Mater. 126: 78–85.

Di Bona, K.R., S. Love, N.R. Rhodes, D. McAdory, S.H. Sinha, N. Kern et al. 2011. Chromium is not an essential trace element for mammals: effects of a "low-chromium" diet. J. Biol. Inorg. Chem. 16: 381–390.

Doble, M. and A. Kumar. 2005. Biotreatment of industrial effluents. Elsevier Inc., Burlington.

Donmez, G. and Z. Aksu. 1999. The effect of copper (II) ions on the growth and bioaccumulation properties of some yeasts. Process Biochem. 35: 135–142.

Elangovan, R. and L. Philip. 2009. Performance evaluation of various bioreactors for the removal of Cr (VI) and organic matter from industrial effluent. Biochem. Eng. J. 44: 174–186.

Fedorovych, D.V., M.V. Gonchar, H.P. Ksheminska, T.M. Prokopiv, H.I. Nechay, P. Kaszycki et al. 2009. Mechanisms of chromate detoxification in yeasts. Microbiol. Biotechnol. 3: 15–21.

Fernández, P.M., M.E. Cabral, O.D. Delgado, J.I. Fariña and L.I.C. Figueroa. 2013. Textile dye polluted waters as an unusual source for selecting chromate-reducing yeasts through Cr(VI)-enriched microcosms. Int. Biodeter. Biodegr. 79: 28–35.

Fernández, P.M., E.L. Cruz, S.C. Viñarta and L.I.C. de Figueroa. 2016. Optimization of culture conditions for growth associated with Cr (VI) removal by *Wickerhamomyces anomalus* M10. Bull. Environ. Contam. Toxicol. 1–7.

Gavrilescu, M. and R.Z. Tudose. 1998. Modelling of liquid circulation velocity in concentric-tube airlift reactors. Chem. Eng. J. 69: 85–91.

Gavrilescu, M. 2004. Removal of heavy metals from the environment by biosorption. Eng. Life Sci. 4: 219–232.

Hawari, A.H. and C.N. Mulligan. 2006. Biosorption of lead (II), cadmium (II), copper (II) and nickel (II) by anaerobic granular biomass. Bioresour. Technol. 97: 692–700.

Hlihor, R.M., L. Bulgariu, D.L. Sobariu, M. Diaconu, T. Tavares and M. Gavrilescu. 2014. Recent advances in biosorption of heavy metals: support tools for biosorption equilibrium, kinetics and mechanism. Rev. Roum. Chim. 59: 527–538.

Iram, S., G. Uzma, S. Rukh and T. Ara. 2013. Bioremediation of heavy metals using isolates of filamentous fungus collected from polluted soil of Kasur, Pakistan. Int. Res. J. Biol. Sci. 66–73.

Ismael, A.R., X. Rodriguez, C. Gutierrez and M.G. Moctezuma. 2004. Biosorption of Chromium(VI) from aqueous solution onto fungal biomasss. Bioinorg. Chem. Appl. 2: 1–7.

Jamnik, P. and P. Raspor. 2003. Stress response of yeast *Candida intermedia* to Cr(VI). J. Biochem. Mol. Toxicol. 17: 316–323.

Jeyasingh, J. and L. Philip. 2005. Bioremediation of chromium contaminated soil: optimization of operating parameters under laboratory conditions. J. Hazard. Mater. 118: 113–120.

Jeyasingh, J., V. Somasundaram, L. Philip and S.M. Bhallamudi. 2012. Pilot scale studies on the remediation of chromium contaminated aquifer using bio-barrier and reactive zone technologies. Chem. Eng. J. 167: 206–214.

Juvera-Espinosa, J., L. Morales-Barrera and E. Cristiani-Urbina. 2006. Isolation and characterization of a yeast strain capable of removing Cr(VI). Enzyme Microb. Technol. 40: 114–121.

Klimaviciute, R., J. Bendoraitiene, R. Rutkaite and A. Zemaitaitis. 2010. Adsorption of hexavalent chromium on cationic cross-linked starches of different botanic origins. J. Hazard. Mater. 181: 624–632.

Konczyk, J., C. Kozlowski and W. Walkowiak. 2010. Removal of chromium (III) from acidic aqueous solution by polymer inclusion membranes with D2EHPA and Aliquat 336. Desalination 263: 211–216.

Krishna, K.R. and L. Philip. 2005. Bioremediation of Cr(VI) in contaminated soils. J. Hazard. Mater. 121: 109–117.

Ksheminska, H., A. Jaglarz, D. Fedorovych, L. Babyak, D. Yanovych, P. Kaszycki et al. 2003. Bioremediation of chromium by the yeast *Pichia guilliermondii*: toxicity and accumulation of Cr(III) and Cr(VI) and the influence of riboflavin on Cr tolerance. Microbiol. Res. 158: 59–67.
Ksheminska, H., D. Fedorovych, L. Babyak, D. Yanovych, P. Kaszycki and H. Koloczek. 2005. Chromium (III) and (VI) tolerance and bioaccumulation in yeast: a survey of cellular chromium content in selected strains of representative genera. Process Biochem. 40: 1565–1572.
Ksheminska, H., D. Fedorovych, T. Honchar, M. Ivash and M. Gonchar. 2008. Yeast tolerance to chromium depends on extracellular chromate reduction and Cr(III) chelation. Food Technol. Biotechnol. 46: 419–426.
Ksheminska, H.P., T.M. Honchar, G.Z. Gayda and M.V. Gonchar. 2006. Extracellular chromate-reducing activity of the yeast cultures. Central Eur. J. Biol. 1: 137–149.
Kulshreshtha, A., A. Agrawal, M. Barar and S. Saxena. 2014. A review on bioremediation of heavy metals in contaminated water. IOSR–J. Environ. Sci. Toxicol. Food Tech. 8: 44–50.
Kumar, P.A., S. Chakraborty and M. Ray. 2008. Removal and recovery of chromium from wastewater using short chain polyaniline synthesized on jute fiber. Chem. Eng. J. 141: 130–140.
Kumar, R., D. Bhatia, R. Singh, S. Rani and N.R. Bishnoi. 2011. Sorption of heavy metals from electroplating effluent using immobilized biomass *Trichoderma viride* in a continuous packed-bed column. Int. Biodeter. Biodegr. 65: 1133–1139.
Marsh, T.L. and M.J. McInerney. 2001. Relationship of hydrogen bioavailability to chromate reduction in aquifer sediments. Appl. Environ. Microbiol. 67: 1517–1521.
Martorell, M.M., P.M. Fernández, J.I. Fariña and L.I.C. Figueroa. 2012. Cr(VI) reduction by cell-free extracts of *Pichia jadinii* and *Pichia anomala* isolated from textile-dye factory effluents. Int. Biodeterior. Biodegradation 71: 80–85.
Mathiyazhagan, N. and D. Natarajan. 2011. Bioremediation on effluents from Magnesite and Bauxite mines using *Thiobacillus* spp. and *Pseudomonas* spp. J. Biorem. Biodegr. 2: 115.
Modoi, O.C., C. Roba, Z. Török and A. Ozunu. 2014. Environmental risks due to heavy metal pollution of water resulted from mining wastes in NW Romania. Environ. Eng. Manag. J. 13: 2325–2336.
Morales-Barrera, L. and E. Cristiani-Urbina. 2006. Removal of hexavalent chromium by *Trichoderma viride* in an airlift bioreactor. Enzyme Microb. Technol. 40: 107–113.
Nikakhtari, H. and G.A. Hill. 2008. Toluene vapours remediation in a continuous packed bed loop bioreactor. Environ. Progress 27: 234–241.
Orozco, A.M.F., E.M. Contreras and N.E. Zaritzky. 2008. Modelling Cr(VI) removal by a combined carbon-activated sludge system. J. Hazard. Mater. 150: 46–52.
Pavel, V.L., M. Diaconu, D. Bulgariu, F. Statescu and M. Gavrilescu. 2012. Evaluation of heavy metals toxicity on two microbial strains isolated from soil: *Azotobacter* sp. and *Pichia* sp. Environ. Eng. Manag. J. 11: 165–168.
Poljsak, B., I. Pócsi, P. Raspor and M. Pesti. 2010. Interference of chromium with biological systems in yeasts and fungi: a review. J. Basic. Microbiol. 50: 21–36.
Quintelas, C., B. Fonseca, B. Silva, H. Figueiredo and T. Tavares. 2009. Treatment of chromium (VI) solutions in a pilot-scale bioreactor through a biofilm of *Arthrobacter viscosus* supported on GAC. Bioresour. Technol. 100: 220–226.
Ramírez-Ramírez, R., C. Calvo-Méndez, M. Ávila-Rodriguez, P. Lappe, M. Ulloa, R. Vázquez-Juárez et al. 2004. Cr(VI) reduction in a chromate-resistant strain of *Candida maltosa* isolated from the leather industry. A. Van Leeuw. 85: 63–68.
Rehman, Z., S. Ahmad, J. Gour and S. Trivedi. 2009. Laboratory scale studies on biological removal of chromium from synthetic wastewater using dual stage process. pp. 182–189. *In*: Proc. of International Conference on Emerging Technologies in Environmental Science and Engineering. October 26–28, 2009 Aligarh Muslim University, Aligarh, India.
Roman, R.V., M. Gavrilescu and V. Efimov. 1992. Evaluation of power consumption for Newtonian system mixing with a various number of Rushton turbines. Hungarian J. Ind. Chem. 20: 155–160.
Rosca, M., R.M. Hlihor, P. Cozma, E.D. Comăniţă, I.M. Simion and M. Gavrilescu. 2015. Potential of biosorption and bioaccumulation processes for heavy metals removal in bioreactors. E-Health and Bioengineering Conference (EHB). DOI: 0.1109/EHB.2015.7391487.
Scott, J.A. and A.M. Karanjkar. 1992. Repeated cadmium biosorption by regenerated *Enterobacter aerogenes* biofilm attached to activated carbon. Biotechnol. Lett. 14: 737–740.
Shen, H. and Y. Wang. 1995. Hexavalent chromium removal in two-stage bioreactor system. J. Environ. Eng. 121: 798–804.
Tekerlekopoulou, A.G., M. Tsiflikiotou, L. Akritidou, A. Viennas, G. Tsiamis, S. Pavlou et al. 2013. Modelling of biological Cr(VI) removal in draw-fill reactors using microorganisms in suspended and attached growth systems. Water Res. 47: 623–636.
Uluozlu, O.D., A. Sari, M. Tuzen and M. Soylak. 2008. Biosorption of Pb(II) and Cr(III) from aqueous solution by lichen (*Parmelina tiliaceae*) biomass. Bioresour. Technol. 99: 2972–2980.

Vergara-Fernández, A.O., E.F. Quiroz, G.E. Aroca and N.A. Alarcón Pulido. 2008. Biological treatment of contaminated air with toluene in an airlift reactor. Electron. J. Biotechnol. 11: 1–7.

Wang, J. and C. Chen. 2006. Biosorption of heavy metals by *Saccharomyces cerevisiae*: A review. Biotechnol. Adv. 24: 427–451.

Wang, J.L. and C. Chen. 2009. Biosorbents for heavy metals removal and their future. Biotechnol. Adv. 27: 195–226.

Williams, P.J., E. Botes, M.M. Maleke, A. Ojo, M.F. DeFlaun, J. Howell et al. 2014. Effective bioreduction of hexavalent chromium-contaminated water in fixed-film bioreactors. Waters S.A. 30: 549–554.

Witek-Krowiak, A. 2013. Kinetics and equilibrium of copper and chromium ions removal from aqueous solutions using sawdust. Environ. Eng. Manag. J. 12: 2125–2135.

Wu, D., Y. Sui, S. He, X. Wang, C. Li and H. Kong. 2008. Removal of trivalent chromium from aqueous solution by zeolite synthesized from coal fly ash. J. Hazard. Mater. 155: 415–423.

Zhao, C., Q. Yang, W. Chen and B. Teng. 2012. Removal of hexavalent chromium in tannery wastewater by *Bacillus cereus*. Can. J. Microbiol. 58: 23–28.

16

Contribution of Genomic and Proteomic Studies toward Understanding Hexavalent Chromium Stress Resistance

Anahí R. Bernal,[1,a] *Elías L. Cruz,*[1,b] *Lucía I. Castellanos de Figueroa*[1,2,c] and *Pablo M. Fernández*[1,3,*]

Introduction

Over the past few decades, heavy metal pollution has become one of the most serious environmental problems (Wang and Chen 2006). These elements deleterious to the environment and therefore to humans who depend on environment products as sources of food. What induces this type of environmental contamination are anthropogenic sources, e.g., mining, smelters and the application of pesticides and fertilizer containing metal, as well as the irresponsible disposal of wastes by various industries (Hookoom and Puchooa 2013, Missaoui et al. 2016).

Actually, the levels of heavy metals are increasing to toxic standards in all environments, such as soil, water, and air. Metal contaminated environments pose a serious threat to human health and ecosystems (Hong et al. 2002, Rorat et al. 2016). Once heavy metals ions enter the cell, their functions are affected by a wide range of actions. The negative impact includes binding of heavy metals ions to sulfhydryl groups of proteins, exchange of essential cations from specific binding sites, leading to enzyme inactivation and production of reactive oxygen species (ROS), generating in oxidative damages to lipids, proteins and nucleic acids (Sharma and Dietz 2009). Among heavy metals, chromium (Cr) is one of the most hazardous environmental pollutants (Shanker et al. 2005, Yıldız and Terzi 2016).

[1] Planta Piloto de Procesos Industriales Microbiológicos PROIMI-CONICET. Av. Belgrano y Caseros (4000), S. M. de Tucumán, Tucumán, Argentina.
[a] Email: anahirbernal@gmail.com
[b] Email: elias-cruz@outlook.com
[c] Email: proimiunt@gmail.com
[2] Microbiología Superior, Facultad de Bioquímica, Química y Farmacia, Universidad Nacional de Tucumán, Ayacucho 450 (4000), S. M. de Tucumán, Tucumán, Argentina.
[3] Calidad de Productos II, Gestión de Empresas Agroindustriales, Universidad de San Pablo – T, Av. Solano Vera y Camino a Villa Nougués, San Pablo, Tucumán, Argentina.
* Corresponding author: pfernandez@proimi.org.ar

However, it represents an essential micronutrient for living organisms as a participant in the maintenance of normal carbohydrate metabolism in mammals and yeasts (Dębski et al. 2004, Lazarova et al. 2014). Chromium (Cr) is a redox active 3d transition metal from group VI-B of the periodic table (Cervantes and Campos-Garcia 2007, Ahemad 2014). Cr naturally occurs in rocks, soils, plants, animals, and volcanic emissions; representing the 7th most abundant element in Earth's crust (Cervantes and Campos-Garcia 2007). It exists in different oxidation states that range from +2 to +6 (Smith et al. 2002, Ahemad 2014). Nevertheless, the most stable, abundant, and ecologically significant forms are the trivalent Cr(III) and the hexavalent Cr(VI) species (Cervantes and Campos-Garcia 2007). Cr(III) forms insoluble complexes because of greater affinity for organics which precipitate generally in the form of hydroxides, oxides, and sulphates (Nickens et al. 2010, Ahemad 2014, Johnson et al. 2016). In contrast, Cr(VI) is exceedingly persistent into the environment due to their considerable hydrosolubility (Barceloux and Barceloux 1999, Ahemad 2014). Cr(VI) is commonly present in solution as the water-soluble chromate (CrO_4^{2-}) or dichromate ($Cr_2O_7^{2-}$) oxyanions and, also, is rapidly reduced to Cr(III) by organic matter and other reducing agents to yield in soil and aquatic settings. Cr(VI) is considered a strong oxidizing agent (McGrath and Smith 1990, Cervantes and Campos-Garcia 2007).

In fact, chromium's relationship with life is still enigmatic and divergent, due to the biological properties of its two prominent oxidation variants, the hexavalent Cr(VI) and the trivalent Cr(III) (Urbano et al. 2012, Johnson et al. 2016). Cr(VI) causes mutagenic and carcinogenic effects on biological organisms because of its high oxidizing potential. The genotoxicity of Cr(VI) is attributed to its intracellular reduction to Cr(III) via reactive intermediates, Cr(VI) does not interact directly with DNA. The resulting types of DNA damage that are produced can be grouped into two categories: (1) oxidative DNA damage and (2) Cr(III)-DNA interactions (Sobol and Schiestl 2012, Joutey et al. 2015).

Traditional methods for debugging metal ions from aqueous solution have been studied in detail, such as chemical precipitation, ion exchange, electrochemical treatment, membrane technologies, adsorption on activated carbon, reverse osmosis, etc. (Hong and Ning 2011, Katsou et al. 2011). All these methods have some limitations as common problems associated with the cost and can be considered techniques not friendly to the environment (Soares and Soares 2012). Bioremediation may employ autochthonous microorganisms or their enzymes to decompose and thus detoxify sites contaminated (Obayori et al. 2009, Chiu et al. 2016). This methodology is characterized by being less expensive and can be considered as an effective and environmentally friendly strategy (Soares and Soares 2012).

Based on this background, this chapter explores concepts of proteomics and genomics, reviewing techniques related to the study of proteins and molecular components that are involved in the removal of heavy metals. A comprehensive study on reduction and repair systems is also provided.

Study of Tools Omics Applied for the Study of Heavy Metals Influence in the Cell

Genomics and Proteomics

The main tools studied in this chapter are genomics (genetic complement) and proteomics (protein synthesis and signaling), but also there are tools other such as transcriptomics (gene expression), metabolomics (concentrations and fluxes of cellular metabolites), and metabonomics (systemic profiling through) (Righetti et al. 2003, Plebani 2005). These kinds of technologies have the potential to find new paths of toxicity and to classify the toxicity of contaminants. In molecular medicine, such technologies (i.e., genomic, transcriptomic, epigenomic, proteomic or metabolomic) are already widely used, and favorable perspective and diagnostic studies have been carried out (Vlaanderen et al. 2010, Baillon et al. 2015). Genetic information (DNA sequence) only indicates the state of the genes, but the genes do not tell us when and how a genetic code occurs, only shows the diverse ways in which cells use their proteins. However, the cell life is a dynamic process. Messenger RNA (mRNA) molecules play an important part in translating genetic information into functional proteins (transcriptomics). The analysis of mRNA shows which genes are active, and which genes exert an influence on an organism. However, mRNA levels often do not correlate with the protein level (Anderson et al. 1998, Plebani 2005). To define protein based gene expression analysis, the concept of the proteome was proposed by Wilkins et al. (1996). A proteome is the entire protein complement expressed by a genome, or by a cell or tissue type (Wilkins et al. 1996, Pennington et al. 1997, Plebani 2005). The genome is simpler than the proteome because any protein may exist in multiple forms that vary within a particular cell or between different cells (Plebani 2005). The concept of the proteome has some differences from that of the genome, the proteome is an entity and has the ability to change in different conditions, and can be several in different tissues of a single organism, as while there is only one definitive genome of an organism (Wilkins et al. 1996). Modifications may derive from different processes such as translational, post-translational, regulatory and degradation processes that modify protein structure, localization, function, and turnover (Plebani 2005). As an extrapolation of the concept of the 'genome project', a 'proteome project' is the research which seeks to identify and characterize the proteins present in a cell or tissue and define their patterns of expression (Wilkins et al. 1996). In other words, proteomics provides unique insights into biological systems which cannot be acquired from genomic or transcriptomic approaches. For example, proteomic has been used extensively to investigate protein expression patterns under heavy metal stress (Labra et al. 2006, Bah et al. 2010, Sharmin et al. 2012, D'Hooghe et al. 2013, Yildiz and Terzi 2016). There are traditionally three types of proteomic approaches: (1) *Protein expression proteomics*, used for the quantitative study of the protein expressions and identify novel proteins; (2) *Structural proteomics*, the main goal of this approach is study to map out the structure of protein complexes or the proteins present in a sub-cellular localization or organelles, such as mitochondria, chloroplasts and nuclei; (3) *Functional proteomics*, for analyze the protein profiles at sub-cellular sites for understanding the functional organization of cells at the molecular level (Lau et al. 2003, Plebani 2005). However, the information initially encoded in the genome is ultimately displayed at the cellular level as cellular traits or phenotypes (Decorosi et al. 2009, Viti et al. 2009). Many studies carried out by genomics and proteomics methods are revealing that there are different mechanisms involved in the Cr(VI)-resistance (Hu et al. 2005, Brown et al. 2006, Decorosi et al. 2009). These technologies allow for the global study of the important macromolecules of cells that convey the information from DNA to RNA to protein.

Microbial Resistance to Cr(VI)

Metal resistance is defined as the capability of a microorganism to survive toxic effects of exposure by means of a detoxification mechanism produced in direct response to the metal species concerned. For its part, tolerance is the ability to survive metal exposure using intrinsic properties and or environmental modification of toxicity (Gadd 1992). Many recent reviews have studied Cr(VI) resistance and tolerance mechanisms and its effects (Ramirez-Diaz et al. 2008, Poljsak et al. 2010). The study focused on the operation of Cr(VI) efflux pumps (Branco et al. 2008), extracellular Cr(VI) reduction to Cr(III) (Gnanamani et al. 2010, Belchik et al. 2011, Chovanec et al. 2012), enzymes involved in cell detoxification caused by ROS (Ackerley et al. 2004a,b, Cheng et al. 2009), repair of DNA lesions (Hu et al. 2005, Miranda et al. 2005, Decorosi et al. 2009), and sulfur metabolism (Viti et al. 2009, Christl et al. 2012). In bacteria, chromium resistance uses pathways such as a specific or nonspecific reduction of Cr(VI), free radical detoxification activities, repair of DNA damage (Morais et al. 2011) and processes related to sulfur or iron homeostasis (Fig. 1) (Ramirez-Diaz et al. 2008).

Several microorganisms possessing Cr(VI)-reducing activities and resistance have been isolated and identified from chromate contaminated environment as well as natural, uncontaminated ecosystems (Fernández et al. 2013). An investigation carried out by Das et al. (2013) revealed that the bacteria isolated from chromite mine soils are resistant towards Cr(VI) along with other heavy metals. It is understood that chromate resistance and reduction are not necessarily interrelated, and not all Cr(VI) resistant bacteria can reduce Cr(VI) to Cr(III). Microbial mechanisms for Cr(VI) resistance generally derive from physiological analysis of wild-type and mutant strains stressed by the metal. The spread of the 'omic' approaches, particularly transcriptomic and proteomic methods, has offered the possibility to carry out studies on the response of global gene and protein expression of microorganisms. In this way, several investigations have shed light on cellular processes that until now were unexplained or poorly highlighted in terms of their involvement in chromate resistance (e.g., enhancement of iron transport and sequestration, enhancement of cysteine biosynthesis, derepression of some central metabolic routes, and phage lytic cycle activation), opening new fields of study to which traditional microbiological methods can contribute (Thatoi and Das 2014).

Because of its structural similarity to sulfate (SO_4^{-2}), (CrO_4^{-2}) maybe crosses the cell membrane in some species bacterial via the sulfate transport system encoded by the chromosomal DNA (Joutey et al. 2015). The chemical analogy of these oxyanions is highlighted by the fact that chromate is a competitive inhibitor of sulfate transport in all bacterial species where it has been tested. In contrast, most cells are impermeable to Cr(III), due to the insolubility of Cr(III) derivatives which forms insoluble compounds in non-acidic aqueous solutions (Cervantes and Campos García 2007).

Once the Cr(VI) enters the cellular interior by the sulfate conveyor is reduced to Cr(III). This intracellular Cr(III) can react with several reducing compounds (such as NAD(P)H, FADH, pentose, cysteine, and antioxidants like ascorbate and glutathione as well as one-electron reducers such as glutathione reductase) while that anaerobic Cr(VI) reduction occurs in the inner membrane through the electron transport pathway by cytochrome b (cyt b) or cytochrome c (cyt c) along the respiratory chains (Ahemad 2014).

These short-lived [Cr(V)/Cr(IV)] intermediates produced by reducing systems generate ROS, including single oxygen (O) and superoxide (O^{2-}) (Cheng et al. 2009, Joutey et al. 2015), hydroxyl (OH) and hydrogen peroxide (H_2O_2) radicals (Ackerley et al. 2004a). These not only negatively affect DNA replication and RNA transcription by damaging DNA but also alter gene expression, proteins and DNA by forming a range of DNA lesions, together with Cr-DNA adducts.

Activation of the SOS response to combat the ROS-generated oxidative stress, include both enzymatic and non-enzymatic systems involved in the cell defense against these damaging oxidants. Repair system is activated in order to repair the damaged DNA, moreover protective metabolic enzymes, superoxide dismutase (SOD), which eliminate the superoxide radical, catalase (CAT), and peroxiredoxin, responsible for the H_2O_2 removal. The most significant non-enzymatic antioxidants involved in direct scavenging of ROS or recycling of oxidized compounds are ascorbate, glutathione, carotenoids, among others. Some outer membrane proteins can also counter the oxidative stress (Ahemad 2014).

Reduced Uptake of Cr(VI)

There are effective systems of protection against the lethal effects of Cr (VI), apparently associated with reduced uptake of Cr(VI). Since chromate ion (CrO_4^{2-}) has structural similarity with tetrahedral sulphate ion (SO_4^{2-}), it can easily pass through cell membranes via SO_4^{2-} transport pathway, with the help of nonspecific anionic carriers (SO_4^{2-}, PO_4^{3-}) (Qi et al. 2000). In a variety of bacterial species, it enters the cells by means of sulfate ABC transporter (Fig. 1) (Cervantes et al. 2001, Aguilar-Barajas et al. 2011). All components of the sulfate ABC transporter (Sbp or CysP, the periplasmic sulphate, and thiosulphate-binding proteins; CysT and CysW, the two inner-membrane transport proteins; CysA, the membrane-associated ATP-binding protein) are generally arranged in operons (Aguilar-Barajas et al. 2011). Through transcriptomic and proteomic studies in *Shewanella oneidensis* MR-1, *Pseudomona putida* F1, *Cupriavidus metallidurans* CH34, and *Arthrobacter* sp. FB24, several authors demonstrated that the sulfate ABC transporter was up-regulated after chromate exposure (Brown et al. 2006, Thompson et al. 2007, 2010, Henne et al. 2009, Monsieurs et al. 2011).

A similar regulation of the sulfate ABC transporter was observed in Cr(VI)-stressed cultures of *Pseudomonas corrugata* 28, a hyper-resistant-Cr(VI) strain (Viti et al. 2007). In this bacterium, *sbp* is located upstream *cysTWA*, but it forms an independent operon with *oscA*, a gene encoding a small protein involved in the utilization of organosulfur compounds (Viti et al. 2009). Both, the transcriptional units *oscA-sbp* and *cysTWA* were overexpressed in the first 30 min after the incorporation of chromate to the medium (Viti et al. 2009). The noted overexpression of the sulfate transporter after chromate exposure suggests that chromate induces sulfur starvation in cells.

Two factors could to explain why chromate induces sulfur limitation: The first include a competition between sulfate and chromate for the transport reduces the bacterial capability to uptake sulfate, while the second consider that the oxidative stress induced by Cr(VI) decreases sulfur availability in cells (Brown et al. 2006). Proteomic analysis of *Arthrobacter* FB24 and *Peusodomonas putida* F1 exposed to a critical chromate stress indicates that these strains up-regulate proteins involved in the uptake and utilization of organic sulfur sources such as aliphatic sulfonates and cysteine (Henne et al. 2009b, Thompson et al. 2010). These results are consistent with the finding reported by Viti et al. (2009), who observed that a Cr(VI)-sensitive mutant of *Pseudomonas corrugata* 28 (a phenotype depending on the activity of the above-mentioned oscA gene) lost the ability to grow on organic sulfur sources such as taurine, methanesulfonic acid, and buthanesulfonic acid. The demand and uptake of organic sulfur sources may be a mechanism used by bacteria to survive sulfur starvation induced by Cr(VI). Several authors (Ohtake et al. 1987, Nies and Silver 1989, Decorosi 2010) observed that the Cr(VI)-minimum inhibitory concentration [Cr(VI)-MIC] is dependent on the concentration of sulfate, proposing the competition of sulfate and chromate for uptake. Decorosi (2010) published that the addition of cysteine or glutathione to the medium driven a strong increase of the Cr(VI)-MIC, hypothesizing that when the cysteine or glutathione are available in the medium, *Pseudomonas corrugata* 28 turns off the sulfate transport, blocking at the same time the uptake of chromate, and activates the transport of these sulfur compounds. In the presence of methionine, sulfate transport may be only partially reduced increasing the chromate-minimum inhibitory concentration [Cr(VI)-MIC)] that remained dependent on the sulfate concentration.

Cr(VI) Reduction

Microbial reduction of hexavalent chromium to Cr(III) can be considered a chromate detoxification mechanism and is usually not plasmid-associated (Cervantes et al. 2001). Two direct Cr(VI) reduction mechanisms have been described: (1) in aerobic conditions associated with soluble chromate reductases (Park et al. 2000) and (2) under anaerobic conditions, Cr(VI) can be used as an electron acceptor in the electron transport chain by some bacteria (Tebo and Obraztova 1998). Cr(VI) can be also reduced indirectly by redox intermediate organic compound (nucleotides, sugars, amino acids, vitamins, glutathione or organic acids) (Myers et al. 2000, Robins et al. 2013). Since the first report of anaerobic Cr(VI) reduction by Romanenko and Koren'Kov (1977) in uncharacterized *Pseudomonas* sp., worldwide researchers have isolated aerobic and anaerobic hexavalent chromium reducing bacteria belonging to a range of genera from different environments (Opperman et al. 2008, He et al. 2009, Alam and Ahmad 2012, Batool et al. 2012, Ge et al. 2013, Narayani and Shetty 2012, Nguema and Luo 2012, Shi et al. 2012). The chromate reducing efficiency is affected by reduction condition (aerobically or anaerobically or both), microbial types (Cervantes et al. 2001, Cervantes and Campos-García 2007), and various factors (Lowe et al. 2003, Alam and Malik 2008, Xu et al. 2013). In general, Cr(VI)-reducing aerobes utilize NADH or NADPH and endogenous cell reserves as cofactors, and Cr(VI) reducing anaerobes use electron transport systems containing cytochromes to reduce Cr(VI) derivatives (Zhu et al. 2008, Dey and Paul 2013). Under the biologically and ecologically point of view, the mechanisms of bacterial Cr(VI) reduction are of specific significance as they transform noxious and mobile chromium into innocuous and immobile reduced species (Daulton et al. 2007, Soni et al. 2013). In fact, the bacterial enzymes which reduce Cr(VI) species are not primarily the Cr(VI) reductases; they may principally act as iron reductases, nitroreductases, glutathione reductase, lipoyl reductase, ferredoxin-$NADP^+$ reductase, or other metal reductases (Kwak et al. 2003, Mazoch et al. 2004, Cervantes and Campos García 2007, Opperman et al. 2008).

Anaerobic conditions

Under anaerobic conditions, many components of the cell's protoplasm such as amino acids, nucleotides, carbohydrates, vitamins, organic acids, glutathione, hydrogen NADH (NADPH in some species), flavoproteins, and hemeproteins reducing Cr(VI) which serves as terminal electron acceptor. Microbial respiration with Cr(VI) as the terminal electron acceptor has never been rigorously shown (Richter et al. 2012). Nevertheless, the global transcriptomic analysis of *Shewanella oneidensis* MR-1, treated with 100 mM Cr(VI) as the sole electron acceptor, showed the up-regulation of genes MtrA, MtrB, MtrC, and OmcA (Bencheikh-Latmani et al. 2005), which are involved in the dissimilatory extracellular reduction of solid ferriciron (hydr)oxides, uranium and technetium (Belchik et al. 2011). The cytochromes MtrC and OmcA of *S. oneidensis* MR-1 were characterized in order to understand their role in Cr(VI) reduction. The data obtained supported the idea that MtrC and OmcA are the terminal reductases of Cr(VI) in *Shewanella oneidensis* MR-1 (Belchik et al. 2011).

Aerobic conditions

In aerobic conditions, NAD(P)H-dependent extracellular soluble reductases are produced purposely by the cell to reduce Cr(VI) that is removed by reacting with functional groups present on the cell surface. For example, *Pseudomonas putida* PRS2000, *Pseudomonas ambigua*

G-1, *Desulfovibrio vulgaris*, and *Escherichia coli* ATCC 33456 have been reported to produce soluble Cr(VI) reductases (Ishibashi et al. 1990, Suzuki et al. 1992, Shen and Wang 1993, Lovley and Phillips 1994, Chen and Hao 1998) which utilize various electron donors and can be located inside or outside the bacterial cell (Chen and Hao 1998). Since reduction mediated by such reductases is an energy-requiring and highly regulated process, these are produced constitutively. To date, the best-studied chromate reductase is ChrR from *P. putida*, a soluble flavin mononucleotide binding enzyme which catalyzes the reduction of Cr(VI) to Cr(III) (Park et al. 2000). The ChrR enzyme functions as a NADH-dependent reductase, having a wide substrate specificity and allowing the NAD(P)H-dependent reduction of quinones, prodrugs, chromate, and hexavalent uranium ions (Barak et al. 2006). ChrR catalyzes a transference of one and two electrons to Cr(VI) with the formation of the unstable species Cr(V) before further reduction to Cr(III). Although a proportion of the Cr(V) intermediate is spontaneously reoxidized generating ROS, its reduction to trivalent chromium through two electron transfer minimizes the production of harmful radicals. Studies with purified *Pseudomonas putida* ChrR revealed that this enzyme has a quinone reductase activity in chromate reduction. Quinols produced by quinone reduction confer tolerance to ROS (Gonzalez et al. 2005). Thus, ROS generated by ChrR activity during Cr(VI) reduction should be neutralized by quinols, which are formed by the quinone reductase activity of the same enzyme (Ackerley et al. 2004a, Cheung and Gu 2007, Ramirez-Diaz et al. 2008). Therefore, though ChrR activity generates ROS during Cr(VI) reduction, it reduces quinones that provide protection against ROS (Ramirez-Diaz et al. 2008). In *Escherichia coli*, a ChrR (ChrR of *Escherichia coli* was formerly called YieF) protein that shares sequence homology with the *Pseudomonas putida* ChrR enzyme was observed (Barak et al. 2006).

Extracellular Cr(VI) Reduction

Extracellular reduction of chromate followed by its binding with functional groups on the bacterial cell surface is another resistance mechanism (Ngwenya and Chirwa 2011) and helps its easy removal from the contaminated environment. Peptidoglycan was found to be a potent binder of Cr(III) (Hoyle and Beveridge 1983). The adsorptive properties are mostly dependent on the distribution of functional groups as carboxyl, amine, hydroxyl, phosphate and sulfhydryl on the cell wall surface of some bacteria (Parmar et al. 2000). Thus when its reduction occurs extracellularly there is no apparent entry of Cr(VI) in the cell. For that reason, biosorption has been also associated with chromate reduction. Fein et al. (2002) showed a nonmetabolic reduction of hexavalent chromium by bacterial surfaces under nutrient limitations as probable results of the oxidation of organic molecules within the cell wall that serve as electron donors. Fungi have the ability to reduce Cr(VI) as a mechanism of detoxification. Filamentous fungi, such as *Aspergillus* sp., *Penicillium* sp., and *Trichoderma inhamatum*, reduce Cr(VI) to Cr(III) by exploiting the reducing power generated by carbon metabolism (Acevedo-Aguilar et al. 2006, Morales-Barrera and Cristiani-Urbina 2008).

Muter et al. (2001) hypothesized that Cr(VI) reduction in *Candida utilis* could be partly dependent on pH changes during the exponential phase or on exo-enzymatic activities during stationary phase. *Candida maltosa*, isolated from a leather factory and characterized by a high tolerance level of chromate, demonstrated the ability to reduce Cr(VI) both in the presence of viable intact cells and in cell-free extracts (Ramirez-Ramirez et al. 2004). This capability was related to NADH-dependent chromate reductase activity associated with soluble proteins and, to a lesser degree, with the membrane fraction (Ramirez-Ramirez et al. 2004). Recently, the reduction to Cr(III) by an enzymatic mechanism has been observed in *Pichia*. Both in intact cells and in cell-free extracts of *Pichia jadinii* M9 and *Pichia anomala* M10 strains Cr(VI) was reduced to Cr(III), suggesting the presence of a chromate reductase activity probably associated with the cytosolic or membrane proteins (Martorell et al. 2012). A non-enzymatic mechanism of chromate reduction has been described for *Aspergillus niger* (Park et al. 2005). The relationship between biosorption and reduction mechanisms could be related to pH values. Low pH provides the protons needed for reduction reaction and increase the protonation level of the adsorbent surface which is thus more positively charged and more attractive for negative Cr(VI) ions (Park et al. 2005, Martorell et al. 2012).

The Cr(III) complexes generated extracellularly may penetrate into the cell and cause intracellular damage. The genotoxicity has been analyzed by means of the SOS test in *Salmonella typhimurium* (Nakamura et al. 1987). The test is based on the capability of genotoxic chemicals to induce *umu* gene expression in the tester strain of *Salmonella typhimurium* TA 1535/pSK1002, in which an *umuC-lacZ* fusion gene has been introduced. The genotoxicity can be quantified reading β-galactosidase activity spectrophotometrically. However, the extracellular reduction of Cr(VI) with different antioxidants did not increase the β-galactosidase activity, which led to determinate that reduction of Cr(VI) to Cr(III) with diverse antioxidants is a detoxification reaction (Poljsak et al. 2005). Additionally, the extracellular reduction of Cr(VI) with antioxidants upped the viability of *Saccharomyces cerevisiae* (Poljsak 2004).

ROS Detoxifying Enzymes/Intracellular Cr(VI) Reduction

During Cr(VI) reduction a short-lived and highly reactive intermediate Cr(V) radical is generated. Thus, Cr(V) is re-oxidized to Cr(VI), giving its electron to dioxygen and ROS. Generation of ROS results in oxidative stress in the bacteria affecting the viability of cells and the efficiency of Cr(VI) reduction. In this process, are also induced the bacterial proteins in the defense against oxidative stress leading to an additional mechanism of chromate resistance (Ramirez-Diaz et al. 2008). However, detoxifying enzymes like glutathione transferase, superoxide dismutase (SOD) and catalase removed the ROS-oxidative stress (Ackerley et al. 2004b). On the basis of the magnitude of oxidative stress generated, the reductase enzymes get reduce Cr(VI) to form the highly unstable Cr(V) intermediate. Cr(V) can get oxidized back to Cr(VI) in a redox cycle, giving its electrons to molecular oxygen and thus, producing a large amount of ROS. The one electron reducers being flavin-dependent enzymes follow the reaction as given below:

$$\text{FMN (Ox.)} + e^- + H^+ \rightarrow \text{FMNH}^+$$

The oxidized flavin nucleotide [FMN(Ox)] accepts one electron flexible semiquinone (a stable free radical) form (FMNH). Some of the identified chromate reducing enzymes is lipoyl dehydrogenase (LpDH) from *Clostridium kluyveri*, ferridoxin-NAD, glutathione reductase and cytochrome c (Shi and Dalal 1990, Ackerley et al. 2004b). The physiological roles of these are to catalyze energetic or biosynthetic reactions (Barak et al. 2006). Otherwise, two electron reduction process to Cr(III) product without forming Cr(V) intermediate. As a result, much smaller amount of ROS is generated during this process than that of one electron reduction. Two electron reducers being NAD(P) dependent enzymes, are characterized by the transfer of H^- (an equivalent of a proton and two electrons) in a reaction as follows:

$$NAD^+ + 2e^- + 2H^+ \rightarrow NADH + H^+$$

$$NADP^+ + 2e^- + 2H^+ \rightarrow NADPH + H^+$$

Some chromate reducing enzymes behaving as two electron reducers includes ChrR from *Pseudomonas putida*, YieF and NfsA from *Escherichia coli* (Barak et al. 2006). The varying one and two electron reducing capability of diverse enzymes have been experimentally verified. Matin (2006) guided an experiment to categorize two chromate reducing enzymes, LpDH and YieF as either one or two electron-reducer. The *in vitro* study, using pure LpDH by 'redox balance' method, revealed the consumption of only 24% of electrons for chromate reduction while remaining 70% electrons for ROS production. The results indicated LpDH as the one electron-reducer. Unlike, an identical test with YieF showed the use of only 25% of electrons for ROS generation while the remaining electrons for Cr(VI) reduction indicating YieF as the two electrons-reducer. Ackerley et al. (2004b) propound a classification of enzymes carrying out Cr(VI) reduction, i.e., Class I or Class II reductases enzymes based on sequence homologies.

DNA Repair Enzymes

Cr(VI) enters the microbial cell which is readily reduced to Cr(III) by the action of various enzymatic or non-enzymatic activities that leads to the generation of ROS. The reactive species such as of Cr(V) and Cr(IV), as soon as formed, can react with DNA, inducing oxidative damage. The ROS generate deleterious effects to DNA, causing base modification, single-strand breaks, and double-strand breaks. Such damages can be repaired by special DNA repair mechanism like the SOS response enzymes (RecA, RecG, RuvB) (Hu et al. 2005). Protection of bacterial cells by DNA repair enzymes of damaged caused by Cr(VI) is another defense shield.

In *Escherichia coli*, Cr(VI) induces the SOS repair system that protects DNA from oxidative damage (Llagostera et al. 1986). Similarly, in *Pseudomonas aeruginosa*, DNA helicases like RecG and RuvB, components of the recombinational DNA repair system, participate in the response to DNA damage caused by chromate (Miranda et al. 2005). Cr-DNA adducts are the most abundant form of DNA damage that causes mutations and chromosomal breaks (Zhitkovich 2011). Cr(III) causes DNA damage and negatively affects DNA topology by inhibiting topoisomerase-DNA relaxation activity (Plaper et al. 2002). These processes explain the DNA alterations observed in those cells exposed to the chromate such as single strand breaks, DNA–DNA interstrand links, DNA-protein cross-links, Cr-DNA adducts, nucleotide oxidation and a basic sites (Stearns and Wetterhahn 1997, Voitkun et al. 1998, Sugden et al. 2001, Salnikow and Zhitkovich 2008). The SOS DNA repair system, initially described in *E. coli* (Radman 1974), is triggered by the treatment of bacteria with DNA-damaging agents. The SOS system, which induces the arrest of DNA replication and cell division, involve more than 40 SOS genes. These genes encode proteins engaged in the protection, replication, repair, mutagenesis, and metabolism of DNA, and it is regulated by the transcriptional repressor LexA and the coprotease RecA that aids the autocatalytic self-cleavage of LexA (Janion 2008). Llagostera et al. (1986) demonstrated that Cr(VI) induces the SOS system in *E. coli*, measuring β-galactosidase activity, using lacZ gene fusions under the control region of different SOS genes.

Two comprehensive sets of metal-responsive genomic profiles were generated following exposure to the metal. One that supplies information on the transcriptional changes associated with metal exposure, and a second that provides information on the relationship between the expression of approximately 4700 non-essential genes and sensitivity to the metal exposure (deletome). Approximately 22% of the genome was affected by exposure to Cr (Jin et al. 2008).

Efflux of Cr(VI) from Cell

Efflux of chromate ions from the bacteria cell cytoplasm mediated by transporters is also considered a resistance mechanism. This prevents the accumulation of toxic ions inside the bacterial cells and is encoded by specific plasmid-borne genes (Ramirez-Diaz et al. 2008). The best understood chromate resistance system involved in chromate efflux mechanism is that conferred by *Pseudomonas aeruginosa* ChrA protein belongs to the chromate ion transporter CHR superfamily. The CHR protein family, which includes putative ChrA orthologs, currently contains about 135 sequences from all three life domains (Ramirez-Diaz et al. 2008).

ChrA is a hydrophobic membrane protein, encoded by plasmids pUM505 of *P. aeruginosa* and pMOL28 from *Cupriavidus metallidurans* (Cervantes et al. 1990, Nies et al. 1990). ChrA protein functions as a chemiosmotic pump that expels chromate from cytoplasm or periplasm to outside propel by the proton motive force (Alvarez et al. 1999). CHR proteins from several bacteria have been demonstrated as involved in chromate resistance by chromate efflux mechanism (Ramirez-Diaz et al. 2008).

Juhnke et al. (2002) reported that chromate efflux in *Cupravidus metallidurans* and *P. aeruginosa* occurring via the ChrA protein and produced resistance levels of 4 and 0.3 mM, respectively. Branco et al. (2008) reported that *Ochrobactrum tritici* 5bvl1 tolerate chromate concentrations of > 50 mM and have the transposon TnOtChr (which contains the group of genes: chrB, chrA, chrC and chrF). The chrB and chrA genes were involved in chromium high resistance of *O. tritici*. The *chr* promoter was induced by chromate or dichromate, but not by Cr(III), oxidants, sulfate, or other oxyanions. Induction of the *chr operon* suppressed accumulation of cellular Cr through the activity of a chromate efflux pump that is encoded by chrA (Branco et al. 2008).

Recently, the *Lysinibacillus fusiformis* ZC1 strain was found to contain large numbers of metal resistance genes, such as the chrA gene, which encodes a chromate transporter that confers chromate resistance. A yieF gene and several genes encoding reductases that were possibly involved in chromate reduction were also found; moreover, the expression of two adjacent putative chromate reduction related genes, nitR and yieF, was regarded to be constitutive (He et al. 2011).

As a structural analog of sulfate (SO_4^{2-}), chromate enters cells through sulfate uptake systems. Cr(VI) will be reduced to Cr(III) if the bacteria possess intracellular chromate reductases. If not, Cr(VI) accumulated inside the cell induces the chr operon and activates the chromate efflux pump that is encoded by chrA. Therefore, the bacterial cell is protected from Cr(VI) toxicity expelling it outside the cell.

ROS Scavenging

Cr(VI) after entering the cell can be reduced to Cr(V) by electron donors like NAD(P)H or some other organic compounds (like glucose). Chromate reductases reduce relative unstable toxic intermediate Cr(V) further to Cr(III) by a two electron transfer (via "semi-tight" mechanism). Sometimes this reaction is not very rapid. However, a portion of Cr(V) intermediate is quickly reoxidized to Cr(VI) thereby generating ROS by a Fenton-like reaction. During the process hydroxyl radicals (•OH) are formed in the microbial cells (Shi and Dalal 1994) as depicted in the equation below:

$$Cr(V) + H_2O_2 \rightarrow Cr(VI) + {}^{\bullet}OH + OH^-$$

During the reduction process, molecular oxygen is reduced to O_2^- radicals, which generates H_2O_2 via dismutation. Hexavalent chromium reacts with H_2O_2 to generate •OH radicals via a Fenton-like reaction. This mechanism is similar to the oxidation of Fe(II) with H_2O_2 in the Fenton reaction as the production of •OH from Fe(II) via Fenton reaction is facilitated greatly by the formation of Fe(II) complexes that have vacant sites for H_2O_2 coordination.

Concluding Remarks

Proteomics and genomics technologies (global analysis of macromolecules) were studied in this chapter to elucidate heavy metals' resistance mechanisms. Different resistance mechanisms were described, Cr(VI) reduction mechanisms have been specified under aerobic conditions (associated with soluble chromate reductases) and in anaerobic conditions (Cr(VI) can be used as a final electron acceptor in the electron transport chain). In the same way, chromate extracellular reduction followed by its binding to functional groups on the bacterial cell surface was described as another resistance mechanism. Additionally, free radical detoxification activities, repair of DNA damage and processes related to sulfur or iron homeostasis. There are effective systems to protect against the lethal effects of Cr (VI), apparently associated with reduced uptake of Cr(VI).

On the other hand, there are an effective protective metabolic systems prepared to combat oxidative stress generated by ROS, such as enzymes superoxide dismutase (SOD), catalase (CAT), and peroxiredoxin. The growing number of studies based on the ability of the microbial reduction of Cr(VI) to Cr(III) is particularly important from the point of view of bioremediation environments contaminated with heavy metals results from greater scientific interest and high biotechnological potential.

Acknowledgments

This work was supported by Agencia Nacional de Promoción Científica y Tecnológica and Fondos para la Invesigación Científica y Tecnológica FONCYT (PICT 1154/2013; PICT 3639/2015); Consejo Nacional de Investigaciones Científicas y Técnicas (PIO CONICET YPF 133201301000022CO); Secretaría de Ciencia, Arte e Innovación Tecnológica de la Universidad Nacional de Tucumán SCAIT UNT (PIUNT D509).

References Cited

Acevedo-Aguilar, F.J., A.E. Espino-Saldana, I.L. Leon-Rodriguez, M.E. Rivera-Cano, M. Vila-Rodriguez, K. Wrobel et al. 2006. Hexavalent chromium removal *in vitro* and from industrial wastes, using chromate-resistant strains of filamentous fungi indigenous to contaminated wastes. Can. J. Microbiol. 52: 809–815.

Ackerley, D.F., C.F. Gonzalez, C.H. Park, R. Blake II, M. Keyhan and A. Matin. 2004a. Chromate-reducing properties of soluble flavoproteins from *Pseudomonas putida* and *Escherichia coli*. Appl. Environ. Microbiol. 70: 873–882.

Ackerley, D.F., C.F. Gonzalez, M. Keyhan, R. Blake and A. Matin. 2004b. Mechanism of chromate reduction by the *Escherichia coli* protein, NfsA and the role of different chromate reductases in minimizing oxidative stress during chromate reduction. Environ. Microbiol. 6: 851–860.

Aguilar-Barajas, E., C. Diaz-Perez, M.I. Ramirez-Diaz, H. Riveros-Rosas and C. Cervantes. 2011. Bacterial transport of sulfate, molybdate, and related oxyanions. Biometals 24: 687–707.

Ahemad, M. 2014. Bacterial mechanisms for Cr(VI) resistance and reduction: an overview and recent advances. Folia Microbial. 59: 321–332.

Alam, M.Z. and A. Malik. 2008. Chromate resistance, transport and bioreduction by *Exiguobacterium* sp. ZM-2 isolated from agricultural soil irrigated with tannery effluent. J. Basic Microbiol. 48: 416–420.

Alam, M.Z. and S. Ahmad. 2012. Toxic chromate reduction by resistant and sensitive bacteria isolated from tannery effluent contaminated soil. Ann. Microbiol. 62: 113–121.

Alvarez, A.H., R. Moreno-Sanchez and C. Cervantes. 1999. Chromate efflux by means of the ChrA chromate resistance protein from *Pseudomonas aeruginosa*. J. Bacteriol. 181: 7398–7400.

Anderson, N.L. and N.G. Anderson. 1998. Proteome and proteomics: new technologies, new concepts, and new words. Electrophoresis 19: 1853–1861.

Bah, A.M., H. Sun, F. Chen, J. Zhou, H. Dai, G. Zhang et al. 2010. Comparative proteomic analysis of *Typha angustifolia* leaf under chromium, cadmium and lead stress. J. Haz. Mat. 184: 191–203.

Baillon, L., F. Pierron, R. Coudret, E. Normendeau, A. Caron, L. Peluhet et al. 2015. Transcriptome profile analysis reveals specific signatures of pollutants in Atlantic eels. Ecotoxicology 24: 71–84.

Barak, Y., D.F. Ackerley, C.J. Dodge, L. Banwari, C. Alex, A.J. Francis et al. 2006. Analysis of novel soluble chromate and uranyl reductases and generation of an improved enzyme by directed evolution. Appl. Environ. Microbiol. 72: 7074–7082.

Barceloux, D.G. and D. Barceloux. 1999. Chromium. J. Toxicol. Clin. Toxicol. 37: 173–194.

Batool, R., K. Yrjälä and S. Hasnain. 2012. Hexavalent chromium reduction by bacteria from tannery effluent. J. Microbiol. Biotechnol. 22: 547–554.

Belchik, S.M., D.W. Kennedy, A.C. Dohnalkova, Y.M. Wang, P.C. Sevinc, H. Wu et al. 2011. Extracellular reduction of hexavalent chromium by cytochromes MtrC and OmcA of *Shewanella oneidensis* MR-1. Appl. Environ. Microbiol. 77: 4035–4041.

Bencheikh-Latmani, R., S.M. Williams, L. Haucke, C.S. Criddle, L.Y. Wu, J.Z. Zhou et al. 2005 Global transcriptional profiling of *Shewanella oneidensis* MR-1 during Cr(VI) and U(VI) reduction. Appl. Environ. Microbiol. 71: 7453–7460.

Branco, R., A.P. Chung, T. Johnston, V. Gurel, P. Morais and A. Zhitkovich. 2008. The chromate-inducible chrBACF operon from the transposable element TnOtChr confers resistance to chromium(VI) and superoxide. J. Bacteriol. 190: 6996–7003.

Brown, S.D., M.R. Thompson, N.C. Verberkmoes, K. Chourey, M. Shah, J. Zhou et al. 2006. Molecular dynamics of the *Shewanella oneidensis* response to chromate stress. Mol. Cell. Proteomics 5: 1054–1071.

Cervantes, C., H. Ohtake, L. Chu, T.K. Misra and S. Silver. 1990. Cloning, nucleotide sequence, and expression of the chromate resistance determinant of *Pseudomonas aeruginosa* plasmid pUM505. J. Bacteriol. 172: 287–291.

Cervantes, C., J. Campos-Garcia, S. Devars, F. Gutierrez-Corona, H. Loza-Tavera, J.C. Torres-Guzman et al. 2001. Interactions of chromium with microorganisms and plants. FEMS Microbiol. Rev. 25: 335–347.

Cervantes, C. and J. Campos-García. 2007. Reduction and efflux of chromate by bacteria. pp. 407–419. *In*: Nies, D. and S. Silver [eds.]. Molecular Microbiology of Heavy Metals. Springer-Verlag, Berlin.

Chen, J. and O. Hao. 1998. Microbial chromium(VI) reduction. Critic. Rev. Environ. Sci. Technol. 28: 219–251.

Cheng, Y.J., Y.M. Xie, J. Zheng, Z.X. Wu, Z. Chen, X.Y. Ma et al. 2009. Identification and characterization of the chromium (VI) responding protein from a newly isolated *Ochrobactrum anthropi* CTS-325. J. Environ. Sci. China 21: 1673–1678.

Cheung, K.H. and J.D. Gu. 2007. Mechanism of hexavalent chromium detoxification by microorganisms and bioremediation application potential: a review. Int. Biodeter. Biodegr. 59: 8–15.

Chiu, J.M., N. Degger, J.Y. Leung, B.H. Po, G.J. Zheng and B.J. Richardson. 2016. A novel approach for estimating the removal efficiencies of endocrine disrupting chemicals and heavy metals in wastewater treatment processes. Mar. Pollut. Bull. 112: 53–57.

Chovanec, P., C. Sparacino-Watkins, N. Zhang, P. Basu and J.F. Stolz. 2012. Microbial reduction of chromate in the presence of nitrate by three nitrate respiring organisms. Front. Microbiol. 3: 416.

Christl, I., M. Imseng, E. Tatti, J. Frommer, C. Viti, L. Giovannetti et al. 2012. Aerobic reduction of chromium (VI) by *Pseudomonas corrugata* 28: influence of metabolism and fate of reduced chromium. Geomicrobiol. J. 29: 173–185.

D'Hooghe, P., S. Escamez, J. Trouverie and J.C. Avice. 2013. Sulphur limitation provokes physiological and leaf proteome changes in oilseed rape that lead to perturbation of sulphur, carbon and oxidative metabolisms. BMC Plant Biology 13: 23.

Das, S., S.S. Ram, H.K. Sahu, D.S. Rao, A. Chakraborty, M. Sudarshan et al. 2013. A study on soil physico-chemical, microbial and metal content in Sukinda chromite mine of Odisha, India. Environ. Earth Sci. 69: 2487–2497.

Daulton, T.L., B.J. Little, J. Jones-Meehan, D.A. Blom and L.F. Allard. 2007. Microbial reduction of chromium from the hexavalent to divalent state. Geochim. Cosmochim. Acta 71: 556–565.

Dębski, B., W. Zalewski, M.A. Gralak and T. Kosla. 2004. Chromium-yeast supplementation of chicken broilers in an industrial farming system. J. Trace Elem. Med. Biol. 18: 47–51.

Decorosi, F., E. Tatti, A. Mini, L. Giovannetti and C. Viti. 2009. Characterization of two genes involved in chromate resistance in a Cr(VI)-hyper-resistant bacterium. Extremophiles 13: 917–923.

Decorosi, F. 2010. Studio di ceppi batterici per il biorisanamento di suoli contaminati da Cr(VI). Ph.D. Thesis, Firenze University Press, Florence.

Dey, S. and A.K. Paul. 2013. Hexavalent chromium reduction by aerobic heterotrophic bacteria indigenous to chromite mine overburden. Braz. J. Microbiol. 44: 307–315.

Fein, J.B., D.A. Fowle, J. Cahill, K. Kemner, M. Boyanov and B. Brunker. 2002. Nonmetabolic reduction of Cr (VI) by bacterial surfaces under nutrient-absent conditions. Geomicrobiol. J. 19: 369–382.

Fernández, P.M., M.E. Cabral, O.D. Delgado, J.I. Fariña and L.I.C. Figueroa. 2013. Textile-dye polluted waters as a source for selecting chromate-reducing yeasts through Cr(VI)-enriched microcosms. Int. Biodeter. Biodegr. 79: 28–35.

Gadd, G.M. 1992. Metals and microorganisms: a problem of definition. FEMS Microbiol. Lett. 100: 197–204.

Ge, S., M. Zhou, X. Dong, Y. Lu and S. Ge. 2013. Distinct and effective biotransformation of hexavalent chromium by a novel isolate under aerobic growth followed by facultative anaerobic incubation. Appl. Microbiol. Biotechnol. 97: 2131–2137.

Gnanamani, A., V. Kavitha, N. Radhakrishnan, G.S. Rajakumar, G. Sekaran and A.B. Mandal. 2010. Microbial products (biosurfactant and extracellular chromate reductase) of marine microorganism are the potential agents reduce the oxidative stress induced by toxic heavy metals. Colloids Surf. B. 79: 334–339.

Gonzalez, C.F., D.F. Ackerley, S.V. Lynch and A. Matin. 2005. ChrR, a soluble quinone reductase of *Pseudomonas putida* that defends against H_2O_2. J. Biol. Chem. 280: 22590–22595.

He, M., X. Li, H. Liu, S.J. Miller, G. Wang and C. Rensing. 2011. Characterization and genomic analysis of a highly chromate resistant and reducing bacterial strain *Lysinibacillus fusiformis* ZC1. J. Hazard. Mater. 185: 682–688.

He, Z., F. Gao, T. Sha, Y. Hu and C. He. 2009. Isolation and characterization of a Cr(VI)-reduction *Ochrobactrum* sp. strain CSCr-3 from chromium landfill. J. Hazard. Mater. 163: 869–873.

Henne, K.L., J.E. Turse and C.D. Nicora. 2009. Global proteomic analysis of the chromate response in *Arthrobacter* sp. strain FB24. J. Prot. Res. 8: 1704–1716.

Hong, K.J., S. Tokunaga and T. Kajiuchi. 2002. Evaluation of remediation process with plant-derived bio-surfactant for recovery of heavy metals from contaminated soils. Chemosphere 49: 379–387.

Hong, S. and C. Ning. 2011. Adsorption of Cr(VI) on FeNi modified bentonites. Environ. Eng. Manag. J. 10: 875–879.

Hookoom, M. and D. Puchooa. 2013. Isolation and identification of heavy metals tolerant bacteria from industrial and agricultural areas in Mauritius. Curr. Res. Microbiol. Biotechnol. 1: 119–123.

Hoyle, B. and T.J. Beveridge. 1983. Binding of metallic ions to the outer membrane of *Escherichia coli*. Appl. Environ. Microbiol. 46: 749–752.

Hu, P., E.L. Brodie, Y. Suzuki, H.H. Mc Adams and G.L. Andersen. 2005. Whole-genome transcriptional analysis of heavy metal stresses in *Caulobacter crescentus*. J. Bacteriol. 187: 8437–8449.

Ishibashi, Y., C. Cervantes and S. Silver. 1990. Chromium reduction in *Pseudomonas putida*. Appl. Environ. Microbiol. 56: 2268–2270.

Janion, C. 2008. Inducible SOS response system of DNA repair and mutagenesis in *Escherichia coli*. Int. J. Biol. Sci. 4: 338–344.

Jin, Y.H., P.E. Dunlap, S.J. McBride, H. Al-Refai, P.R. Bushel and J.H. Freedman. 2008. Global transcriptome and deletome profiles of yeast exposed to transition metals. PLoS Genet. 25: 4-1000053.

Johnson, A.J., F. Veljanoski, P.J. O'Doherty, M.S. Zaman, G. Petersingham, T.D. Bailey et al. 2016. Revelation of molecular basis for chromium toxicity by phenotypes of *Saccharomyces cerevisiae* gene deletion mutants. Metallomics 8: 542–550.

Joutey, N.T., H. Sayel, W. Bahafid and N. El Ghachtouli. 2015. Mechanisms of hexavalent chromium resistance and removal by microorganisms. pp. 45–69. *In*: Whitacre, D.M. [ed.]. Reviews of Environmental Contamination and Toxicology Volume 233. Springer International Publishing.

Juhnke, S., N. Peitzsch, N. Hubener, C. Grosse and D.H. Nies. 2002. New genes involved in chromate resistance in *Ralstonia metallidurans*strain CH34. Arch. Microbiol. 179: 15–25.

Katsou, E., S. Malamis and K.J. Haralambous. 2011. Industrial wastewater pre-treatment for heavy metal reduction by employing a sorbent-assisted ultrafiltration system. Chemosphere 82: 557–564.

Kwak, Y.H., D.S. Lee and H.B. Kim. 2003. *Vibrio harveyi* nitroreductase is also a chromate reductase. Appl. Environ. Microbiol. 69: 4390–4395.

Labra, M., S. Regondi, F. Grassi and E. Agradi. 2006. *Zea mays* L. protein changes in response to potassium dichromate treatments. Chemosphere 62: 1234–1244.

Lau, A.T., Q.Y. He and J.F. Chiu. 2003. Proteomic technology and its biomedical applications. Acta Biochim. Biophys. Sin. 35: 965–975.

Lazarova, N., E. Krumova, T. Stefanova, N. Georgieva and M. Angelova. 2014. The oxidative stress response of the filamentous yeast *Trichosporon cutaneum* R57 to copper, cadmium and chromium exposure. Biotechnol. Biotechnol. Equip. 28: 855–862.

Llagostera, M., S. Garrido, R. Guerrero and J. Barbe. 1986. Induction of SOS genes of *Escherichia coli* by chromium compounds. Environ. Mutagen. 8: 571–577.

Lovley, D.R. and J.P.E. Phillips. 1994. Reduction of chromate by *Desulfovibrio vulgaris* and its C3 cytochrome. Appl. Environ. Microbiol. 60: 726–728.

Lowe, K.L., W. Straube, B. Little and J. Jones-Meehan. 2003. Aerobic and anaerobic reduction of Cr(VI) by *Shewanella oneidensis*: effects of cationic metals, sorbing agents and mixed microbial cultures. Acta Biotechnol. 23: 161–178.

Martorell, M.M., P.M. Fernandez, J.I. Farina and L.I.C. Figueroa. 2012. Cr(VI) reduction by cell-free extracts of *Pichia jadinii* and *Pichia anomala* isolated from textile-dye factory effluents. Int. Biodeter. Biodegr. 71: 80–85.

Matin, A.C. 2006. Development of combinatorial bacteria for metal and radionuclide bioremediation, Grant No. ER63627-1021953-0009581, Final Technical Report, Reporting period: 09/15/2003–06/15/2006.

Mazoch, J., R. Tesařík, V. Sedláček, I. Kučera and J. Turánek. 2004. Isolation and biochemical characterization of two soluble iron(III) reductases from *Paracoccus denitrificans*. Eur. J. Biochem. 271: 553–562.

McGrath, S.P. and S. Smith. 1990. Chromium and nickel. pp. 125–150. *In*: Alloway, B.J. [ed.]. Heavy Metals in Soils. Wiley, New York.

Miranda, A.T., M.V. Gonzalez, G. Gonzalez, E. Vargas, J. Campos-Garcia and C. Cervantes. 2005. Involvement of DNA helicases in chromate resistance by *Pseudomonas aeruginosa* PAO1. Mutat. Res. 578: 202–209.

Missaoui, A., I. Said, Z. Lafhaj and E. Hamdi. 2016. Influence of enhancing electrolytes on the removal efficiency of heavy metals from Gabes marine sediments (Tunisia). Mar. Pollut. Bull. 113: 44–54.

Monsieurs, P., H. Moors, R. Van Houdt, P.J. Janssen, A. Janssen, I. Coninx et al. 2011. Heavy metal resistance in *Cupriavidus metallidurans* CH34 is governed by an intricate transcriptional network. Biometals 24: 1133–1151.

Morais, P.V., R. Branco and R. Francisco. 2011. Chromium resistance strategies and toxicity: what makes *Ochrobactrum tritici* 5bvl1 a strain highly resistant. Biometals 24: 401–410.

Morales-Barrera, L. and E. Cristiani-Urbina. 2008. Hexavalent chromium removal by a Trichoderma inhamatum fungal strain isolated from tannery effluent. Water Air Soil Pollut. 187: 327–336.

Muter, O., A. Patmalnieks and A. Rapoport. 2001. Interrelations of the yeast *Candida utilis* and Cr(VI): metal reduction and its distribution in the cell and medium. Process Biochem. 36: 963–970.

Myers, C.R., B.P. Carstens, W.E. Antholine and J.M. Myers. 2000. Chromium(VI) reductase activity is associated with the cytoplasmic membrane of anaerobically grown *Shewanella putrefaciens* MR-1. J. Appl. Microbiol. 88: 98–106.

Nakamura, S., Y. Oda, T. Shimada, I. Oki and K. Sugimoto. 1987. SOS-inducing activity of chemical carcinogens and mutagens in *Salmonella typhimurium* TA1535/pSK1002: examination with 151 chemicals. Mutat. Res. 192: 239–246.

Narayani, M. and K.V. Shetty. 2012. Characteristics of a novel *Acinetobacter* sp. and its kinetics in hexavalent chromium bioreduction. J. Microbiol. Biotechnol. 22: 690–698.

Nguema, P.F. and Z. Luo. 2012. Aerobic chromium(VI) reduction by chromium-resistant bacteria isolated from activated sludge. Ann. Microbiol. 62: 41–47.

Ngwenya, N. and E.M.N. Chirwa. 2011. Biological removal of cationic fission products from nuclear wastewater. Water Sci. Technol. 63: 124–128.

Nickens, K.P., S.R. Patierno and S. Ceryak. 2010. Chromium genotoxicity: a double-edged sword. Chem. Biol. Interact. 188: 276–288.

Nies, A., D.H. Nies and S. Silver. 1990. Nucleotide sequence and expression of plasmid-encoded chromate resistance determinant from *Alcaligenes eutrophus*. J. Biol. Chem. 265: 5648–5653.

Nies, D.H. and S. Silver. 1989. Metal ion uptake by plasmid-free metal-sensitive *Alcaligenes eutrophus*. J. Bacteriol. 171: 4073–4075.

Obayori, O.S., M.O. Ilori, S.A. Adebusoye, G.O. Oyetibo and O. Amund. 2009. Degradation of hydrocarbons and biosurfactant production by *Pseudomonas* sp. strain LP1. World J. Microbiol. Biotechnol. 25: 1615–1623.

Ohtake, H., C. Cervantes and S. Silver. 1987. Decreased chromate uptake in *Pseudomonas fluorescens* carrying a chromate resistance plasmid. J. Bacteriol. 169: 3853–3856.

Opperman, D.J. and E. van Heerden. 2008. A membrane-associated protein with Cr(VI)-reducing activity from *Thermus scotoductus* SA-01. FEMS Microbiol. Lett. 280: 210–218.

Park, C.H., M. Keyhan, B. Wielinga, S. Fendorf and A. Matin. 2000. Purification to homogeneity and characterization of a novel *Pseudomonas putida* chromate reductase. Appl. Environ. Microbiol. 66: 1788–1795.

Park, D., Y.S. Yun, J.H. Jo and J.M. Park. 2005. Mechanism of hexavalent chromium removal by dead fungal biomass of *Aspergillus niger*. Water. Res. 39: 533–540.

Parmar, A.N., T. Oosterbroek, S. Del Sordo, A. Segreto, A. Santangelo, D. Dal Fiume et al. 2000. Broad-band BeppoSAX observation of the low-mass X-ray binary X 1822-371. Astron. Astrophys. 356: 175–180.

Pennington, S.R., M.R. Wilkins, D.F. Hochstrasser and M.J. Dunn. 1997. Proteome analysis: from protein characterization to biological function. Trends Cell. Biol. 7: 168–173.

Plaper, A., S. Jenko-Brinovec, A. Premzl, J. Kos and P. Raspor. 2002. Genotoxicity of trivalent chromium in bacterial cells. Possible effects on DNA topology. Chem. Res. Toxicol. 15: 943–949.

Plebani, M. 2005. Proteomics: the next revolution in laboratory medicine? Clin. Chim. Acta 357: 113–122.

Poljsak, B. 2004. Pro-oxidative vs. antioxidative properties of ascorbic acid and trolox in Cr(VI) induced damage. Ph.D. Thesis, Nova Gorica Polytechnic, School of Environmental Science.

Poljsak, B., Z. Gazdag, S. Jenko-Brinovec, S. Fujs, M. Pesti, J. Belagyi et al. 2005. Pro-oxidative vs. antioxidative properties of ascorbic acid in chromium(VI)-induced damage: an *in vivo* and *in vitro* approach. J. Appl. Toxicol. 25: 535–548.

Poljsak, B., I. Pocsi, P. Raspor and M. Pesti. 2010. Interference of chromium with biological systems in yeasts and fungi: a review. J. Basic Microbiol. 50: 21–36.

Qi, W., R.J. Reiter, D.X. Tan, J.J. Garcia, L.C. Manchester, M. Karbownik et al. 2000. Chromium (III)-induced 8-hydroxydeoxyguanosine in DNA and its reduction by antioxidants: comparative effects of melatonin, ascorbate, and vitamin E. Environ. Health Perspect. 108: 399–402.

Radman, M. 1974. Phenomenology of an inducible mutagenic DNA repair pathway in *Escherichia coli:* SOS repair hypothesis. pp. 128–142. *In*: Sherman, S., M. Miller, C. Lawrence and W.H. Tabor [eds.]. Molecular and Environmental Aspects of Mutagenesis.

Ramírez-Díaz, M.I., C. Díaz-Pérez, E. Vargas, H. Riveros-Rosas, J. Campos-García and C. Cervantes. 2008. Mechanisms of bacterial resistance to chromium compounds. Biometals 21: 321–332.

Ramirez-Ramirez, R., C. Calvo-Mendez, M. Avila-Rodriguez, P. Lappe, M. Ulloa, R. Vazquez-Juarez et al. 2004. Cr(VI) reduction in a chromate-resistant strain of *Candida maltosa* isolated from the leather industry. Anton. Leeuw. 85: 63–68.

Richter, K., M. Schicklberger and J. Gescher. 2012. Dissimilatory reduction of extracellular electron acceptors in anaerobic respiration. Appl. Environ. Microbiol. 78: 913–921.

Righetti, P.G., A. Castagna, F. Antonucci, C. Piubelli, D. Cecconi, N. Campostrini et al. 2003. The proteome: anno Domini 2002. Clin. Chem. Lab. Med. 41: 425–438.

Robins, K.J., D.O. Hooks, B.H.A. Rehm and D.F. Ackerley. 2013. *Escherichia coli* NemA is an efficient chromate reductase that can be biologically immobilized to provide a cell free system for remediation of hexavalent chromium. PLoS ONE 8: 59200.

Romanenko, V.I. and V.N. Koren'kov. 1977. A pure culture of bacteria utilizing chromate and dichromate as hydrogen acceptors in growth under anaerobic conditions. Mikrobiologiya 46: 414–417.

Rorat, A., D. Wloka, A. Grobelak, A. Grosser, A. Sosnecka, M. Milczarek et al. 2016. Vermiremediation of polycyclic aromatic hydrocarbons and heavy metals in sewage sludge composting process. J. Environ. Manage. 187: 347–353.

Salnikow, K. and A. Zhitkovich. 2008. Genetic and epigenetic mechanisms in metal carcinogenesis and cocarcinogenesis: nickel, arsenic, and chromium. Chem. Res. Toxicol. 21: 28–44.

Shanker, A.K., C. Cervantes, H. Loza-Tavera and S. Avudainayagam. 2005. Chromium toxicity in plants. Environ. Int. 31: 739–753.

Sharmin, S.A., I. Alam, K.H. Kim, Y.G. Kim, P.J. Kim, J.D. Bahk et al. 2012. Chromium-induced physiological and proteomic alterations in roots of *Miscanthus sinensis*. Plant Sci. 187: 113–126.

Sharrma, S.S. and K.J. Dietz. 2009. The relationship between metal toxicity and cellular redox imbalance. Trends Plant Sci. 14: 43–50.

Shen, H. and Y.T. Wang. 1993. Characterization of enzymatic reduction of hexavalent chromium by *Escherichia coli* ATCC 33456. Appl. Environ. Microbiol. 59: 3771–3777.

Shi, Y., L. Chai, Z. Yang, Q. Jing, R. Chen and Y. Chen. 2012. Identification and hexavalent chromium reduction characteristics of *Pannonibacter phragmitetus*. Bioprocess Biosyst. Eng. 35: 843–850.

Shi, X. and N.S. Dalal. 1994. Generation of hydroxyl radical by chromate in biologically relevant systems: role of Cr (V) complexes versus tetraperoxochromate (V). Environ. Health Perspect. 102: 231–236.

Shi, X.L. and N.S. Dalal. 1990. Evidence for a fenton-type mechanism for the generation of OH radicals in the reduction of Cr (VI) in cellular media. Arch. Biochem. Biophys. 281: 90–95.

Smith, W.A., W.A. Apel, J.N. Petersen and B.M. Peyton. 2002. Effect of carbon and energy source on bacterial chromate reduction. Bioremed. J. 6: 205–215.

Soares, E.V. and H.M. Soares. 2012. Bioremediation of industrial effluents containing heavy metals using brewing cells of *Saccharomyces cerevisiea* as a green technology: a review. Environ. Sci. Pollut. Res. 19: 1066–1083.

Sobol, Z. and R.H. Schiestl. 2012. Intracellular and extracellular factors influencing Cr(VI) and Cr(III) genotoxicity. Environ. Mol. Mutagen. 53: 94–100.

Soni, S.K., R. Singh, A. Awasthi, M. Singh and A. Kalra. 2013. *In vitro* Cr(VI) reduction by cell-free extracts of chromate-reducing bacteria isolated from tannery effluent irrigated soil. Environ. Sci. Pollut. Res. Int. 20: 1661–1674.

Stearns, D.M. and K.E. Wetterhahn. 1997. Intermediates produced in the reaction of chromium(VI) with dehydroascorbate cause single-strand breaks in plasmid DNA. Chem. Res. Toxicol. 10: 271–278.

Sugden, K.D., C.K. Campo and B.D. Martin. 2001. Direct oxidation of guanine and 7,8-dihydro-8-oxoguanine in DNA by a high-valent chromium complex: a possible mechanism for chromate genotoxicity. Chem. Res. Toxicol. 14: 1315–1322.

Suzuki, T., N. Miyata, H. Horitsu, K. Kawai, K. Takamizawa, Y. Tai et al. 1992. NAD(P)H-dependent chromium(VI) reductase of *Pseudomonas ambigua* G-1: Cr(VI) intermédiate is formed during the reduction of Cr(VI) to Cr(III). J. Bacteriol. 174: 5340–5345.

Tebo, B.M. and A.Y. Obraztsova. 1998. Sulfate-reducing bacterium grows with Cr(VI), U(VI), Mn(IV), and Fe(III) as electron acceptors. FEMS Microbiol. Lett. 162: 193–199.

Thatoi, H., S. Das, J. Mishra, B.P. Rath and N. Das. 2014. Bacterial chromate reductase, a potential enzyme for bioremediation of hexavalent chromium: a review. J. Environ. Manage. 146: 383–399.

Thompson, D.K., K. Chourey, G.S. Wickham, S.B. Thieman, N.C. Verberkmoes, B. Zhang et al. 2010. Proteomics reveals a core molecular response of *Pseudomonas putida* F1 to acute chromate challenge. BMC Genomics 11: 311.

Thompson, M.R., N.C. Verberkmoes, K. Chourey, M. Shah, D.K. Thompson and R.L. Hettich. 2007. Dosage-dependent proteome response of *Shewanella oneidensis* MR-1 to acute chromate challenge. J. Proteome Res. 6: 1745–1757.

Urbano, A., L.M.R. Ferreira and M.C. Alpoim. 2012. Molecular and cellular mechanisms of hexavalent chromium-induced lung cancer: an updated perspective. Curr. Drug. Metab. 13: 284–305.

Viti, C., F. Decorosi, E. Tatti and L. Giovannetti. 2007. Characterization of chromate-resistant and -reducing bacteria by traditional means and by a high-throughput phenomic technique for bioremediation purposes. Biotechnol. Prog. 23: 553–559.

Viti, C., F. Decorosi, A. Mini, E. Tatti and L. Giovannetti. 2009. Involvement of the oscA gene in the sulphur starvation response and in Cr(VI) resistance in *Pseudomonas corrugata* 28. Microbiology 155: 95–105.

Vlaanderen, J., L.E. Moore, M.T. Smith, Q. Lan, L. Zhang and C.F. Skibola. 2010. Application of OMICS technologies in occupational and environmental health research: current status and projections. Occup. Environ. Med. 67: 136–143.

Voitkun, V., A. Zhitkovich and M. Costa. 1998. Cr(III)-mediated crosslinks of glutathione or amino acids to the DNA phosphate backbone are mutagenic in human cells. Nucleic Acids Res. 26: 2024–2030.

Wang, J. and C. Chen. 2006. Biosorption of heavy metals by *Saccharomyces cerevisiae*: a review. Biotechnol. Adv. 24: 427–451.

Wilkins, M.R., J.C. Sanchez, A.A. Gooley, R.D. Appel, I. Humphery-Smith, D.F. Hochstrasser et al. 1996. Progress with proteome projects: why all proteins expressed by a genome should be identified and how to do it. Biotechnol. Genet. Eng. Rev. 13: 19–50.

Xu, W.H., Y.G. Liu, G.M. Zeng, X. Li and W. Zhang. 2013. Promoting influence of organic carbon source on chromate reduction by *Bacillus* sp. Adv. Mat. Res. 610–613: 1789–1794.

Yıldız, M. and H. Terzi. 2016. Proteomic analysis of chromium stress and sulfur deficiency responses in leaves of two canola (*Brassica napus* L.) cultivars differing in Cr (VI) tolerance. Ecotoxicol. Environ. Saf. 124: 255–266.

Zhitkovich, A. 2011. Chromium in drinking water: sources, metabolism and cancer risk. Chem. Res. Toxicol. 24: 1617–1629.

Zhu, W., L. Chai, Z. Ma, Y. Wang, H. Xiaoa and K. Zhao. 2008. Anaerobic reduction of hexavalent chromium by bacterial cells of *Achromobacter* sp. strain Ch1. Microbiol. Res. 163: 616–623.

17

Effects of Environmental Factors on Bioremediation of Co-Contaminated Soils Using a Multiresistant Bacterium

Juan D. Aparicio,[1,2,a] *María Z. Simón Solá,*[1,2,b] *Claudia S. Benimeli,*[1,3,c] *Lucía Bigliardo*[1,2,d] and *Marta A. Polti*[1,4,*]

Introduction

Increasing soil pollution problems, resulting from industrialization and urbanization, has caused worldwide concerns. This problem is extremely acute in areas where the presence of different families of organic pollutants is accompanied by heavy metals, in concentrations exceeding permissible levels (Chen et al. 2015). Both organic and inorganic compounds when entering the soil pose a huge threat to human health and natural ecosystem. For instance, according to data from the Environmental Protection Agency (EPA), 40% of hazardous waste sites included in the National Priority List are contaminated with organic compounds and heavy metals (Olaniran et al. 2013). The treatment of this simultaneous contamination, known as co-contamination, represents a real current challenge.

In particular, chromium contamination in soil and water has been detected in association with industrial areas (Benimeli et al. 2003, Nie et al. 2010, Srinivasa Gowd et al. 2010). Cr(VI) is a dangerous pollutant, classified as carcinogenic by EPA (Bagchi et al. 2002). Similarly, residues of the gamma isomer of hexachlorocyclohexane (γ-HCH), commercially known as lindane, have been detected in soil, water, air, plants and animals, because of its

[1] Planta Piloto de Procesos Industriales Microbiológicos (PROIMI-CONICET). Av. Belgrano y Pasaje Caseros. 4000 Tucumán, Argentina.
[a] Email: dani_aparicio@hotmail.com.ar
[b] Email: zoleicas@hotmail.com
[c] Email: cbenimeli@yahoo.com.ar
[d] Email: luciabigliardo@hotmail.com
[2] Facultad de Bioquímica, Química y Farmacia, Universidad Nacional de Tucumán. Ayacucho 491. 4000. Tucumán, Argentina.
[3] Facultad de Ciencias Exactas y Naturales, Universidad Nacional de Catamarca. Esquiú 799. 4700. Catamarca, Argentina.
[4] Facultad de Ciencias Naturales e Instituto Miguel Lillo, Universidad Nacional de Tucumán. Miguel Lillo 205. 4000. Tucumán, Argentina.
* Corresponding author: mpolti@proimi.org.ar

unselective use, mainly in agriculture practices (Fuentes et al. 2011). Lindane is extremely recalcitrant with numerous health effects, including neurological problems and cancer (Saez et al. 2012). Mixed pollution by chromium and lindane has been detected around the world in sediment and soil, at concentrations up to 140 mg kg^{-1} and 400 mg kg^{-1}, for chromium and lindane, respectively (Benimeli et al. 2003, Maggi et al. 2012, Arienzo et al. 2013, Coatu et al. 2013).

The treatment of soils with mixed pollution is a complex issue because the remediation technologies are different for each class of pollutants (Puzon et al. 2002, Dong et al. 2013). In this context, bioremediation is a low cost technology which simultaneously allows the degradation of organic compounds and the removal or stabilization of metals into non-toxic or less toxic forms (Owabor et al. 2011).

Actinobacteria are a group of bacteria with a cosmopolitan distribution in different ecosystems. They have demonstrated the ability to remove several organic and inorganic pollutants (Benimeli et al. 2008, Polti et al. 2009, Albarracín et al. 2010, Alvarez et al. 2012). In this context, bioaugmentation with actinobacteria to degrade organic compounds and remove or stabilize heavy metals represents a promising approach to bioremediate co-polluted environments. In fact, Polti et al. (2014) previously demonstrated the potential of *Streptomyces* sp. M7 to clean up artificially contaminated non-sterile soils with hexavalent chromium and lindane. However, the bioremediation efficacy depends on several factors and their interactions, including the temperature, the humidity and the initial contaminant concentrations (Owabor et al. 2011). For proper study of the influence of these factors, it is necessary to apply experimental design methods, which determine the effective factors and their interactions, as well as to model and optimize the whole system. The full factorial design generates maximum information regarding the factors, identifies the interactions between separate experimental factors and predicts the effect that such interactions could have on the experimental response (Antony 2003, Mason et al. 2003). Consequently, bioremediation effectiveness could be improved.

On the other hand, to assess whether bioremediation processes are acceptable, it is mandatory to investigate toxic effects of microbial metabolites produced during the pollutant removal (Repetto et al. 2001). Bioindicators change their response in front of changes in environmental pollution. *Lactuca sativa* is one of the recommended species for this purpose (Charles et al. 2011), since it allows evaluating lethal and sublethal effects and can be used in samples with high turbidity, reducing pretreatment interference. Furthermore, it has high sensitivity, so it requires a reduced exposure time, has low cost and does not require sophisticated equipment (Sobrero and Ronco 2004).

This chapter will show a full factorial design like a useful tool to study the effect of environmental factors on a bioremediation process. Further, it will be demonstrated how the efficiency of this bioprocess could be proved by using *L. sativa* as bioindicator.

Streptomyces sp. M7, a Multiresistant Bacterium

Previously, Benimeli et al. (2003) have isolated *Streptomyces* sp. M7 from sediments contaminated with organochlorine pesticides and heavy metals. First, *Streptomyces* sp. M7 showed pesticide degradation ability, under aerobic conditions in minimal medium supplemented with lindane (Benimeli et al. 2007a). Then, studies conducted in a soil extract liquid medium (Benimeli et al. 2007b) and sterile soil samples (Benimeli et al. 2008) showed that inoculation with *Streptomyces* sp. M7 was an effective alternative in the remediation of lindane polluted sites.

Recently, *Streptomyces* sp. M7 was not only demonstrated to be a potential agent for treating lindane contaminated soils, but also this strain was able to degrade others pesticides and a mixture of them (Fuentes et al. 2013). Even the inoculation with this actinobacterium has been demonstrated to be the most appropriate strategy for bioremediation of Cr(VI) and lindane co-contaminated soils (Polti et al. 2014). This is possible due to *Streptomyces* sp. M7 being a metal-resistant microorganism with the ability to biodegrade the organic pollutant in co-contaminated soils, reducing the bioavailability of the metal (Polti et al. 2014).

Optimal Inoculum Concentration in the Bioremediation Process

The optimization of a biotechnological process involves the evaluation of different parameters affecting the effectiveness and profitability of the process. It is, therefore, essential to optimize the quantity of required microbial biomass, since the production of the same significantly affects the costs of the process (Wolski et al. 2006).

Aparicio et al. (2015) determined the optimal inoculum concentration of *Streptomyces* sp. M7 to bioremediate soil contaminated with Cr(VI) and lindane. Different inoculum concentrations were evaluated: 0.5, 1, 2, and 4 g kg^{-1}. The *Streptomyces* sp. M7 inoculum was obtained by cultivating the strain in Tryptic Soy Broth, containing (in g l^{-1}): tryptone, 15; soy peptone, 3; NaCl, 5; K_2HPO_4, 2.5; and glucose, 2.5, during 3 d at 30°C (200 rpm) (Polti et al. 2009). After two washes, co-contaminated soil samples were inoculated with the biomass in the previously specified concentrations.

The soils samples (SS) employed in the mentioned assay were collected from a non-polluted experimental area. Glass pots were filled with 200 g of soil, and contaminated with 25 μg kg^{-1} of lindane and 50 mg kg^{-1} of Cr(VI) as $K_2Cr_2O_7$. The humidity content was fixed at 20% using distilled water. The pollutants' concentration and humidity selection was based on values preliminary obtained (Polti et al. 2014).

The glass pots were incubated during 14 d. Also, non-inoculated SS contaminated with both toxic compounds were used as controls. In all cases, samples were taken at the end of each assay to determine both lindane and bioavailable chromium residual concentrations. The extraction procedure for γ-HCH in soil was performed using 5 g of soil and mixed with 10 ml of a 4:1:5 water-methanol-hexane solution. The tubes were hermetically sealed and shaken during 10 min in order to allow the extraction of lindane from soil to the organic phase, and then centrifuged (2,500 × *g* during 10 min) for separation of the organic and aqueous phases. Organic phase was evaporated to dryness. The residues were suspended in hexane and analyzed in a Gas Chromatograph Agilent 7890A equipped with a HP5 capillary column (30 m × 0.53 mm × 0.35 m) and 63Ni micro-electron capture detector, a split/splitless Agilent 7693B injector and Agilent Chem-Station software (Fuentes et al. 2011).

Potentially bioavailable chromium in the soil was extracted by a physical method: 100 g of soil was centrifuged at 5,050 × *g* during 60 min, in order to reproduce the maximum plant suction (soil water potential: 1,500 kPa, conventional wilting point) (Csillag et al. 1999). It is important to highlight that chromium is bioavailable only as Cr(VI) (Polti et al. 2011). Supernatant was recovered, filtered and analyzed by atomic absorption spectrometry for determining Cr content (Polti et al. 2011).

Cr(VI) and lindane removals in SS inoculated with different concentrations of *Streptomyces* sp. M7 inoculum, are shown in Table 1. The lowest Cr(VI) removal (54.6%) occurred when SS were inoculated with 0.5 g kg^{-1} of *Streptomyces* sp. M7, while the highest removal (89.5%) was achieved with 4 g kg^{-1} of *Streptomyces* sp. M7. However,

Table 1. Contaminants' removal in soil samples inoculated with different *Streptomyces* sp. M7 inoculum concentrations, after 14 d of incubation at 30°C. Different letters indicate significantly differences between pollutant removal means.

Inoculum concentration ($g\ kg^{-1}$)	Pollutant removal (%)	
	Cr(VI)	Lindane
0	0.0 ± 0.0^{a}	$0.0 \pm 0.0^{a'}$
0.5	54.6 ± 15.0^{b}	$17.7 \pm 5.7^{b'}$
1	69.2 ± 12.5^{b}	$38.1 \pm 6.1^{c'}$
2	76.4 ± 10.4^{b}	$27.6 \pm 6.2^{b'c'}$
4	89.5 ± 1.2^{b}	$6.0 \pm 4.2^{a'}$

no significant differences were observed in Cr(VI) removal achieved by the different inoculum concentrations tested. The lowest lindane removal (6%) was achieved with 4 g kg^{-1} of *Streptomyces* sp. M7, showing no statistical differences ($p < 0.05$) in comparison to the uninoculated SS. The maximum lindane removal (38%) was obtained when 1 g kg^{-1} of *Streptomyces* sp. M7 was inoculated in SS, and it was significantly greater than that obtained with 0.5 and 4 g kg^{-1}. However, significant differences in comparison with 2 g kg^{-1} of *Streptomyces* sp. M7 were not observed.

Based on the statistical analysis, Aparicio et al. (2015) selected the inoculum of 1 g kg^{-1} of *Streptomyces* sp. M7 for further assays, since it was the lowest inoculum concentration which allowed the highest simultaneous removal of Cr(VI) and lindane.

Similarly, Benimeli et al. (2008) observed that the highest inoculum concentration was not optimum for lindane removal by using *Streptomyces* sp. M7.

Previously, Polti et al. (2014) reported that chromium and lindane removal did not occur simultaneously. First, metal is reduced, and then the pesticide is degraded. The Cr(VI) reduction to Cr(III) is a process which uses NADH from bacterial metabolism and, therefore, any process that affects its production affects the reduction of Cr(VI) (Polti et al. 2010). Also, electron acceptors significantly affect lindane degradation (Robles-González et al. 2012). It is a co-metabolic process which improves with an additional energy source (Benimeli et al. 2006, 2008). Soil energy sources could be used primarily to obtain NADH for Cr(VI) reduction, and residual carbon sources could be used for lindane removal. A large inoculum size could use all the energy sources for cell reproduction; therefore, the residual energy for lindane degradation would be lower and thus, a significant removal of the pesticide would not be achieved.

Influence of Environmental Factor in the Bioremediation Process

Environmental conditions, such as humidity and temperature, influence the bioremediation process through the modification of microbial activity and pollutants' mobility, solubility, and bioavailability.

There are several reports that high levels of moisture cause inhibition of microbial activity through diminishing of gas diffusion in soil. This could be the reason for the decreased organic toxic compounds' biodegradation in the soil (Lahel et al. 2016). Also, the moisture influences the contaminated soil chemistry and could affect the mobility of numerous dissolved inorganic compounds, including chromium. In this context, it is possible to avoid chromium mobility, and hence toxicity, by regulating matrix moisture content (Shahid et al. 2017).

Walworth and Reynolds (1995) have demonstrated that soil temperature had a pronounced influence on organic compounds' biodegradation rates. Modern pesticides (insecticides, fungicides, herbicides), mostly organic compounds, are subject to biological decomposition. If bioremediation process is carried out considering factors like proper temperature, the degradation of different organic compounds present in the soil can be improved. Lahel et al. (2016) proved the removal efficiency of lindane (81.36%) when the temperature was increased. Due to heat implementation, pesticides and other organic and inorganic pollutants readily dissolve in the aqueous phase and are readily available to the microbes. On the other hand, microbial ability to remove pollutants is strongly affected by soil temperature. In this sense, this factor also has to be included as an experimental parameter.

In order to evaluate the importance and the interactions of four independent variables, initial Cr(VI) concentration, initial lindane concentration, temperature and humidity, Aparicio et al. (2015) performed an experimental design and statistical analysis using a 2^4 full factorial design (four factors, each at two levels) plus a centre point (Table 2). The levels were selected according to what was previously reported (Polti et al. 2014). The methodology followed was the same as described above, employing the optimal inoculum concentration and using the combinations of initial Cr(VI) and lindane concentrations, temperature and humidity given by the factorial design. Non-inoculated SS contaminated with both toxics and incubated under the same factorial design conditions were used as controls. The experimental design and their results were performed using Minitab®17 (PA, USA) statistical software. Differences were accepted as significant when $p < 0.05$.

The evaluated responses were residual Cr(VI) and residual lindane concentrations. All assays were performed in triplicate and the results were presented as the mean value ± standard deviation. Analysis of variance (ANOVA) for residual Cr(VI) and lindane concentrations showed that residual Cr(VI) concentrations in the inoculated soil samples

Table 2. Removal of Cr(VI) and lindane by *Streptomyces* sp. M7. Combinations obtained from full factorial design with center point.

Conditions	Humidity (%)	Initial Cr(VI) (mg kg^{-1})	Initial Lindane (μg kg^{-1})	Temperature (°C)	Final Cr(VI) (mg kg^{-1})		Final Lindane (μg kg^{-1})	
					Non-Inoculated	Inoculated	Non-Inoculated	Inoculated
A	10	20	10	25	11.0 ± 0.9	8.8 ± 0.2	10.1 ± 0.0	8.0 ± 0.3
B	10	20	10	35	13.5 ± 0.5	0.0 ± 0.0	10.0 ± 0.2	4.7 ± 0.4
C	10	20	40	25	10.2 ± 1.3	5.4 ± 0.3	36.2 ± 0.3	11.2 ± 0.0
D	10	20	40	35	13.7 ± 0.9	2.2 ± 0.3	37.1 ± 0.2	9.2 ± 1.0
E	10	80	10	25	42.3 ± 0.2	24.9 ± 0.3	10.2 ± 1.0	7.6 ± 0.0*
F	10	80	10	35	43.7 ± 0.1	4.5 ± 0.2	13.6 ± 0.6	2.1 ± 0.5
G	10	80	40	25	41.3 ± 0.0	5.8 ± 0.5	37.2 ± 0.8	13.1 ± 0.1
H	10	80	40	35	40.9 ± 0.4	6.4 ± 0.1	36.0 ± 0.7	4.9 ± 0.6
I	20	50	25	30	15.5 ± 0.1	3.3 ± 0.2	21.2 ± 1.3	9.4 ± 0.8
J	30	20	10	25	14.3 ± 0.1	4.8 ± 0.6	8.9 ± 0.1	3.5 ± 0.0
K	30	20	10	35	15.3 ± 1.1	5.2 ± 0.4	17.1 ± 0.3	5.3 ± 0.4*
L	30	20	40	25	16.7 ± 0.2	10.0 ± 0.2	37.0 ± 0.9	15.4 ± 1.7
M	30	20	40	35	14.2 ± 0.3	8.7 ± 0.0*	37.8 ± 0.6	34.7 ± 0.9*
N	30	80	10	25	41.8 ± 0.5	36.9 ± 1.2	13.2 ± 0.4	5.3 ± 1.1
O	30	80	10	35	43.0 ± 0.6	6.8 ± 0.9	12.9 ± 0.6	5.1 ± 0.1
P	30	80	40	25	43.3 ± 2.7	34.1 ± 0.4	36.3 ± 0.8	11.8 ± 0.6
Q	30	80	40	35	39.3 ± 0.1	6.2 ± 0.1	36.0 ± 1.4	7.8 ± 2.8

* non-significant differences in comparison with non-inoculated control.

were significantly lower than those found in uninoculated controls, at 16 evaluated conditions (Table 2). On the other hand, residual lindane concentrations in the inoculated soil samples were significantly lower than that obtained in uninoculated controls, at 14 evaluated conditions (Table 2).

In conventional experimental design, every process parameter changes one by one to investigate how those affect the bioremediation process (Mason et al. 2003). Contrary to this, the experimental design employed by Aparicio et al. (2015) provides both residual Cr(VI) and lindane concentrations responses at a time. The factorial design was a useful tool to identify the factors affecting the Cr(VI) and lindane removals and the interaction effects of these on the evaluated responses. The factorial design helped to identify if there was a factor that showed more effect and influence on the others factors. This was important because usually, the interactions between all the factors are neglected with conventional experiments, which makes it difficult to achieve real optimal performance conditions.

The 2^4 factorial experimental design included one-, two-, three- and four-factor combinations of the parameters. The factors were assigned as: (A) initial Cr(VI) concentration, (B) initial lindane concentration, (C) temperature, and (D) humidity.

The validation of both responses (residual Cr(VI) and lindane concentrations) was carried out. The normality and constant variance assumptions were satisfactorily checked using Normal Probability of the residuals and plotting the residual versus the fit of the residuals (data not shown).

A lineal model was assembled, obtaining the best regression equations for residual concentrations of Cr(VI) (1) and lindane (2):

$$(1)\ \textit{Final Cr(VI) concentration} = -11.2 + 0.03686\,A - 0.0592\,B + 0.082\,C + 0.0979\,D - 0.000009\,A * B - 0.000180\,A * C + 0.000021\,A * D + 0.000302\,B * C + 0.000066\,B * D - 0.000682\,C * D - 7.41\,Ct\,Pt$$

A high correlation was observed between residual Cr(VI) concentration, experimental values, and those predicted by the statistical model. The r^2 value was 0.8788. Moreover, predicted r^2 was 0.7186, indicating that the model could explain the 87.88% of the observed data and could predict more than 70% of new data.

$$(2)\ \textit{Final Lindane concentration} = 19.0 + 0.02189\,A - 0.0122\,B - 0.1129\,C - 0.1575\,D - 0.00010\,A * B - 0.000094\,A * C - 0.000016\,A * D + 0.000136\,B * C + 0.000096\,B * D + 0.000899\,C * D + 0.42\,Ct\,Pt$$

The fit of this model was verified with the value of r^2 0.8151, indicating that 81.51% of the variability in this response could be explained by the variation of the analyzed factors and their interactions. However, predicted r^2 was 0.5702, suggesting that the model will not predict new observations nearly as well as it fitted the existing data (Aparicio et al. 2015).

The main effects and significant 2-way interactions of the factors were identified based on the p-values with < 5% of significance level (data not shown). Only the initial Cr(VI) concentration had statistically significant effect on the residual Cr(VI) concentrations. The p-values for all 2-factor combinations was lower than 0.05 for all the interactions where initial Cr(VI) concentration was present. This means that the individual effect of initial Cr(VI) concentration on the residual Cr(VI) concentration change in the presence of the other factors.

Initial Cr(VI) concentration and humidity had statistically significant effect on the "residual lindane" response. The p-values were lower than 0.05 in all the interactions where initial Cr(VI) concentration and humidity were present. This means that the individual effect of initial Cr(VI) concentration and humidity on the residual lindane concentration change in the presence of the other factors.

A response optimizer is a method that allows compromise among various responses. All possible combinations were considered, in terms of initial concentrations of Cr(VI) and lindane, which could be present in a natural co-contaminated environment.

The optimum conditions for bioremediation were 35°C and 10% of humidity, in three of the four evaluated scenarios (high initial concentration of at least one of the pollutants), while at low initial concentrations of both pollutants, the optimum bioremediation conditions were 25°C and 30% of humidity.

Aparicio et al. (2015) observed that, in general, low humidity levels favored Cr(VI) removal; however, there was a particular situation where the highest Cr(VI) removal was achieved at high humidity levels. In this case, both contaminants' initial concentrations were low. These observations could indicate that at high concentrations of pollutants, the major removal mechanism was physicochemical, including low mobility and high adsorption to soil particles whereas, at low contaminant concentrations, Cr(VI) removal could result from microbial activity.

Experimental data indicated that lindane has high affinity for organic matter which is due to its hydrophobic nature (Robles-González et al. 2006). Therefore, lindane mobility decreases by increasing humidity (Willett et al. 1998). In order to extract lindane from soil, water-immiscible solvents with affinity for hydrophobic compounds are necessary. This can help in attracting the contaminants' molecules adsorbed onto soil, transferring the contaminant into the solvent phase and, afterwards, facilitating the exchange of contaminant between the solvent and the aqueous phase where the microorganisms can finally degrade the pollutant (Robles-González et al. 2012). This agrees with our results, since lindane removal was increased at low humidity level.

Temperature affects several processes involved in the accumulation of organochlorine pesticides and heavy metals in the soil. This effect may be direct, by modifying the adsorption/desorption, diffusion, volatilization and degradation/chemical reduction of the compounds, or indirect, by increasing soil microbial activity, favoring biological removal processes (Dhal et al. 2013). The way these factors affect the bioremediation process is complex, and may have opposite effects and interactions. The temperature and humidity affect microbial activity, and thereby facilitate or inhibit the degradation of lindane (Ali et al. 2014). Furthermore, Cr(VI) could be toxic to the microorganisms involved in the process; however, lindane removal was higher with high Cr(VI) initial concentrations. Possible mechanisms of resistance to Cr(VI) in this actinobacterium conserve its metabolic activity and therefore, its ability to remove lindane (Ali et al. 2014). Aparicio et al. (2015) observed positive effects of temperature on Cr(VI) removal. According to optimizer response, higher temperature conditions promote contaminants' removal when one of them was at high initial concentration, which could be related to non-biological activity. Several studies indicate that the presence of metals in a contaminated site affects microbial population development, decreasing the growth rate and inhibiting degradative metabolism (Hong et al. 2007, Liu et al. 2007, Moreira et al. 2013).

Bioremediation strategies should aim at causing the least possible disturbance in the polluted area, using low-cost technologies. With this premise, the optimum temperature and humidity for the bioremediation of Cr(VI) and lindane contaminated environments were determined (Aparicio et al. 2015). The technologies used to achieve increased temperature involve assembling piles of soil, and subsequent coverage. In contrast, the temperature decrease is performed by removing the soil, mixing/uprising material, which, besides decreasing temperature, promotes soil aeration (Kauppi et al. 2011, Kalantidou et al. 2012).

The humidity can be regulated by watering, manually or mechanically. On the other hand, humidity content can be decreased by exposing the material to the sun, evaporating water gradually. These strategies could be employed to change the temperature and humidity of the contaminated environment during treatment with actinobacteria, for enhancing their bioremediation activity.

Lactuca sativa as a Bioindicator of the Efficacy of Soil Bioremediation

After a bioremediation process, it is necessary to perform ecotoxicology to evaluate the safety of the process (Fuentes et al. 2013). Ecotoxicology test could be performed using bioindicators. According to Holt and Miller (2011), bioindicators include biological processes, species, or communities that can be used to assess the quality of the environment and its changes. In this context, Charles et al. (2011) have demonstrated that plant biomonitoring test using *Lactuca sativa* was a useful tool to evaluate the toxicity of industrial effluents presenting trace metal polycontamination. Similarly, Fuentes et al. (2013) and Saez et al. (2015) confirmed that lettuce seeds (*L. sativa*) are sensitive to organochlorine pesticides and their toxic intermediates. Lindane has shown phytotoxicity through the disruption of vital processes in plants, affecting germination, cell growth and photosynthesis (Bidlan et al. 2004, Calvelo Pereira et al. 2006, Saez et al. 2014). On the other hand, high availability of metals such as Cr(VI), can induce physiological and biochemical changes in plants, such as inhibition of root growth and interveinal chlorosis with chlorophyll reduction (Sharma et al. 2003, Polti et al. 2011).

Aparicio et al. (2015) evaluated the *L. sativa* sensitivity in a mixture of Cr(VI) and lindane, to establish if it is useful as a tool to assess the bioremediation success. For this purpose, three parameters were considered on lettuce seedlings: germination, root elongation and hypocotyl elongation. The seeds were placed into sterile Petri plates containing 15 g of soil sample bioremediated by *Streptomyces* sp. M7, then plates were sealed and incubated at 22 ± 2°C in darkness, for 5 d. At the end of the incubation period, the number of germinated seeds was registered and the length of roots and hypocotyl was measured. Vigour index (VI) ((mean root length + mean hypocotyl length) × (percent germination/ 10)) was also calculated (Bidlan et al. 2004, Fuentes et al. 2013, Saez et al. 2014). In order to evaluate the bioremediation efficacy, three conditions (D, Q and I, Table 3) were randomly selected among the ones which showed significant differences between contaminated and non-bioremediated soil. Tested seeds were adversely affected by the addition of the contaminated soil. Moreover, an inverse relationship between the toxic effects observed in seedlings and Cr(VI) initial concentration ($r^2 = 0.9154$) was determined.

Roots and hypocotyls lengths and the VI were significantly lower in contaminated SS compared to non-contaminated soils (Table 3). At D and I combinations, significant differences were not observed in roots' length of seedlings grown on bioremediated and non-bioremediated SS. However, at Q combination, significant differences were observed between root lengths developed on bioremediated and non-bioremediated SS. On the other hand, at all assayed conditions, VI and hypocotyls length were significantly higher in bioremediated SS than in non-bioremediated ones. Furthermore, at D and I condition, VI and hypocotyls lengths did not show significant differences between bioremediated and non-contaminated SS (Table 3).

At Q condition, lettuce seeds were not able to germinate before bioremediation. However, germination was greater than 50% after the treatment with *Streptomyces* sp. M7. These results are in accordance with the contaminants' removal achieved.

Table 3. Factors and levels of selected conditions. Vigour Index (VI) of lettuce seedlings.

Condition	Treatment	Initial Cr(VI) ($mg\ kg^{-1}$)	Initial Lindane ($\mu g\ kg^{-1}$)	Temperature (°C)	Humidity (%)	*Streptomyces* sp. M7	Residual Cr(VI) ($mg\ kg^{-1}$)	Residual Lindane ($\mu g\ kg^{-1}$)	VI
D	Bioremediated	20	40	35	10	+	2.7	9.4	22^{cd*}
	Non bioremediated	20	40	35	10	-	18.1	20.0	9^{be}
	Non contaminated	-	-	35	10	-	-	-	27^{c}
I	Bioremediated	50	25	30	20	+	3.8	9.6	11^{ab}
	Non bioremediated	50	25	30	20	-	13.4	25.7	2.2^{ef}
	Non contaminated	-	-	30	20	-	-	-	18^{ad}
Q	Bioremediated	80	40	35	30	+	6.2	7.8	8^{be}
	Non bioremediated	80	40	35	30	-	26.6	21.9	0^{f}
	Non contaminated	-	-	35	30	-	-	-	22^{cd}

* Means with different letters were significantly different ($p < 0.05$). VI = (mean root length + mean shoot length) × percentage of germination/10.

Toxic effects of these contaminants on lettuce seedlings was observed, confirming that it is an appropriate indicator for studying the process efficiency. Moreover, at all tested conditions, properties of bioremediated soil improved significantly. Any relationship between residual toxic concentrations was not found, indicating that factors affected the system as a whole. According to these results, Aparicio et al. (2015) inferred that an effective reduction of lindane and Cr(VI) bioavailability in co-contaminated soil samples was carried out, with generation of less toxic or non-toxic metabolites.

Concluding Remarks

This chapter focused on the feasibility of applying bioremediation protocols in mixed contaminated soils. In this context, the work done by Aparicio et al. (2015) represents an important advancement on this subject.

The use of the actinobacteria *Streptomyces* M7 led to a successful bioremediation of co-contaminated soils under different operating conditions, revealing the metabolic versatility of this microorganism. The statistical approach allowed the generation of a mathematical model that efficiently predicted the operation of the system, knowing the initial conditions of the same. Moreover, the success of bioremediation was confirmed by using a suitable bioindicator to evaluate the toxicity of the mixture of contaminants.

This study represents a significant advancement in the bioremediation of co-contaminated sites, and the next step would be to scale the process to achieve bioremediation of larger environments suffering from mixed pollution.

Acknowledgements

The authors gratefully acknowledge financial support of Secretaría de Ciencia y Técnica, Universidad Nacional de Tucumán (SCAIT), Agencia Nacional de Promoción Científica y Tecnológica (ANPCyT) and Consejo Nacional de Investigaciones Científicas y Técnicas (CONICET).

References Cited

Albarracín, V.H., M.J. Amoroso and C.M. Abate. 2010. Bioaugmentation of copper polluted soil microcosms with *Amycolatopsis tucumanensis* to diminish phytoavailable copper for *Zea mays* plants. Chemosphere 79: 131–137.

Ali, M., A.A. Kazmi and N. Ahmed. 2014. Study on effects of temperature, moisture and pH in degradation and degradation kinetics of aldrin, endosulfan, lindane pesticides during full-scale continuous rotary drum composting. Chemosphere 102: 68–75.

Alvarez, A., C.S. Benimeli, J.M. Saez, M.S. Fuentes, S.A. Cuozzo, M.A. Polti et al. 2012. Bacterial bio-resources for remediation of hexachlorocyclohexane. Int. J. Mol. Sci. 13: 15086–15106.

Antony, J. 2003. Full factorial design. pp. 54–72. *In*: Antony, J. [ed.]. Design of Experiments for Engineers and Scientists. Elsevier, New York.

Aparicio, J.D., M.Z. Simón Solá, C.S. Benimeli, M.J. Amoroso and M.A. Polti. 2015. Versatility of *Streptomyces* sp. M7 to bioremediate soils co-contaminated with Cr(VI) and lindane. Ecotoxicol. Environ. Saf. 116: 34–39.

Arienzo, M., A. Masuccio and L. Ferrara. 2013. Evaluation of sediment contamination by heavy metals, organochlorinated pesticides and polycyclic aromatic hydrocarbons in the Berre coastal lagoon (Southeast France). Arch. Environ. Contam. Toxicol. 65: 396–406.

Bagchi, D., S.J. Stohs, B.W. Downs, M. Bagchi and H.G. Preuss. 2002. Cytotoxicity and oxidative mechanisms of different forms of chromium. Toxicology 180: 5–22.

Benimeli, C.S., M.J. Amoroso, A.P. Chaile and G.R. Castro. 2003. Isolation of four aquatic streptomycetes strains capable of growth on organochlorine pesticides. Bioresour. Technol. 89: 133–138.

Benimeli, C.S., G.R. Castro, A.P. Chaile and M.J. Amoroso. 2006. Lindane removal induction by *Streptomyces* sp. M7. J. Basic Microbiol. 46: 348–357.

Benimeli, C.S., G.R. Castro, A.P. Chaile and M.J. Amoroso. 2007a. Lindane uptake and degradation by aquatic *Streptomyces* sp. strain M7. Int. Biodeterior. Biodegrad. 59: 148–155.

Benimeli, C.S., J. González, P. Chaile and M.J. Amoroso. 2007b. Temperature and pH effect on lindane removal by *Streptomyces* sp. M7 in soil extract. J. Basic Microbiol. 47: 468–473.

Benimeli, C.S., M.S. Fuentes, C.M. Abate and M.J. Amoroso. 2008. Bioremediation of lindane-contaminated soil by *Streptomyces* sp. M7 and its effects on *Zea mays* growth. Int. Biodeterior. Biodegrad. 61: 233–239.

Bidlan, R., M. Afsar and H.K. Manonmani. 2004. Bioremediation of HCH-contaminated soil: Elimination of inhibitory effects of the insecticide on radish and green gram seed germination. Chemosphere 56: 803–811.

Calvelo Pereira, R., M. Camps-Arbestain, B. Rodríguez Garrido, F. Macías and C. Monterroso. 2006. Behaviour of α, β, γ and δ-hexachlorocyclohexane in the soil-plant system of a contaminated site. Environ. Pollut. 144: 210–217.

Charles, J., B. Sancey, N. Morin-Crini, P.M. Badot, F. Degiorgi, G. Trunfio et al. 2011. Evaluation of the phytotoxicity of polycontaminated industrial effluents using the lettuce plant (*Lactuca sativa*) as a bioindicator. Ecotoxicol. Environ. Saf. 74: 2057–2064.

Chen, M., P. Xu, G. Zeng, C. Yang, D. Huang and J. Zhang. 2015. Bioremediation of soils contaminated with polycyclic aromatic hydrocarbons, petroleum, pesticides, chlorophenols and heavy metals by composting: Applications, microbes and future research needs. Biotechnol. Adv. 33: 745–755.

Coatu, V., D. Ţigănuş, A. Oros and L. Lazăr. 2013. Analysis of hazardous substance contamination of the marine ecosystem in the Romanian black Sea coast, part of the Marine Strategy Framework Directive (2008/56/EEC) implementation. Rech. Mar. 43: 144–186.

Csillag, J., G. Pártay, A. Lukács, K. Bujtás and T. Németh. 1999. Extraction of soil solution for environmental analysis. Int. J. Environ. Anal. Chem. 74: 305–324.

Dhal, B., H.N. Thatoi, N.N. Das and B.D. Pandey. 2013. Chemical and microbial remediation of hexavalent chromium from contaminated soil and mining/metallurgical solid waste: A review. J. Hazard. Mater. 250–251: 272–291.

Dong, Z.Y., W.H. Huang, D.F. Xing and H.F. Zhang. 2013. Remediation of soil co-contaminated with petroleum and heavy metals by the integration of electrokinetics and biostimulation. J. Hazard. Mater. 260: 399–408.

Fuentes, M.S., J.M. Saez, C.S. Benimeli and M.J. Amoroso. 2011. Lindane biodegradation by defined consortia of indigenous *Streptomyces* strains. Water Air Soil Pollut. 222: 217–231.

Fuentes, M.S., G.E. Briceño, J.M. Saez, C.S. Benimeli, M.C. Diez and M.J. Amoroso. 2013. Enhanced removal of a pesticides mixture by single cultures and consortia of free and immobilized *Streptomyces* strains. Biomed. Res. Int. 2013: 392573.

Holt, E. and S. Miller. 2010. Bioindicators: using organisms to measure environmental impacts. Nat. Educ. Knowl. 3: 8.

Hong, H., I. Nam, Y. Kim and Y. Chang. 2007. Effect of heavy metals on the biodegradation of dibenzofuran in liquid medium. J. Hazard. Mater. 140: 145–148.

Kalantidou, A., A.M. Tang, J.M. Pereira and G. Hassen. 2012. Preliminary study on the mechanical behaviour of heat exchanger pile in physical model. Géotechnique. 62: 1047–1051.

Kauppi, S., A. Sinkkonen and M. Romantschuk. 2011. Enhancing bioremediation of diesel-fuel-contaminated soil in a boreal climate: Comparison of biostimulation and bioaugmentation. Int. Biodeterior. Biodegrad. 65: 359–368.

Lahel, A., A.B. Fanta, N. Sergienko, M. Shakya, M.E. López, S.K. Behera et al. 2016. Effect of process parameters on the bioremediation of diesel contaminated soil by mixed microbial consortia. Int. Biodeterior. Biodegradation 113: 375–385.

Liu, T., C. Sun, N. Ta, J. Hong, S. Yang and C. Chen. 2007. Effect of copper on the degradation of pesticides cypermethrin and cyhalothrin. J. Environ. Sci. 19: 1235–1238.

Maggi, C., A. Ausili, R. Boscolo, F. Cacciatore, A. Bonometto, M. Cornello et al. 2012. Sediment and biota in trend monitoring of contaminants in transitional waters. TrAC—Trends Anal. Chem. 36: 82–91.

Mason, R.L., R.F. Gunst and J.L. Hess. 2003. Statistical Design and Analysis of Experiments. 2nd ed. John Wiley & Sons, Inc., New Jersey.

Moreira, I., C. Amorim, M. Carvalho and A. Ferreira. 2013. Effect of the metals iron, copper and silver on fluorobenzene biodegradation by *Labrys portucalensis*. Biodegradation 24: 245–255.

Nie, M., N. Xian, X. Fu, X. Chen and B. Li. 2010. The interactive effects of petroleum-hydrocarbon spillage and plant rhizosphere on concentrations and distribution of heavy metals in sediments in the Yellow River Delta, China. J. Hazard. Mater. 174: 156–161.

Olaniran, A.O., A. Balgobind and B. Pillay. 2013. Bioavailability of heavy metals in soil: Impact on microbial biodegradation of organic compounds and possible improvement strategies. Int. J. Mol. Sci. 14: 10197–10228.

Owabor, C.N., O.C. Onwuemene and I. Enaburekhan. 2011. Bioremediation of polycyclic aromatic hydrocarbon contaminated aqueous-soil matrix: effect of co-contamination. J. Appl. Sci. Environ. Manag. 15: 583–588.

Polti, M.A., R.O. Garcia, M.J. Amoroso and C.M. Abate. 2009. Bioremediation of chromium(VI) contaminated soil by *Streptomyces* sp. MC1. J. Basic Microbiol. 49: 285–292.

Polti, M.A., M.J. Amoroso and C.M. Abate. 2010. Chromate reductase activity in *Streptomyces* sp. MC1. J. Gen. Appl. Microbiol. 56: 11–18.

Polti, M.A., M.C. Atjian, M.J. Amoroso and C.M. Abate. 2011. Soil chromium bioremediation: Synergic activity of actinobacteria and plants. Int. Biodeterior. Biodegrad. 65: 1175–1181.

Polti, M.A., J.D. Aparicio, C.S. Benimeli and M.J. Amoroso. 2014. Simultaneous bioremediation of Cr(VI) and lindane in soil by actinobacteria. Int. Biodeterior. Biodegrad. 88: 48–55.

Puzon, G.J., J.N. Petersen, A.G. Roberts, D.M. Kramer and L. Xun. 2002. A bacterial flavin reductase system reduces chromate to a soluble chromium(III)-NAD+ complex. Biochem. Biophys. Res. Commun. 294: 76–81.

Repetto, G., A. Jos, M.J. Hazen, M.L. Molero, A. Del Peso, M. Salguero et al. 2001. A test battery for the ecotoxicological evaluation of pentachlorophenol. Toxicol. Vitr. 15: 503–509.

Robles-González, I.V., E. Ríos-Leal, J. Galíndez-Mayer, S. Caffarel-Méndez, J. Barrera-Cortés, F. Esparza-García et al. 2006. Comportamiento adsortivo-desortivo del lindano en un suelo agrícola. Interciencia. 31.

Robles-González, I.V., E. Ríos-Leal, I. Sastre-Conde, F. Fava, N. Rinderknecht-Seijas and H.M. Poggi-Varaldo. 2012. Slurry bioreactors with simultaneous electron acceptors for bioremediation of an agricultural soil polluted with lindane. Process Biochem. 47: 1640–1648.

Saez, J.M., C.S. Benimeli and M.J. Amoroso. 2012. Lindane removal by pure and mixed cultures of immobilized actinobacteria. Chemosphere 89: 982–987.

Saez, J.M., A. Alvarez, C.S. Benimeli and M.J. Amoroso. 2014. Enhanced lindane removal from soil slurry by immobilized *Streptomyces*. Int. Biodeterior. Biodegrad. 93: 63–69.

Saez, J.M., J.D. Aparicio, M.J. Amoroso and C.S. Benimeli. 2015. Effect of the acclimation of a *Streptomyces* consortium on lindane biodegradation by free and immobilized cells. Process Biochem. 50: 1923–1933.

Shahid, M., S. Shamshad, M. Rafiq, S. Khalid, I. Bibi, N.K. Niazi et al. 2017. Chromium speciation, bioavailability, uptake, toxicity and detoxification in soil-plant system: A review. Chemosphere 178: 513–533.

Sharma, D.C., C.P. Sharma and R.D. Tripathi. 2003. Phytotoxic lesions of chromium in maize. Chemosphere 51: 63–68.

Sobrero, M. and A. Ronco. 2004. Ensayo de toxicidad aguda con semillas de lechuga (*Lactuca sativa* L.). pp. 71–79. *In*: Castillo, G. [ed.]. Ensayos toxicológicos y métodos de evaluación de calidad de aguas. Estandarización, intercalibración, resultados y aplicaciones. IMTA, México.

Srinivasa Gowd, S., M. Ramakrishna Reddy and P.K.K. Govil. 2010. Assessment of heavy metal contamination in soils at Jajmau (Kanpur) and Unnao industrial areas of the Ganga Plain, Uttar Pradesh, India. J. Hazard. Mater. 174: 113–121.

Willett, K.L., E.M. Ulrich and R.A. Hites. 1998. Differential toxicity and environmental fates of hexachlorocyclohexane isomers. Environ. Sci. Technol. 32: 2197–2207.

Walworth, J.L. and C.M. Reynolds. 1995. Bioremediation of a petroleum-contaminated cryic soil: Effects of phosphorus, nitrogen, and temperature. J. Soil Contam. 4: 299–310.

Wolski, E.A., S.E. Murialdo and J.F. Gonzalez. 2006. Effect of pH and inoculum size on pentachlorophenol degradation by *Pseudomonas* sp. Water SA. 32: 93–98.

Index

P

R

S

T

U

V

W

X